International Conference on Metal Material Processes and Manufacturing (ICMMPM 2018)

AA004061

MATEC Web of Conferences Volume 207 (2018)

Jeju, South Korea
19 – 20 July 2018

Editor:

Dong-Won Jung

ISBN: 978-1-5108-7180-9

Printed from e-media with permission by:

Curran Associates, Inc.
57 Morehouse Lane
Red Hook, NY 12571

Some format issues inherent in the e-media version may also appear in this print version.

This work is licensed under a Creative Commons Attribution 4.0 International Licence. Licence details: http://creativecommons.org/licenses/by/4.0/.

Printed by Curran Associates, Inc. (2018)

For additional information, please contact EDP Sciences – Web of Conferences
at the address below.

EDP Sciences – Web of Conferences
17, Avenue du Hoggar
Parc d'Activité de Courtabœuf
BP 112
F-91944 Les Ulis Cedex A
France

Phone: +33 (0) 1 69 18 75 75
Fax: +33 (0) 1 69 28 84 91

contact-edps@webofconferences.org

Additional copies of this publication are available from:

Curran Associates, Inc.
57 Morehouse Lane
Red Hook, NY 12571 USA
Phone: 845-758-0400
Fax: 845-758-2633
Email: curran@proceedings.com
Web: www.proceedings.com

TABLE OF CONTENTS

ICMMPM2018 Committee

Conference Chair

Prof. Dong-Won Jung, Jeju National University, South Korea

Scientific Editor

Program Committee

Prof. Marcin Barburski, Lodz University of Technology, （Poland）

Dr. Roya Darabi, Jeju National University, （South Korea）

Dr. Wu-Le Zhu, Kyoto University（Japan ）

Prof. Hwang Yi, Ajou University（South Korea）

Dr. Oscar Hui, University of East Anglia（UK）

Prof. Ing. Abdullah Al-Janabi, Sultan Qaboos University（German）

Prof. Fatma Yalcinkaya, Technical University of Liberec（Turkey）

Prof. Huan-Liang Tsai, Da-Yeh University（Taiwan）

Dr. Kim Si jin, Samsung Display Company（South Korea）

Dr. BROURI Adil, Moulay Ismail University（Morocco）

Prof. GHORBEL Elhem, University of Cergy-Pontoise（France）

Dr. Abdeen Mustafa Omer, Energy Research Institute (ERI), Nottingham （UK）

Dr. Hugo Miguel Andrade Lopes Figueiredo da Silva, University of Minho（Portugal）

Prof. DEBILI MOHAMED YACINE, Badji-Mokhtar University（Algeria）

Prof. Mohamed Abdel Moneim Deyab, Egyptian Petroleum Research Institute (EPRI), Cairo（Egypt）

Prof. Ayssar Nahlé, University of Sharjah（Sharjah）

Dr. Zh. Algazy, Satbayev University（Kazakhstan）

Dr. RAJIV KUMAR, Goswami Ganesh Dutt Sanatan Dharam College（India）

Dr. Subhasis Roy, University of Calcutta（India）

Dr. Siamak Hoseinzadeh, Islamic Azad University（Iran）

Dr. Yasin Polat, Erciyes University（Turkey）

Prof. N.ETHIRAJ, Dr. M.G.R Educational and Research Institute（Chennai）

Prof. Farid ABED-MERAIM, Laboratory of Microstructure Analysis and Mechanics of Materials（French）

Prof. Osman ADIGUZEL, Ankara University（Turkey）

Dr. Yasin Polat, Nevşehir Hacı Bektaş Veli University（Turkey）

Assoc.Prof. Achanai Buasri, Silpakorn University（Thailand）

Dr. Antonio Riveiro Rodríguez, University of Vigo（Spain）

Prof. Ji Shijun, Jilin University（China）

Dr. Brahim Safi, University M'hamed Bougara of Boumerdes（Algeria）

Prof. Arnulfo Luévanos-Rojas, Autonomous University of Coahuila（México）

Dr. Yanjun Li, Florida Atlantic University（USA）

Assoc.Prof. Dr. Soumya Mukherjee, Amity University（India）

Dr. Beddiaf ZAIDI, University of Batna 1（Algeria）

Prof. Vineet Jain, Amity University Haryana（Gurgaon）

Sponsor

Jeju National University, South Korea

X-academy

Preface

Dear Distinguished Authors and Guests,

This proceeding book consists of 48 papers from 'International Conference on Metal Material Processes and Manufacturing (ICMMPM2018), which was held on July 19th and 20th, 2018 in Jeju Island, South Korea. This conference focused on recent scientific research and engineering technology in the field of metal material processes, manufacturing and engineering. The main topics of the ICMMPM2018 were 'Metal Material Processes', 'Manufacturing', and other materials related issues, which cover the whole area ranging from metal material process to manufacturing engineering.

All papers were accepted through a peer-review process by the editorial committee and international reviewers, and were categorized into 4 chapters.

On behalf of the organizing committee, the editor would like to acknowledge the technical comments from the reviewers and thank all of the authors for their contributions to this proceeding book. With our warmest regards,

Editor of ICMMPM2018

edp sciences

MATEC Web of Conferences

Statement of Peer review

In submitting conference proceedings to MATEC Web of Conferences, I certify to the Publisher that I adhere to the **Policy on Publishing Integrity** of the journal in order to safeguard good scientific practice in publishing.

1. All articles have been subjected to peer review administered by the proceedings editors.
2. Reviews have been conducted by expert referees, who have been requested to provide unbiased and constructive comments aimed, whenever possible, at improving the work.
3. Proceedings editors have taken all reasonable steps to ensure the quality of the materials they publish and their decision to accept or reject a paper for publication has been based only on the merits of the work and the relevance to the journal.

Title, date and place of the conference

International Conference on Metal Material processes and Manufacturing (ICMMPM 2018)
July 19-20, 2018
Jeju Hana Hotel, Jeju, Korea

Proceedings editor(s):

Dong-Won Jung, Jeju National University, Korea

Date and editor's signature

August 13, 2018

Using Support Vector Machine to Improve Ultrasonic Pulse Velocity Test for Concrete

Loan Ngo Tu Quynh[1,a] and Yu-Ren Wang[1]

[1] *Department of Civil Engineering, National Kaohsiung University of Applied Sciences, Taiwan*

Abstract. In the construction industry, to evaluate the compressive strength of concrete, destructive and non-destructive testing methods are used. Non-destructive testing methods are preferable due to the fact that those methods do not destroy concrete samples. However, they usually give larger percentage of error than using destructive tests. Among the non-destructive testing methods, the ultrasonic pulse velocity test is the popular one because it is economic and very simple in operation. Using the ultrasonic pulse velocity test gives 20% MAPE more than using destructive tests. This paper aims to improve the ultrasonic pulse velocity test results in estimating the compressive strength of concrete using the help of artificial intelligent. To establish a better prediction model for the ultrasonic pulse velocity test, data collected from 312 cylinder of concrete samples are used to develop and validate the model. The research results provide valuable information when using the ultrasonic pulse velocity tests to the inputs data in addition with support vector machine by learning algorithms, and the actual compressive strengths are set as the target output data to train the model. The results show that both MAPEs for the linear and nonlinear regression models are 11.17% and 17.66% respectively. The MAPE for the support vector machine models is 11.02%. These research results can provide valuable information when using the ultrasonic pulse velocity test to estimate the compressive strength of concrete.

1 Introduction

In recent years, support vector machines have been widely used in various sectors. Non-destructive testing (NDT) methods used in the construction industry to examine the compressive strength of concrete are not only important alternatives but also feasible and economical as well. When using destructive methods in the lab, the concrete compressive strength test results in some cases do not represent the in-situ cast concrete because the strength and the quality of concrete could be affected through the transportation process, the placement, tamping and curing. Moreover, taking the core samples on site is not always feasible since the drilling process might damage the structure. Therefore, non – destructive tests are more suitable in measuring the concrete strength in cases where destructive tests are not. Typically, ultrasonic pulse velocity (UPV) is a non-destructive testing method to estimate concrete compressive strength without damaging the concrete structure. The testing method can effectively evaluate the uniformity and relative quality of concrete structures. Besides, UPV has some advantages such as low cost, easier to operate and convenient to carry.

The actual concrete compressive strength is a self-compacted that refers to the unconfined compressive strength in geotechnical term [1]. It is a kind of self-flowing cementitious material, also known as the flowable fill, the un-shrinkable fill, the controlled density fill, and other names [2]. UPV testing is done by measuring the travelling period and distance of an ultrasonic pulse, the ultrasonic wave velocity can be obtained through the travelling time and distance. It is shown in [3] that the better the concrete quality, the higher the ultrasonic wave velocity.

To predict the concrete compressive strength, a prediction model is constructed. The author of [4] stated that Multiple Linear Regression cannot give a precise prediction for compressive strength. In [5], it is said that using Artificial Neural Network is more suitable to give the estimated displacement in concrete reinforcement building; in addition, the bearing capacity of shallow foundations also be determined by neuro-fuzzy models.

The investigation constructed in this paper is a process using simulation, which aims to improve the accuracy of the results of ultrasonic pulse wave velocity mode when measuring concrete strength. Then with the use of the neural network, the support vector machine SVMs, the prediction and the generation of strength will be enhanced and predict the ultrasonic pulse velocity, this research aims to get the intensity value closer to the actual compressive strength value. In addition, this research will provide proof that using SVMs can improve the regression results and artificial models created predict good results.

[a] Corresponding author: loandecor88@gmail.com

© The Authors, published by EDP Sciences. This is an open access article distributed under the terms of the Creative Commons Attribution License 4.0 (http://creativecommons.org/licenses/by/4.0/).

https://doi.org/10.1051/matecconf/201820701001

2 Proposed diagnostic method

2.1 Ultrasonic Pulse Velocity

UPV is used to evaluate the property of concrete. UPV test is done based on the principle of measuring the speed of a pulse of ultrasonic wave passing through concrete knowing the traveling time and the traveling distance [6]. Paper [7] has the full report on the correlation between the compressive strength, the pulse velocity and the elastic modulus. In [8], a relationship between pulse velocity and compressive strength is also stated. [9] proposes an empirical equation on the relationship between the unconfined compressive strength (UCS) and the pulse velocity. The equation as below:

$$UCS = a.e^{bV} \tag{1}$$

where the properties of material will determine a and b, and V is the speed of ultrasonic pulse.

2.2 UPV Experimental Setup

In this study, the instrument used for ultrasonic pulse velocity test is a TICO concrete ultrasonic detector developed by Process Company in Zurich, Switzerland. As shown in Figure. 1, the instrument can perform the calculation and evaluation on functions of concrete uniformity, column hole, crack depth, elastic modulus, and concrete strength. The ultrasonic pulse velocity test is conducted in this study as follows: the measurement will be done four times for each of the cylindrical concrete sample with a circular top view of diameter 12 cm and height of 24 cm (refer to Figure. 2).

The ultrasonic pulse velocity tests are conducted for each of the 312 test samples. The travelling distance is the distance L between two ends of the transducer. The travelling time T is the time between the excitation to the moment that the first pressure wave is received by the probe. The ultrasonic pulse velocity is the quotient of distance and time.

The ultrasonic equipment consists of three parts: the ultrasonic pulser, the receive amplifier and the time measurement device that indicates the time of ultrasonic wave travel through the medium. The detected data is displayed and recorded in the control Host.

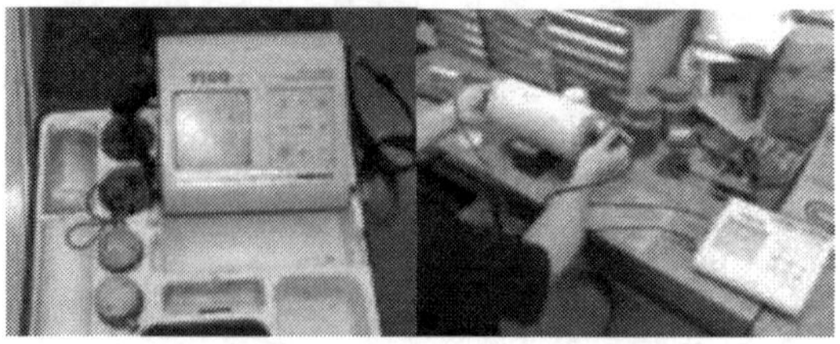

Figure 1. Ultrasonic pulse velocity test and TICO ultrasonic pulse velocity

To determine the velocity of ultrasonic, two quantities are measured: the transmission distance and the duration ultrasonic pulses transmitted. The ultrasonic velocity (v) was determined as follows:

$$v = \frac{l}{t}10^3 \tag{2}$$

where, l is the specimen length (peat–cement–sand brick with length measured in millimeters (mm)), v is the velocity of ultrasound (m/s) and t is the actual travelling time through medium of the ultrasonic pulse, measured in microseconds (μs).

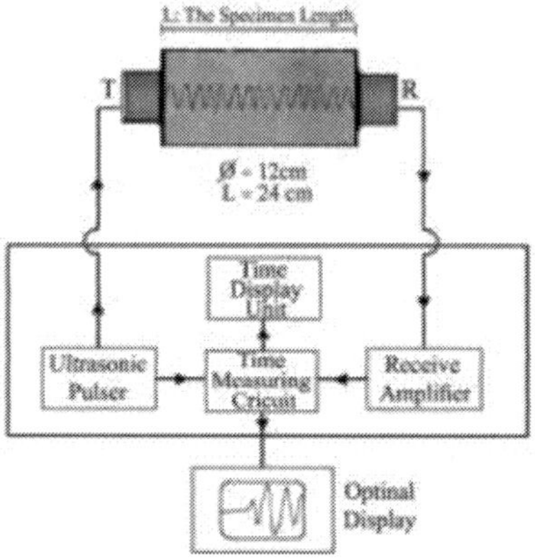

Figure 2. Model of schematic diagram of pulse velocity testing circuit

According to different settings of transducers location, there are three possible ways of measuring pulse of ultrasonic pulse velocity tests [10]. Measurements can be made by the following methods:

- Direct transmitter method and receiver are on two parallel surfaces was placed on the opposite sides of the concrete, pulse transmission path and the concrete surface perpendicular to the path length is the distance between two parallel transducers
- Indirect method or surface transmission where both transducers are on the same surface, pulse transmission path as the right triangle triangular oblique edge, the path length of the two electro-acoustical transducers center point distance.

Semi-direct Methods, transmitters and receivers are on two mutually perpendicular surfaces or two different surfaces, need to put two electro-acoustical transducers at different distances repeated measurements, and then measured The return time is the average pulse wave velocity.

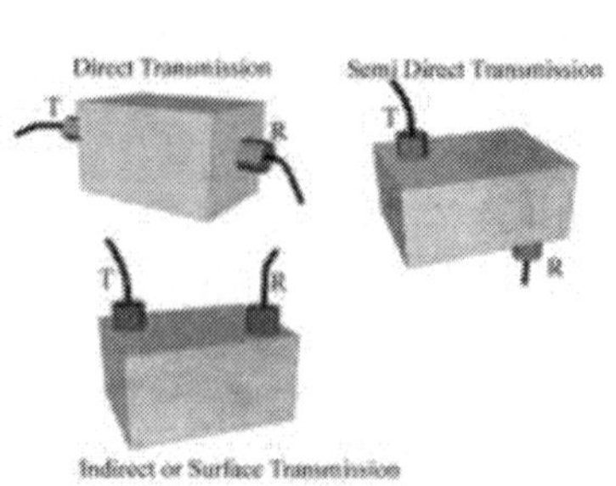

Figure 3. Ultrasonic Pulse Velocity test arrangements

2.3. Support Vector Machines

Developed base on statistical learning theory, Support Vector Machines (SVMs) is an algorithm often used for small scale, nonlinear or high dimensional problems. The foundations of SVMs have been developed by Vapnik [11]. It is not only well founded theoretically, but also superior in application domains including object recognition problems, prediction formula, protein structure prediction, functional classification and also been successfully used in the construction-related areas in recent years [12]. SVMs in this research is used to create the prediction model for concrete compressive strength. The predicted result then will be used in comparison with the real test result.

In SVMs, the upper bound of generalization error has been minimized. This enables the ability to process data better than other artificial intelligence techniques even for unseen data [13]. Some basic concepts of SVMs model will be described as follow [14].

A dataset $G = \{(x_u, d_i)\}_i^n$ is given with input vector x_u, target value di, and the size of the dataset n. By mapping x into high-dimensional feature space, the nonlinear regression $f(x)$ in low dimensional space is represented as $f(x) = \omega.\Phi(x) + b$ with ω is the weight vector, $\Phi(x)$ is the high dimensional feature space and b the bias of the hyper plane. In SVMs, by minimizing the risk penalty function, the parameters above can be found.

$$\hat{f}_{n,l}(x) = \arg\min_{f \in H_n} R_{emp}[f] \quad (3)$$

The empirical risk minimizing makes sense only if: $\lim_{l \to \infty} R_{emp}[f] = R[f]$

Using the law of large numbers: $\lim_{l \to \infty} \min_{f \in H_n} R_{emp}[f] = \min_{f \in H_n} R[f]$

Next, this condition is less intuitive and requires that the minima also converge:

$$R[f] \le R_{emp}[f] + \sqrt{\frac{h\ln(\frac{2l}{h}+1) - \ln(\frac{\delta}{4})}{l}}$$

By introducing the Lagrange multipliers: a_i and a_i^*, the SVMs decision function becomes:

$$f(x, a_i, a_i^*) = \sum_{i=1}^{n} (a_i - a_i^*) K(x, x_i) + b$$

where $K(x, x_i)$ is defined as the Kernel Function. It is the inner product of x and x_i in the feature spaces $\psi(x_i)$ and $\psi(x_j)$ respectively. There are four basic Kernel functions used in Support Vector Machines as follows [15].

Linear Kernel: $K(x_i, x_j) = x_i^T x_j$

Polynomial Kernel: $K(x_i, x_j) = (\gamma x_i^T x_j + r)^d, \gamma > 0$

Sigmoid Kernel: $K(x_i, x_j) = \tanh(\gamma x_i^T x_j + r)^d$

Radial Basis Function (RBF) Kernel:

$$K(x_i, x_j) = \exp(-\gamma \| x_i - x_j \|^2), \gamma > 0$$

With γ, r and d are Kernel parameters.

2.4 Regression SVM

Regression is performed by using support vector regression SVR, this supervised learning method generates an SVM by input-output mapping functions from a labeled training dataset. This function solves both classification and regression problems. Abubakar et al. (2013) stated that in a regression SVM, the functional dependence of the dependent variable y on a set of independent variables x has to be estimated. It assumes, like other regression problems, that the relationship between the independent and dependent variables is given by a deterministic function f plus the addition of some additive noise [16]:

$$y = f(x) + noise \quad (4)$$

The task is then to find a functional form for f that can correctly predict new cases that the SVM has not been presented with before. This can be achieved by training the SVM model on a sample set, i.e., training set, a process that involves, like classification (see above), and the sequential optimization of an error function [17,18]. Depending on the definition of this error function, two types of SVM models can be recognized:

For this type of SVM the error function is.

$$\frac{1}{2} w^T w + C \sum_{i=1}^{N} \xi_i + C \sum_{i=1}^{N} \xi_i^*$$

The error function was minimized subject to:

$$w^T \phi(x_i) + b - y_i \le \varepsilon + \xi_i^*$$
$$y_i - w^T \phi(x_i) - b_i \le \varepsilon + \xi_i$$
$$\xi_i, \xi_i^* \ge 0, i = 1, \ldots, N$$

There are several numbers of kernels that can be used in Support Vector Machines models. These include linear, polynomial, radial basis function (RBF) and sigmoid:

$$K(X_i, X_j) = \begin{Bmatrix} X_i . X_j & \text{Linear} \\ (\gamma X_i . X_j + C)^d & \text{Polynomial} \\ \exp(-\gamma |X_i - X_j|^2) & \text{RBF} \\ \tanh(\gamma X_i . X_j + C) & \text{Sigmoid} \end{Bmatrix}$$

Where $K(X_i, X_j) = \phi(X_i) \bullet \phi(X_j)$

That is, the kernel function, represents a dot product of input data points mapped into the higher dimensional feature space by transformation O. Gamma is an adjustable parameter of certain kernel functions. The RBF is by far the most popular choice of kernel types used in Support Vector Machines. This is mainly because of their localized and finite responses across the entire range of the real x-axis.

The SVR uses ε parameter (insensitive loss) to compute a linear regression function for a dimensional feature space while concurrently minimizing $\| \omega \|^2$ to reduce model complexity.

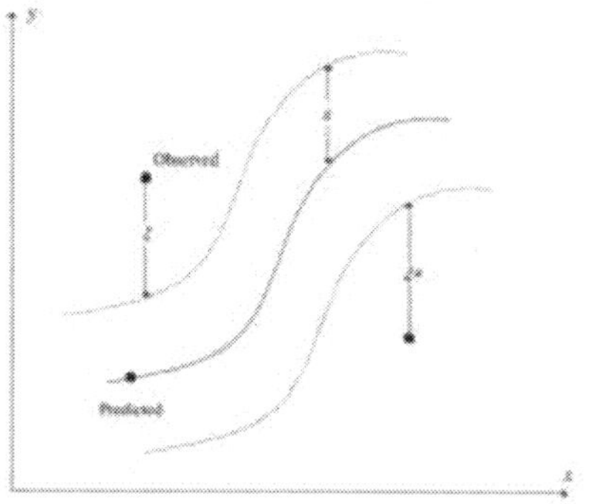

Figure 4. Support vector regression [19]

Two new parameters ξ_i, ξ_i^* are introduced to solve the minimization problem as show figure.4. Thus, SVR is formulated as minimizing following constraint-optimization problem:

Minimize $\frac{1}{2}\|\omega\|^2 + C\sum_{i=1}^{l}\xi_i + \xi_i^*$ subject to

$$\begin{cases} y_i - <\omega, x_i> - b \le \varepsilon + \xi_i \\ <\omega, x_i> + b - y_i \le \varepsilon + \xi_i^* \\ \xi_i, \xi_i^* \ge 0 \end{cases}$$ where the constant C ≥ 0

determine the trade-off between the flatness of f(x,w) and the tolerance for deviations larger than ε .

2.5 Least squares support vector machine (LSSVM)

The least square support vector machine (LSSVM) have been developed and introduced by Suykens [20]. By using the equality constraints instead of the inequality ones, the solution of LSSVM follows from solving a set of linear equations and the computational complexity is rather lower than that of SVM.

The LSSVM helps the process of programming for constrained optimization in SVMs less heavy. The error square is used as a measure of the error condition, which represents the error between the output and the correct result. So the LSSVM guidelines is represented as

$$\min_{w,\xi} Q(w) = \frac{1}{2}w^2 + \gamma\frac{1}{2}\sum_{i=1}^{n}\xi_i^2 \quad , \quad \text{constrain:}$$

$$y_i\left(w.\Phi(x_i)+b\right) = 1 - \xi_i ?? = \ ,\ldots,n$$

where γ is the same as the regular parameter in the traditional SVMs.

The aim is also to margin maximization the spacing with error minimization between the transactions ξ_i is the wrong variable.

The difference between the LMSVM and the traditional SVM is that the LSSVM is an equation and the SVM is an inequality, and the optimal solution for the LSSVM is not as sparse as the traditional SVM (Kruif & Vries, 2003). And solve the problem of (4) optimization is the same as using the Lagrangian method, so LSSVM optimization problem to solve all the above linear equations of the problem, reducing the computational complexity. The solving process of LSSVM is mainly composed of two steps. Firstly, the final solution to the right weight vector can be gotten by the gradient descent algorithm. Secondly, for the fixed right weight vector in each iteration, an analytical solution to the left weight vector can be gained. The detailed description about at LSSVM can be found in [21]. Because the search is not for the sparse solution, so LSSVM does not like SVM to find some influential information as a support vector (support vector), but all the data as a support vector.

In this research a total of 312 cylinder concrete samples are collected in collaborating with a local material testing laboratory, With 312 test samples collected in the study, 252 of them are randomly chosen to be the training dataset. 60 remaining samples are used as the test group data. The model calculates the mean absolute percentage error (MAPE) and uses the Microsoft Excel 2013 and MATLAB 2013 suite of software to construct the LSSVM model. The model predicts the concrete strength of the concrete column will be compared with the actual compressive strength of the concrete column specimen, and further verify the model is used to predict the accuracy of the actual compressive strength of the concrete cylindrical specimen, and whether the result is more accurate and the construction process of the model is as follows figure 5.

Step1: Import training group and test group data.

The prediction model will use the ultrasonic wave velocity of the ultrasonic wave to obtain the average velocity V value as the input, and the compressive strength of the concrete is the output. In the 312 cylinder concrete samples, the average wave velocity (X) and the concrete compressive strength Y) data as the LSSVM model training. The optimal prediction model is found by the model training process. The remaining 60 remaining samples are set as the best prediction model for the test group (Xt) to calculate the average absolute error percentage. The model calculates the mean absolute percentage error (MAPE) and uses the Microsoft Excel 2013 and MATLAB 2013 suite of software to construct the LSSVM model as follows:

X=xlsread('X.xlsx') (5)

Y=xlsread('Y.xlsx') (6)

Xt=xlsread('Xt.xlsx') (7)

Step2: Adjust the values of C (sigma) and γ (gama) to establish the LSSVM prediction model.

In the LSSVM model analysis to adjust the value of C (sigma) and γ (gama), repeatedly test, adjust and record the relationship between sig and gam, the input item data set through the training group instruction training, and then use the test group instructions the prediction results are predicted and the predicted results are compared with the actual compressive strength and the average absolute error percentage is calculated until a best combination of the parameters is found. The import command syntax is as follows:

Training group instruction

[alpha, b] = trainlssvm({X,Y,'f',gam, sig2}) (8)

Test group instructions

Yt = simlssvm({X,Y,'f',gam, sig2},{alpha,b},Xt) (9)

In this research, a revolution of support vector machine called least square support vector machine (LSSVM) is used to import data into MATLAB to predict the model for properties of concrete after testing those properties in the lab. LSSVM is modified from existed SVMs and then the mean absolute percentage error (MAPE) is calculated. In figure 6, MATLAB gives the prediction strength data. The value for C and Y need to be setup differently for a better prediction model. Support vector instruction inputs are shown in figure 7.

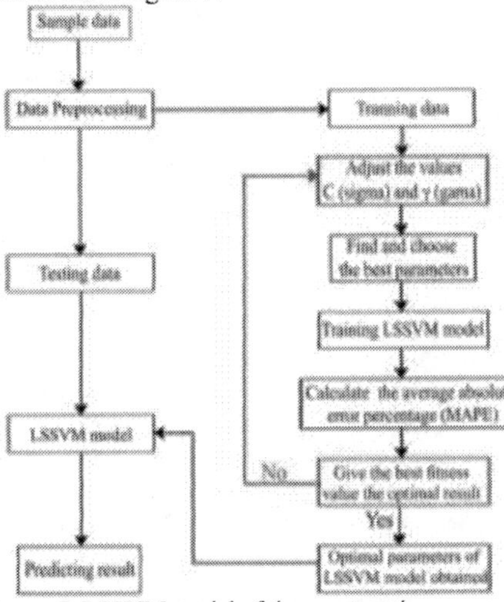

Figure 5. LSSVM model of the construction process

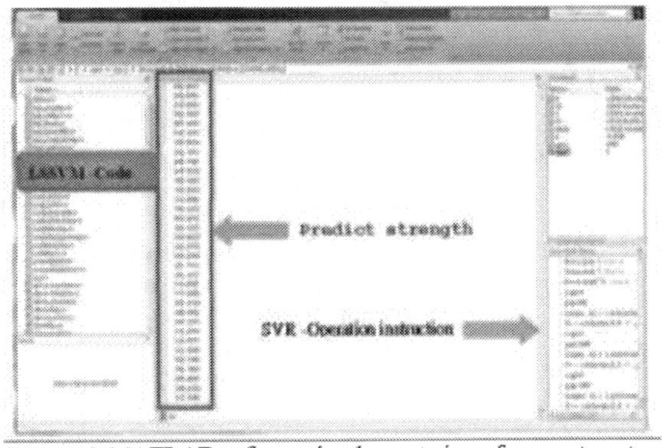

Figure 6. MATLAB software implementation of support vector regression

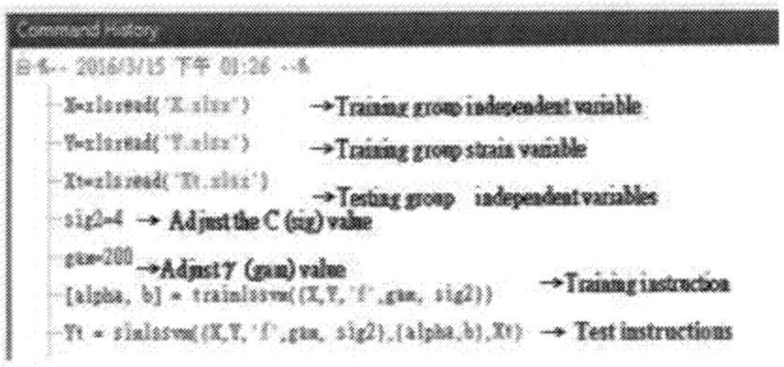

Figure 7. Support vector instruction inputs.

3 Results Analysis

SVM is developed and highly valued for combined NDTs (SonReb) for using UPV tests to the inputs data. The actual compressive strengths are set as the target output data to train the model. Total 312 test samples are collected in the lab experiment, which 252 of them are randomly selected as the training dataset and 60 test patterns and reloaded to the testing dataset. The 60 samples in the testing dataset are "unseen" that remains in the data will not be seen or new data for the support vector machine model is trained.

The prediction accuracy is measured by the mean absolute percentage error (MAPE) illustrated in the equation below:

$$MAPE = \frac{1}{n}\sum_{i=1}^{n}\left|\frac{A_i - P_i}{A_i}\right| \tag{10}$$

where, A_i is the actual concrete sample compressive strength and P_i is the predicted strength.

MAPE is calculated by comparing the data of the test group into the equation with the actual compressive strength. The outputs predict the estimated concrete compressive strength are compared directly to the actual compressive strength obtained. The result for model prediction are shown in Table 1.

In UPV test, to detect the ultrasonic wave velocity of the ultrasonic wave, the setup are as follows: the transmitting probe and the receiving probe are placed on the upper and lower sides of the concrete cylindrical test specimen (distance 24cm); the speed of transmission of the pulse in the concrete specimen then calculated as mentioned in the introduction section. MAPE was calculated by comparing the data of the test group into the equation with the actual compressive strength. The linear and non-linear regression are illustrated as in figure 8 and 9, respectively.

Table 1. Comparison of Linear and Nonlinear Regression Models.

Forecasting model	Prediction formula	Standard deviation	MAPE (%)
Linear regression	$fc = 0.3713\ V - 1162.8$	*16.80*	*11.17*
Nonlinear regression	$fc = 0.8648 exp\ (0.0015V)$	*23.13*	*17.66*

Shown in figure 8, the linear regression model is

$y = 0.3713x - 1162.8$, where x is the average ultrasonic wave velocity and y is the actual compressive strength, and using 60 test group data bands. The MAPE calculated by comparing the predicted formula with the actual compressive strength is 11.17%. Similarly for the non-linear regression in figure 9, the model is

$y = 0.8648e\ 0.0015x$. MAPE calculated is 17.66%.

As shown in Table 1, the compressive strength of concrete predicted by linear and nonlinear regression model is 11.17% and 17.66% respectively. It can be found that the linear regression model is better than the non-linear regression one. In this study, we try different kinds of artificial intelligence methods to correct the inaccuracy of ultrasonic wave velocity prediction, and to establish different types of concrete strength estimation system for improving the reliability of concrete strength when using the ultrasonic wave velocity.

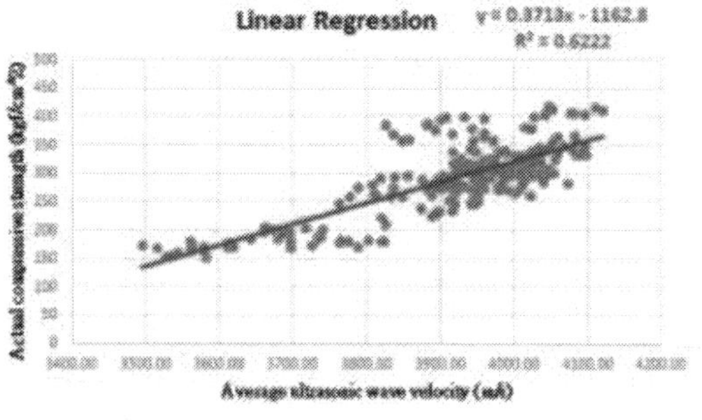

Figure 8. Linear regression graphs.

In this research a total of 312 cylinder concrete samples are collected in collaborating with a local material testing laboratory, 252 cylinder of them are randomly chosen to be the training dataset. 60 cylinder remaining samples are used as the test group data. The model calculates the mean absolute percentage error (MAPE) and uses the Microsoft Excel 2013 and MATLAB 2013 suite of software to construct the LSSVM model. And the data according to the program required to be organized into the variable X (252 cylinder training group data) the average velocity and strain Y (252 cylinder training group data of the actual strength) xlsx. File into the Matlab software by adjusting the C (sig) and γ (gam) value to get the best predictive model, and finally Xt (the remaining 60 test group data) as a model Of the test group data into the best prediction model and calculate the average absolute error percentage.

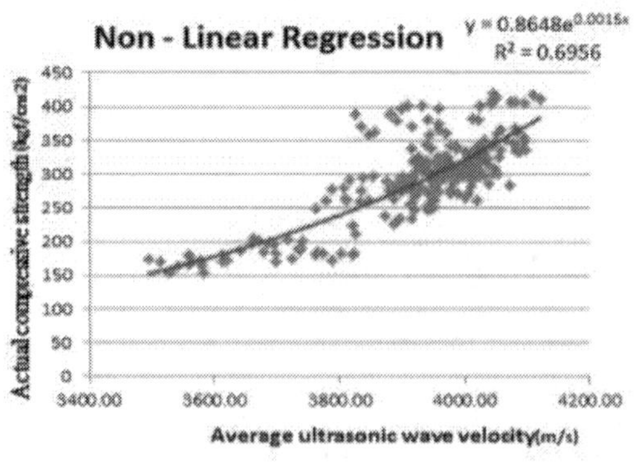

Figure 9. Non-linear regression graphs.

In the majority of the routing stations, the settings of input variable were randomly selected as the test samples for the least square support vector machine (LSSVM) methods. For this models, the actual compressive strength is set up as input variable. In the case of trial and error, we set the γ (Gamma) value change from 200 to 100000, we adjust the C (Sigma) value from 2 to 512 in order to searching and comparing with the actual compressive strength to get the best parameter (Table 2).

A total of 312 samples, the settings of input variable were randomly selected as the test samples for the support vector machine (SVM) methods. For this models, the actual compressive strength is set up as output variable. The model prediction results are summarized in Table 3.

The results show that the best MAPE values of 11.02% respectively was obtained by SVMs test with the remaining 60 data as the model. The results show that the prediction results are improved by means of artificial intelligence. The model does improve the prediction accuracy of concrete compressive strength. When attempting to increase the input of the model, the results show that the maximum error rate of the concrete is less than 9.996% and 10.495% respectively for SVMs methods. The results are better than the predictions of the one input and two input, which proves that the increase in the input of the model can reduce the mean absolute percentage error.

4 Conclusion

To improve the result of non-destructive tests, SVMs is used in order to predict the concrete compressive strength. Within 312 test samples, 252 test samples are randomly chosen to be the train data set for SVMs. The remained 60 test samples are set as the data set to evaluate the prediction model accuracy. The model the support vector machine is developed

The results shows that MAPE is about 20%, and the compressive strength of the concrete predicted by the linear and nonlinear regression model is 11.17% and 17.66% respectively. Therefore, the study show the establishing a predictive strength model with the goal for the intensity of ultrasonic wave velocity detection method becomes more reliable, and also increasing the prediction accuracy. Comparing the prediction of three models mentioned, the best MAPE values of 11.02% is obtained by the SVM test. In addition, the prediction results from one input (average ultrasonic pulse wave velocity) and two input (average ultrasonic pulse wave velocity and standard deviation) show that the maximum error rate of the concrete is less than 9.996% and 10.495% the mean absolute percentage error of the model. Which imply that the results are better than the predictions of the three input (3 velocity ultrasound V values). For the collected sample, the research results have shown that the increase in the input of the model can reduce the absolute error percentage.

Table 2. The LSSVM models prediction accuracy (one input, two input, three input).

	One input		Two input		Three input	
γ	C	MAPE	C	MAPE	C	MAPE
200	2	9.595%	2	9.604%	2	10.981%
200	4	9.685%	4	9.592%	4	10.874%
200	8	9.700%	8	9.687%	8	11.104%
200	16	9.772%	16	9.797%	16	11.082%
200	32	9.873%	32	9.891%	32	11.100%
200	64	9.955%	64	9.981%	64	11.193%
200	128	9.996%	128	10.096%	128	11.248%
200	256	10.029%	256	10.179%	256	11.256%
200	512	10.051%	512	10.195%	512	11.244%
200	2	9.595%	4	9.592%	4	10.874%
400	2	9.565%	4	9.686%	4	10.872%
800	2	9.535%	4	9.665%	4	10.860%
1000	2	9.486%	4	9.631%	4	10.855%
1500	2	9.452%	4	9.617%	4	10.889%
3000	2	9.476%	4	9.565%	4	10.889%
6000	2	9.471%	4	9.597%	4	10.978%
20000	2	9.483%	4	9.574%	4	11.099%
50000	2	9.489%	4	9.613%	4	11.518%
60000	2	9.634%	4	9.634%	4	11.879%
75000	2	9.719%	4	9.719%	4	12.025%
100000	2	9.720%	4	9.720%	4	12.225%

Table 3. The results of the test group data into the SVM model.

SVM Model	C	γ	Standard deviation	Training MAPE *(%)*	Testing MAPE *(%)*
One input	2	1500	14.41	9.452	10.495
Two input	4	3000	14.11	***9.565***	9.996
Three input	4	1000	16.38	***10.855***	11.02

References

1. S. T. Bhat, C. W. Lovell, *Use of Coal Combustion Residues and Foundry Sands in Flowable Fill*. Pub FHWA/IN/JHRP-96/02: J. H. R. P. (1996).
2. ACI-229R. *Controlled-Low Strength Materials*, Farm. Hills. (MI) (2005).
3. K. Komlos, S. Popovics, T. Nürnbergerova, B. Babal, and J. S. Popovics, *Ultrasonic Pulse Velocity Test Of Concrete Properties As Specified In Various Standards*. Cem. Concr. Compos. 18 (1996).
4. F. Khademi, S. M. Jamal, *Predicting the 28 Days Compressive Strength of Concrete using Artificial Neural Network*. i-Manager's J. Civ.Eng. 6 (2), (2016).
5. M. Nikoo, P. Zarfam, M. Nikoo, *Determining Displacement in Concrete Reinforcement Building with using Evolutionary Artificial Neural Networks*. World Appl. Sci. J. 16(12) (2012).
6. J. H. Bungey, J. H. Millard, and M. G. Grantham. *Testing of Concrete in Structures*. NY: Taylor and Francis, (2006).
7. Info: http://www.engineeringcivil.com/rebound-hammer-test.html.
8. Info: http://www.enkaymachine.com/proceq6.htm
9. L. Y. Yu, and S. Yong. *The Relationship Between The Compressive Strength of Concrete Slabs and Rebound Hammer Number of Research*. Taiwan NKUAST. 37(2) (2008).
10. Info: http://www.spectrogroup.com/ultrasonic-pulse-velocity.aspx
11. Vapnik, V.N. *The Nature of Statistical Learning Theory*, Springer Science & Business Media, (2013).
12. Noble, W. S. In B. Schoelkopf, K. Tsuda, & J.-P. Vert (Eds.), *Kernel methods in computational biology* (pp. 71–92). Cambridge, MA. (2004).
13. S. X. Wei, *ANFIS Applied to Concrete Compression Strength Prediction Mode*, NKUAST., Master's thesis, Kaohsiung, (2012).
14. G. Ye, P. Lura, K. Van Breugel, A. Fraaij, *Study on the Development of the Microstructure in Cement-Based Materials by Means of Numerical Simulation and Ultrasonic Pulse Velocity Measurement*, Concr. Compos. 26(5) (2014).
15. Hsu, C.W.; Chang, C.C.; Lin, C.J. *A Practical Guide to Support Vector Classification*. NTU: Taiwan, (2003).
16. O. Chapelle, V. Vapnik, O. Bousquet, S. Mukherjee, *Choosing multiple parameters for support vector machines*, Mach. Learn. 46 (2002).
17. Abubakar S. Magaji, Audu Isah, Victor Onomza Waziri, K.R. Adeboye, *A conceptual Nigeria stock exchange prediction: implementation using support vector machines-SMO model*, W. Com. Sci. Inf. Tech. J. (WCSIT) 3 (4) (2013).
18. S. Lukaszyk, *A New Concept Of Probability Metric And Its Applications In Approximation Of Scattered Data Sets*, Com. Mech. 33 (2004).
19. J.-S. Chou, D.-K. Bui, *Modeling heating and cooling loads by artificial intelligence for energy-efficient building design*, Energy and Buildings 82 (2014).
20. Suykens, J.A.K.; Van Gestel, T.; De Moor, B.; Vandewalle, *J. Basic Methods of Least Squares Support Vector Machines. In Least Squares Support Vector Machines*. W. Sci. Pub. (2002).
21. Kruif, B. J., de Vries, T. J. A. *Pruning Error Minimization In Least Squares Support Vector Machines*. IEEE Trans on Neural N. 14 (2016)

Analysis of masonry arches reinforced with FRP sheets: experimental results and numerical evaluations

Stefano Galassi[1,a]

[1] *Department of Architecture, University of Florence, Italy*

Abstract. In recent years fiber-reinforced polymers (FRP) have been widely used for strengthening masonry structures. In particular, in the case of masonry arches, the use of FRP sheets increases load-bearing capacity by limiting or preventing the occurrence of tensile cracks that could activate collapse mechanisms. The effectiveness of the strengthening intervention depends on the bond between FRP and substrate, due to the shear and normal stresses that occur in the bond interface, so much so that the typical failure mode of an arch reinforced with narrow FRP sheets at the intrados is exactly delamination. In this paper a predictive numerical procedure of the combined mode I and mode II failure is proposed. Numerical results provided by this procedure are compared to the experimental results on in-scale arch models taken from a recent work of the author.

1 Introduction

The need for preserving, restoring and consolidating architectural and monumental heritage [1, 2], as well as ruins in archaeological sites [3-6], is the reason for the development of innovative materials and techniques for strengthening and repairing them [7-14]. In this framework, masonry arches, vaults and domes are typical and recurrent structural elements of architectural heritage that need to be preserved. For the analysis of these types of structures, new numerical tools [15-27] have been recently presented.

A very widespread technique for consolidating and retrofitting masonry structures is the use of fiber-reinforced polymer sheets (FRP) bonded to the substrate by an adhesive layer (often an epoxy matrix). In order to effectively increase the load carrying capacity and the ductility of FRP bonded joints mechanical anchors can also be used [28, 29]. The main reason for the success of the strengthening intervention using FRP materials is due to their compatibility with masonry [30] as well as other advantages that they allow, compared to traditional materials and techniques, such as lightness, high resistance, high elastic modulus and stiffness and rapid installment in the structure.

However, the effectiveness of the reinforcement intervention strongly depends on the bond performance between FRP and substrate. In the bond interface shear and normal stresses (i.e. adhesion and peeling forces) are generated during load transfer from the masonry to the reinforcement. The peeling forces, that are provoked due to the curved shape of the arch, are tension if the reinforcement is placed at the intrados. The combination of peeling and shear forces are responsible for the typical delamination failure, that is the detachment of the sheet from the masonry. Experimental investigations have shown that, unlike FRCM composites in which failure occurs exactly in the fiber-matrix interface [31-33], in FRPs failure occurs at the composite-substrate interface affecting the superficial layer of the masonry [34-39], due to the higher strength of the adhesive as respect to that of masonry.

In this paper a novel mechanical model of an arch reinforced with intrados FRP sheets is presented and a non-linear numerical procedure for investigating the bond behaviour is also proposed. The algorithm checks the shear and peeling forces in the bond interfaces and monitors the full delamination process, step by step, during the increase of an external load.

2 Reinforced arch model

Two main failure modes are considered: the occurrence of cracks in the mortar joints due to tensile forces and the delamination of the sheet from the substrate due to peeling and adhesion forces in the bond interfaces. In order to address both failure modes, the FRP reinforced arch (Fig. 1a) is modeled as follows. The masonry arch is composed of n rigid-blocks assembled by elastic mortar joints. In the mortar joints, all the elasticity (i.e.: deformability) of the arch is concentrated. Each mortar joint is, in turn, represented by a device composed of four links orthogonal to the mid-surface of the joint, plus an additional link arranged along that surface (hereafter referred to as tangential link).

The reinforcement is represented through $n+1$ additional links, placed at the intrados of the arch. Such

[a] Corresponding author: stefano.galassi@unifi.it

© The Authors, published by EDP Sciences. This is an open access article distributed under the terms of the Creative Commons Attribution License 4.0 (http://creativecommons.org/licenses/by/4.0/).

https://doi.org/10.1051/matecconf/201820701002

links connect, in series, the midpoints of the intrados of adjacent blocks.

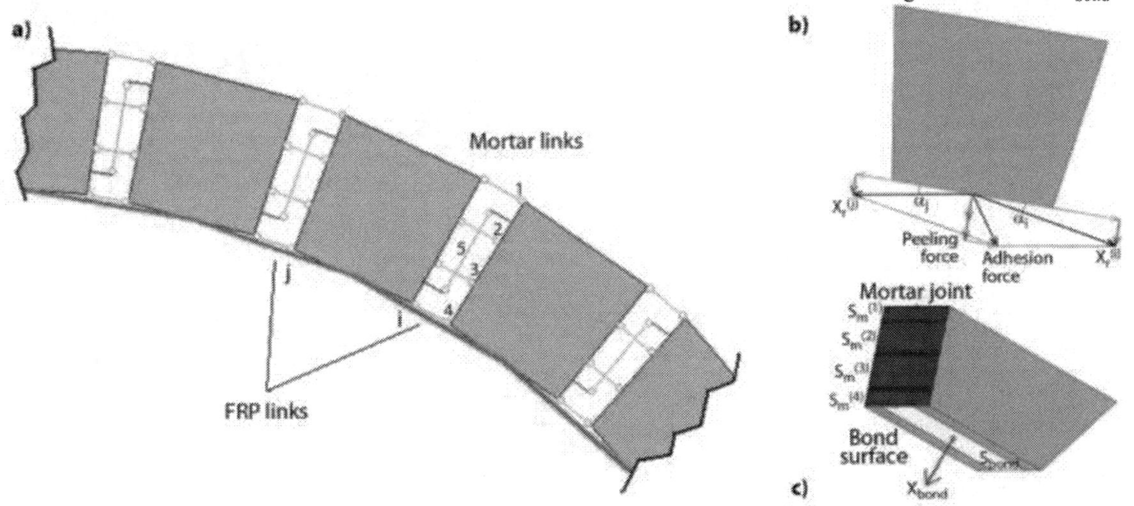

Figure 1. a) The reinforced arch model; b) Decomposition of the FRP link axial forces into the adhesion and peeling forces; c) Peeling force and bond surface in the generic arch block; area of the axial links in a mortar joint.

The links orthogonal to the mid-surface of the mortar joints measure the axial force. If the axial force is a tension higher than the tensile strength of mortar, the corresponding link fails and a crack occurs. Instead, the tangential link that measures the shearing force in the mortar joint is considered to be infinitely rigid and strong, in order to avoid sliding joints. As for the bond behaviour between FRP and masonry, each couple of reinforcement links (*i* and *j* in Fig. 1a) provides, at the intrados midpoint of the generic block of the arch, a resultant force directed towards the centre of the arch, whose components measure both the adhesion and peeling forces (Fig. 1b). If such forces exceed the tensile strength of masonry, delamination occurs and, in the mechanical model, links *i* and *j* are substituted by a new reinforcement link that connects the midpoint (start node of link *i*) of the previous block to the midpoint (end node of link *j*) of the next one.

3 Solution algorithm

The two failure criteria formulated above are represented by Eq. (1), Fig. 1b,c:

$$\begin{cases} X_t \leq \overline{X}_t \\ X_{bond} = X_r^{(i)} \sin\alpha_i + X_r^{(j)} \sin\alpha_j \leq \overline{X}_{bond} \end{cases}$$

$$\text{with} \quad \begin{cases} \overline{X}_t = S_m \cdot \overline{\sigma}_t^{(m)} \\ \overline{X}_{bond} = S_{bond} \cdot \overline{\sigma}_t^{(b)} \end{cases} \qquad (1)$$

where X_t is the tensile force in a specific link orthogonal to a mortar joint, $X_r^{(i)}$ and $X_r^{(j)}$ the axial forces in the couple of external links that simulate the intrados reinforcement in correspondence to a generic block, α_i and α_j the slope of these vectors. $\overline{X}_t$ is the threshold value of the mortar tensile force in the generic link and $\overline{X}_{bond}$ is the debonding force (herein assumed to be coincident with the peeling force, Fig. 1c). $\overline{X}_t$ and $\overline{X}_{bond}$ are computed by multiplying the tensile strength of mortar $\overline{\sigma}_t^{(m)}$ by the cross sectional area of the axial link S_m and the tensile strength of the block $\overline{\sigma}_t^{(b)}$ by the bond surface S_{bond} under the generic block respectively.

To perform the analysis of the reinforced rigid block arch, the equilibrium equation and the elastic-kinematic equation are formulated in Eq. 2 in matrix form:

$$\begin{cases} AX = \lambda F \\ \tilde{A}x + KX = \Delta \end{cases} \qquad (2)$$

where {F} is the vector of the loads and λ the incremental load factor, {X} is the vector of the unknown forces in the links of the mortar and FRP joints, [A] is the equilibrium matrix, {x} is the vector of the displacements of the block centroids and [K] is the matrix whose entries are the deformability of mortar and FRP links. Vector {Δ}, the "impressed distortion vector", is used to simulate, in the model, the mortar tensile failure, i.e. the failure of some links orthogonal to the joint.

Eq. 3 is the solution of Eq. 2, that provides the forces both in the mortar and FRP links. At the beginning of the analysis, vector {Δ}={0} because the masonry arch and the reinforcement are assumed to be undamaged when values of λ are low:

$$X = K^{-1}\tilde{A}(AK^{-1}\tilde{A})^{-1}\cdot\lambda F + \left(I - K^{-1}\tilde{A}\left(AK^{-1}\tilde{A}\right)^{-1} A\right)\cdot\Delta \quad (3)$$

During the analysis, the value of λ is slowly increased and, in correspondence to each value of λ, the two failure criteria in Eq. (1) are checked. According to Eq. (1), the algorithm introduces a distortion coefficient [40] in the mortar link whose force, at the current step of the incremental analysis, does not respect the first of the inequalities in Eq. (1). The distortion provokes the failure of that link. Instead, as for the debonding failure of the reinforcement, it occurs when the second of Eq. 1 is not respected, in which $\overline{X}_{bond}$ is computed considering the strength of masonry, where the failure occurs (ripping), rather than the strength of the adhesive. Delamination is simulated by changing the reinforcement pattern as described in the previous section, without introducing further distortion coefficients in the numerical model.

At the end of the incremental nonlinear analysis, both the value of the load factor λ and the failure mode of the arch are captured.

4 Experimental evidence and numerical results

In the experimental campaign performed in [39] on in scale 1:2 models of masonry arches reinforced with FRP composite materials at the intrados, it was observed that the tested specimens with narrow FRP sheets failed due to delamination of the reinforcement, while specimens reinforced with wider sheets failed due to sliding blocks and/or masonry crushing.

Such models reproduced a segmental arch with a 150 cm spam, an internal radius corresponding to 86.5 cm, with fixed imposts and inclined by 30° and a 10x10 cm square cross section.

The experimental trial took place by impressing an increasing vertical displacement at the keystone using a hydraulic jack. In correspondence to a load of 1.98 kN the first crack at the keystone appeared and then, later, at a load of 2.63 kN at the haunches. In correspondence to a load of 4.32 kN the delamination process occurred, and, starting from the keystone sections this process was enlarged little by little to the sections near the left haunch taking away part of the masonry (Fig. 2).

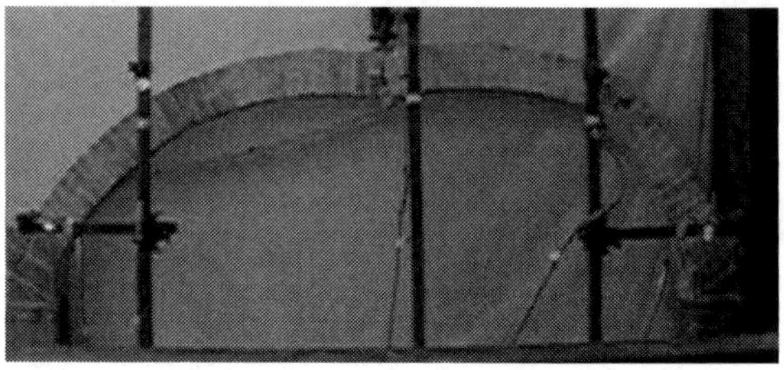

Figure 2. Experimental trial. (Taken from [39]).

In the numerical model the action of the hydraulic jack at the keystone of the arch was simulated by inputting two equal point forces at the centre of gravity of both the keystone blocks. During the analysis, these forces were slowly increased.

The numerical model showed (Fig. 3), step by step, the delamination process which also occurred in the laboratory. It highlighted a peak load of the same level as the one achieved experimentally as well as the cracking pattern, which shows the failure of mortar joints due to tensile forces, and also the position of the sheet detached from many intrados interfaces. It is worth noting that the configuration drawn by the computer program, in which the numerical procedure is implemented, is the mirror image of that obtained in the laboratory.

5 Conclusions

In this paper a mechanical model and the related numerical procedure for the analysis of masonry arches reinforced with FRP sheets at the intrados, has been presented.

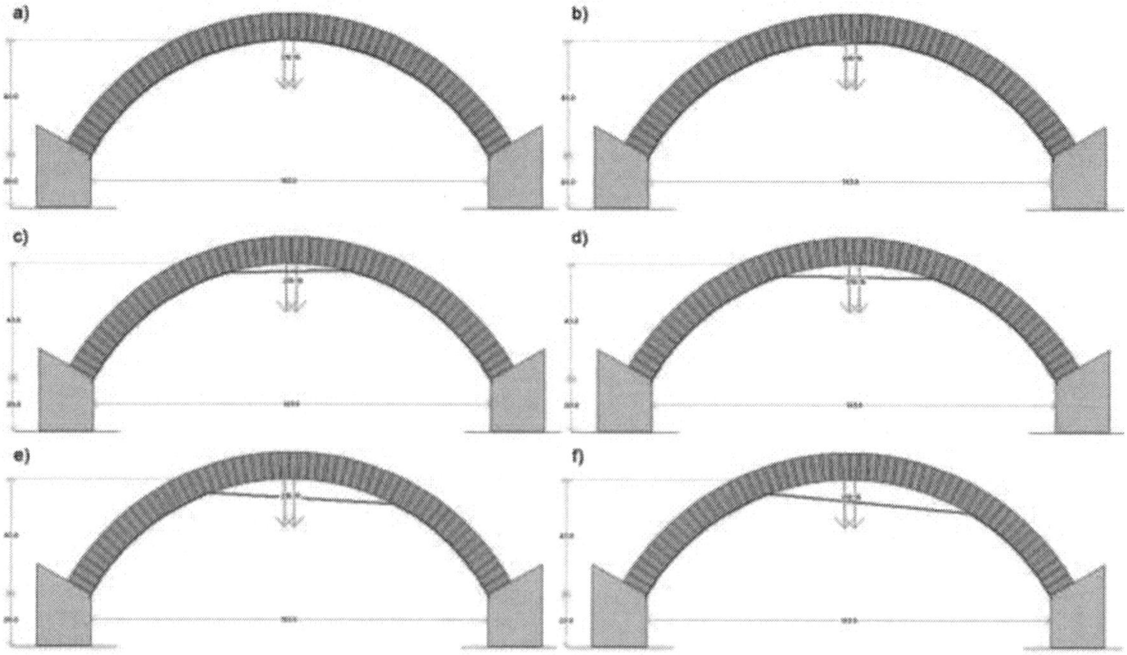

Figure 3. Delamination process computed by the proposed algorithm. (Taken from [39]).

Even if a reinforced arch can fail in different ways, the delamination of the reinforcement from the arch that, generally, follows or occurs simultaneously to a phase of micro-cracking in the mortar joints due to tensile forces, is herein proposed to be assumed as the reference failure mode for designing the reinforcement intervention. Indeed, according to the strength hierarchy criterion, the failure due to delamination is more preferable to masonry crushing, which spoils the structure without any remedy, and also to FRP rupture, that causes the arch to collapse almost immediately.

Therefore, the numerical procedure discussed in this paper is proposed in order to provide guidance for designing a targeted reinforcement system in order to avoid undesirable failure modes.

The agreement between numerical results and the results of the experimental arches tested in the laboratory confirms the reliability of the proposed procedure.

Acknowledgments

The author wishes to thank Architect Daniela Sinicropi, that performed the experimental trial in our Laboratory.

References

1. S. Galassi, L. Dipasquale, N. Ruggieri, G. Tempesta, Andalusian timber roof structure in Chefchaouen, Northern Morocco: construction technique and structural behavior, J. Archit. Eng. **24**, 3, 04018012 (2018) 10.1061/(ASCE)AE.1943-5568.0000315
2. M. Paradiso, S. Galassi, S. Benedetti, Il caratteristico Forte di San Fernando de Bocachica, Colombia, in: Luis Villegas, Ignacio lombillo, Clara Liano, Haydee Blanco (Eds.), *Patologia della Costruccion, Tecnologia de la Rehabilitacion y Gestion del Patrimonio*, Proc. Conf. Latinoamericano Patologie del costruito, tecniche di riabilitazione e gestione del patrimonio (Rehabend 2014), Santander, Spain, Gráficas Iguña, S.A. (2014)
3. N. Ruggieri, S. Galassi, G. Tempesta, Pompeii's Stabian Baths. Mechanical behaviour assessment of selected masonry structures during the 1st century seismic events, Int. J. Archit. Herit. **12**, 5, 859-878 (2018) 10.1080/15583058.2017.1422571
4. M. Sassu, M. Andreini, C. Casapulla, A. De Falco, Archaeological consolidation of UNESCO masonry structures in Oman: The Sumhuram citadel of Khor Rori and the Al Balid Fortress, Int. J. Archit. Herit. 7,4, 339-374 (2013)
5. R. Aguilar, R. Marques, K. Sovero, C. Martel, F. Trujillano, R. Boroschek, Investigations on the structural behaviour of archaeological heritage in Peru: From survey to seismic assessment, Eng. Struct. **95**, 94-111 (2015)
6. A. Albuerne, M. S. Williams, Structural appraisal of a Roman concrete vaulted monument: the Basilica of Maxentius, Int. J. Archit. Herit. **11**,7, 901-912 (2017)
7. M. Paradiso, S. Galassi, A. Borri, D. Sinicropi, "Reticolatus": an innovative reinforcement for irregular masonry. A numeric model, in: Paulo J. S. Cruz (Ed.), *Structures and Architecture: Concepts, Applications and Challenges*, CRC Press / Balkema, Taylor and Francis Group, London, Proc. 2nd Int. Conf. on Structure & Architecture ICSA2013, Guimaraes, Portugal (2013)
8. V. Alecci, F. Focacci, L. Rovero, G. Stipo, M. De Stefano, Intrados strengthening of brick masonry arches with PBO–FRCM composites: experimental and analytical investigations, Compos. Struct. **176**, 898-909, (2017) 10.1016/j.compstruct.2017.06.023
9. V. Alecci, M. De Stefano, F. Focacci, R. Luciano, L. Rovero, G. Stipo, Strengthening masonry arches with lime-based mortar composite, Buildings **7**, 2, 49 (2017) 10.3390/buildings7020049
10. M. Fagone, G. Ranocchiai, An Experimental Analysis of the Mechanical Behaviour of Anchored CFRP-to-Masonry Reinforcements Loaded by Out-of-Plane Actions, Key Eng. Mat. **747**, 204–11 (2017) 10.4028/www.scientific.net/KEM.747.204
11. L. Jurina, Experimental tests on consolidation of masonry bridges using "ram-reinforced arch method", *Proc. IIX International Conference on Arch Bridges*, 5-7 October 2016, Wroclaw, Poland (2016)
12. A. Borri, G. Castori, M. Corradi, Intrados strengthening of brick masonry arches with composite materials, Compos. Part B-Eng. **42**, 5, 1164-1172 (2011)
13. M. Corradi, A. Borri, G. Castori, R. Sisti, The Reticulatus method for shear strengthening of fair-faced masonry, Bulletin of Earthquake Engineering **14**, 12, 3547–3571 (2016) 10.1007/s10518-016-0006-5
14. G. Castori, A. Borri, M. Corradi, Behavior of thin masonry arches repaired using composite materials, Compos. Part B-Eng. **87**, 311-321 (2016) 10.1016/j.compositesb.2015.09.008
15. S. Galassi, G. Misseri, L. Rovero, G. Tempesta, Equilibrium analysis of masonry domes. On the analytical interpretation of the Eddy-Lévy graphical method, Int. J. Archit. Herit. **11**, 8, 1195-1211 (2017) 10.1080/15583058.2017.1372823
16. S. Galassi, G. Tempesta, Analysis of Masonry Block Structures with Unilateral Frictional Joints, in: *Proc. Int. Conf. on Electronic, Control, Automation and Mechanical Engineering (ECAME 2017)*, 19-20 November 2017, Sanya, China, DEStech Transactions on Engineering and Technology Research, 246-51, 10.12783/dtetr/ecame2017/18396
17. F. Pugi, S. Galassi, 2013 Seismic analysis of masonry voussoir arches according to the Italian building code, Ing. Sismica-Ital. **30** ,3, 33-55 (2013)
18. M. Paradiso, S. Galassi, S. Benedetti, Stabilità di archi e volte in muratura: metodi tradizionali e calcolo automatico, in: Luis Villegas, Ignacio lombillo, Clara Liano, Haydee Blanco (Eds.), *Patologia della Costruccion, Tecnologia de la Rehabilitacion y Gestion del Patrimonio*, Proc. Conf Latinoamericano Patologie del costruito, tecniche di riabilitazione e gestione del patrimonio (Rehabend 2014), Santander, Spain, Gráficas Iguña, S.A. (2014)
19. V. Alecci, M. De Stefano, S. Galassi, M. Lapi, M. Orlando, An assessment of American criterion for detecting plan irregularity, in: *Proc. of Eighth European Workshop on the Seismic Behaviour of*

Irregular and Complex Structures (8EWICS), 19-20 October 2017, Bucharest, Romania (2017)
20. S. Briccoli Bati, M. Fagone, T. Rotunno, Lower Bound Limit Analysis of Masonry Arches with CFRP Reinforcements: A Numerical Method, J. Compos. Constr. **17**, 366 (2013) 10.1061/(asce)cc.1943-5614.0000350
21. A. Caporale, R. Luciano, L. Rosati, Limit analysis of masonry arches with externally bonded FRP reinforcements, Comput. Method Appl. M. **196**, 247-260 (2006)
22. S. Galassi, N. Ruggieri, G. Tempesta, A novel numerical tool for seismic vulnerability analysis of ruins in archaeological sites, Int. J. Archit. Herit. (2018), 10.1080/15583058.2018.1492647
23. S. Galassi, G. Misseri, L. Rovero, G. Tempesta, Failure modes prediction of masonry coussoir arches on moving supports, Eng. Struct. **173**, 706-717 (2018) 10.1016/j.engstruct.2018.07.015
24. S. Galassi, M. Paradiso, BrickWORK Software-Aided Analysis of Masonry Structures, IERI Procedia **7**, 62-70 (2014), Special Issue 2013 International Conference on Applied Computing, Computer Science, and Computer Engineering (ICACC 2013), USA, December 2013, Elsevier B.V., ISSN: 2212-6678, 10.1016/j.ieri.2014.08.011
25. G. Salerno, How to recognize the order of infinitesimal mechanisms: A numerical approach, International Journal for Numerical Methods in Engineering **35**, 7, 1351-1395 (1992) 10.1002/nme.1620350702
26. G. Salerno, A. Billotta, F. Porco, A finite element with micro-scale effects for the linear analysis of masonry brickwork, Comput. Methods Appl. Mech. Eng. **190**, 34, 4365-4378 (2001) 10.1016/S0045-7825(01)00165-7
27. G. Uva, G. Salerno, Towards a multiscale analysis of periodic masonry brickwork: A FEM algorithm with damage and friction, Int. J. Solids Struct. **43**, 13, 3739-3769 (2006) 10.1016/j.ijsolstr.2005.10.004
28. S. Briccoli Bati, M. Fagone, G. Ranocchiai, The effects of mortar joints on the efficiency of anchored CFRP sheets reinforcements of brick-masonry, Key Eng. Mater. **624**, 575–83 (2015) 10.4028/www.scientific.net/KEM.624.575
29. Fagone M, Ranocchiai G, Briccoli Bati S. An experimental analysis about the effects of mortar joints on the efficiency of anchored CFRP-to-masonry reinforcements, Compos. Part B-Eng. **76**, 133–48 (2015) 10.1016/j.compositesb.2015.01.050
30. D. Sinicropi, E. Perria, S. Galassi, M. Paradiso, A. Borri, Artificial Ageing of Mortar Prisms Reinforced through Steel, Glass and Organic Fibers, Journal of Key Eng. Mat., A. Di Tommaso, C. Gentilini and G. Castellazzi (Eds.) **624**, 4, 542-50 (2014) 10.4028/www.scientific.net/KEM.624.542
31. V. Alecci, M. De Stefano, R. Luciano, L. Rovero, G. Stipo, Experimental Investigation on Bond Behavior of Cement-Matrix–Based Composites for Strengthening of Masonry Structures, J. Compos. Constr. **20**, 1 (2016)
32. C. Pellegrino, T. D'Antino, Experimental behaviour of existing precast prestressed reinforced concrete elements strengthened with cementitious composites, Compos. Part B-Eng. **55**, 31-40 (2013)
33. V. Alecci, S. Briccoli Bati, G. Ranocchiai, Concrete columns confined with CFRP wraps, Mater. Struct. Constr. **47**, 3, 397–410 (2014) 10.1617/s11527-013-0068-7
34. H. Yuan, J.C. Teng, R. Seracini, Z.S. Wu, J. Yao, Full-range behavior of FRP-to-concrete bonded joints, Eng. Struct. **26**, 553-65 (2004)
35. M.R. Valluzzi et al., Round robin test for composite-to-brick shear bond characerization. Mater. Struct. **45**, 12, 1761-91 (2012)
36. L. Rovero, F. Focacci, G. Stipo, Structural behavior of arch models strengthened using FRP strips of different lengths, J. Compos. Constr. **17**, 2, 249-58 (2013) 10.1061/(ASCE)CC.1943-5614.0000325
37. T. Rotunno, L. Rovero, U. Tonietti, S. Briccoli Bati, Experimental study of bond behavior of CFRP-to-brick joints, J. Compos. Constr. 04014063 (2015) 10.1061/(ASCE)CC.1943-5614.0000528
38. S. Briccoli Bati, L. Rovero, Towards a methodology for estimating strength and collapse mechanism in masonry arches strengthened with fibre reinforced polymer applied on external surfaces, Mater. Struct. **41**, 7, 1291-1306 (2008)
39. S. Galassi, A numerical procedure for failure mode detection of masonry arches reinforced with fiber reinforced polymeric materials, IOP Conf. Ser.: Mater. Sci. Eng. **369**, 012038 (2018) 10.1088/1757-899X/369/1/012038, Proc. of 5th Global Conference on Polymer and Composite Materials (PCM 2018) (10-13 April 2018, Kitakyushu City, Japan)
40. G. Colonnetti, *Scienza delle Costruzioni* (Edizioni Scientifiche Einaudi, Torino, 1955)

Economic-engineering design for logistics infrastructure routing

Weerapol Namboonruang[1a] , Noppadon Amdee[2] , Nutthanun Suphadon[3] and Boontham Krisonsri[4]

[1]*Department of Civil Technology and Architecture, Faculty of Industrial Technology, Muban Chombueng Rajabhat University, MCRU 70150, Thailand*
[2]*Department of Manufacturing Technology, Faculty of Industrial Technology, Muban Chombueng Rajabhat University, MCRU 70150, Thailand*
[3]*National Metal and Materials Technology Center 12120, Thailand*
[4]*Bureau of Location and Design, Department of Highway, Ministry of Transport 10400, Thailand*

Abstract. The existing transport route starting from Lamchabang seaport - Bangkok - Nakhonpathom province was studied in the terms of logistic and supply chain. This work, the new route starting from Ratchaburi-Suanphuang-Chong-Tagobon border will be studied which will be promoted to be the new transportation and transfer of goods route from Thailand to Myanmar via Chong-Tagobon border checkpoint, Ratchaburi province to the Highway No.9 of Myanmar. The study focuses on the possibility of infrastructure and logistic system of Ratchaburi province selected as the center of goods distribution. There are two optional routes that are studied. The first route starts from km. 0+0.000 – km. 91+174 with a total distance of 91.174 kilometers. The second route starts from km. 0+0.000 – km. 93+219 with a total distance of 93.219 kilometers. For the analysis in format of engineering and economics found that Route No.1 is more appropriate to use as the transport route for connecting the economics between Thailand-Myanmar borders. The value of the investment in logistic infrastructure (road) is approximately 3,612 million baths.

1 Introduction

In 2015, Thailand and the nine other neighboring countries have joined the AEC (Asean Economics Community). Then, it is necessary to study routes that can be linked to the benefits of trade and economic development together. Chong - Tagobon, Ratchaburi province is one of area where has the border line with Myanmar which is currently used as barter, labor checkpoint and the western border trade of Thailand. Also, it is located at the main highway network (No.4) and Asia highway No.2(AH2).

The distance from the main highway network is just 90 km. Also, this route is connected to the highway No.9 of Myanmar where can be used as alternative route to the Dawei deep seaport. Although the expanded infrastructure is just part of managing congestion, especially the expansion of the road, which requires the state's fiscal austerity measures and need to be carefully considered [1]. The regulatory and legal economy is a priority for the development of infrastructure effectively and could lead to transport efficiency, as well as control legislation. Management of the economic activity [2] was developed and deployed in the field of control of state infrastructure. This could be a study of the economic dimension of transport and mechanisms related to the impact of regulatory measures (on the development of transport infrastructure [3-11]. Furthermore, it also covers the financing of efficient transport infrastructure [12]. For the development of country, if Chong-Tagobon could be developed as the main economic border, it would be one of the most important the economic border of Thailand.

2 Methodology

2.1 Design details of logistic infrastructure using Asphalt Institute Method

The asphalt road structure can be designed by various ways which are widely separated into 2 methods: using experience and analysis. The design using Asphalt Institute Method Vol.8 (1970) [13,14] is one of the design using experience method which has been selected by the department of highway to design asphalt road for 25 years because it is convenient, fast, and non-complex testing method. Therefore, this guide is designed to interesting people and officials who are responsible for the structural design of asphalt to understanding of the principles and processes of pavement roads in Thailand. The structural design of the pavement road by the Asphalt Institute No. 8 is experiential method which brings information taken from the results of the road testing by following AASHTO Road Test, WASHO Road Test, British Road Test and real experience to analyze the relationship between all factors and the depth of

[a] Corresponding author: bannork_civil@hotmail.com

© The Authors, published by EDP Sciences. This is an open access article distributed under the terms of the Creative Commons Attribution License 4.0 (http://creativecommons.org/licenses/by/4.0/).

pavement. Materials in pavement layer following this relationship will be asphalt material called the Full-Depth Asphalt Thickness (T_A).

However, the cost of Full-Depth Asphalt Pavement is high and cannot be replaced by any local materials. Then, the Asphalt institute has set the constant parameter called the substitution ratio to convert the depth of asphalt to the depth of the replaced material which allows designer can select other cheaper local materials and design the pavement structure as multi-layers. According to the study found that the compressive force in the pavement structure relates to the depth as shown in Fig.1. It clearly shows that the compressive force shows the maximum value at the area of the road surface and decreases with increasing the depth. Therefore, to reduce the construction cost, it is often designed using the best quality materials close to the road surface area and inferior quality materials used in deeper layers.

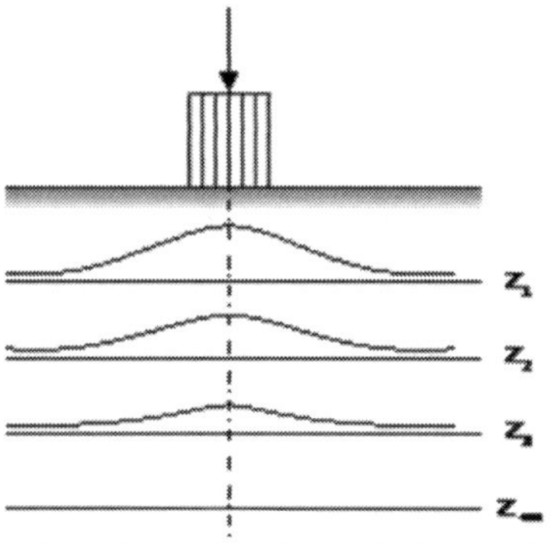

Figure 1. Show the vertical force distribution from vehicle weight at different depths[15].

2.2 Element of design

The primary information used to design of the asphalt road construction following the Asphalt Institute method No.8 are the traffic volume and the load bearing of the soil foundation which are measured in terms of the design Traffic Number (DTN) and CBR, Plate Bearing k-value or R-value, respectively. Details of these values will be explained in Chapter 2.3, 2.4 and 2.5. In addition, the ratio of the strength of materials used in each layer (Substitution Ratios) is another important factor that is used to convert the thickness of full depth asphalt pavement road as a multi structure layer by substitution ratios.

2.3 Traffic volume

Traffic volume used to design following the asphalt institute No.8 is used in the term of design traffic number (DTN) which is the average number of trips per day, the number of vehicles with axle weight standards equivalent to 18,000 pounds which are converted from different car types and axle weights expected to use in traffic lane in all designed duration. To evaluate the DTN value is dependent on various variables such as duration, number of the heavy trucks and the increasing rate of number of heavy trucks etc.

2.4 The strength of the soil foundation

The design of pavement by the Asphalt Institute No.8. [16,17] presents that factors related to the strength of the soil are shown in the form of CBR, Plate Bearing K-value, or R value. However, comparison to all factors, CBR is the most widely used because it is fast and convenient method. Then the department of highway has selected the percentage of CBR method to design the pavement structure following Asphalt Institute No.8. [16,17]. In addition, this work will explain <u>only</u> the CBR testing method.

2.5 Design of asphalt thickness

As mentioned earlier, the designed thickness is the Full Depth Asphalt Thickness (T_A) and then converted to the multi-layer thickness. This thickness design according to the Asphalt Institute No. 8 can be easily calculated which there are only two important parameters related such as DTN and the strength of the soil. Full Depth Asphalt Thickness (T_A) can be obtained by using basic equations developed from experiments or Monograph generated from the above equation. The basic equations used in the design are comprised of three equations as shown in equation (1)-(3). It is dependent on the designer that what information they have. For example, if selects the strength of soil foundation in term of percentage of CBR, equation 1 will be chosen. If there is R-value, equation (2) or (3) can be selected depending on DTN value and so on.

$$T_A = \frac{9.19 + 3.97 \log DTN}{(CBR)^{0.4}} \tag{1}$$

$$T_A = 6.37 + 2.75 \log DTN - 0.0893\ DTN^{0.119}(R - 12) \tag{2}$$

$$T_A = 6.37 + 2.75 \log DTN - 0.117\ DTN^{0.0279}(R - 12) \tag{3}$$

When: T_A = Thickness of asphalt layer

2.6 Research Tools

The analysis and engineering design are following by standards of ASTM, ASSHTO, Asphalt Institute Method, the Department of Highways, Ministry of Transport and geographic information systems and engineering information. The data are collected from engineering property testing as well as the use of geographic information systems for engineering design and analysis. In this study, results are analyzed using data from engineering testing and geographic information systems. An engineering design for the infrastructure and the laboratory tests are processing by the computer as a tool for analysis, including statistics related to the test such as Mean values and a percentage and so on.

3 Results and Discussions

3.1 Topographic and Route Survey

Survey work and the design of road construction starting from Ratchaburi province to the border of Thailand and Myanmar is the construction project promoted for the border business. The beginning point starts at the conjunction of the highway No.4 at km.102+713 through Ratchaburi capital district, Chom-buang district and to meet the border of Thailand - Myanmar at Suan-Phung district. The total distance is approximately 90 km. Due to the initial design, data are only imported form Geographic Information Systems Engineering which comprised of the satellite images and contour line as shown in Figure 2 and 3.

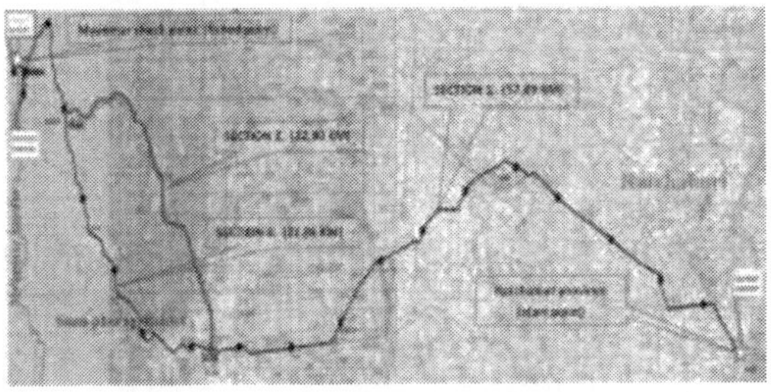

Figure 2. Map of the project line

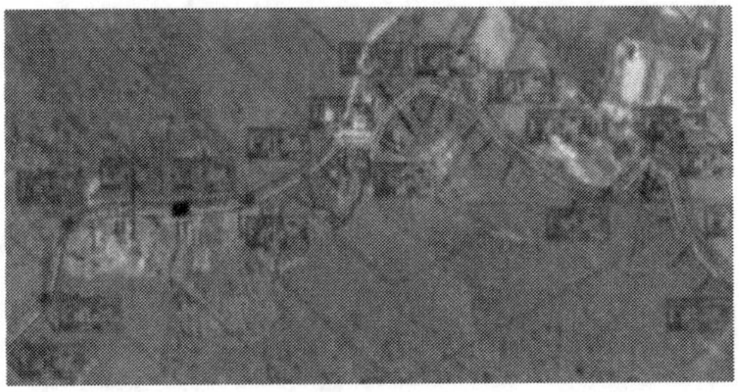

Figure 3. Show the satellite geographic image and contour line to apply for the project design

3.2 Traffic volume survey

By counting the traffic volume of the department of highway in 2015 on the Highway No. 3208, mine is the road that overlaps with the studied route project. Then, a total of three points are inspected. First point is at km.3+000 which has the ADT value of 12, 422 vehicles per day. For second point (km. 19+714) and third point (km. 51+823), the ADT values are 6,755 and 5,893 vehicles per day, respectively.

3.3 Detail of intersection design

Typically, the type of intersections in different places often varies according to the amount of traffic, the direction of turns, and the complexity of the traffic flow. For example, Intersections, where have the traffic volume on the secondary road less than 500 vehicles per day, it would take the typical intersections which is often used to a general link. If inter section has a traffic volume of more than 1,000 vehicles per day, it would be considerably managed the channelization at the conjunction and the traffic light would be considerably used in case a traffic volume of more than 1,500 vehicles per day or other emergency case, for example, the statistics of accidents occurring at intersections that are relatively high and so on. Therefore, the design of intersections at the same level will consider as a type of traffic lights and no traffic lights. Consideration of the various elements mentioned above will refer to the provisions of the HCM (TRB) as the first priority and the design of shape of the geometry will follow the standards of the Department of Rural Roads together with the standard by AASHTO-A policy on Geometric Design of Highway and Streets, 2004 [15] and the visible safety range in other cases can be designed by following to the same standard.

3.4 Analysis of route selection feasibility

The study will start by selecting the most appropriate route. The study will be based on two main factors for instance economics and engineering. In figure 4, there are 2 routes which are compared. In selecting the appropriate route, the data of section 3 of route alternative No.1 and the section 2 of route alternative No.2 are compared since they are difference between the two alternatives.

Table 1 shows the comparison of alternative route No.1(km.0+0.000 – km.91+174). and No.2 (km.0+0.000 – km.93+219). It is found that the first alternative's distance is shorter than the second alternative route about 2.04 kilometers and it is considered that it shows a better geometry because it has less curve which is more convenient and safer for the driver. However, by measuring from both routes have nearly an equal number of vertical curves, the second alternative route has a greater slope and it cuts through the water with a longer distance. It is measured that the total bridge length is 360 meters and the route is more winding and steep. Consideration by economic analysis in construction details, it found that the stability of the road and the prevention of erosion are quite major factor resulting on higher cost of construction of the route No.2. By comparison of engineering factors, it clearly shows that route No.1 is suitable to use as the logistic route in this study.

Table 1. Show the comparison of engineering factors for route No.1 (section 3) and route No.2 (section 2)

Route No.	Horizontal curve (N)	Vertical curve (N)	Slope (%)	Bridge (m.)
1 (Section 3)	60	15	6.2	240
2 (Section 2)	135	14	11.1	360

Table 2. and Figure 2 show a comparison of construction costs of logistic infrastructure designed by using material from Ratchaburi sources. For example, the cost of soil reclamation including transportation, operation, profit and tax is about 110 baths per cubic meter and so on. Therefore, comparing the cost of all materials used for road construction, it is found that the

routes No.2 show a higher value which makes the cost of overall construction projects is also high. Then, if considers only the investment in infrastructure, it can be concluded that the route No.1 is more appropriate than route No.2 which would be used as logistic route from Laemchabang seaport-Nakhonpathom-Highway No.4-Ratchaburi- Jedeehak intersection- Highway No. 3087 and Chong-Takobon border checkpoint.

Table 2 shows the cost of construction infrastructure, starting from Khoa-Ngu intersection (km 0 + 0.000 intersect with Highway No.4, AH2) along the route No.1. to Chong-Takobon checkpoint (km 91. +174) with overall distance of 91.174 kilometers. The total costs calculated only the road and the bridge construction based on economic investment from the government is 3,612 million baths.

Table 2. Show the cost of the logistic infrastructure investment compared between the route No.1 (section 3) and route No.2 (section 2)

Route No.	Cost of Pavement (average 40 million baths/kilometer)	Cost of bridge (average 320,000 baths/meter)	Total (million baths)
1 (section 3)	874.40	7.68	882.08
2 (section 2)	956.00	11.52	967.52
Route No 1. (km.0+0.000-91+174) Total distance of 91.174 km.			3,612.00
Route No 2. (km.0+0.000-93.219) Total distance of 93.219 km.			3,693.00

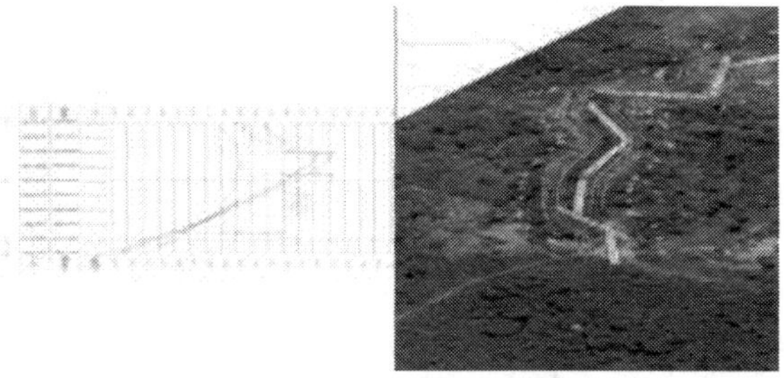

Figure 4. Show the engineering profile design of infrastructure on the Chong-Takobon border (Thailand side) and the border where connecting to the Highway No. 9. (Myanmar side)

3.5 The selection of the routes by the project analysis

According to study above, the appropriate route to use for logistic transportation considered by engineering infrastructure factor is the route No.1. Next, environmental effects that will be analyzed and considered the suitability to start the construction process including the implementation of the project in the next step. By counting the traffic volume of Department of Highway, it is found that the maximum traffic volumes is 12,422 vehicles per day, and by the standards of the Department of Highways, the road should be designed as four lanes (special class). For crossing the natural channel, the short bridge will be used and for mountain section, the suitable structure design will be considered to fit with the shape of the terrain which might cut deep or fill high. Moreover, the design will add special lanes for trucks to climb steep slopes (Climbing Lane), so four-lane roads with lane width of 3.50 meters and shoulder lane width of 2.50 meters are designed. Including the middle island with Painted Median type of 1.60 meters in width is used. In addition, for the area, where accesses to the intersection, the middle mountain is designed as a raised is mountain (raised median) of 4.60 meters in width because it is convenience and safety for vehicles waiting to turn as shown in Figure 5. For the road at the hillside area, the road is designed as 4 lanes with the width of 3.50 meters and shoulder with the width of 2.50 meter. the Barrier is used to divide the traffic direction at the curve area to reduce an accident. The size of the middle island is 1.60 meters in width. The road is also designed to prevent the erosion of the slope. The cross-section picture is shown in Figure 6.

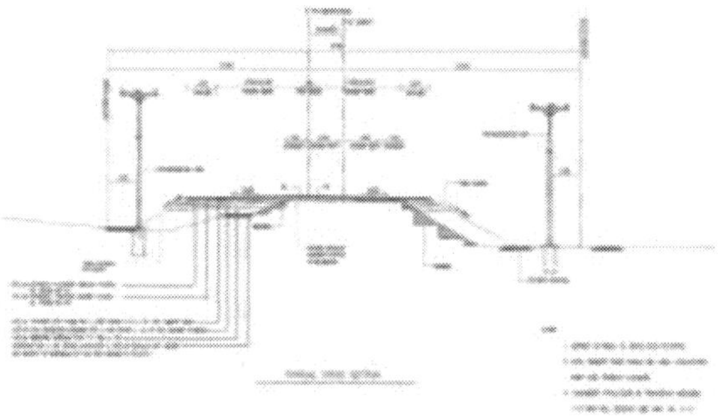

Figure 5. Show the cross-section of the road structure

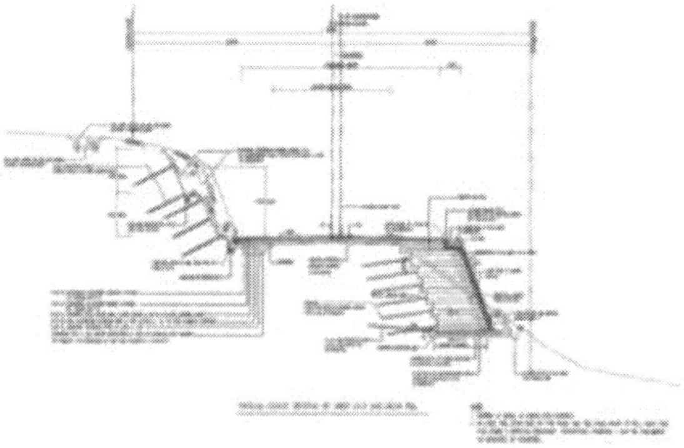

Figure 6. Show the cross-section of the road structure at hillside area

4 Conclusions

4.1 Consideration by various factors, it can conclude that road design can be shown in Figure 6 and 7 according to engineering format. For economic analysis, it can be seen that route No.1 is the most appropriate choice with the value of the investment approximately 3,612 million baths. Also, such analysis indicates that if the government of Thailand decides to invest in this project, it is not only for Thailand side but also they should consider investing in the important infrastructure in the border area including at Myanmar side to support the connection of the two countries in economy.

4.2 Continuous development format and the link with the project which the government should support. The existing transportation route starting from Lamchabang seaport - Bangkok - Nakhonpathom province was studied in the terms of logistic and supply chain. This work, the extended route Ratchaburi - Suanpuang - Chong Tagobon will be studied. This route will be constructed to be the new transportation and transfer of goods route from Thailand to Myanmar via Chong Tagobon, Ratchaburi province to the highway No.9 of Myanmar. Therefore, to develop this project continuously.

4.3 Develop the tourist attraction widely known as Suan-Phuang district. The development should be appropriate and study and environmental impact including the forest conservation together with the development of preliminary infrastructure. Develop and review, including the clarity of the border open at the level of international cooperation (Thailand - Myanmar).

4.4 The cooperation of the public and private sectors of Thailand - Myanmar in service trucks connection between the dock and the industrial estates of Thailand (Laem-Chabang Industrial Estate - Klong - Toei seaport) and the Myanmar (Dawei industrial estate).

4.5 Economics, social and stability should be discussed and cooperated in the development of Highway No. 9, where is the main transportation of Chong-Takobon border to Dawei industrial estate.

References

1. Grant-Muller, S.M., Xu, M.,Transp. Rev. **34(2)**, (2014)
2. Posner, R.A., BellJ. Econ. **5(2)**, (1974)
3. Edwards, H., Berglund, M., European Commission DGVII, (1999)
4. Guasch, J.L.,Hahn,R.W., World Bank Res. Observ.**14(1)**, (1999)
5. Estache, A., World Bank Res. Observ. **16(1)**, (2001)
6. Harris, C., World Bank Working Paper 5.World Bank Washington, DC.(2003).
7. Fan,S.,Chan-Kang,C., Transp. Policy. **15(5)**, (2008).
8. Xu,M.,Ceder,A.,Gao,Z.Y.,Guan,W., Transportation, **37(5)**, (2010)
9. Yang, H.J., Zhang, A.M., J.Transp. Econ. Policy, **46(3)**, (2012)
10. NERA Economic Consulting., URN09D-596, UK, (2009)
11. Wang, G.M., Gao, Z., Xu, M., Sun, H. J., Transp. Res. Part A. **67**, (2014).
12. Helm, D., Thompson, K., J.Transp.Econ.Policy. **25**, (1991)
13. AASHTO-A policy on Geometric Design of Highway and Streets, (2004).
14. Thickness Design-Asphalt Pavements for Highways and Streets, The Asphalt Institute, College Park, Md., (1981)
15. Handbook Asphalt Institute Method. Department of Highway, Ministry of Transport. **8**, (1970)

A Review on Early Age and Long-term Compressive Strength of High-volume Fly Ash Concrete

Mu Li[1,a]

[1] *Beijing University of Civil Engineering and Architecture, China*

Abstract. Fly ash is a by-product of the combustion of the coal-fired electric power stations, and disposal of fly ash has been one of the environmental challenges. Much of the studies have been focused on the mechanical property of fly ash concrete. It is no doubt that the use of high-volume fly ash as a partial replacement of cement is also one of the effect way to utilize fly ash. It is known that the compressive strength of fly ash concrete is lower than that of ordinary concrete at early age, especially for high-volume fly ash concrete. It is urgent for engineers to consider the compressive strength of high-volume fly ash concrete at different curing age. In this review, the compressive strength of high-volume fly ash concrete in various literature was reported and then analyzed. Furthermore, the proposal of the utilization of high-volume fly ash concrete is provided.

1 Introduction

Concrete is the most common building material around the world, and more than 10 billion tons of concrete are made each year[1]. Portland cement is the necessary component of the concrete. The manufacturing process for Portland cement is a highly energy-intensive process that emits a large amount of green-house gas[2]. The green-house gas is definitely a contributor to the greenhouse effect and global warming. Additionally, the amount of fly ash, released by factories and thermal power plants, has increased throughout the whole world, and the disposal of this large amount of fly ash has become a serious environmental challenge. The role of fly ash(FA) in the blended cement and concrete has been investigated widely in the recent years due to the benefit of reducing emission of CO2 and improving the performance of concrete as well as reducing the cost of construction.

High-volume fly ash(HVFA) concrete with the properties of high-performance concrete was first developed by the Canadian Centre for Mineral and Energy Technology (CANMET) in 1985[3]. High-volume fly ash concrete has emerged as useful construction material as it can consume much more fly ash than FA concrete, and HVFA concrete is also defined as concrete which contains 50% or more than 50% fly ash by weight of total binder materials[4, 5]. The heat of hydration of HVFA concrete is lower than that of ordinary concrete and therefore the thermal cracking of HVFA concrete is less than that of ordinary concrete. Thus, it can be used in massive concrete structure.

As manufacturing concrete and building up the structure, compressive strength of concrete is the property most valued by structure designers and constructors. Although HVFA concrete has positive effect on durability of concrete[6-10], the compressive strength of HVFA concrete is lower than that of ordinary concrete at early age, the difference between the long-term compressive strength of HVFA concrete and ordinary concrete is still needed to be considered. In this review, the compressive strength of HVFA concrete is analyzed with previous literatures in order to utilize HVFA concrete better in the practical engineering.

2 Fly ash

According to American Society for Testing and Materials ASTM-C618[11], FA is classified into two sorts: Class F and Class C, where Class F fly ash has pozzolanic properties, Class C fly ash has pozzolanic properties and also some cementitious properties. The CaO content of Class F FA is nearly equal to 10% or less than 15%, in the meantime, the CaO content of Class C FA ranges from 15% to 40%[12]. Low-calcium FA (Class F) reacts slowly compared with high-calcium FA (Class C). Without description in this review, the default type of fly ash is Class F.

3 Early age and long-term compressive strength

Compressive strength development is one of the important mechanical properties of hardening concrete. It relates to the mold turnover and the construction quality of the building.

HVFA concrete with 50% Class C FA replacement level by mass showed lower compressive strength than fly ash concrete with 25% Class C FA replacement level by mass

[a] Corresponding author: limu26@126.com

© The Authors, published by EDP Sciences. This is an open access article distributed under the terms of the Creative Commons Attribution License 4.0 (http://creativecommons.org/licenses/by/4.0/).

in early age, however, this trend reversed at the curing age of 90 days[13].

Rafat Siddique et al[14] found that 53.4%, 39.8%, and 39.1% reduction in the compressive strength of concrete at curing age of 7 days, 28 days and 56 days by partially replacing cement with 50% FA, respectively, at water/binder ratio of 0.44. The compressive strength of HVFA concrete with 50% Class C FA replacement was 22.6%, 21.7% and 18.3% less than that of ordinary concrete, respectively, for 3, 14 and 28 days curing[15]. When the curing period is 28 days, compressive strength of HVFA concrete increased by nearly 20% as the FA content increased from 50% to 70% by mass of total cementitious materials[16]. Michelle R. Nokken et al[17] found nearly 47.6% and 85.7% reduction of compressive strength, respectively, for 60% and 80% FA replacement level by mass at the curing age of 28 days. Min-Hong Zhang and Jahidul Islam[18] found that the compressive strength of HVFA concrete with 50% FA replacement level and water-binder ratio of 0.45 was 19.5MPa, 28.1MPa, 43MPa and 50.2MPa, respectively, for 3 days, 7days, 28days and 91days.

As the replacement level of Class C FA by mass is 75%, the difference of compressive strength between HVFA concrete and ordinary concrete is the highest for 3 days, and the lowest for 28 days at the water-binder ratio of 0.4[19]. Faiz U.A. Shaikh and Steve W.M. Supit[20] found that the difference of compressive strength between HVFA concrete with 60% FA replacement level and ordinary concrete is the lowest for 7 days compared with the experimental values of 3 days, 28 days, 56 days and 90 days at the water-binder ratio of 0.4. For 50% FA replacement level, this difference is the highest for 14 days at the water-binder ratio of 0.4[21].

W. Chalee et al[22] found that the difference of compressive strength between HVFA concrete with 50% FA replacement by mass and ordinary concrete is the lowest for the water-binder ratio of 0.65 at the curing age of 28 days among the water-binder ratio of 0.45, 0.55 and 0.65. A. Durán-Herrera et al[23] also found the same trend that the difference of compressive strength between HVFA concrete with 60% and 75% FA replacement by mass and ordinary concrete is the lowest for the water-binder ratio of 0.6 at the curing age of 7days, 21 days, 28 days and 56 days among the water-binder ratio of 0.5, 0.55 and 0.6.

C.D. Atiş[24] found a reduction in compressive strength of HVFA concrete with 50% and 70% FA replacement level by mass. The reduction in compressive strength was 53.3%, 17.1%, 22.9%, 6.2%, 7.4%, 2.6% and 4.8%, respectively, at curing period of 1 day, 3 days, 7 days, 28 days, 3 months, 6 months and 1 year for HVFA concrete with 50% FA replacement level, while corresponding reduction was 85.4%, 57.4%, 51.3%, 45.3%, 37.3%, 38.6% and 36.6% for HVFA concrete with 70% FA replacement level. It was also shown that the difference of compressive strength between HVFA concrete with 50% FA replacement level and reference concrete is only 3.4MPa as the curing time is 1 year.

Guðmundur Hannesson et al[25] found that the compressive strength of HVFA concrete with 60% Class C FA replacement level was higher than that of reference concrete at the curing age of 56 days, while the corresponding curing age was 168 days for HVFA concrete with 80% Class C FA replacement level. For Class F FA, the compressive strength of HVFA concrete with 60% Class F FA replacement level was higher than that of reference concrete at the age of 84 days, while the compressive strength of HVFA concrete with 80% Class F FA replacement level was lower than that of reference concrete at all curing ages.

Chung-Ho Huang et al[26] found that the compressive strength of HVFA concrete with 60% FA replacement level nearly approached that of reference concrete at the curing age of 91 days, while the corresponding curing age was 182 days for HVFA concrete with 80% FA replacement level. Compared with the experimental values at the curing age of 7 days, the compressive strength gain of the control concrete was less than that of HVFA concrete at the curing ages of 28 days, 56 days, 91 days, 182 days and 365 days.

Aliakbar Gholampour and Togay Ozbakkaloglu[27] replaced the high early strength cement(HESC) with ordinary Portland cement(OPC), and the main difference between HESC and OPC is the value of specific surface area(Specific surface area is 330 m2/kg for OPC, while 390 m2/kg for HESC.). Compared with compressive strength of HVFA concrete with 50% FA replacement level by mass with 3 days curing, the values of compressive strength were increased by 36%, 156% and 203%, respectively, for 7 days, 28 days and 90 days. In the same way, for HVFA concrete with 60% and 90% FA replacement level, the values of compressive strength were increased by 7% and 67%, 166% and 378%, 263% and 1089%, respectively, for 7 days, 28 days and 90 days. The higher the FA replacement level was, and consequently the more compressive strength gain of HVFA concrete. Katherine Kuder et al[28] got the same trend of the compressive strength gain of HVFA concrete compared with the experimental values of 7 days, and additionally the fly ash type is Class C.

Xiao-Yong Wang and Ki-Bong Park[29] found that the compressive strength of HVFA concrete with 55% FA replacement level might not surpass that of control concrete at later age when the ratio of water to binder is higher, but the compressive strength of HVFA concrete could be higher than that of control concrete as the water to binder ratio was lower. It could be seen that a lower water to binder ratio is more suitable for HVFA concrete.

Seyoon Yoon[30] found that when the curing period is 365 days and the amount of the binder is fixed in all of the concrete mixtures(450kg/m3), the compressive strength of HVFA concrete with 50% FA replacement level is higher than that of ordinary concrete as the ratio of water to binder is 0.3, while the trend reversed as water to binder ratios were 0.35 and 0.4. The compressive strength of HVFA concrete with 60% fly ash was lower than that of ordinary concrete in all of the water to binder ratios (0.3, 0.35 and 0.4), and the differences of compressive strength between HVFA concrete with 60% FA replacement level and ordinary concrete were 4.2MPa, 13.9MPa and 10.9MPa, respectively, for water to binder ratio of 0.3, 0.35 and 0.4, in the meantime, for HVFA concrete with 50% fly ash replacement, the corresponding differences were -7.2MPa, 14.5MPa and 8.5MPa. Thus, it can be seen that the lower the water to binder ratio was, the lower the difference between HVFA concrete and ordinary concrete.

4 Conclusion

HVFA concrete has been used in the study of the concrete component[31] and the field application[32]. The application of HVFA concrete in the real structure is on the way. It is necessary for engineers to understand the influence of time-dependent effect on the compressive strength of HVFA concrete in relation to the analysis and design of concrete structures. Utilizing the long-term compressive strength of HVFA concrete is preferable for construction reality if the early age compressive strength of HVFA concrete is not emphasized in field construction. The compressive strength gain of HVFA concrete is higher than that of ordinary concrete at different curing ages. The higher the ratio of water to binder is, the lower the difference of compressive strength of HVFA concrete and ordinary concrete.

References

[1] C. Meyer, The greening of the concrete industry, Cement and Concrete Composites 31(8) (2009) 601-605.
[2] J. Yu, C. Lu, C.K.Y. Leung, G. Li, Mechanical properties of green structural concrete with ultrahigh-volume fly ash, Construction and Building Materials 147 (2017) 510-518.
[3] A. Bilodeau, V.M. Malhotra, High-Volume Fly Ash System: Concrete Solution for Sustainable Development, Aci Materials Journal 97(1) (2000) págs. 41-48.
[4] P.K. Mehta, High-performance, high-volume fly ash concrete for sustainable development, 2004 Symposium on cement concrete technology and sustainable development 24 (2002).
[5] H.A. Alaka, L.O. Oyedele, High volume fly ash concrete: The practical impact of using superabundant dose of high range water reducer, Journal of Building Engineering 8 (2016) 81-90.
[6] E.G. Moffatt, M.D.A. Thomas, A. Fahim, Performance of high-volume fly ash concrete in marine environment, Cement and Concrete Research 102 (2017) 127-135.
[7] P. Jiang, L. Jiang, J. Zha, Z. Song, Influence of temperature history on chloride diffusion in high volume fly ash concrete, Construction and Building Materials 144 (2017) 677-685.
[8] O. Karahan, Transport properties of high volume fly ash or slag concrete exposed to high temperature, Construction and Building Materials 152 (2017) 898-906.
[9] F.U.A. Shaikh, S.W.M. Supit, Chloride induced corrosion durability of high volume fly ash concretes containing nano particles, Construction and Building Materials 99 (2015) 208-225.
[10] P. Van den Heede, E. Gruyaert, N. De Belie, Transport properties of high-volume fly ash concrete: Capillary water sorption, water sorption under vacuum and gas permeability, Cement and Concrete Composites 32(10) (2010) 749-756.
[11] ASTM-C618, Standard Specification for Coal Fly Ash and Raw or Calcined Natural Pozzolan for Use in Concrete, ASTM International.
[12] P.J.M. Monteiro, Concrete : Microstructure, Properties, and Materials / P.K. Mehta, P.J.M. Monteiro, McGraw-Hill Professional (1993).
[13] R.U.D. Nassar, P. Soroushian, T. Ghebrab, Field investigation of high-volume fly ash pavement concrete, Resources Conservation & Recycling 73(2) (2013) 78-85.
[14] R. Siddique, K. Kapoor, E.H. Kadri, R. Bennacer, Effect of polyester fibres on the compressive strength and abrasion resistance of HVFA concrete, Construction & Building Materials 29(4) (2012) 270-278.
[15] Y. Khodair, B. Bommareddy, Self-consolidating concrete using recycled concrete aggregate and high volume of fly ash, and slag, Construction and Building Materials 153 (2017) 307-316.
[16] I. Ignjatović, Z. Sas, J. Dragaš, J. Somlai, T. Kovács, Radiological and material characterization of high volume fly ash concrete, Journal of Environmental Radioactivity 168 (2017) 38-45.
[17] M. Nokken, Standardized Testing for Determining the Durability of High Volume Fly Ash Mixtures, Journal of Materials in Civil Engineering 97(26) (2012) 206-210.
[18] M.-H. Zhang, J. Islam, Use of nano-silica to reduce setting time and increase early strength of concretes with high volumes of fly ash or slag, Construction and Building Materials 29 (2012) 573-580.
[19] Y. Khodair, M. Raza, Sustainable self-consolidating concrete using recycled asphalt pavement and high volume of supplementary cementitious materials, Construction and Building Materials 131 (2017) 245-253.
[20] F.U.A. Shaikh, S.W.M. Supit, Compressive strength and durability properties of high volume fly ash (HVFA) concretes containing ultrafine fly ash (UFFA), Construction and Building Materials 82 (2015) 192-205.
[21] X. Wei, H. Zhu, G. Li, C. Zhang, L. Xiao, Properties of high volume fly ash concrete compensated by metakaolin or silica fume, Journal of Wuhan University of Technology-Mater. Sci. Ed. 22(4) (2007) 728-732.
[22] W. Chalee, P. Ausapanit, C. Jaturapitakkul, Utilization of fly ash concrete in marine environment for long term design life analysis, Materials & Design 31(3) (2010) 1242-1249.
[23] A. Durán-Herrera, C.A. Juárez, P. Valdez, D.P. Bentz, Evaluation of sustainable high-volume fly ash concretes, Cement and Concrete Composites 33(1) (2011) 39-45.
[24] C.D. Atiş, High-Volume Fly Ash Concrete with High Strength and Low Drying Shrinkage, Journal of Materials in Civil Engineering 15(2) (2003) 153-156.
[25] G. Hannesson, K. Kuder, R. Shogren, D. Lehman, The influence of high volume of fly ash and slag on the compressive strength of self-consolidating concrete, Construction & Building Materials 30(30) (2012) 161-168.
[26] C.-H. Huang, S.-K. Lin, C.-S. Chang, H.-J. Chen, Mix proportions and mechanical properties of concrete containing very high-volume of Class F fly ash, Construction and Building Materials 46 (2013) 71-78.

[27] A. Gholampour, T. Ozbakkaloglu, Performance of sustainable concretes containing very high volume Class-F fly ash and ground granulated blast furnace slag, Journal of Cleaner Production 162 (2017) 1407-1417.
[28] K. Kuder, D. Lehman, J. Berman, G. Hannesson, R. Shogren, Mechanical properties of self consolidating concrete blended with high volumes of fly ash and slag, Construction & Building Materials 34(34) (2012) 285-295.
[29] X.-Y. Wang, K.-B. Park, Analysis of compressive strength development of concrete containing high volume fly ash, Construction and Building Materials 98 (2015) 810-819.
[30] S. Yoon, P.J.M. Monteiro, D.E. Macphee, F.P. Glasser, M.S.-E. Imbabi, Statistical evaluation of the mechanical properties of high-volume class F fly ash concretes, Construction and Building Materials 54 (2014) 432-442.
[31] S.-W. Yoo, G.-S. Ryu, J.F. Choo, Evaluation of the effects of high-volume fly ash on the flexural behavior of reinforced concrete beams, Construction and Building Materials 93 (2015) 1132-1144.
[32] M. Reiner, K. Rens, High-Volume Fly Ash Concrete: Analysis and Application, Practice Periodical on Structural Design & Construction 11(1) (2006) 58-64.

MATEC Web of Conferences **207**, 02001 (2018) https://doi.org/10.1051/matecconf/201820702001
ICMMPM 2018

Peak Discharge and Hydrograph Assessments Induced by Heavy Rainfall Events Using Tank Model

Yen-Kun Hsu[1], Szu-Hsien Peng[1,a] and Chiao-Wen Tsai[1]

[1]*Department of Spatial Design, Chienkuo Technology University, Changhua City 500, Taiwan, R.O.C.*

Abstract. Tank Model is a kind of simulation of rainfall movement in soil horizon. With the runoff and piping rate, the peak discharge could be effectively calculated. Having 17 rain gauge stations in 13 debris flow events during 1996-2010 as the studied cases, the peak discharge at 12 control points along Chenyulan River is simulated. Furthermore, the data in Neimaopu discharge station is established parameters of Tank Model to estimate the peak discharge in Shenmu Village. By comparing with the parameters of Shueili Station and Japanese Granite, the mean error of the parameter in this study is 51.0%, which is better than those of Japanese Granite 189% and Shueili discharge stations 251%. The parameter in this study appears the highest in allowance analysis, showing that it is more suitable for simulating the peak discharge than the other two. In spite that the percentage of the three parameters is still low, Shenmu Village could be ignored as it locates in the sub-basin of Chenyulan River with few factors. The parameters of Tank Model are applied to transform average rainfall into hydrograph so as to solve the problem of no discharge records when analysing the areas with various debris flow simulation programs.

1 Introduction

Rainfall and peak discharge are the factors causing sediment disasters in slopeland. Since the Unit Hydrograph for simulating effective rainfall and hydrograph was proposed [1-2], it has been largely studied by researchers. In addition to effective rainfall causing direct runoff, the rainfall mobile mechanism of rainfall infiltrating to soil horizon and further effusing as runoff becomes the blind spot for effective rainfall and hydrograph. Not until the rainfall moving in soil with a four-layer tank was simulated [3], the proposed Tank Model became one of the major methods to simulate rainfall and peak discharge in hydrology. With a three-layer tank and eleven parameters, the study [4] calculated hydrograph by dividing sub-basin, analysed the relations between the rainfall and runoff in the watersheds of Kohira, Minamihatajiki, Tsukigase, Mino, and Kiyohorobashi, established five Tank Model parameters for the five watersheds, and confirmed four parameters of Tank Model from the geological distribution in the watersheds (see Table 1). Up to then, Japanese researchers followed such a framework for further studies. In 1990, Japan Meteorological Agency Forecast applied Automated Meteorological Data Acquisition System (AMEDAS) and 80% detectivity to national verification and found the difference between simulation and actual measurement merely several percentage [5] when utilizing the parameters of granite and other geology. As a consequence, the parameter of granite was utilized for forecast.

Although Tank Model has been used for years in Japan, it is still under research and promotion in Taiwan. The previous study [6] estimated the effective rainfall and direct runoff of rainfall events with soil water index Φ infiltration and further simulated the rainfall runoff in sub-watersheds with the built-in kinematic-wave based geomorphic IUH model, triangular Unit Hydrograph, linear reservoir model, dimensionless hydrograph method, and Tank Model. A hydrologic process-based rainfall-runoff model was also established for the earth surface and subsurface [7-10]. The subsurface runoff model is structured with two in-line Tank Models to simulate interflow and groundwater flow; the total is the runoff. Another research [11] increased vegetation covering and established rainfall-runoff model with a five-layer tank. Nonetheless, most researchers still focused on direct runoff and effusion from soil horizon caused by water after rainfall; the total runoff was not discussed the water concentrating on channel flow. According to Chingualiaochiao, Chinshan, Xinanchiao, Shueili, and Tukuchiao discharge stations, the parameter of Tank Model was established by applying the Least Square Method [12]. Based on the research framework of this research, the peak discharge simulation is preceded in this study.

The researchers [13] proposed that Tank Model is a conceptual model which simulates the moving behaviour of water in the soil layers including runoff, infiltration and percolation, because the sum of rainfall depth as "depth of rainfall retention in soil layer" in three tank of tank model. Therefore, they referred to as a "soil water

[a] Corresponding author: shpeng@cc.ctu.edu.tw

© The Authors, published by EDP Sciences. This is an open access article distributed under the terms of the Creative Commons Attribution License 4.0 (http://creativecommons.org/licenses/by/4.0/).

index (SWI)," which as the national criteria of disaster preparedness phase in Japan. Besides, Tank Model is also employed to simulate groundwater [14-16], and many optimization methods [17-19] are proposed to raise the accuracy of this model. Therefore, the Tank Model is still one of the commonly used models to estimate the relationship of rainfall and runoff.

Table 1. Parameters of Tank Model in Japan and Taiwan.

Area	L_1	L_2	L_3	L_4	a_1	a_2	a_3	a_4	b_1	b_2	b_3
Japan, Ishihara (1979)											
Granite	15	60	15	15	0.10	0.15	0.05	0.01	0.12	0.05	0.01
Volcanic rocks	30-40	60-75	15	15	0.10	0.15	0.05	0.01	0.12	0.08	0.01
Paleozoic	30	75	5	15	0.10	0.15	0.05	0.01	0.12	0.04	0.01
Quaternary	15	40	5	15	0.10	0.15	0.05	0.01	0.12	0.04	0.01
Taiwan, Chen (2011)											
Chingualiaochiao	15	30	5	5	0.15	0.3	0.03	0.02	0.15	0.03	0.02
Chinshan	20	75	20	20	0.05	0.2	0.05	0.01	0.40	0.20	0.05
Xinanchiao	10	30	15	15	0.10	0.4	0.10	0.01	0.10	0.05	0.05
Shueili	15	60	15	15	0.10	0.2	0.10	0.05	0.10	0.05	0.01
Tukuchiao	5	30	15	15	0.10	0.2	0.10	0.01	0.10	0.05	0.01
Suggested parameters in this study											
Chenyulan River	30	100	100	100	0.1	0.2	0.02	0.01	0.3	0.02	0.01

Equation

$R_1=(H_1-L_1)\times a_1$
$R_2=(H_1-L_2)\times a_2$
$R_3=(H_2-L_3)\times a_3$
$R_4=(H_3-L_4)\times a_4$
$F_1=H_1\times b_1$
$F_2=H_2\times b_2$
$F_3=H_3\times b_3$

L_1: First tank Layer 1 discharge depth
L_2: First tank Layer 2 discharge depth
L_3: Second tank discharge depth
L_4: Third tank discharge depth
a_1: First tank Layer 1 discharge rate
a_2: First tank Layer 2 discharge rate
a_3: Second tank discharge rate
a_4: Third tank discharge rate
b_1: First tank infiltration rate
b_2: Second tank infiltration rate
b_3: Third tank infiltration rate
R_1: First tank Layer 1 discharge
R_2: First tank Layer 2 discharge
R_3: Second tank discharge
R_4: Third tank discharge
H_1: First tank depth
H_2: Second tank depth
H_3: Third tank depth

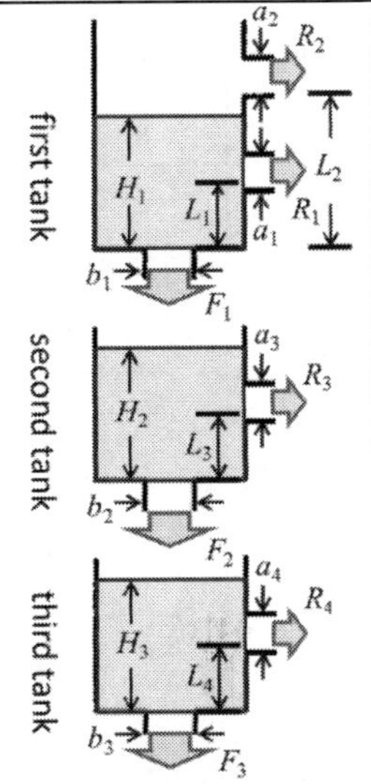

a_1~a_4 & b_1~b_3 unit: no unit

L_1~L_4, R_1~R_4 & H_1~H_4 unit: mm (millimeter)

2 Materials and methods

2.1 Research area

Chenyulan River watershed in Nantou County is one of the areas with the most debris flow in Taiwan, where sixteen debris flows appeared in Shenmu Village during 1996-2010. Neimaopu discharge station is the only one still working in Chenyulan River watershed; however, it is far away from Shenmu Village that the peak discharge is hard to estimate. Taking the thirteen debris flow events during 1996-2010 as the cases, the peak discharge at twelve control spots along Chenyulan River is simulated. Data of Neimaopu discharge stations are established the parameter of Tank Model for estimating the peak discharge of Shenmu Village. The thirteen debris flow events occurred in Shenmu Village, located along Huosa River and Chushuei River, the upstream of Chenyulan River, is simulated the discharge.

2.2 Research method

The hourly precipitation around rain gauge stations is first collected for calculating the average hourly precipitation. With Kriging method in ArcGIS is utilized for analyses in order to draw the hourly discharge distribution.

Chenyulan River watershed is divided into thirty sub-basins, which are numbered as Figure 1. According to these sub-basins in Figure 1, the main tributary control spots, depending on the afflux, are set twelve, from *A* to *L*, where the control spot *I* and Neimaopu discharge station are overlapped.

Table 2. Debris flow events in Shenmu Village.

No.	Event	Date
1	Typhon Herb	1996/08/01
2	Heavy Rainfall	1999/05/28
3	Typhon Toraji	2001/07/30
4	Typhon Mindulle	2004/07/02

5	Typhon Haitang	2005/07/19
6	Typhon Matsa	2005/08/04
7	Typhon Talim	2005/09/01
8	Heavy Rainfall	2006/06/09
9	Heavy Rainfall	2007/08/13
10	Typhon Sepat	2007/08/18
11	Typhon Krosa	2007/10/06
12	Typhon Kalmaegi	2008/07/18
13	Typhon Morakot	2009/08/08

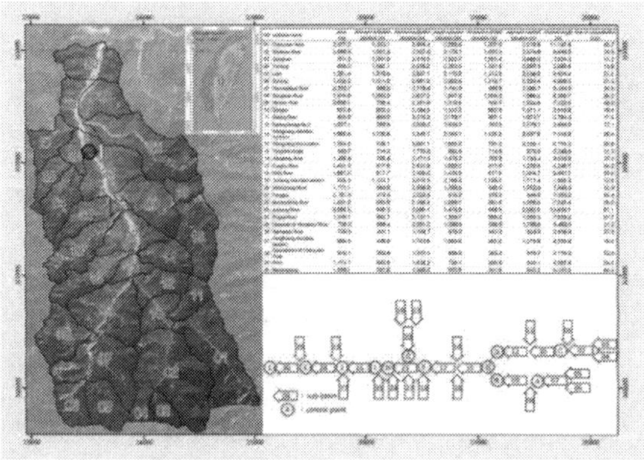

Figure 1. Distribution of sub-basin in the upstream and discharge balance of Chenyulan River watershed.

The discharge of each control spot is calculated by summing up the discharge of each sub-basin during the time of concentration at the control spots. The time of concentration (t_C) is the sum of inflow time of slope (t_1) and the outflow time of channel (t_2). The equations for the time of concentration at each sub-basin are shown as follows [20].

$$t_C = t_1 + t_2 \quad (1)$$

$$t_1 = L_1/v \quad (2)$$

$$t_2 = L_2/W \quad (3)$$

$$W = 72(H/L_2)^{0.6} \quad (4)$$

where, t_C is the time of concentration (hr), t_1 the inflow time of slope (hr), t_2 the outflow time of channel (hr), L_1 spreading length (m), L_2 the length of river channel (km), v the overland flow velocity (0.3-0.6 m/sec in general), W the channel velocity (km/hr), and H the elevation difference of river channel (m).

The difference in the time of concentration between each control spot and each sub-basin is regarded as the time difference between the peak discharges of each sub-basin to each control spot. The peak discharge of each sub-basin to the control spot is summed every hour to establish the hydrograph of each control spot.

Allowance analysis aims to understand the closeness between the simulated parameter of Tank Model to the observed value in discharge stations. The structure of allowance analysis shows (1) the important information in deducting the observed error from the simulated parameter, (2) certain limitation on accuracy being acceptable as the model cannot be perfect, and (3) the accuracy of the parameter being definable to understand the limitation of the model.

3 Results and discussions

3.1. Peak discharge simulation results

From the discharge simulation results, the average hourly precipitation and the peak discharge of the thirty sub-basin are shown as Figure 2.

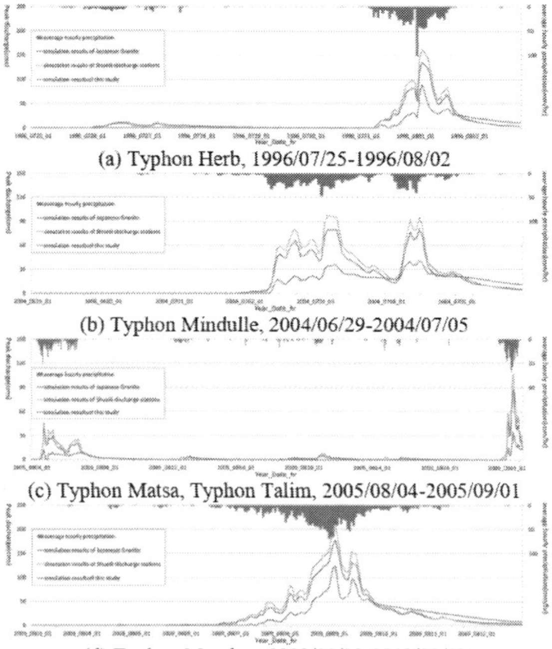

(a) Typhon Herb, 1996/07/25-1996/08/02

(b) Typhon Mindulle, 2004/06/29-2004/07/05

(c) Typhon Matsa, Typhon Talim, 2005/08/04-2005/09/01

(d) Typhon Morakot, 2009/08/03-2009/08/12

Figure 2. Average hourly precipitation and simulated peak discharge of Lulin sub-basin in Shenmu Village.

The hydrographs simulated by the parameters of Japanese Granite and Shueili discharge station appear large difference with the observed value of Neimaopu discharge station using the data of Typhoon Mindulle as the training sample. With the parameters to simulate the hydrograph at the control spot *I* in the thirteen hydrologic events, they are further compared with the observed values of Neimaopu discharge station, Figure 3. The discharge simulated by the parameters of Japanese granite and Shueili discharge station is obviously doubled the observed discharge of Neimaopu discharge station.

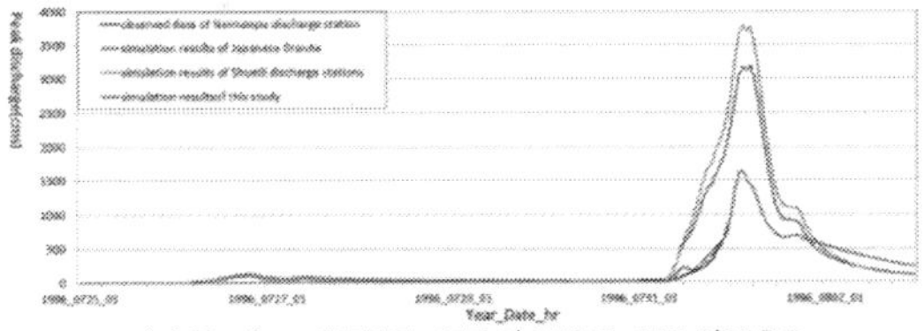

(a) Typhon HERB, 1996/ 07/25~1996/08/02

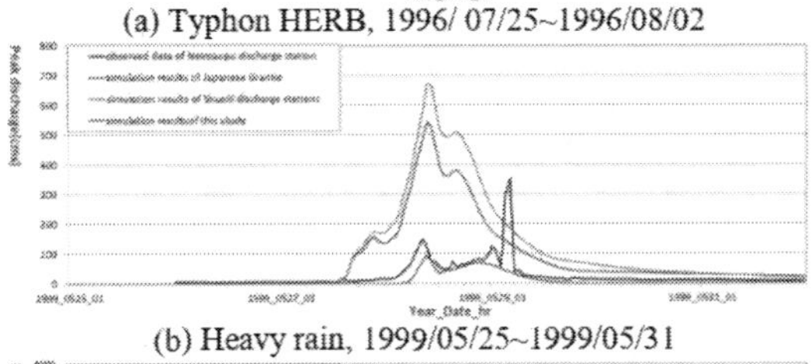

(b) Heavy rain, 1999/05/25~1999/05/31

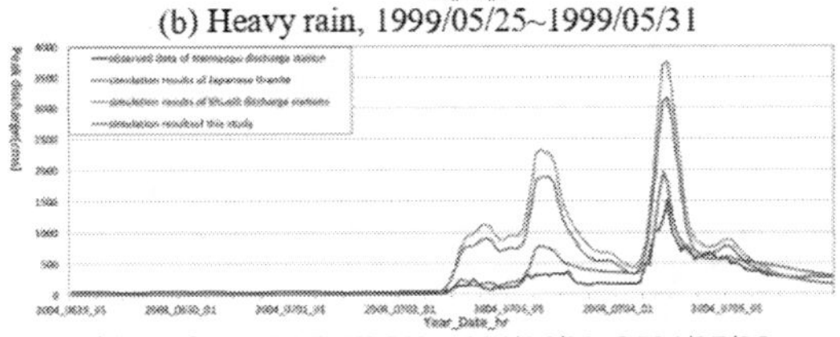

(c) Typhon MINDULLE, 2004/06/29~2004/07/05

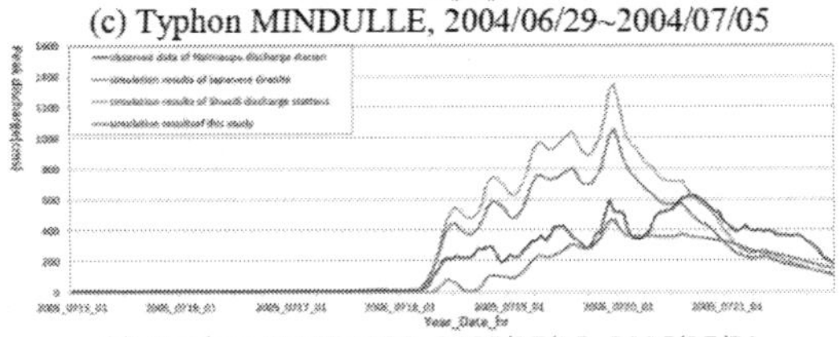

(d) Typhon HAITANG, 2005/07/15~2005/07/21

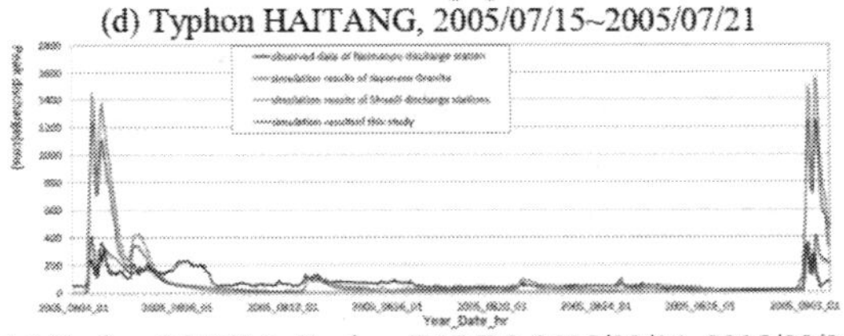

(e) Typhon MATSA, Typhon TALIM, 2005/08/04~2005/09/01

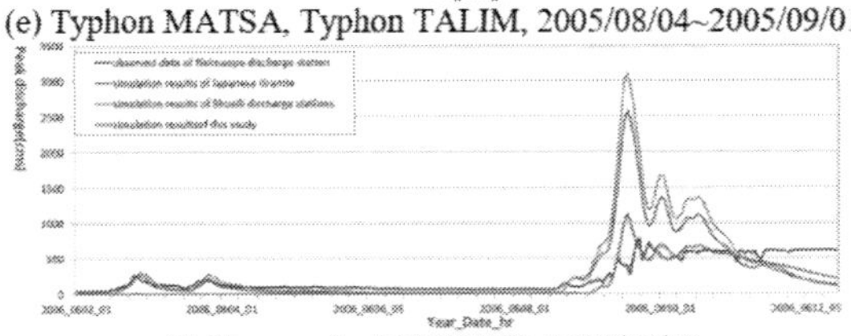

(f) Heavy rain, 2006/06/02~2006/06/12

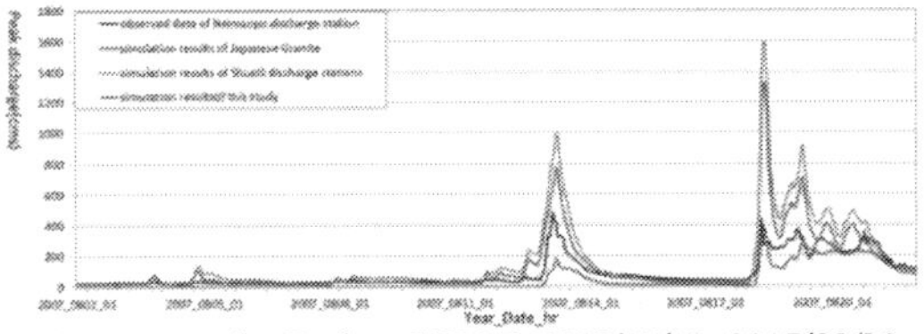

(g) Heavy rain, Typhon SEPAT, 2007/08/02~2007/08/21

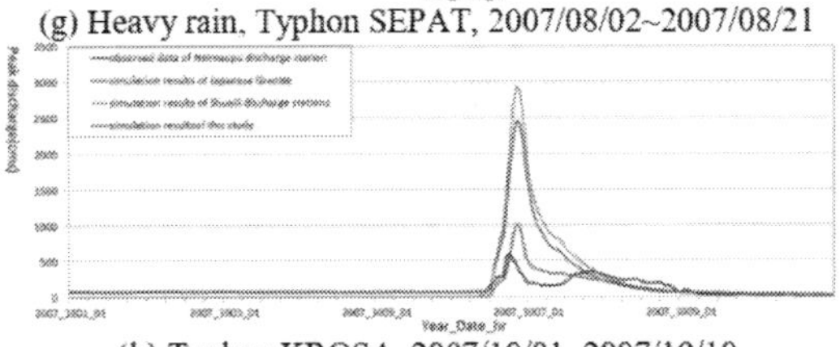

(h) Typhon KROSA, 2007/10/01~2007/10/10

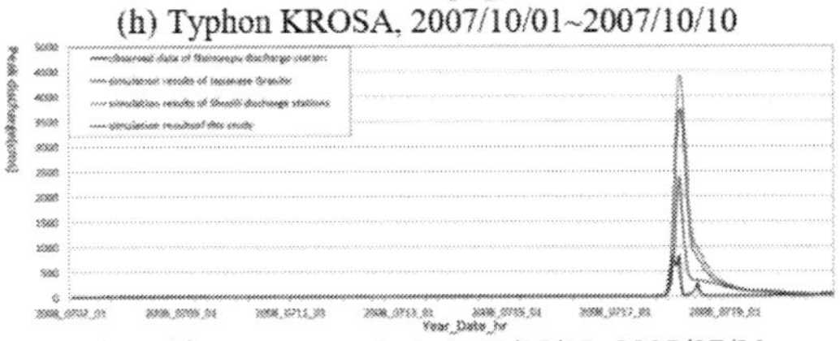

(i) Typhon KALMAEGI, 2008/07/07~2008/07/20

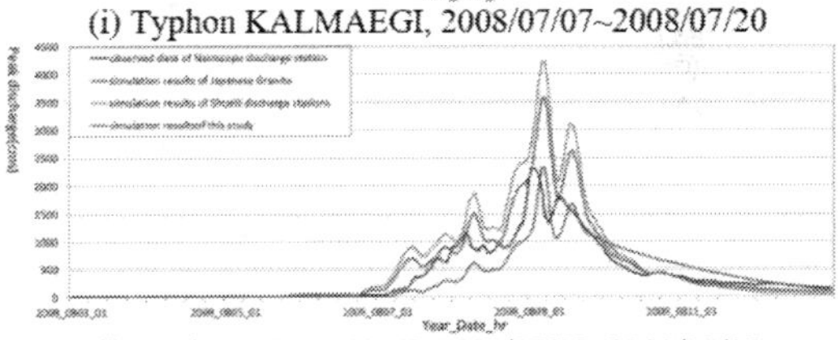

(j) Typhon MORAKOT, 2009/08/03~2009/08/12

Figure 3. Hydrograph at the control spot *I* and Neimaopu discharge station.

3.2 Comparison of simulated discharges and observed values

According to the peak discharge of each hydrologic event, the largest error appears on the mean error of the parameter of Shueili Station 251%, and followed by it of the parameter of Japanese granite 189%. The mean error of the parameter 51.0% in this study is favourable.

From Figure 4(a), the linear relationship between the measured discharge (x) and the simulated discharge (y) appears $y = a \times x$, where both the measured discharge and the simulated discharge should be equal. For this reason, when variable a approaches 1, the parameter simulation is better. The parameter in this study receives a = 0.8542, close to 1, and R^2 = 0.71; the parameter of Japanese Granite acquires a = 1.4648 and R^2 = 0.60; and, the parameter of Shueili Station obtains a = 1.759 and R^2 = 0.59, presenting that the parameter in this study is more suitable for this area.

Finally, the allowance of the three parameters are analyzed, Figure 4(b), where the parameter in this study appears the highest allowance percentage, that it is more suitable for simulating the peak discharge in this area. When the allowance percentage is set lower than 50%, the percentage of the parameter of Shueili

Station is higher than it of the parameter of Japanese granite; however, it shows the opposite situation when the allowance percentage is higher than 50%. In fact, the percentage could reach up to 80% when the allowance is set 40%. The three parameters therefore present low percentage, as Neimaopu discharge station is located in the downstream that the factors in discharge contain rainfall distribution, discharge of each sub-basin, afflux time, and the velocity and discharge at each section. Tank Model cannot overcome such problems, that other theories and models are required. Nevertheless, Shenmu Village is located at the sub-basin of upstream Chenyulan River, the factors are few so that they are ignored. The parameter of Tank Model is utilized for transforming average rainfall into hydrograph so as to solve the problem of no discharge when simulating various debris flows.

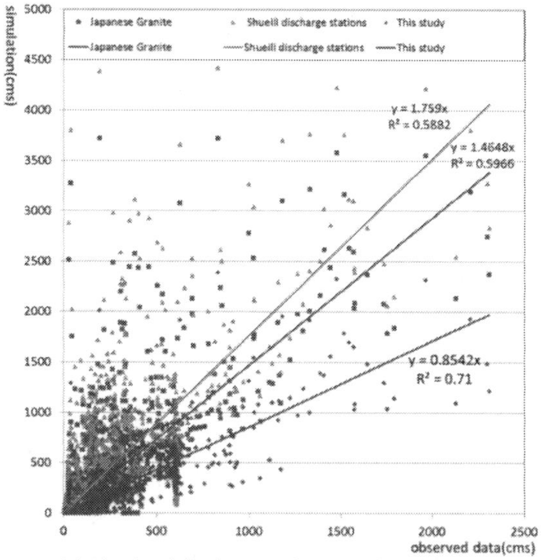

(a) Simulated discharge and measured discharge

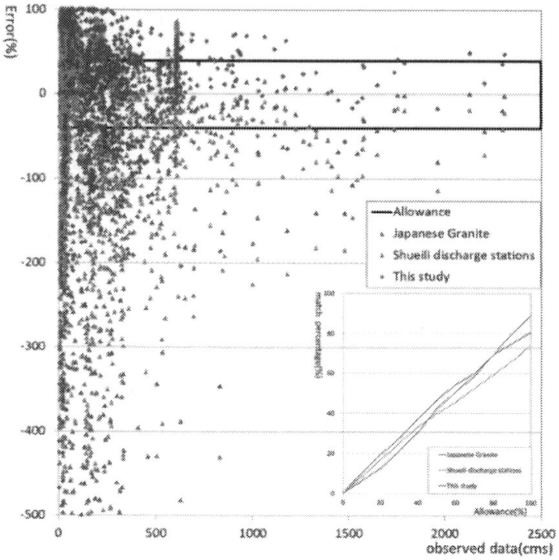

(b) Allowance analysis of parameter errors

Figure 4. Comparison and allowance analysis of simulated and observed data.

4 Conclusions

Thirteen debris flow events around seventeen rain gauge stations within 1996-2010 are regarded as the cases for simulating the peak discharge at twelve control spots along Chenyulan River. The data in Neimaopu discharge station are established the parameter of Tank Model to estimate the peak discharge in Shenmu Village. The parameters of Shueili Station and Japanese Granite are further compared that the mean errors of the discharge simulated by the parameters achieve 251% and 189%, respectively. The mean error of the parameter in this study 51.0% is favourable, showing the simulated hydrograph is close to the measured discharge in Neimaopu Station. Shenmu Village locates at the sub-basin on the upstream Chenyulan River that the few factors could be neglected. The parameter of Tank Model therefore could be applied to transforming average rainfall into hydrograph to solve the problem of no discharge when simulating various debris flows.

Acknowledgements

Thanks for the parts of support from the Project (NSC 99-2622-B-270-001-CC3) of Ministry of Science and Technology, Taiwan.

References

1. L.K. Sherman, Engineering News Record **108,** 501-5 (1932)

2. J.C. Dooge, Journal of Geophysical Research **64**(2), 241-56 (1959)
3. M. Sugawara, E. Ozaki, I. Watanabe, Y. Katsuyama, Tank model and its application to Bird Creek, Wollombi Brook, Bikin River, Kitsu River, Sanaga River and Nam Mune. Report of National Research Center for Disaster Prevention (1974)
4. Y. Ishihara, S. Kobatake, Bulletin of the Disaster Prevention Research Institute **29**(1), 27-43 (1979)
5. K. Okada, Weather Service Bulletin **69**(5), 67-97 (2002)
6. K.T. Lee, Comprehensive watershed management planning and Flooding information system building plan, Research Report of Water Resources Planning Institute, Water Resources Agency, Taiwan (2007) (in Chinese)
7. L.C. Pi, A Study on Surface and Subsurface Rainfall-Runoff Model. Doctoral Dissertation, Department of Civil Engineering, National Chung Hsing University (2004) (in Chinese)
8. S.K. Yu, Application on Rainfall-Runoff Analysis for Fa-Tzu River by Simplified Tank Model. Master Thesis, Department of Civil Engineering, National Chung Hsing University (2007) (in Chinese)
9. Y.N. Chan, Study on Rainfall-Runoff Analysis for Urban Area of Fa-Tzu River by Simple Tank Model and Their Application. Master Thesis, Department of Civil Engineering, National Chung Hsing University (2008) (in Chinese)
10. S.W. Ahmad, Journal of Physics: Conference Series **846**(1), 012019 (2017)
11. G.Q. Wang, J.H. Liu, The Digital Watershed Model. Science Press, Beijing (2005) (in Chinese)
12. S.C. Chen, The application of the tank model and the soil water index for the debris flow disaster evacuation time, Research Report of Soil and Water Conservation Bureau, Taiwan (2011) (in Chinese)
13. C.W. Tsai, S.C. Chen, M.C. Chen, H.Y. Yin, C.Y. He, Journal of Disaster Management **2**(2), 35-53. (2013) (in Chinese)
14. S.W. Aqili, N. Hong, T. Hama, Y. Suenaga, Y. Kawagoshi, Journal of Water and Environment Technology **14**(2), 57-66 (2016)
15. N. Hong, T. Hama, Y. Suenaga, S.W. Aqili, X. Huang, Q. Wei, Y. Kawagoshi, Science of The Total Environment **541**(Supplement C), 383-90 (2016)
16. W. Nie, Y.C. Liang, L. Chen, W. Shao, Geofluids **2017**, 11 (2017)
17. V.A. Cooper, V.T.V. Nguyen, J. A. Nicell, Water Science and Technology **36**(5), 53-60 (1997)
18. S.K. Chen, R.S. Chen, T.Y. Yang, Hydrological Sciences Journal **59**(5), 1020-31 (2014)
19. B.J. Amiri, N. Fohrer, J. Cullmann, G. Hörmann, F. Müller, J. Adamowski, Water Resources Management **30**(14), 5065-85 (2016)
20. SWCB, Technical Manual for Soil and Water Conservation. Soil and Water Conservation Bureau, Council of Agriculture, Taiwan (2005) (in Chinese)

Correction and accuracy improvement of non-parallel shear zone model

Yaoke Wang[1], Meng Kou[1], Wei Ding[1], Huan Ma[1] and Liangshan Xiong[1,2,a]

[1]*Huazhong University of S & T, School of Mechanical Science and Engineering, Luoyu 1037, Wuhan, China*
[2]*Huazhong University of S & T, State Engineering Research Center of Digital Manufacturing and Equipment, Luoyu 1037, Wuhan, China*

Abstract. When applying the non-parallel shear zone model to predict the cutting process parameters of carbon steel workpiece, it is found that there is a big error between the prediction results and the experimental values. And also, the former approach to obtain the relevant cutting parameters of the non-parallel shear zone model by applying coordinate transformation to the parallel shear zone model has a theoretical error – it erroneously regards the determinant ($|J|$) of the Jacobian matrix (J) in the coordinate transformation as a constant. The shape of the shear zone obtained when $|J|$ is not constant is drew and it is found that the two boundaries of the shear zone are two slightly curved surfaces rather than two inclined planes. Also, the error between predicted values and experimental values of cutting force and cutting thrust is slightly smaller than that of constant $|J|$. A corrected model where $|J|$ is a variable is proposed. Since the specific values of inclination of the shear zone (α, β), the thickness coefficient of the shear zone (a_s) and the constants related to the material (f_0, p) are not given in the former work, a method to obtain the above-mentioned five constants by solving multivariable constrained optimization problem based on experimental data was also proposed; based on the obtained experimental data of AISI 1045 steel workpiece cutting force, cutting thrust, chip thickness, the results of five above-mentioned model constants are obtained. It is found that, compared with prediction from uncorrected model, the cutting force and cutting thrust of AISI 1045 steel predicted by the corrected model with the obtained constants has a better agreement with the experimental values obtained by Ivester.

1 Introduction

To build cutting analysis model is an effective way to quantitatively analyze different kinds of phenomena and output (cutting force, cutting temperature, chip deformation and workpiece quality) in the cutting process and has long been paid attention by many researchers [1-4].

In some sense, the development of cutting theory is to establish new, more accurate analysis model which is more consistent with the actual situation to replace the old, poorly-accurate analysis model which is relatively inconsistent with the actual situation. In order to make the cutting analysis model more accurate in predicting the cutting process, the predecessors have conducted various observations, analyses, and assumptions on the physical processes and geometric features in metal cutting. Then, they have proposed a series of new cutting analysis models accordingly.

An analysis and prediction theory based on parallel plane shear zones proposed by Oxley and his co-workers [5] provides a mathematical model suitable for the prediction of the cutting process parameters of carbon steel and aluminum alloy workpieces [6]. In this theory, the shear zone is regarded as a strip zone bounded by two parallel planes and bisected by the primary shear plane, which is more consistent with the actual situation than the Merchant's shear plane model [1]. Also, since the effect of work hardening and temperature softening on workpiece deformation is taken in to consideration, a more accurate prediction result can be obtained.

Astakhov [7,8] and Leopold [9] obtained and analyzed the metallographic photographs of the chips obtained from orthogonal cutting of stainless steel and carbon steel. It was found that the shear zone consisted of two sections of unequal thickness (the first section was wide, where the chip speed changes slowly; the second one was narrow, where the chip speed changes rapidly) rather than the symmetrical strip area bisected by the primary shear plane. Based on this finding, Tounsi etc. [10] proposed a method for solving the coefficients of constitutive equation of workpiece material based on the cutting experiment data obtained under orthogonal cutting conditions. Li [11, 12] further established unequal division shear-zone model which assumed that the flow velocity of the cut material in the shear zone obeyed a piecewise power-law distribution, and also, applied the thermal control equation in the shear zone to predict the cutting process parameters under orthogonal cutting conditions. The validity of this model has been verified in stainless steel orthogonal cutting experiments: the predicted cutting force and cutting thrust share consistent trends with the experimental data while feed rate, cutting speed and rake angle changes, but there was a big error between the predicted results and the results of cutting experiments (Up

[a] Liangshan Xiong: liangsx@hust.edu.cn
© The Authors, published by EDP Sciences. This is an open access article distributed under the terms of the Creative Commons Attribution License 4.0 (http://creativecommons.org/licenses/by/4.0/).

to 54%).

`By analyzing the metallography from orthogonal cutting on carbon steel and aluminum alloy workpiece and the relationship between shear strain and shear angle, Jesper [13] found that the two boundaries of shear zone (the initial shear plane and ultimate shear plane) are not parallel, and the initial shear plane and the ultimate shear plane have an inclination of about 0° to 9° relative to the main shear plane. On the basis of this, Zhou [14] applied the coordinate transformation method to the unequal division shear zone model, which is based on parallel shear planes, and cleverly obtained the formula of non-parallel shear zone model from a calculation formula of the unequal division shear zone model by using a set of transformation matrices. Then, the solution of the cutting force and cutting thrust of the non-parallel shear zone model is obtained. Unfortunately, although the prediction accuracy of the non-parallel shear zone model is higher than the unequal division shear zone model, the prediction results are still far from the experimental data (Error up to 28%).

When learning and applying the solution of Zhou, it was found that, in order to simplify the process of solving, this solution artificially regarded the determinant ($|J|$) of the Jacobian matrix (J) used in calculating the coordinate value of the boundary of the transformed shear zone as a constant. Obviously, this approach greatly weakened the mathematic rigor of the proposed solution and may increase the error of the results obtained.

This paper first proves that $|J|$ used by Zhou in coordinate transformation and solution is not a constant, and its values are different at different stages of the solution process. Then, on basis of the inverse method to obtain the theoretical model coefficients based on the experimental data proposed by the predecessors [15-19], a method to obtain the coefficients in non-parallel shear zone model based on experimental values of the cutting force, cutting thrust and chip thickness are proposed. By using optimization techniques, in this paper, a set of reversely determined coefficients were obtained by a set of orthogonal cutting experimental data of cutting force, cutting thrust and chip thickness of a AISI 1045 steel workpiece. Finally, in order to verify the validation of the method and the accuracy of the results obtained, this set of inversed coefficients is applied in the corrected non-parallel shear zone model. The obtained results of cutting force and cutting thrust of AISI 1045 steel workpiece are compared with the experimental results of Ivester etc. [25]. The results show that the prediction error of the corrected model is smaller than that of Zhou [14].

2 Defect and Correction of Non-Parallel Shear Zone Model

2.1 Introduction of Non-Parallel Shear Zone Model and its solving method

Figure.1 is a schematic diagram of non-parallel shear zone model. The calculation formula can be obtained from the unequal division shear zone model shown in Figure.2 by establishing a certain one-to-one correspondence to relate the four boundary points (C'D'F'E') of the former and four boundary points (CDFE) of the latter.

Assume the mapping relationship between the x-y coordinate values and the ξ-η coordinate values of the boundary points is:

$$\begin{cases} x_i = a_0 + a_1\xi_i + a_2\eta_i + a_3\xi_i\eta_i \\ y_i = b_0 + b_1\xi_i + b_2\eta_i + b_3\xi_i\eta_i \end{cases} \quad (i=1,2,3,4) \tag{1}$$

From the mapping theorem, for any function (the distribution function of the shear strain rate in Zhou's model), its partial differential mapping relationship is:

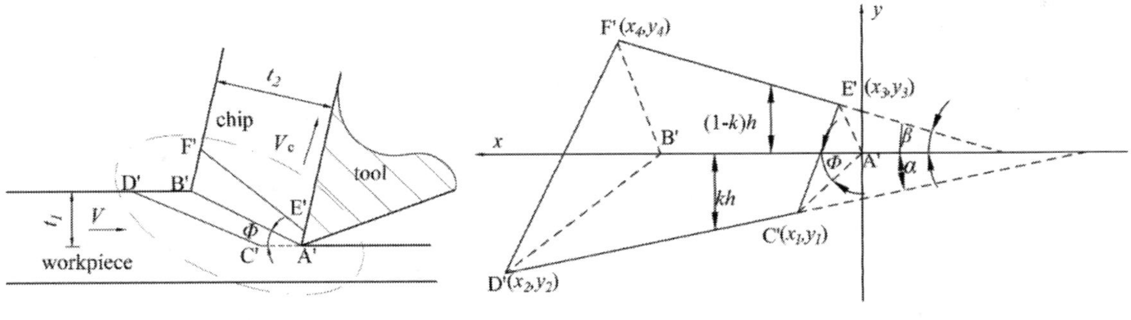

Figure 1. Non-Parallel Shear Zone Model.

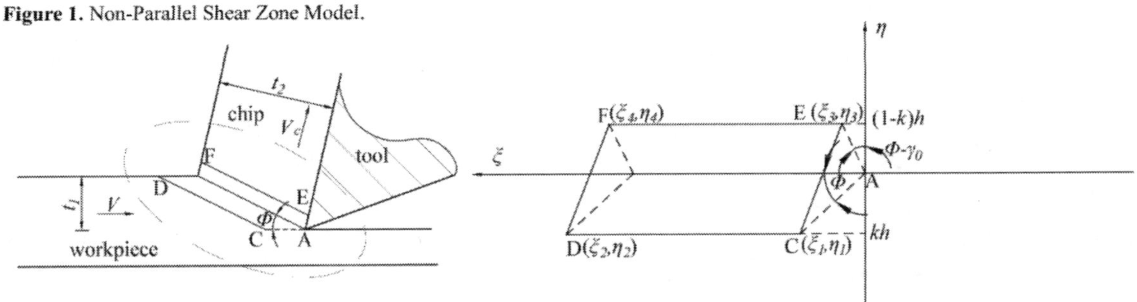

Figure.2 Unequal Division Shear Zone Model

$$\begin{bmatrix} \dfrac{\partial \psi}{\partial x} \\ \dfrac{\partial \psi}{\partial y} \end{bmatrix} = J^{-1} \begin{bmatrix} \dfrac{\partial \psi}{\partial \xi} \\ \dfrac{\partial \psi}{\partial \eta} \end{bmatrix}, \quad J = \begin{bmatrix} \dfrac{\partial x}{\partial \xi} & \dfrac{\partial y}{\partial \xi} \\ \dfrac{\partial x}{\partial \eta} & \dfrac{\partial y}{\partial \eta} \end{bmatrix} \tag{2}$$

J is the Jacobian matrix that represents the transformation process.

Through the coordinate transformation determined by equations (1) and (2), the parallel plane shear zone can be transformed to the non-parallel plane shear zone, and the calculation formula of non-parallel shear zone model can be obtained by the calculation formula applied in parallel plane shear zone.

2.2 Proof that $|J|$ is not a Constant

In the process of transforming the parallel plane shear zone to the non-parallel plane shear zone, in order to realize the calculation of the shear strain rate at each point in the shear zone, the value of $|J|$ needs to be calculated. In the derivation process, Zhou [14] took the value of $|J|$ at the center of gravity of the quadrilateral CDFE as the value of $|J|$ at any point in the entire shear region:

$$|J| = 1 + f(\alpha, \beta, \Phi, \gamma_0, k) \tag{3}$$

which means that $|J|$ is a constant unrelated to ξ and η.

From Equations (1) ~ (2), the original expression of $|J|$ is:

$$|J| = a_1 b_2 - b_1 a_2 + (b_2 a_3 - b_3 a_2)\eta + (b_3 a_1 - b_1 a_3)\xi \tag{4}$$

Then when the coefficients of ξ and η are not 0, $|J|$ is a function of ξ, η, which means its value changes as the coordinate values changes. Actually, the value of $|J|$ does change with the position of the point (ξ, η) in the shear zone (the quadrilateral CDFE), and is not a constant. The proof is as follows.

The necessary condition for the formulas (3) and (4) is

$$\begin{cases} b_2 a_3 - b_3 a_2 = 0 \\ b_3 a_1 - b_1 a_3 = 0 \end{cases} \tag{6}$$

Which equals to:

$$\frac{b_3}{a_3} = \frac{b_2}{a_2} = \frac{b_1}{a_1} = Const. \tag{7}$$

Then, that means:

$$\frac{y_i - b_0}{x_i - a_0} = \frac{b_1\xi_i + b_2\eta_i + b_3\xi_i\eta_i}{a_1\xi_i + a_2\eta_i + a_3\xi_i\eta_i} = Const. \tag{8}$$

Equation (8) is actually an equation of a line passing through (a_0, b_0), which means, if the value of $|J|$ in the equation is constant, the four known points E, F, C, and D must be at the same straight line. This obviously contradicts the fact that CDFE is a quadrilateral. Thus, the value of $|J|$ should not be a constant for this coordinate transformation.

2.3 Influence when $|J|$ is taken as a constant

It has been calculated and verified that although the relative error of $|J|$ obtained when $|J|$ is taken as a constant is not large, only about 1%%, the model is still affected by the following three aspects.

(1) The calculation of $|J|$ is simplified. If $|J|$ is not a constant, the computational flow of the entire model can be represented by Figure 3, which is more complicated. When $|J|$ is a constant, the "point-by-point calculation" process of $|J|$ in Figure. 3 is omitted, and the integral term (proportional to the shear strain rate) in the integration process to obtain the temperature field distribution becomes conventional power functions, which makes the integration calculation process relatively simple.

(2) The distortion of shape of the transformed shear zone was caused. By further calculation, it is found that if $|J|$ is not a constant, the boundary of the non-parallel shear zone is not a plane, but an outward convex surface which is shown in Figure 4.

(3) The prediction error of the model is increased. When $|J|$ is regarded as a constant and a non-constant

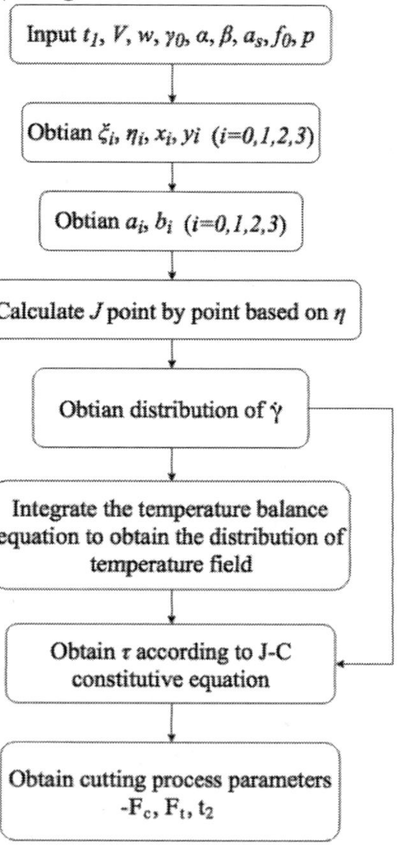

Figure.3 Computational Flow

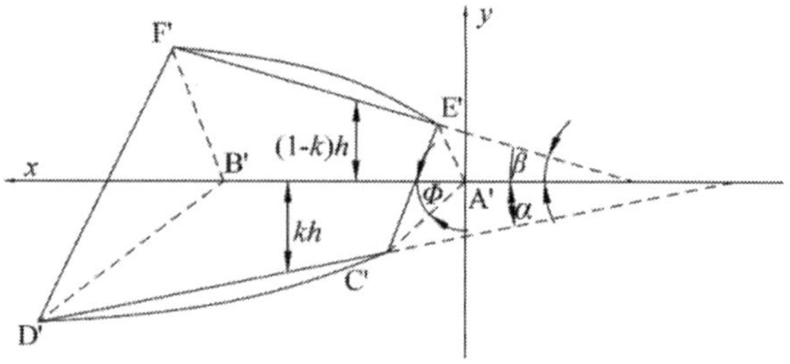

Figure.4 Actual Shear Zone Diagram

number, the model's forecast results under a particular cutting condition are shown in Figure 5. It can be seen that, in most cases, using the constant $|J|$ actually artificially increases the prediction error of the model (For example, when the cutting conditions are V=187.2m/min, t_l=0.2mm, γ_0=0°, the prediction error of the cutting force increased by 7.6%; the prediction error of the cutting thrust increased by 3.7%)

It can be seen from Figure.5 that correcting Zhou's mistake and taking the value of $|J|$ as a non-constant number improves the prediction accuracy of the model although the forecasting error is still large.

In terms of narrative convenience, the non-parallel shear zone model where the value of $|J|$ is changed to non-constant is called "corrected non-parallel shear zone model" and referred to as "corrected model" for short.

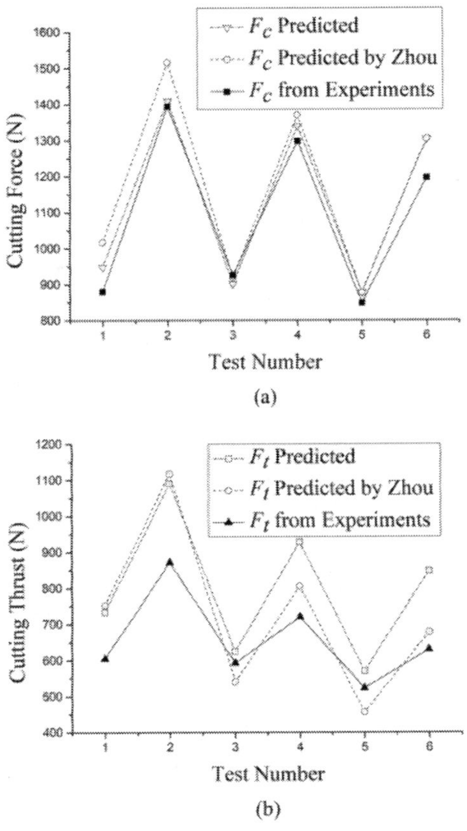

Figure.5 Predicted and Experimental Cutting Parameters (a) Comparison on Cutting Force; (b) Comparison on Cutting Thrust

3 Coefficient and Constant Inverse of Corrected Model

3.1 Strategy of Coefficient and Constant Inverse

The reason why the non-parallel shear zone model prediction accuracy is not exceptional is due to many factors, one of which is that many coefficients and constants used in the model, especially the constants defining the shape of the non-parallel shear zone, the coefficients related to the degree of chip deformation and the state of knife-chip friction (α, β, a_s, f_0, p) are all artificially given. These constants and coefficients are difficult to completely agree with the actual situation.

It is not difficult to understand that the values of the constants α, β, a_s, f_0, p should be determined by the cutting process according to its inherent laws. Only when the values of these constants meet the inherent laws of the cutting process, the output parameters of the cutting process, such as F_c, F_t, and t_2, are consistent with the actual situation. Conversely, if cutting force F_c, cutting thrust F_t, the ratio between cutting force and thrust F_c / F_t and chip thickness t_2 predicted by the non-parallel shear zone model are consistent with the actual situation, the corresponding α, β, a_s, f_0, p is the parameter that conforms to the inherent laws of the cutting process. This provides an idea for determining the values of the constants α, β, a_s, f_0, p indirectly by measuring the force and the thickness of the chip, and further improving the prediction accuracy of the non-parallel shear zone model. The constants α, β, a_s, f_0, p which make the forecast result consistent with the experimental value are the actual values of α, β, a_s, f_0, p which are in line with the laws in metal cutting.

3.2 Influence of Each Coefficient and Constants on Cutting Process Parameters

It should be noted that the premise of using this idea to inversely determine the constants above is that the output parameters of the cutting process (F_c, F_t, F_c/F_f and t_2) which are determined by α, β, a_s, f_0, p change monotonously as α, β, a_s, f_0, p changes. So, is the actual situation not the case?

3.2.1 *Influence of Inclination (α, β) on Output Parameters of Cutting Process*

According to the study of Jaspers [13], the inclination α, β varies from 0 to 9°.

When the α and β change from 0° to 9° and the other input coefficients are constant, influence on the cutting force F_c, the cutting thrust F_t, calculated by the non-parallel shear zone model is shown in Figure 6. Obviously, within a given range of α, β, the cutting force F_c and the cutting thrust F_t are monotonous as α, β increase or decrease. In addition, from the numerical calculation, the chip thickness t_2 and the ratio of cutting force and thrust F_c/F_t will not change when α, β change, which also represents a monotonous relation.

3.2.2 *Influence of the Thickness Coefficient of Shear Zone (a_s) on Output Parameters of Cutting Process*

In order to predict the cutting force, the shear zone thickness is also one of the input conditions. Grzesik [21] proposed a relation between the thickness of the shear zone and the cutting thickness:

$$h = \frac{t_2}{10 \sin \Phi} \tag{9}$$

 https://doi.org/10.1051/matecconf/201820702002

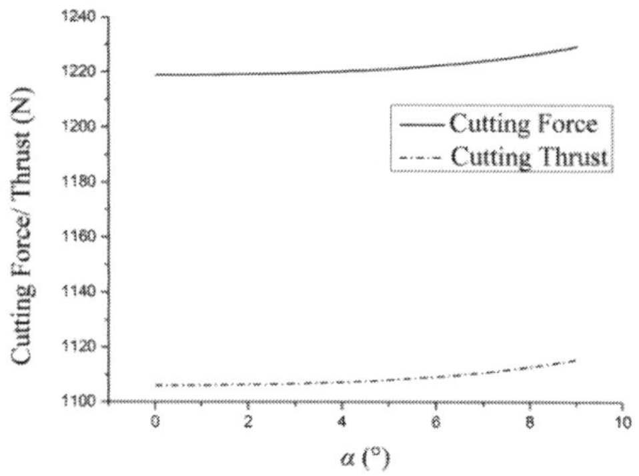

(a)

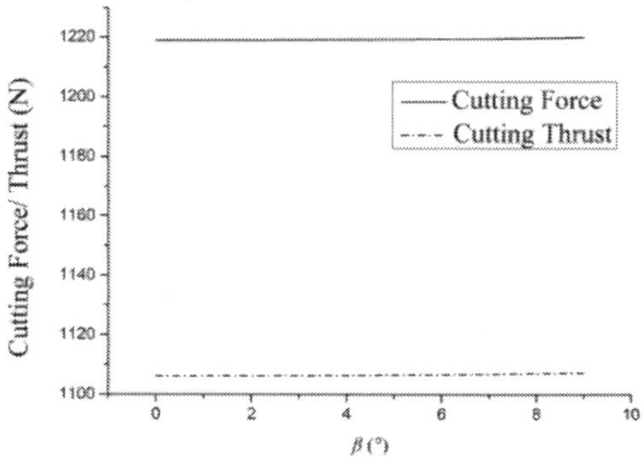

(b)

Figure.6 F_c, F_t Calculated when α, β Change
(a) α's Influence on Cutting Force and Cutting Thrust; (b) β's Influence on Cutting Force and Cutting Thrust

where Φ is the shear angle and t_2 is the cutting thickness. According to the Merchant theory and the actual situation, Φ must be taken within the interval (5°, 45°). Assume that:

$$h = a_s \times t_2 \tag{10}$$

Thus, when $\Phi \in (5°,45°)$, from equation (9):

$$a_s \in (0.14,1.14) \tag{11}$$

To increase the reliability of the analysis process, set that $a_s \in (0.1,1.5)$.

When a_s changes from 0.1 to 1.5 and other input coefficients are constant, by numerical simulation, it is found that F_c, F_t, t_2, F_c/F_t calculated by the non-parallel shear zone model are monotonically related to a_s, which is similar to the relations with α, β.

3.2.3 *Influence of the Knife-Chip Friction Coefficients* (f_0, p) *on Output Parameters of Cutting Process*

The coefficients f_0 and p are mainly related to the friction coefficient. Dudzinski [22] proposed that the average chip friction coefficient between the chip and knife can be expressed as a power function with respect to the chip speed V_c:

$$f = f_0 V_c^p \tag{12}$$

where f_0 is a constant and p <0. It can be seen that the friction coefficient between the chips and knife is a non-constant number. In this equation, f_0 represents the reference friction coefficient measured at a certain cutting speed and p shows the effect of cutting speed on friction.

When f_0 varies from 0.6 to 1.5, p varies from -0.4 to 0, and other input coefficients are constant, by numerical simulation, it is found that the corresponding F_c, F_t, t_2, F_c/F_t are all monotonically related to f_0, p,

3.2.4 *Summary of Influence of* (α, β, a_s, f_0, p)

In summary, F_c, F_t, t_2 and F_c/F_t all monotonically change with changes in coefficient / constants (α, β, a_s, f_0, p) which are to be determined. Assume that the functions representing these relations are respectively f_1, f_2, f_3 and f_4, with the results of numerical simulation mentioned above, it can be concluded that:

$$\begin{cases} F_c = f_1(+f_0,-p,+\alpha,-a_s,+\beta) \\ F_t = f_2(+f_0,-p,+\alpha,-a_s,+\beta) \\ \dfrac{F_c}{F_t} = f_3(-f_0,+p) \\ t_2 = f_4(+f_0,-p) \end{cases} \tag{13}$$

In the equation, the positive sign before each parameter indicates that the function is positively related to the desired parameter; the negative sign indicates that the function is negatively related to the desired parameter.

3.3 Inverse of Coefficients and Constants of the Model

3.3.1 *Design and Data of Cutting Experiment*

Based on the strategy introduced above, modern optimization techniques are used to retrieve the above coefficients and constants from the experimental data of the cutting process parameters F_c, F_t, F_c/F_f, and t_2.

To obtain the forces F_{c0}, F_{t0} and the chip thickness t_{20} of the orthogonal cutting, the experiment was carried out on a Mazak CNC lathe. The tool was sharpened YT15 carbide turning tool; the workpiece material was AISI 1045 steel, the workpiece shape was a circular tube shaped from bar stock, the outer diameter was 80mm, and the inner diameter was determined by required cutting width w; the Kistler 9257B three-way dynamometer is used to measure the cutting forces F_{c0} and F_{t0}, and a truncated caliper is used to measure t_{20}. In order to eliminate the influence of factors such as built-up edge and size effect on the test results, the cutting speed U≥100m/min and the undeformed cutting thickness t_i≥0.1mm are guaranteed during the experiments. In order to reduce the number of experiments, the designed cutting experiment was a 4 factor 5 level orthogonal cutting experiment and a 5-6 orthogonal table is applied. In order to eliminate the error caused by accidental factors as much as possible, each cutting condition is repeated twice, and the average value of the two measured cutting forces is taken as the actual value of the cutting force. The thicknesses of the chip were measured five times at different positions where the cutting in and cutting out points of the tool are avoided, and the average value was taken as the measured value of the cutting thickness. Then, the average value of values measured in two cutting experiments was taken as the final measured value.

Experimental Data is shown in Table 1:

Table 1. Experimental Data

	γ_0	t_1	w	V	F_{c0}	F_{t0}	t_{20}
U.	N	mm	mm	m/min	N	N	*Mm*
1	-9.83	0.10	1.00	100	383	346	0.350
2	-9.83	0.15	1.31	150	631	524	0.420
3	-9.83	0.20	1.50	200	877	648	0.490
4	-9.83	0.25	1.81	250	1140	761	0.560
5	-9.83	0.30	1.90	300	1462	909	0.630
6	-4.67	0.10	1.31	200	446	417	0.280
7	-4.67	0.15	1.50	250	707	595	0.370
8	-4.67	0.20	1.81	300	94	685	0.470
9	-4.67	0.25	1.90	100	1354	915	0.630
10	-4.67	0.30	1.00	150	80	488	0.440
11	-0.50	0.10	1.50	300	477	416	0.250
12	-0.50	0.15	1.81	100	779	578	0.370
13	-0.50	0.20	1.90	150	1023	664	0.470
14	-0.50	0.25	1.00	200	591	315	0.550
15	-0.50	0.30	1.31	250	839	410	0.640
16	4.00	0.10	1.81	150	522	401	0.285
17	4.00	0.15	1.90	200	754	524	0.357
18	4.00	0.20	1.00	250	469	280	0.423
19	4.00	0.25	1.31	300	676	366	0.532
20	4.00	0.30	1.50	100	1037	542	0.669
21	10.67	0.10	1.90	250	487	313	0.236
22	10.67	0.15	1.00	300	332	185	0.303
23	10.67	0.20	1.31	100	555	264	0.455
24	10.67	0.25	1.50	150	798	356	0.536
25	10.67	0.30	1.81	200	1013	392	0.615
26	-6.50	0.10	1.00	100	343	295	0.306
27	-6.50	0.15	1.31	150	593	451	0.401
28	-6.50	0.20	1.50	200	827	572	0.486
29	-6.50	0.25	1.81	250	1164	755	0.563
30	-6.50	0.30	1.90	300	1383	840	0.636
31	3.17	0.10	1.31	200	373	273	0.258
32	3.17	0.15	1.50	250	570	364	0.346
33	3.17	0.20	1.81	300	845	495	0.425
34	3.17	0.25	1.90	100	1142	635	0.587
35	3.17	0.30	1.00	150	661	315	0.643
36	6.50	0.10	1.50	300	395	275	0.238
37	6.50	0.15	1.81	100	706	443	0.380
38	6.50	0.20	1.90	150	893	502	0.453
39	6.50	0.25	1.00	200	538	255	0.521
40	6.50	0.30	1.31	250	798	364	0.585
41	8.83	0.10	1.81	150	488	337	0.256
42	8.83	0.15	1.90	200	680	405	0.34
43	8.83	0.20	1.00	250	432	212	0.415
44	8.83	0.25	1.31	300	663	309	0.484
45	8.83	0.30	1.50	100	950	431	0.647
46	14.00	0.10	1.90	250	458	286	0.230
47	14.00	0.15	1.00	300	319	158	0.309
48	14.00	0.20	1.31	100	567	267	0.449
49	14.00	0.25	1.50	155	744	325	0.506
50	14.00	0.30	1.81	200	1013	421	0.566

3.3.2 Objective Function and Constraint function

According to the strategy above, the values of α, β, a_s, f_0, p that make the prediction result closest to the experimental data are the values that match the cutting rule best. Then, the values of α, β, a_s, f_0, p, which is the closest to the predicted cutting process parameters and the experimental results as a whole are the demanded values.

Construct the objective function:

$$\min_x F(x)=\sum_{i=1}^{n}\left[\left(\frac{t_{2i}}{t_{20i}}-1\right)^2+\left(\frac{F_{ti}}{F_{t0i}}-1\right)^2+\left(\frac{F_{ci}}{F_{c0i}}-1\right)^2+\left(\frac{F_{ci}}{F_{ti}}\Big/\frac{F_{c0i}}{F_{t0i}}-1\right)^2\right]$$

$$s.t.\begin{cases}0.25t_{1i}-h_i<0\\ hi-t_{1i}<0\\ 0.6-f<0\\ f-1.5<0\end{cases} \tag{13}$$

In the objective function above, x is the variable to be optimized (x=[α, β, a_s, f_0, p]); t_{20} is the experimentally measured chip thickness; t_2 is the predicted chip thickness; F_{c0} is the experimentally measured cutting force; F_c is the predicted cutting force; F_{t0} is the experimentally measured cutting thrust; F_t is the cutting thrust; h is the thickness of the shear zone; t_1 is the cutting thickness; f is the average friction coefficient between the knife and chips.

3.3.2 Optimization Algorithm

The experimental data can be used to optimize the objective function above using the optimization functions ***ga*** and ***fmincon*** provided by the Matlab2014a software platform to obtain the desired model coefficients and constants α, β, a_s, f_0, p.

Specific Process :

(1) First use the genetic algorithm ***ga*** to search in a larger range to get a rough solution x_0;

(2) Set x_0 as the center and rebuild a smaller search interval. If the original interval width is B, the upper limit of the original interval is UB and the lower limit is LB, then the new interval is [max(X_0-$0.5aB$, LB), min(X_0-$0.5aB$, UB)] (the coefficient a ranges from 0 to 1 ($0<a<1$);

(3) Taking x_0 as the initial value and also taking the new

interval as the range, the ***fmincon*** function is used to obtain a smaller objective function value in the new search interval. The corresponding solution x_1 is the demanded solution.

3.3.4 The result of the solution

Through the process above, according to the experimental data in Table 1, the following optimization results can be obtained:

$$\alpha=0.9596°; \beta=0.6142°; a_s=1; f_0=0.703; p=-0.175$$

For convenience of narrative, coefficients obtained above is referred to as optimized coefficients.

4 Prediction Accuracy of Revised Model with Optimized Coefficients

4.1 Experimental Data for Comparison

To verify the validity of the proposed method, the two boundaries of the cutting forces' and cutting thrusts' reference values based on four groups of AISI 1045 steel cutting processes which were measured by R. W. Ivester etc [19] are used. The cutting forces and cutting thrusts were calculated for AISI 1045 steel orthogonal cutting under the exact cutting conditions.

4.2 Prediction Accuracy of Revised Model With Optimized Coefficients

The prediction results of cutting process parameters of Zhou's model and revised model after coefficients/constants re-determination are shown in Figure 7. As can be seen from the figures, most of the prediction accuracy after correction is higher than that of Zhou's. Only a few data have lower prediction accuracy than Zhou's after eliminating several groups of data that neither models can predict preferably. After calculation, the cutting forces' average prediction accuracy predicted by revised model is only increased by 0.4%, but the cutting thrust forces' prediction accuracy is improved from 22% to 18%. Hence, it can be concluded that the overall prediction accuracy of the corrected non-parallel shear zone model is higher than Zhou's prediction accuracy, the predicted forces are closer to experimental data.

5 Conclusions

The determinant of Jacobian matrix in Zhou's paper is not a constant value, conversely, it should be a variable.

Setting $|J|$ as constant value could result in the deformation of shear region in non-parallel shear region model and the increasement of cutting force's prediction error.

Revised model to solve the above problem is proposed.

A method using modern optimization methods to reversely calculate the five coefficients (α, β, a_s, f_0, p) with experimental values (cutting force, cutting thrust and chip thickness) in corrected model is proposed.

Designed and finished a set of cutting experiments based on AISI 1045 Steel workpieces orthogonal cutting to obtain cutting force, cutting thrust and chip thickness.

According to the methods and experimental data above, the specific values of five optimized coefficients are obtained while applied revised model in AISI 1045 steel workpieces; The average cutting thrust error is decreased from 22% of the erroneous models proposed by predecessor to 18% after applying optimized coefficients in corrected model.

6 Acknowledges

This research work was supported by the National Natural Science Foundation of China (No: 51675203) and the Undergraduate Research Opportunity Program of Huazhong University of Science and Technology (No: 20160205Z). The relevant cutting experiments were completed using the machine tools and test instruments of the State Key Laboratory of Digital Manufacturing Equipment & Technology.

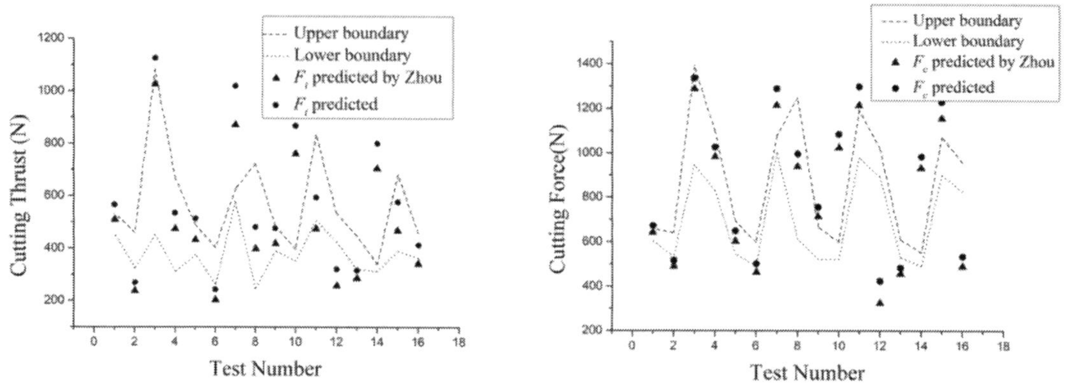

(a) (b)

Figure.7 Cutting Process Parameters Predicted by Zhou's Model and revised model with optimized coefficients.
(a) Comparison on Cutting Thrust; (b) Comparison on Cutting Force;

References

1. H. Ernst, M.E. Merchant, *T. Am. Soc. Metal.*, **ASM 29,** 299-378 (1941)
2. E.H. Lee, B.W. Shaffer, *J. Appl. Mech.*, **18**, 405–413 (1951)
3. S. Kobayashi, E.G. Thomsen, *J. Eeg. Ind.*, **84**, 71–80, (1962)
4. E.J.A. Armarego, *Mach. Sci. Technol.*, **2(2)**, 191–211 (1998)
5. P.L.B Oxley, Mechanics of Machining, Chichester: Ellis Horwood (1989)
6. M.R.V. Sereshk, P. Heydarizadeh, The Evaluation of Analytical Models for Orthogonal Cutting Process, International Conference on Electrical, Automation and Mechanical Engineering (2015)
7. V.P. Astakhov, M.O.M. Osman, M.T. Hayajneh, *Int. J. Mach. Tool Manu.*, **41(3)**, 393-418(2001)
8. V.P. Astakhov, Metal Cutting Mechanics, Boca Raton: CRC Press, 1998.
9. J. Leopold, Mechanical and physical models of machining, Proceedings of the Second CIRP Wordshop on Modeling of Machining Operation (1999)
10. N. Tounsi, J. Vincenti, A. Otho, et al, *Int. J. Mach. Tool Manu.*, **42(12)**, 1373-1383 (2002)
11. B. Li, X. Wang, Y. Hu, et al, *Int. J. Adv. Manuf. Tech.*, **54(5-8)**, 431-443 (2011)
12. B. Li, Y. Hu, X. Wang, et al., *Mach. Sci. Technol.*, **15(4)**, 453-484 (2011)
13. S.P.F.C. Jaspers, J.H. Dautzenberg, *J. Mater. Process Tech.*, **121**, 123-135 (2002)
14. F. Zhou, X. Wang, Y. Hu, *J. Appl. Mech. Mater.*, **229**, 503-506 (2012)
15. T. Dereli, I. H. Filiz, A. Baykasoglu, *Int. J. Prod. Res.*, **39(15)**, 3303-3328 (2001)
16. P. Asokan, R. Saravanan, K. Vijayakumar, *Int. J. Adv. Manuf. Tech.*, **21(1)**,1-9 (2003)
17. P.E. Amiolemhen, A.O. A Ibhadode, *Int. J. Mach. Tool. Manu.*, **44(12–13)**,1403-1412 (2004)
18. R.Q. Sardiñas, M.R. Santana, *Eng. Appl. Artif. Intel.*, **19(2)**, 127-133 (2006)
19. L. Xiong, J. Wang, Y. Gan, et al., *Int. J. Adv. Manuf. Tech.*, **77(1-4)**, 1-13 (2015)
20. R.W. Ivester, M. Kennedy, M. Davies et al., *Mach. Sci. Technol.*, **4(3)**, 511-538 (2000)
21. W. Grzesek, Advanced Machining Processes of Metallic Materials, London: Elsvier (2008)
22. D. Dudzinski, A. Molinari, *Int. J. Mech. Sci.*, **39(4)**, 369-389 (1997)

 https://doi.org/10.1051/matecconf/201820702003

Design Analysis of Overhead Crane for Maintenance Workshop

Siva Sitthipong[1,a], Chaiyoot Meengam[2] Suppachai Chainarong[2] and Prawit Towatana[3]

[1]*Physical Testing Laboratory, Thailand Institute of Scientific and Technological Research, Samut Prakan, Thailand*
[2]*Faculty of Industrial Technology, Songkla Rajabhat University, Songkhla, Thailand*
[3]*Marine and Coastal Resources Institute, Prince of Songkla University, Songkhla, Thailand*

Abstract. This research aimed to study the design of the overhead crane of a small fishing boat maintenance factory according to the building and functional requirements of the project based on the ASME B30.2-2005 Standard. The results of the study showed that the design of the runway with steel structure BS: 5950: 2000 grade s460 provided vertical and horizontal deflection values of 4.96 and 16.62 respectively that did not exceed the allowed deflection. It is strong enough for use in construction when the stresses on the beam bridge were analysed by the finite element program compared with the strength of the reinforced steel providing a safety value of 1.83.

1 Introduction

Utilization of the maintenance buildings possesses specification for the overhead crane which has less than 600 square meters in size, 20 meter wide bridge beam and 30 meter running distance as shown in Fig 1. The height of the crane to the floor does not exceed. 15 meters and the crane must be able to carry a load of 20 tons and the design must comply with the requirements. The structure must be strong, stable, and cost-effective.

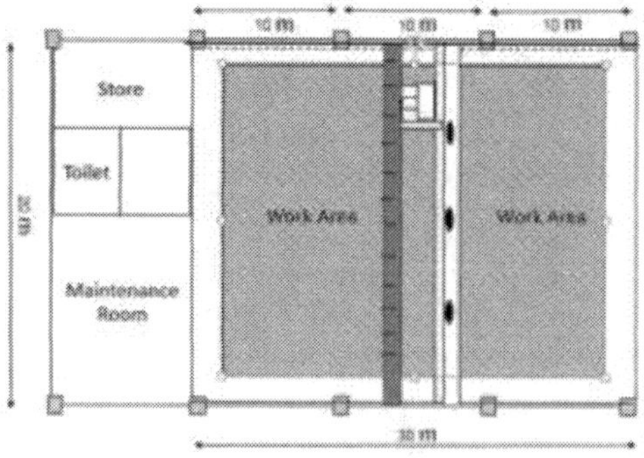

Fig. 1: Install Area of Overhead Crane

2 Crane Design Method

2.1 Design Guidelines

The main structure of the crane is iron with BS: 5950: 2000 grade S460 and has high strength per weight unit. The design is in accordance with ASME B30.2-2005. The figure is shown in Figure 2. The primary data required for the design calculation is shown in Table 1., requirements. The structure must be strong, stable, and cost-effective.

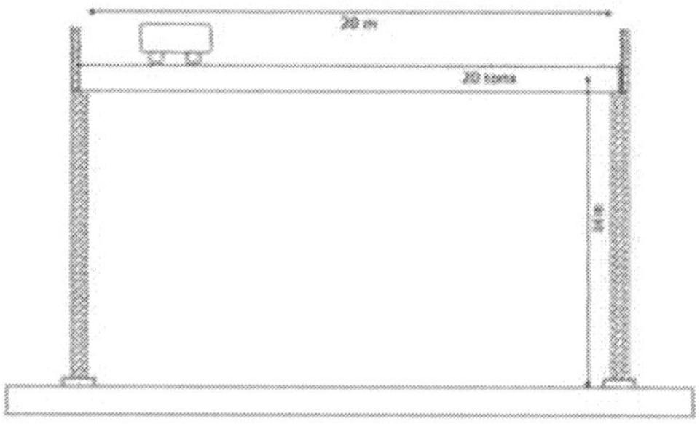

Fig. 2: Overhead Crane

Table 1. Primary data designated in the design

Parameters	Symbol	Value
Crane Capacity	W_{cap}	200 kN
Weight of Crane Bridge and End Carriages	W_c	122 kN
Weight of Crab	W_{cb}	15.6 kN
Span Beetween Crane Rails	L_c	20 m
Minimum Hook Approach	ah	0.92 m
Wheel Centres in End Carriages	s	4 m
No. of Wheel Flanges	-	2
Height of Rail	HR	50 mm
Span of Crane Girder	L	10 m
Self Weight of Crane Girder	W_g	12 kN
Stiff Bearing Length	b1	150 mm
Beam end to Stiff Bearing Edge	be	150 mm
Flange Restrains For Position and Direction	--	1
Design Strength	P_Y	440 N/mm^2
Specific Thickness	t	20.2 mm

[a] Corresponding author: siva@tistr.or.th

© The Authors, published by EDP Sciences. This is an open access article distributed under the terms of the Creative Commons Attribution License 4.0 (http://creativecommons.org/licenses/by/4.0/).

2.2 Runway Deflection Analysis

The vertical and horizontal actual deflection was calculated from (1) and (2), respectively, which was acceptable only if the result was no more than the permitted deflection as calculated from equation (3)

Vertical Actual Deflection = Wus L^3 [3(a+c)/L – 4(a^3 + c^3)/L^3]/[48E Ix] (1)

Horizontal Actual Deflection = FR L^3/(48E Iyfc) (2)

Permitted Deflection = Span / 600 (3)

The design of the beam cross-section used for deflection analysis is as follows:
The figure is shown in Fig. 3:

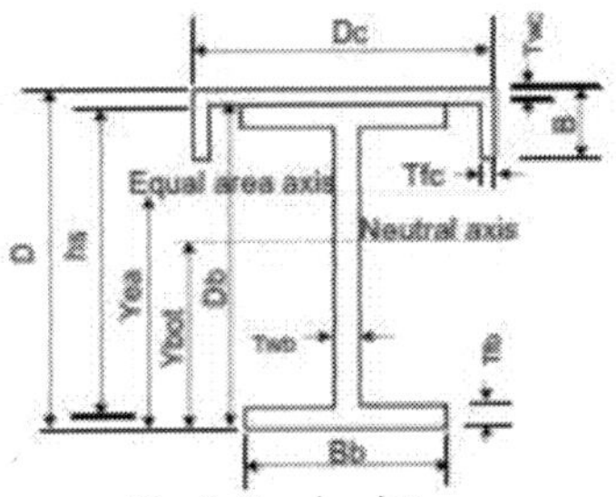

Fig. 3: Overhead Crane.

Table 2. Main Section

Mass	200.9
Ab cm^2	256
Db mm	903
Bb mm	303.3
Tfb mm	20.2
Twb mm	15.1
Ixb cm^4	325254
Iyb cm^4	9423
Jb cm^4	291
r mm	19.1
d mm	824.4

Table 3. Top Section

Mass kg/m	67.510
Ac cm^2	86
Dc mm	430
Bc,DL mm	20
Tfc mm	0
Twc cm	20
Ixc cm^4	13251.167
Iyc cm^4	28.667
Jc cm^3	114.667
Cy cm	1

Table 4. Forming a Compound Crane Girder Section.

M kg/m	A cm^2	D mm	B mm	Ytop mm	Ybot mm	Ix cm^4	Iy cm^4	Iyfc cm^4	Iyft cm^4	ry cm
268.410	342.000	923.000	430.000	355.450	567.550	462388	22674	13251	4697	8.142
Zx top cm^3	**Zxbot cm^3**	**Zyfc cm^3**	**Sx cm^3**	**J cm^4**	**hs cm**	**X const**	**DL mm**	**sai**	**N ratio**	
13009	8147	616	10962	405.667	89.280	46.398	20.000	0.385	0.738	

Table 5. Equivalent Area Axis & Plastic Modulus of the Combined Section.

strip	Width	Height	y=0	Area	Dist comp	Dist ten	Area comp
A1	430	20	Y=20	8600	20	0	8600
A2	303.3	20.2	Y=40.2	812.66	20.2	0	6126.66
A3	15.1	0	Y=40.2	0	0.0000	0.0000	0
A4	15.1	862.6	Y=902.8	13025.26	146.5317881	716.0682119	221263
A5	303.3	20.2	Y=923	6126.66	0	20.2	0
		923		33878.58		923	16939.29

Area ten	yc top	yt top	Sx cm^3
0	10.000	186.732	-86.000
0	30.100	186.732	-184.412
0	40.200	186.732	0.000
1081263	113.466	544.766	5639.294
6126.66	186.732	912.900	5693.028
16939.29	10.000	186.732	10961.909

2.3. Strength Analysis of Bridge Beam Box

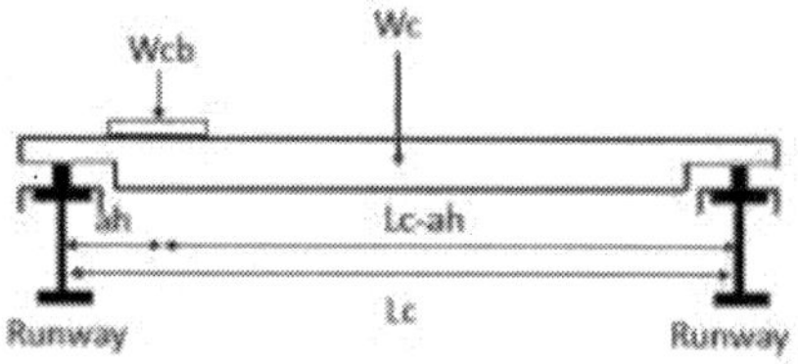

Fig. 4: Bridge Crane.

In processing, the finite element method with the solid work simulation program was employed in the 8 following main steps:

(1) Check the integrity of the CAD file

(2) Define the boundary conditions of the grip and support the actual working conditions.

(3) Determine the force load to determine both position and size. The value derived from the calculation.

(4) Specify the type of material as standard iron steel with BS: 5950: 2000 grade 460.

(5) Determine the type and shape of the elements that are appropriate for the complex structure work.

(6) Create mesh especially using the fine mesh in the complex structure area.

(7) Use the workpiece command and then the obtained results is analyzed for the accuracy.

3 Results and Discussion

3.1 Result of Design Analysis of Runway Overhead Crane

The results obtained from the calculation of the overhead crane design were shown in Table 6. Fig. 5-6 showed the position of the analysis. Fig. 7-10 showed the maximum bending moment. The forces acting on the support points were shown in Fig. 11-12.

Table 6. Result of Design Analysis

Maximum Vertical Moments			
Distances	a = 2 m	S = 4 m	C = 4 m
Due to Dead Load	R_a = 6 kN	R_b = 6 kN	M = 14.4 kNm
Due to Crane Load	R_a = 208 kN	R_b = 138 kN	M = 554.7 kNm
Momement Mx1 = 1.4 Dead Load + 1.6 Crane Load with Impact			M=907.7 kNm
Momement Mx2 = 1.4 Dead Load + 1.4 Crane Load with Impact			M=796.7 kNm
Maximum Horizontal Moments			
Due to Crane Surge	R_a = 6.5kN	R_b = 4.3kN	M = 17.2 kNm
Moment My = 1.4 Crane Load due to Surge			M = 24.1 kNm
Due to Crane Crabbing	My = 1.4FR.L/4		75.8 kNm
Vertical Deflection Check			
Distances	a = 3m	S = 4m	C =3 m
Max Actual Deflection			4.96 mm
Permit Deflection			16.67 mm
Horizontal Deflection Check			
Distances	a = 3m	S = 4m	C =3 m
Surge Deflection			6.99 mm
Crabbing Deflection			16.62 mm
Max Actual Deflection			16.62 mm
Permit Deflection			20.00 mm
Shear, Bending, Compression Check			
Shear			3599.7 kN
Shear Ratio 0.126			452.16
Bending (1.4 Dead Load + 1.6 Vertical Crane Load)			0.483
Maximum Bending 1.4 (Dead Load + Vertical Crane Load + Horizontal Crabbing)			0.703
Compression Under Wheel			101.8 N/mm^2

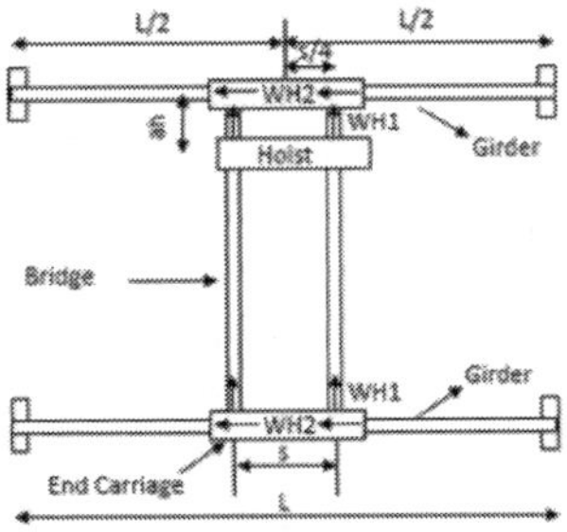

Fig. 5: Crane layout & transverse surge force for maximum horizontal moment

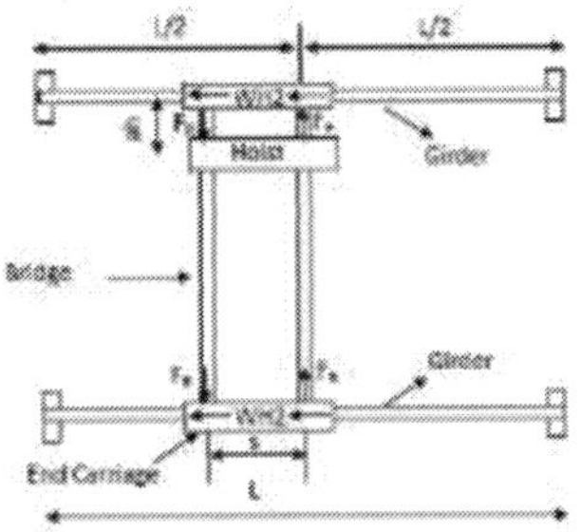

Fig. 6: Crane layout & crabbing forces for maximum horizontal moment

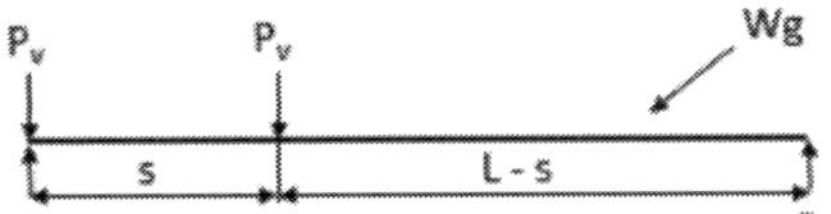

Fig. 7: Vertical Load Positions Causing Maximum Vertical Shear

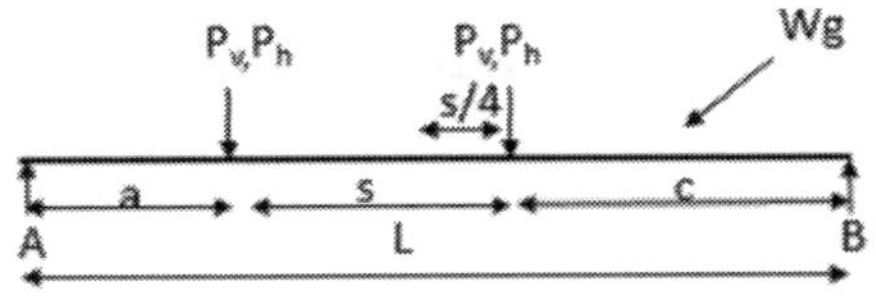

Fig. 8: Load Positions Causing Maximum Vertical & Horizontal Moment

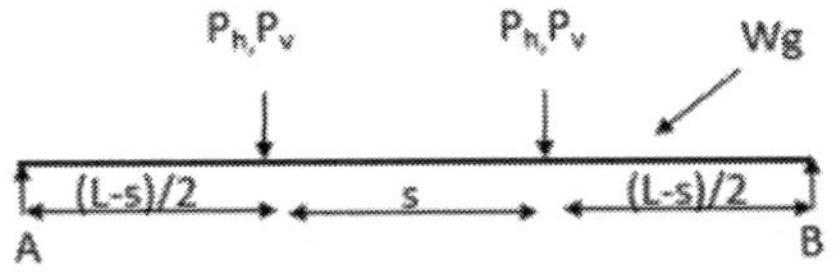

Fig. 9: Load Positions Causing Maximum Vertical & Horizontal Deflection

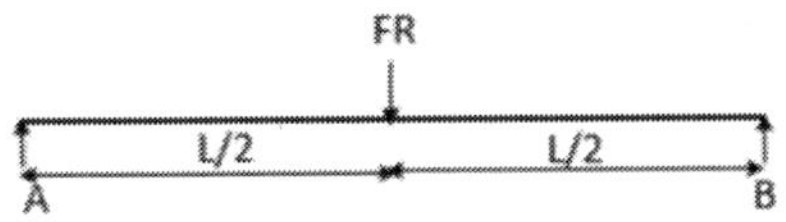

Fig. 10: Crabbing Load Position Causing Maximum Horizontal Moment & Deflection

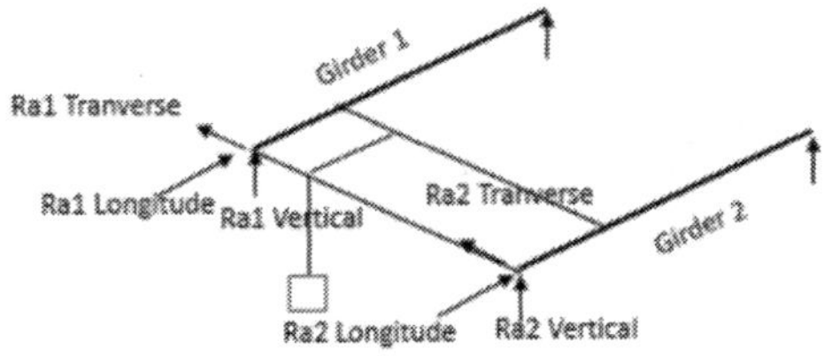

Fig. 11: Reactions on to the Support Structure with Crane Crabbing

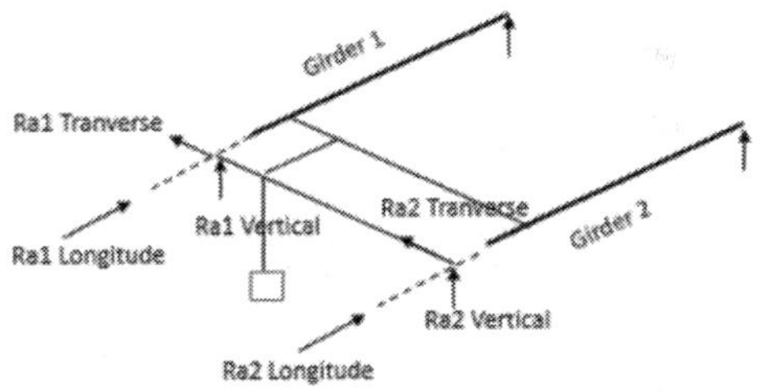

Fig. 12: Reactions on to the Support Structure with Crane Crabbing

3.2 Results of Strength Analysis of Bridge Beam Box

The stress analysis of the beams with finite element method [1-14] is shown in Fig. 13. The maximum stress of 338 MPa gives a safety value of 1.83.

Fig.13: Result of Stress Analysis.

4 Summary

Crane design analysis, which satisfies ASME B30.2-2005 standards, is highly accurate and reliable. Therefore, the cost of construction can be reduced to 1.83., and provides 20 tons of cranes as desired. It is suitable for use within designated areas. The data of this design will be employed to construct the overhead crane of the local fishing boat maintenance workshop.

References

1. Haniszewski T., Strength analysis of overhead traveling crane with use of finite element method, J trans Prob. Vol. 9(1) (2014), p. 19-26.
2. Pratik R, Patel V.K., A review on structural analysis of overhead crane girder using FEA technique, Int. j. innov. res. sci. eng technol. Vol. 2(4) (2013), p. 41-44.
3. Alkin C, Imrak C.E, Kocabas H.J., Solid modeling and finite element analysis of an overhead crane bridge, Acta Polytech. Vol. 45(3) (2005), p. 61-67.
4. Sankar V.V, Vijayan P. Ashraf M.Y., Reducing the structural mass of a real-world double girder overhead crane, Int. j. adv. eng. technol. Vol. 8(2) (2015), p. 150-162.
5. Yifei T, Wei Y, Zhen Y, Dongbo L, Xiangdong L., Research on multidisciplinary optimization design of bridge crane, Math. probl. eng. (2013), p. 1-10.
6. Khalidurfeasif P, Bharti D., Finite element analysis of 450 t eot crane box girder, Int. j. adv. eng. res. dev. Vol. 2(11) (2015), p. 171-178.
7. Kapadni R, Ganiger S.G., A review paper on design and structural analysis of simply supported gantry crane beam for eccentric loading, Int. res. j. eng. Technol. Vol. 2(8) (2015), p. 1622-1626.
8. Vlada G, MIlorad M, Zoran., FEA Implementation in moving load problem at bridgr cranes, Mach des. (2010), p. 25-30.
9. Vankatesh A, Vignesh S, Iyappan S, Kumar P, Tamilvanan G, Sarathy R., Design of an over head plate gantry girder, Int. j. dev. Res. Vol. 5(6) (2016), p. 7821-7823.
10. Patel K, Jani V.K., J. inf. knowl. Design and dynamic analysis of 70 t double girder electrical overhead crane, Res. mech. eng. Vol. 2(2) (2013), p. 496-503.
11. Nenad M, Rodolju., Analysis of overhead travelling crane's motion in horizontal plane, Int. j. eng. (2016), p. 21-24.
12. Bhutta M, Misbahuddin M, Qureshi M, Ahmad S, Mughal H., Using finite element analysis for design optimization of twin beam crane arm, Sci. int. Vol. 28(1) (2016), p. 131-138.
13. Sitthipong S, Muangjunburee P, Detchvayukul, Totarat N, Stress analysis of swing shaft in the transmission system of electric shovel, Naresuan Phayao J. Vol. 9(4) (2016), p. 21-24.
14. Sitthipong S, Muangjunburee P, Detchvayukul, Totarat N., Stress Analysis of Final Drive Pinion Shaft of Electric Shovel Through The Use of FEM, Thaksin. Uni. J. Vol. 20(1) (2017), p. 57-66.

Design and Analysis of a Wind Turbine Blade with Dimples to Enhance the Efficiency through CFD with ANSYS R16.0

M. Rajaram Narayanan[a], S. Nallusamy , M. Ragesh Sathiyan

Department of Mechanical Engineering, Dr. M G R Educational and Research Institute,
Chennai-600095, Tamilnadu, India

Abstract. In the global scenario, wind turbines and their aerodynamics are always subjected to constant research for increasing their efficiency which converts the abundant wind energy into usable electrical energy. In this research, an attempt is made to increase the efficiency through the changes in surface topology of wind turbines through computational fluid dynamics. Dimples on the other hand are very efficient in reducing air drag as is it evident from the reduction of drag and increase in lift in golf balls. The predominant factors influencing the efficiency of the wind turbines are lift and drag which are to be maximized and minimized respectively. In this research, surface of turbine blades are integrated with dimples of various sizes and arrangements and are analyzed using computational fluid dynamics to obtain an optimum combination. The analysis result shows that there is an increase in power with about 15% increase in efficiency. Hence, integration of dimples on the surface of wind turbine blades has helped in increasing the overall efficiency of the wind turbine.

1 Introduction

In the modern scenario energy crisis and fossil fuel depletion has urged many countries to switch to other forms of renewable sources of energy. Among various forms of renewable energy, hydel and wind energy plays vital role in energy generation because of their surplus availability and higher efficiencies of energy transformation compared with tidal or solar. The amount of installed wind power is increasing every year and many nations have made plans to make large investments in wind power in the near future. High efficiency wind turbines are being designed on the chase of energy conversion and the analysis showed that the maximum extraction of energy possible from a turbine rotor is about 59% of the incoming kinetic energy [1-4]. The general public associates' wind turbines with horizontal axis wind turbines and are unaware of the several other technologies based on the vertical axis wind turbines which are less frequently used [5-8]. The energy absorbed from the wind is converted into mechanical energy by using wind turbines that is used to drive an electrical generator and then converted into electrical energy [9-13]. Based on our requirement airfoil blades are selected from the national renewable energy laboratory S-series, thicker than the ones that are seen on airplanes and for the generation of about 1.5MW power at wind speeds of 12-20m/s, the optimal blade was found to have a span of 40 meters [14-17]. The turbine blade is joined with the hub with the help of a cylindrical root section and blade profile is divided into three sections. Three different profiles such as S818, S825 and S826 are included in each section of the blade starting from the root. The blade was modelled with 3D CAD software and airfoils modelled with the point data which defines a particular airfoil with the blade dimensions including twist, chord length and span with respect to 40 meters blade length [18-21]. Earlier the scientists used to create scaled prototypes of the aerofoil to be analyzed and then place them in large wind tunnels to simulate the real world working conditions [22-25]. In modern times the process of testing and analysis has been highly computerized and is solved using computational fluid dynamics (CFD) [26-29]. The geometry of the problem here is the turbine blade which has to be analysed. The blade is modelled using the solid works modelling software and then imported to the CFD analysis software and a domain is created for the study which acts as a virtual boundary for the fluid [30-33]. The behaviour of a wind turbine rotor in a flow field may conveniently be analyzed by introducing the actuator disc principle [32-35]. The basic idea of the actuator disc principle in connection with rotor aerodynamic calculations is to replace the real rotor with a permeable disc of equivalent area where the forces from the blades are distributed on the circular disc. The basic assumption for a classical blade by element

[a] Corresponding author: drmrajaram@yahoo.com

© The Authors, published by EDP Sciences. This is an open access article distributed under the terms of the Creative Commons Attribution License 4.0 (http://creativecommons.org/licenses/by/4.0/).

 https://doi.org/10.1051/matecconf/201820702004

momentum method is the annual stream tubes are independent [36-39]. Thus by using this methodology along with the Reynolds averaged navier-stokes equations a near to actual result can be obtained.

1.1 Dimples

Dimples are small depressions on the surface which is usually permanent like in case of a golf ball. The presence of dimples increases the amount of energy and linear momentum that are available to those particles of fluid that move in a tiny region near to the surface of the ball called the boundary layer as shown in Figure 1. The existence and importance of the boundary layer are due to the peculiar behaviour of fluid friction near the surface of a solid body that is in relative motion with the surrounding fluid. In case of a golf ball, dimples trip air particles that are moving close to them. This disturbance causes the particles to jiggle sideways while they travel forward, instead of staying in lanes of traffic that are perfectly parallel to each other at all times, as expected in laminar flow. This jiggling forces particles in adjacent lanes to bump into each other, causing linear momentum to be transferred through bumping. When bumping is vigorous enough, its net result is that the bulk flow of air in the boundary layer becomes turbulent. Based on the above an attempt was made to increase the efficiency by changes in surface topology of wind turbine blades by CFD and ANSYS R 16.0.

1.2 CFD Analysis

Initially, CFD analysis was carried out without dimple done in order to find out the degree of accuracy to obtained results and match with the actual results and also to differentiate the efficiency between the blades with and without dimples. Blade alignment, domain dimensions, mesh settings and boundary conditions are considered to be the same for both the blades. Since the blades of the wind turbine are identical and are symmetrically arranged at angular gaps of 1200, it is enough to analyse only one blade with a domain in the shape a sector having 1200 angle. The blade is positioned 1m away from the axis of rotation to compensate for the hub. The wind speed considered for analysis is 12 m/s with the blade rotating at the speed of 2.2 rad/sec or 21 rpm. The relative motion between the blade tip and the flow stream results in a relative velocity which approaches the blade at an angle of 960 where as this angle is obtained by using the velocity vectors of the wind as well as the blade and then finding their resultant. Hence, the blade is also rotated in order to make its tip incline at the same angle. This causes the tip of the blade to have an almost zero angle of attack. Based on the actuator disc principle, a fluid domain is created for the wind turbine with dimensions of inlet face radius is 120 m, inlet face distance from rotor is 90m, outlet face radius is 240 m and outlet face distance from rotor is 180 m. The above mentioned radii are that of the sector which will cover only one blade, using the symmetric boundary condition the entire wind mill is analysed. The domain is created using the ANSYS R16.0 design modular software. The blade profile is then subtracted from the blade to create a void which acts as a virtual boundary. The speed of rotation is given as 2.2 rad/s which is the maximum speed with which the windmill will rotate in its optimum conditions with wind speeds of 12-20m/s. K-omega equation model is chosen for the analysis and hybrid initialization carried out.

2 Results and Discussion

The plain blade was first analysed with the above mentioned settings and the following results were obtained. The pressure (Pr) contour of the top side of the blade as shown in the Figure 2 shows the low pressure created at the top surface of the wind turbine blade. The pressure contour of the bottom side of the blade is shown in the Figure 3. The pressure developed at the bottom is significantly higher when compared to that of the top side of the blade which confirms that lift is generated. From Figure 4 it is clear that a high pressure is created at the bottom surface just near the leading edge of the blade and it creates a resistance to the rotation of the turbine and is otherwise known as drag. So, from this the position at which the drag force acts and is maximum visible. This data is then used to approximate a region where the dimples would be most effective and this region is found to be on the bottom face of the blade in the mid region near the trailing edge. Only the mid region is selected as the region near the root is too slow for drag to have a considerable effect and the region near tip is too fast for dimples to have any considerable effect hence dimples are designed and shown in Figure 5.

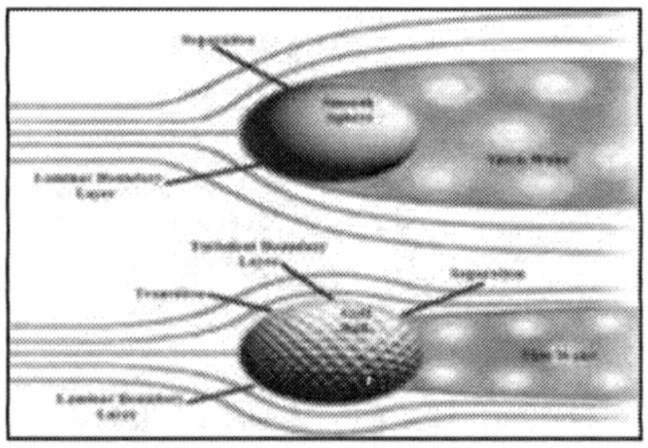

Figure 1. Dimple Effects on Golf Ball

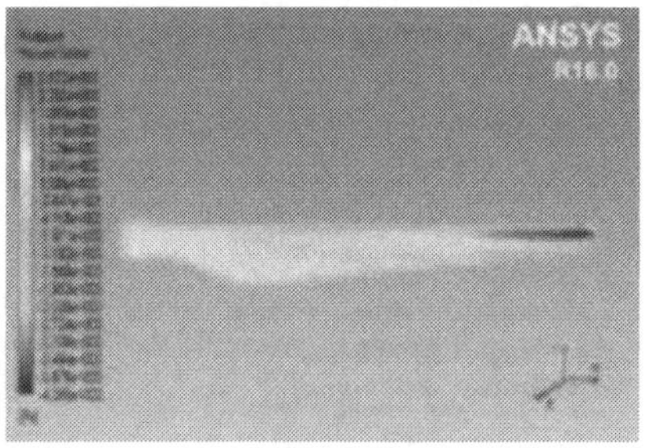

Figure 2. Pressure of Plain Blade Top View

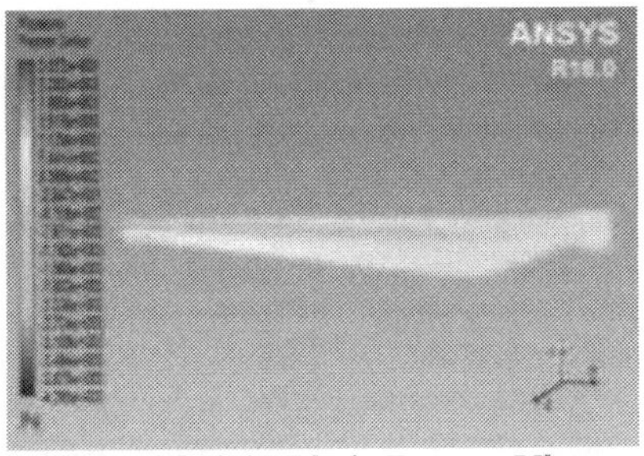

Figure 3. Pressure of Plain Blade Bottom View

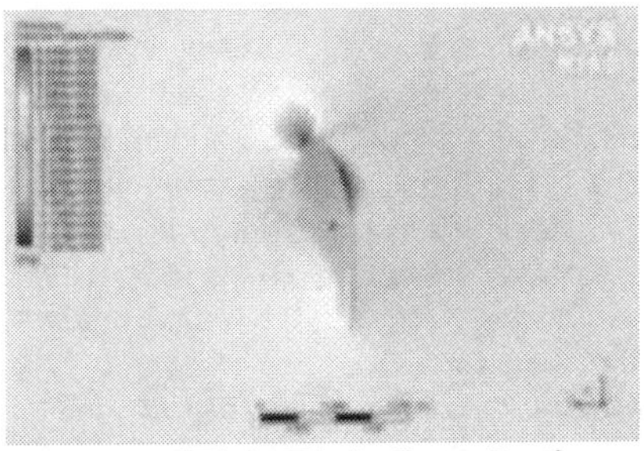

Figure 4. Pressure of Plain Blade Cross Section

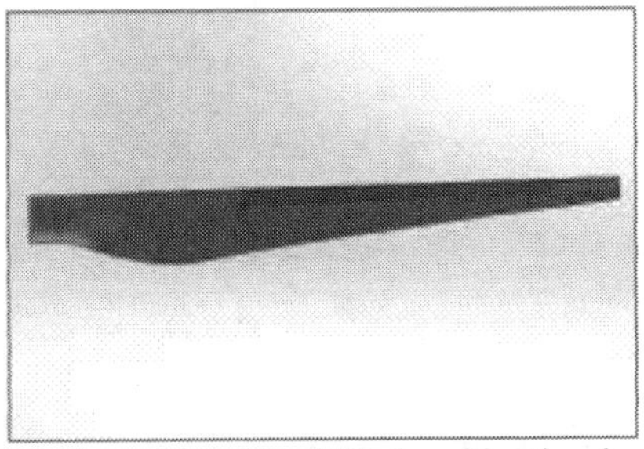

Figure 5. Design of Modified Blade with Dimples

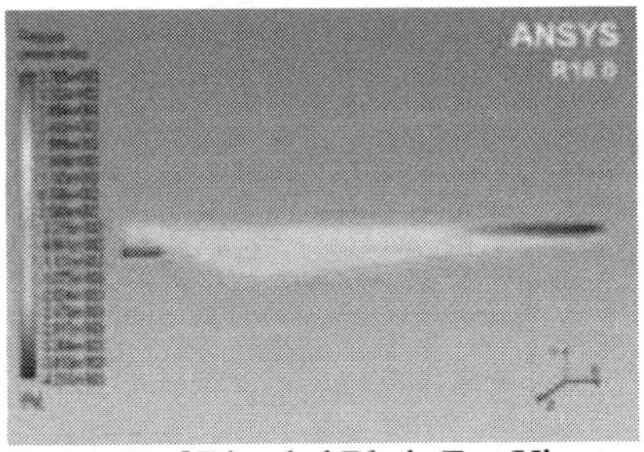

Figure 6. Pressure of Dimpled Blade Top View

After the dimples have been included in the design of the blade, it is then analyzed with the above mentioned settings and the following results are obtained. The pressure contour of the top side of the dimpled blade is shown in the Figure 6 which shows that the low pressure that is created at the top surface of this blade is similar to that of the plain blade. This implies that there is no effect of the dimples on the pressures of the top surface. The pressure contour of the bottom side of the dimpled blade is shown in the Figure 7.

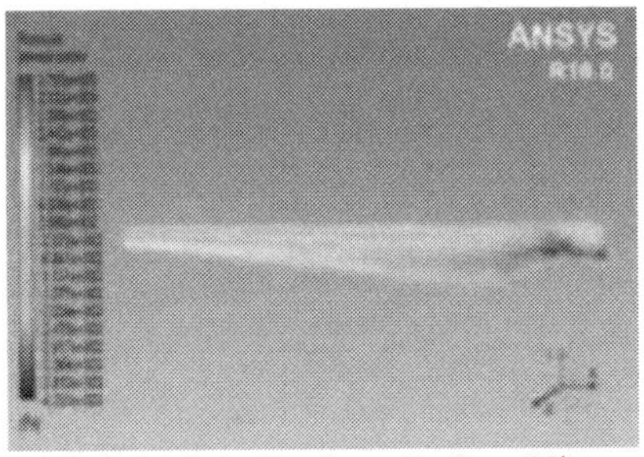

Figure 7. Dimpled Blade Pressure Bottom View

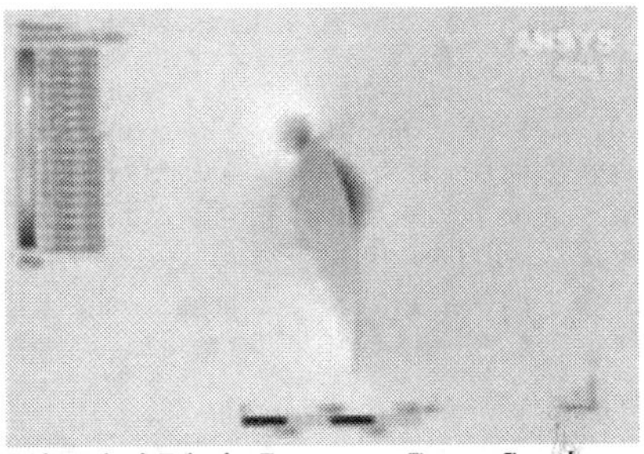

Figure 8. Dimpled Blade Pressure Cross Section

The pressure developed at the bottom is significantly higher when compared to that of the top side of the blade. This high pressure region however is much lesser compared to the plain blade at the region where the dimples have been placed. This indicates that the dimples have effectively reduced the pressure in the region by creating turbulence and hence reduced the drag. By comparing the pressure contour of the cross-section of the dimpled blade as shown in Figure 8 to that of the plain blade, it is seen that the highest pressure as described in the legends have a negligible difference but the region of high pressure is significantly larger in the plain blade. This implies that due to a smaller region of high pressure, the drag in dimpled blades is lesser.

2.1 Analysis of Torque and Power

The torque produced by the wind turbine is calculated by means of the function calculator of ANSYS R16.0 CFD post. Only the torque produced by the blade is calculated about the global z-axis as it is the defined axis of rotation in this model. For the plain blade, the torque produced by single blade is found to be 258895 Nm and so the torque produced by the turbine is 776715 Nm which can be converted to power using the formula,

$$P = 2\pi NT/60,$$

Where, P is the power, N is the seed of rotation in rpm and T is the torque produced

The power is thus found to be 1.70 MW and this power is transmitted to the generator and during the transmission about 20% loss occurs. This makes a power Of 1.36 MW to the generator which is much closer to the actual value. For the dimpled blade the torque produced by single blade is found to be 296615 Nm and so the torque produced by the turbine is 890054 Nm which when converted to power is 1.9 MW. After the transmission loss the power available to the generator is about 1.5MW. While comparing the power outputs, it is inferred that there is about 15% increase of power in dimpled blade turbine when compared to the plain blade turbine.

3 Conclusions

The above study on wind turbine blade designed with dimples was done for reducing the air drag in order to increase the efficiency of the blades. From the results the following conclusions were arrived.

- The numerical investigation focused on the aerodynamic effects contributing to torque which eventually improved the performance levels due to the inclusion of dimples.
- The S-series blade profile was chosen for optimum performance of generating a target of 1.5MW power under wind speed range of 12-20 m/s by combining three profiles together.
- The combination proved to have more strength and was found to be thick using actuator disc method without any modifications on the surface.
- Nearly about 1.4 MW power was found to be generated when the dimples were included which is a good level of accuracy when compared with actual value.
- The results showed an increase in power and also 15% increase in efficiency with the addition of dimples on the surface of wind turbine blades.

References

[1] Lin Wang et al., *State of the art in the Aeroelasticity of wind turbine blades: Aeroelastic modelling*, Renewable and Sustainable Energy Reviews, **64**, 195-210, (2016).

[2] S. Nallusamy, *Thermal conductivity analysis and characterization of copper oxide nanofluids through different techniques*, Journal of Nano Research, **40**, 105-112, (2016).

[3] Suman Kumar Jha, S. Nallusamy and Sinha Akshat Diwakar, *Future energy scenario in India - An essential overview*, Pollution Research, **35(3)**, 595-599, (2016).

[4] S. Nallusamy, et al., *Analysis of design features for inspection robot make use of concrete structures-An assessment*, International Journal of Engineering Research in Africa, **17**, 74-81, (2015).

[5] S. Nallusamy, *Characterization of epoxy composites with TiO_2 additives and E-glass fibers as reinforcement agent*, Journal of Nano Research, **40**, 99-104, (2016).

[6] Mohamed Khaled et al., *Aerodynamic design and blade angle analysis of a small horizontal-axis wind turbine*, American Journal of Modern Energy, **3(2)**, 23-37, (2017).

[7] S. Nallusamy and N. Manikanda Prabu, *Heat transfer enhancement analysis of Al_2O_3-water nanofluid through parallel and counter flow in shell and tube heat exchangers*, International Journal of Nanoscience, **16(5&6)**, 01-07, (2017).

[8] S. Nallusamy and A. Karthikeyan, *Synthesis and wear characterization of reinforced glass fiber polymer composites with epoxy resin using granite powder*, Journal of Nano Research, **49**, 01-09, (2017).

[9] Karthikeyan and S. Nallusamy, *Experimental analysis on sliding wear behaviour of aluminium-6063 with SiC particulate composites*, International Journal of Engineering Research in Africa, **31**, 36-43, (2017).

[10] S. Nallusamy and G. Majumdar, *Enhancement of overall equipment effectiveness using total productive maintenance in a manufacturing industry*, Int. J. of Per. Engg., **13(2)**, 01-16, (2017).

[11] S. Nallusamy, *Productivity enhancement in a small scale manufacturing unit through proposed line balancing and cellular layout*, Int. Journal of Performability Engg., **12(6)**, 523-534, (2016).

[12] Murtagh, Basu and Broderick, *Along-wind response of a wind turbine tower with blade coupling subjected to rotationally sampled wind loading*, Engineering Structures, **27**, 1209-1219, (2005).

[13] S. Nallusamy and Saravanan, *Enhancement of overall output in a small scale industry through VSM, line balancing and work standardization*, Int. J. of Engg. Res. in Africa, **26**, 176-183, (2016).

[14] S. Nallusamy, *Efficiency enhancement in CNC industry using value stream mapping, work standardization and line balancing*, Int. Journal of Performability Engg., **12(5)**, 413-422, (2016).

[15] Suman Kumar Jha, S. Nallusamy and Sinha Akshat Diwakar, *Future energy scenario in India-An essential overview*, Pollution Research, **35(3)**, 595-599, (2016).

[16] S. Nallusamy, *Enhancement of productivity and efficiency of CNC machines in a small scale industry using total productive maintenance*, Int. J. of Engg. Res. in Africa, **25**, 119-126, (2016).

[17] Weijun Tian et al., *Bionic design of wind turbine blade based on long-eared owl's airfoil*, Applied Bionics Biomechanics, **2017**, DOI: 10.1155/2017/8504638, (2017).

[18] S. Nallusamy, Manoj Babu and N. M. Prabu, *Investigation on carbon nanotubes over review on other heat transfer nano fluids*, Int. Journal of Applied Engg. Research, **10(62)**, 112- 117, (2015).

[19] S. Nallusamy, M.R. Narayanan and J. Logeshwaran, *Synthesis and machining characterization of copper-multiwalled carbon nanotubes-graphene hybrid composite using SEM and ANOVA*, Journal of Nano Research, **50**, 105-115, (2017).

[20] Xin Jin et al., *Investigation into parameter influence of upstream deflector on vertical axis wind turbines output power via 3-dimensional CFD simulation*, Renewable Energy, **115**, 41-53, (2018).

[21] S. Nallusamy et al., *Reliability analysis of passenger transport vehicles in public sector undertaking*, International Journal of Applied Engineering Research, **10(68)**, 843-850, (2015).

[22] S. Nallusamy, *A proposed model for sustaining quality assurance using TQM practices in small and medium scale industries*, Int. Journal of Engineering Research in Africa, **22,** 184-190, (2016).

[23] Xinzi Tang et al., *A direct approach of design optimization for small horizontal axis wind turbine blades*, Procedia CIRP, **36**, 12-16, (2015).

[24] S. Nallusamy and Adil Ahamed, *Implementation of lean tools in an automotive industry for productivity enhancement - A case study*, Int. J. of Engg. Research in Africa, **29**, 175-185, (2017).

[25] S. Nallusamy, *Frequency analysis of lean manufacturing system by different critical issues in Indian automotive industries*, Int. Journal of Engineering Research in Africa, **23**, 181-187, (2016).

[26] P J. Schubel and Crossley, *Wind turbine blade design*, Energies, **5**, 3425-3449, (2012).

[27] S. Nallusamy, *A review on the effects of casting quality, microstructure and mechanical properties of cast Al-Si-0.3Mg alloy*, Int. Journal of Performability Engg., **12(2)**, 143-154, (2016).

[28] S. Nallusamy and A. Karthikeyan, *Analysis of wear resistance, cracks and frictional properties of metal matrix composites with SiC additives and Al2O3 as reinforcement*, Indian Journal of Science and Technology, 9(35), 01-06, (2016).

[29] Okulov et al., *Experimental investigation of the wake behind a model of wind turbine in a water flume*, J. of Physics, Conference Series, **555**, DOI: 10.1088/1742-6596/555/1/012080, (2014).

[30] S. Nallusamy, *Analysis of MRR and TWR on OHNS die steel with different electrodes using electrical discharge machining*, International Journal of Engineering Research in Africa, **22**, 112-120, (2016).

[31] Galih Bangga et al., *CFD studies on rotational augmentation at the inboard sections of a 10 MW wind turbine rotor*, Journal of Renewable and Sustainable Energy, **9(2),** 676-681, (2017).

[32] S. Nallusamy, et al., *Analysis of static stress in an alloy wheel of the passenger car*, International Journal of Engineering Research in Africa, **16**, 17-25, (2015).

[33] S. Nallusamy and G. Majumdar, *Effect of stacking sequence and hybridization on mechanical properties of jute-glass fiber composites*, Int. Journal of Performability Engg., **12(3)**, 229-239, (2016).

[34] Nikolaos Simisiroglou, Simon-Philippe Breton and Stefan Ivanell, *Validation of the actuator disc approach using small scale model wind turbines*, Journal of Physics, **753**, 01-10, (2016).

[35] S. Nallusamy, *Analysis of welding properties in FSW aluminium 6351 alloy plates added with silicon carbide particles*, Int. Journal of Engineering Research in Africa, **21**, 110-117, (2015).

[36] S. Nallusamy, *Characterization of Al_2O_3/water nanofluid through shell and tube heat exchangers over parallel and counter flow*, Journal of Nano Research, **45**, 155-163, (2017).

[37] M. Rajaram Narayanan and S. Nallusamy, *An experimental analysis of aluminium alloy with tungsten carbide metal matrix composite through in-situ method using SEM*, Rasayan Journal of Chemistry, **11(1)**, 355-360, (2018).

[38] S. Nallusamy, et al., *Experimental analysis on reinforced aluminium metal matrix with boron carbide, graphite and fly ash composites*, Rasayan Journal of Chemistry, **10(4)**, 1368-1373, (2017).

[39] Sarkar and Behera, *Wind turbine blade efficiency and power calculation with electrical analogy*, International Journal of Scientific and Research Publications, **2(2)**, 01-05, (2012).

Structural Failure Trends of the Assembly Welding Fixture for the Front Sidewall on the basis of Finite Element Analysis

Duo Sun [1,a], Tianmin Guan[2] and Lei Lei [1]

[1, 2, 3] *School of Mechanical Engineering, Dalian Jiaotong University, Dalian 116028, Liaoning, China*

Abstract. The operating conditions of the assembly welding fixture for the front sidewall have been analyzed in this paper, and all the typical conditions have been applied with the nonlinear finite element analysis of the contact, aiming to ensure the safe utilization of the assembly welding fixture for the front sidewall of tooling equipment for the production of rail vehicles. The potentially risky component locations are predicted and the weak spots are discovered on the basis of the analysis results of the static strength. The anticipation of the component failure trends shows a guiding role in the failure prediction in actual production and facilitates accident prevention.

1 Introduction

The assembly welding fixture for the front sidewall of buses as the processing and tooling equipment in a machinery factory, Ltd., has been applied for the assembly welding for the front sidewall of buses, and generally installed on the positioner of 10T. The finite element analysis has been applied on the assembly welding fixture for the front sidewall, aiming to ensure the safe, scientific and reasonable operation of the equipment. The component failure trends are predicted and the weak spots are discovered on the basis of the static strength simulation and stiffness calculation under various operating loads, providing guidelines for maintenance and facilitating accident prevention.

2 Nonlinear finite element analysis of the contact

As to the contact problem of highly nonlinear characteristics, the contact zone is unknown before solving the problem, and it is also indefinite that the surfaces are in contact or separate. Such surfaces may be suddenly varied depending on the loads, materials, boundary conditions and other factors. Besides, the friction needs to be calculated in most of the contact problems, and a few types of calculation methods and models are applicable to compute the friction. However, all of them are involved with nonlinear properties, making the solving of the problem difficult to converge.

The finite element software of ANSYS as a multi-purpose finite element calculation program, can be utilized to solve structural, fluid, electric, electromagnetic and collision problems. The ANSYS supports three contact ways: point-point, point-surface and surface-surface, and the contact elements for each contact way are suitable for certain types of problems. To create a model of the contact problem, it's significant to firstly figure out which parts of the model may be contacted to each other. If one of the interactions is a point, the corresponding element of the model shall be a node. If one of the interactions is a surface, the corresponding element of the model shall be a cell, such as the beam element, shell element or solid element. The finite element model can be applied to identify potential contact pairs by means of the specified contact element.

3 The finite element model for the assembly welding fixture of the front sidewall

The 3D drawing software of PRO/E is applied to model the three-dimensional assembly welding fixture for the front sidewall, and the HyperMesh v10.0 is employed to carry out finite element modeling on the equipment. The large-scale, general-purpose finite element analysis software of ANSYS is utilized to conduct the static strength analysis on the assembly welding fixture for the front sidewall. Such a fixture has been represented in terms of three-dimensional physical model, as shown in Fig. 1, in accordance with its drawing data and the relationship between various components on the drawings.

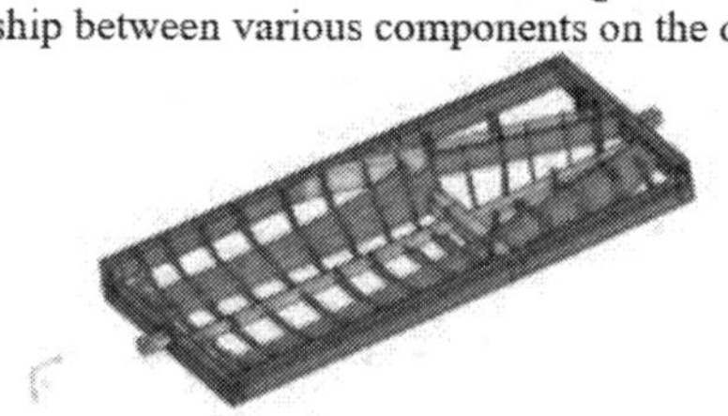

a. General assembly drawing

[a] Duo Sun: sddjtu@163.com

© The Authors, published by EDP Sciences. This is an open access article distributed under the terms of the Creative Commons Attribution License 4.0 (http://creativecommons.org/licenses/by/4.0/).

b. Parts drawing

Figure 1. Three-dimensional model for the assembly welding fixture for the front sidewall

Then the three-dimensional model is imported into HyperMesh for finite element meshing and developing a finite element model. Any structure that contributes to the overall stiffness and local strength of the assembly welding fixture for the front sidewall has been taken into account. The finite element model for the assembly welding fixture of the front sidewall as shown in Fig. 2, is mainly composed of hexahedral solid cells for the accuracy of computing.

Figure 2. Overall view of the finite element model for the assembly welding fixture for the front sidewall

The contact pairs shall be defined after the finite element modeling for all components. Two methods are available to be selected for the surface-surface contact pairs: one is to select the solid-solid, and the other is to produce two shells on the two contact surfaces, applied to define the contact relationship between the surfaces of the two shell elements. Taking into account the large size of the equipment parts, the solid-solid contact increases the number of operations, hence it is reasonable to define the contact relationship between the surface shell elements and perform operations so as to facilitate the calculation. The defined contact surface is displayed in Fig. 3.

Figure 3. Schematic of the defined contact surface

4 Load and boundary conditions

The calculation results for each operating condition are accessible by virtue of the post-processing of ANSYS, and the deformation nephogram and Von. Mises stress nephogram can be obtained respectively under each operating condition. The positions for the second largest stress can be easily discovered after removing the maximum stress point, and the stress value can be displayed. Five stress points are selected for each operating condition in order to find as many of potentially risky locations as possible. The calculation results of static strength for the assembly welding fixture for the front sidewall are summarized in Table 2.

Table 1. The partial load conditions for the assembly welding fixture for the front sidewall

No.	1	2	3	4
Load conditions	The rated load condition at 0°	The rated load condition at 30°	The rated load condition at 90°	The rated load condition at 180°
Labels and description	FZ	FZ	FZ	FZ
	Each assembly welding fixture for the front sidewall	Each assembly welding fixture for the front sidewall	Each assembly welding fixture for the front sidewall	Each assembly welding fixture for the front sidewall
Load values	7840 kN	7840 kN	7840 kN	7840 kN
Positions of the loads applied	Chassis center	Chassis center	Chassis center	Chassis center

The Fig. 4 shows the positions of the displacement constraints and load conditions applied on the assembly welding fixture for the front sidewall under the rated load conditions at 0°, 30°, 90° and 180°.

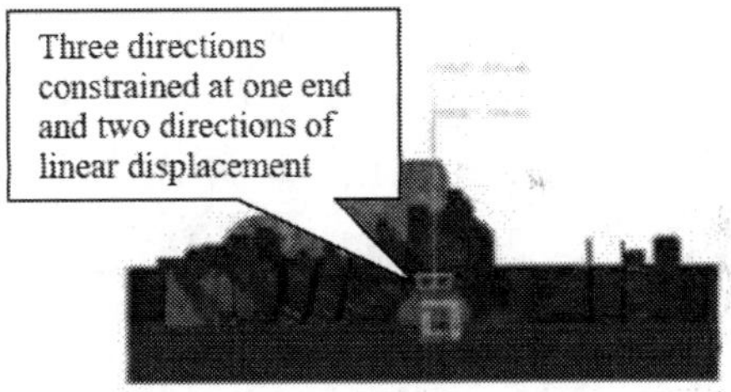

a.Rated load condition at 0°

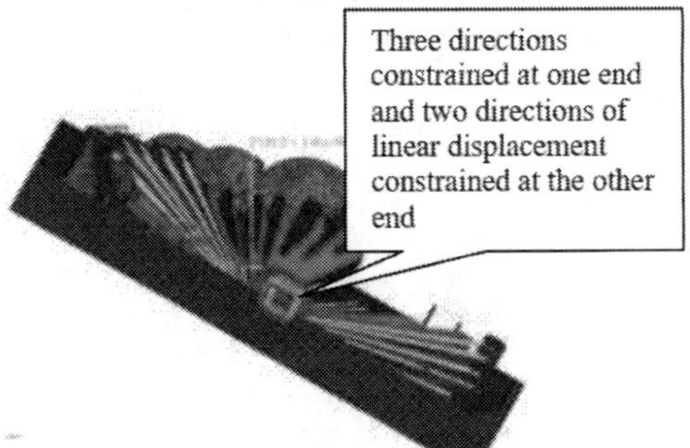

b. Rated load condition at 30°

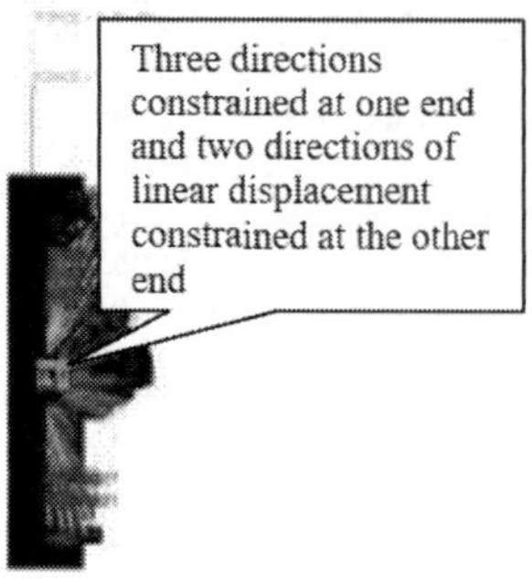

c. Rated load condition at 90°

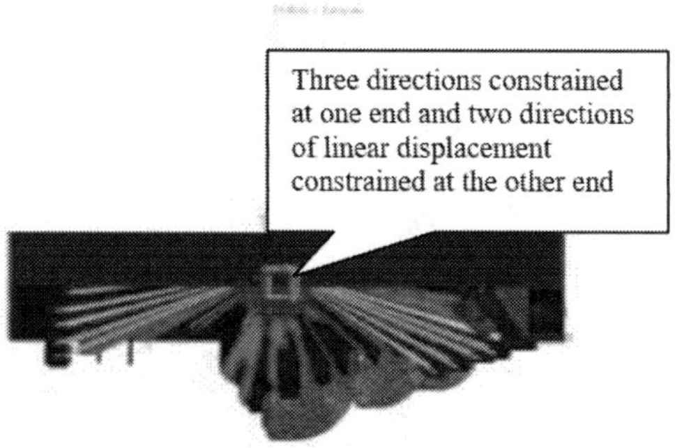

d. Rated load condition at 180°

Figure 4. Schematic of the loading positions on the assembly welding fixture for the front sidewall

5 Calculation results of static strength

The calculation results for each operating condition are accessible by virtue of the post-processing of ANSYS, and the deformation nephogram and Von. Mises stress nephogram can be obtained respectively under each operating condition. The positions for the second largest stress can be easily discovered after removing the maximum stress point, and the stress value can be displayed. Five stress points are selected for each operating condition in order to find as many of potentially risky locations as possible. The calculation results of static strength for the assembly welding fixture for the front sidewall are summarized in Table 2.

From the results in Table 2, it can be seen that the operating condition 3 is the riskiest one since the deformation and stresses at each location are the maximum values under each operating condition. Thus the operating condition 3 shall be avoided as much as possible during the usage of the equipment.

The maximum stress of the equipment takes place in the elliptical hole at the extended end, hence the surface of the elliptical hole shall be monitored.

The main stress point appears at the welding positions between the extended end and the crossbeam in the operating state. Based on the comprehensive consideration of strain and stress, the welding positions between the extended end and the crossbeam, rib plate as vulnerable parts tend to cause equipment failure, thus such positions ought to be conducted with 100% UT inspection. The followed are the welding positions between various supporting plates and crossbeam, longitudinal beam.

From the analyzed strain and stress nephogram it can be seen that significant stresses and deformation also take place at the welding positions between small longitudinal beam and crossbeam, supporting plate, consequently special attention shall be paid to their overhauling and maintenance.

Table 2. Calculation summary of the nonlinear static strength for the assembly welding fixture for the front sidewall

Calculation conditions		Operating condition 1	Operating condition 2	Operating condition 3	Operating condition 4
Maximum deformation (mm)		0.076	0.091	0.170	0.095
Von. Mises stress (MPa)	Maximum stress	14.3	17.2	17.6	12.6
	Second largest stress 2	10.7	12.8	12.8	9.7
	Second largest stress 3	8.6	10.3	10.4	9.6
	Second largest stress 4	8.2	9.8	10.2	9.1
	Second largest stress 5	5.9	7.1	10.2	8.2

The key monitoring positions are summarized for the assembly welding fixture for the front sidewall in Table 3.

Table 3 Summary of the key monitoring positions for the assembly welding fixture of the front

No	Key monitoring positions
1	Elliptical hole at the extended end
2	Location between the bottom plate at the extended end and the rib plate of crossbeam
3	Welding positions between the longitudinal supporting plate and crossbeam
4	Welding positions between small longitudinal beam and crossbeam

6 Conclusion

Taking into account ensuring the safe, scientific and reasonable operation of the assembly welding fixture for the front sidewall, the operating conditions have been selected for the equipment in this paper, and the nonlinear finite element analysis of the contact has been applied for all the typical conditions. The potentially risky component locations take place in the elliptical hole at the extended end, the same positions as the actual fault, on the basis of the analysis results of the static strength. The weak spots are discovered and the component failure trends are predicted, providing guidelines for the failure prediction in actual production and facilitating accident prevention. In the meanwhile, the anticipation of the component failure trends may pave the way for creating a database of the equipment reliability analysis.

References

1. Qingwu Wang, Fang Zuo, Renxi Hu, et al. Advanced Applications of Mechanical Design by ANSYS 10.0 (second edition) [M]. China Machine Press, 2006, 1
2. Hongyun Li, Shexu Zhao, Yan Sun. ANSYS 10.0 Foundations and Applications in Engineering [M]. China Machine Press, 2008
3. Feifei Zhang, Hongwei Ma, Peng Chai. Optimization Design of FSW Fixtures based on Structural Static Analysis [J]. Tooling Equipment, 2010, 1

Kinematics Analysis of Multi-link Mechanism for Miniature Servo Press Based on Matlab

Xinjian Lu[1,2,3]

[1] *School of Mechanical Engineering, Nanjing Institute of Technology, Nanjing 211167,China*
[2] *Research Department of Intelligent Manufacturing Equipment, Nanjing Institute of Technology, Nanjing 211167, China*
[3] *Yangzhou SHUANGYING Metal Forming Machine Co.,Ltd. Yangzhou 225000,China*

Abstract. The degree of freedom for a multi-link drive mechanism of a 50kN miniature servo press was calculated. The kinematics equations of the mechanism was established based on the rod group method. The choice or design basis for rod length and some hinge points were proposed. The multi-linkage mechanism's line segment model was constructed in SolidWorks and a set of design parameters was obtained that met the requirements. The M file of kinematic equations were wrote and calculated in Matlab and the displacement, velocity and acceleration curve of the slider were obtained. Compared with the crank-slider mechanism, the mechanism has the features of quick return, small return speed, slightly smaller speed and larger acceleration near B.D.C. Therefore, the speed and acceleration characteristics of the mechanism were inferior to the crank-slider mechanism in the case of reverse rotation and higher speed.

1 Introduction

Multi-link drive mechanism is widely used in mechanicalpresses. For example, the presses with eight-link (1) or the six-link (double-action press) (2) were mainly used for the body cover's stamping for automobiles, the presses with double crank mechanism (3) mechanism were mainly used for stamping parts of home appliances and auto parts industry, mainlymulti-link drive systems for micro-motor and micro-electronic stamping industry are: (4)symmetrical toggle lever(Kyori and Fair Oaks adopted),(5)double-link (Yamada Dobby adopted) and (5)leverage(Bruderer adopted) and so on. By using multi-link mechanism the contact speed and acceleration near bottom dead center(B.D.C) can be reduced in punching, and the life of molds can be improved also. There are few researches on the driving mechanism of mini press (the nominal force is usually less than 100kN)and crank-slider mechanism was usually used as working mechanism. With the popularity of servo drive technology, more and more servo drives are used in large presses. The multi-link drive's using in a mini presswill help to reduce the torque required during the blanking process, reduce the servo motor's power and increase the flexibility of the press. In this paper, a 50kN micro servo press (for precision electronic connector stamping) drive system has been studied. It's kinematic equations were derived and the slider displacement, velocity and acceleration curve are calculated based on Matlab. and curve shape's influencing factors were studied also. Research results for reference only.

2 Multi-link drive mechanism

The 50kN mini servo press multi-link drive mechanism was shown in Figure 1. The mechanism is composed of crank rocker mechanism rocker slider mechanism in series. By rational design the mechanism can save torque(6).That is to say, the same torque can produce greater stamping force. The freedom of this body is:

$$f = 3x5 - 2x7 = 1 \quad (1)$$

Therefore, the mechanism has a definite movement. The movement of the mechanism is as follows: The crank 2 rotates around the center O of the coordinate system and is hinged with the connecting rod 3 to the point P_{23}, the connecting rod 3 drives the connecting rod 4 to swing within a certain angle range, the inner part of the connecting rod 4 and the frame are hinged at point P_{14}, one end of the connecting rod 4 is hinged with the connecting rod 3 at point P_{34} and the other end is hinged with the connecting rod 5 at point P_{45} , the slider 6 reciprocates up and down in the rail of the rack 1, and is hinged with the connecting rod 5 through the hinge point P_{56}.

[a] Corresponding author: jw41281997@tom.com

© The Authors, published by EDP Sciences. This is an open access article distributed under the terms of the Creative Commons Attribution License 4.0 (http://creativecommons.org/licenses/by/4.0/).

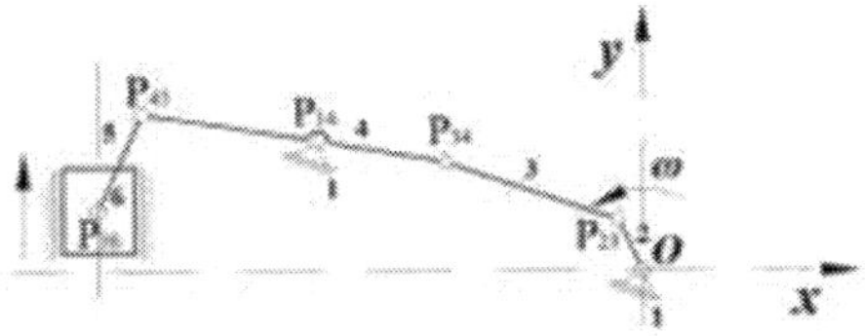

Figure 1. Multi-link drive mechanism

3 Kinematics equations

The coordinate system of the multi-link drive mechanism was shown in Figure 1. The origin of the coordinates is located at the center of rotation O of the crank 2, with the x-axis horizontal to the right and the y-axis vertically upwards. Kinematic equations of this mechanism were established by using rod group analysis. The mechanism is broken down into one side link and two II grade rod groups as shown in Figure 2. The location equations of each hinge point were established. Speed and acceleration for hinge points were obtained by calculating the first and second derivatives of the position equation over time.

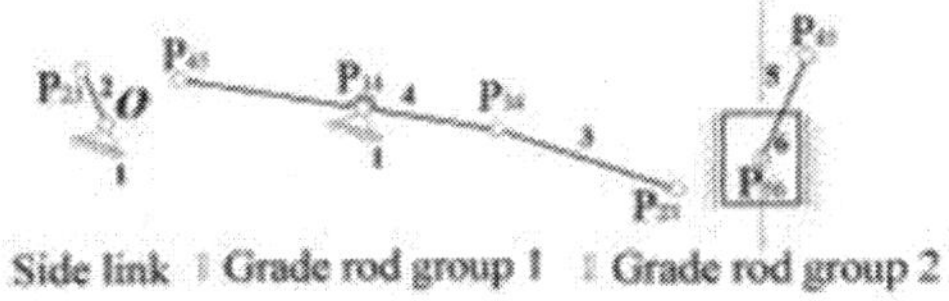

Figure 2. Side link and II grade rod groups

3.1 Side link

As shown in Figure 3, the radius of the crank 2 is r, the rotation angle is φ, the angular velocity is ω , and the displacement of the hinge point P_{23} between the crank 2 and the connecting rod is:

$$\left.\begin{aligned} x_{23} &= \mathrm{r}\cos\varphi \\ y_{23} &= \mathrm{r}\sin\varphi \end{aligned}\right\} \tag{2}$$

By calculating the first and second derivatives of equation (2) over time the velocity and acceleration of the hinge point P_{23} were obtained (irrespective of the fluctuation of the crank rotation speed during punching, and the angular acceleration of the crank when the machine was started and braked).

$$\left.\begin{aligned} \dot{x}_{23} &= -\omega \mathrm{r}\sin\varphi \\ \dot{y}_{23} &= \omega \mathrm{r}\cos\varphi \end{aligned}\right\} \tag{3}$$

$$\left.\begin{aligned} \ddot{x}_{23} &= -\omega^2 \mathrm{r}\cos\varphi \\ \ddot{y}_{23} &= -\omega^2 \mathrm{r}\sin\varphi \end{aligned}\right\} \tag{4}$$

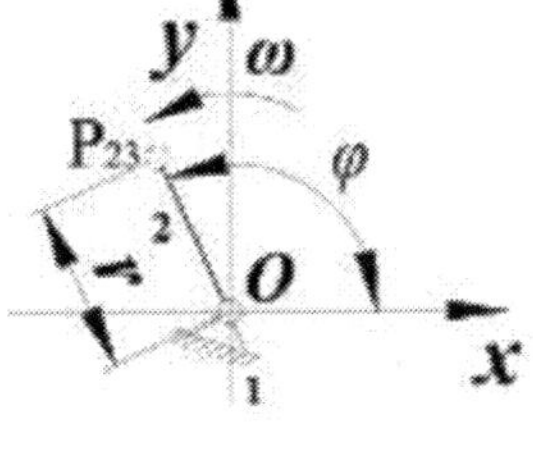

Figure 3. Side link

3.2 Grade rod group 1

Rod group 1 was shown in Figure 4. The length of the connecting rod 3 was set to be l_3. The x coordinate and the y coordinate of the hinge point P_{14} were respectively x_{14} and y_{14}, which were known quantities. The distance between the hinge points P_{34} and P_{45} was l_4, and the distance between the hinge points P_{34} and P_{14} was l_{41}.

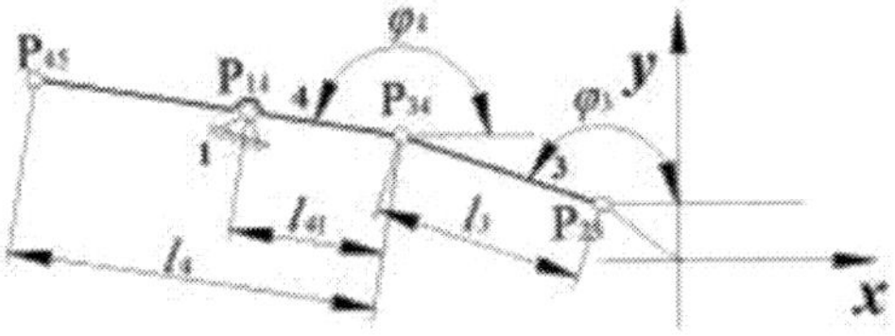

Figure 4. II grade rod group 1

The displacement equations for hinge point P_{34} between the connecting rod 3 and the connecting rod 4 was:

$$\left.\begin{aligned} x_{34} &= x_{23} + l_3\cos\varphi_3 \\ y_{34} &= y_{23} + l_3\sin\varphi_3 \end{aligned}\right\} \tag{5}$$

By calculating the first and second derivatives of equation (5) over time the velocity and acceleration of the hinge point P_{34} were obtained.

$$\left.\begin{aligned} \dot{x}_{34} &= \dot{x}_{23} - \dot{\varphi}_3 l_3\sin\varphi_3 \\ \dot{y}_{34} &= \dot{y}_{23} + \dot{\varphi}_3 l_3\cos\varphi_3 \end{aligned}\right\} \tag{6}$$

$$\left.\begin{aligned} \ddot{x}_{34} &= \ddot{x}_{23} - l_3(\dot{\varphi}_3^2\cos\varphi_3 + \ddot{\varphi}_3\sin\varphi_3) \\ \ddot{y}_{34} &= \ddot{y}_{23} + l_3(\ddot{\varphi}_3\cos\varphi_3 - \dot{\varphi}_3^2\sin\varphi_3) \end{aligned}\right\} \tag{7}$$

The displacement equation of the hinge point P_{14} between the connecting rod 4 and the body 1 was:

$$\left.\begin{aligned} x_{41} &= x_{34} + l_{41}\cos\varphi_4 \\ y_{41} &= y_{34} + l_{41}\sin\varphi_4 \end{aligned}\right\} \tag{8}$$

The velocity of the hinge point P_{14} was obtained by calculating the first derivation of equation (8) over time. Since the hinge point is a fixed point, the speed value should be zero.

$$\left.\begin{aligned} 0 &= \dot{x}_{34} - \dot{\varphi}_4 l_{41}\sin\varphi_4 \\ 0 &= \dot{y}_{34} + \dot{\varphi}_4 l_{41}\cos\varphi_4 \end{aligned}\right\} \tag{9}$$

Solving Equations (5) and (8).

$$\varphi_4 = 2\pi + 2\mathrm{arctg}\frac{B-\sqrt{B^2+A^2-C^2}}{A+C} \tag{10}$$

$$\varphi_3 = \mathrm{atan2}(y_{34}, x_{34}) \tag{11}$$

In the above formula,

$$A = x_{41} - \mathrm{r}\cos\varphi$$

$$B = y_{41} - \mathrm{r}\sin\varphi$$

$$C = -\frac{l_3^2 - l_{41}^2 - \mathrm{r}^2 - x_{41}^2 - y_{41}^2 + 2(x_{41}\mathrm{r}\cos\varphi + y_{41}\mathrm{r}\sin\varphi)}{2l_{41}}$$

The angular velocity and angular acceleration for φ_4 were:

$$\dot{\varphi}_4 = \frac{\dot{C} - \dot{A}\cos\varphi_4 - \dot{B}\sin\varphi_4}{B\cos\varphi_4 - A\sin\varphi_4} \tag{12}$$

$$\ddot{\varphi}_4 = \frac{\ddot{C} - \ddot{A}cos\varphi_4 + \dot{A}sin\varphi_4\dot{\varphi}_4 - \ddot{B}sin\varphi_4 - \dot{B}cos\varphi_4\dot{\varphi}_4}{Bcos\varphi_4 - Asin\varphi_4} - \frac{(\dot{C} - \dot{A}cos\varphi_4 - \dot{B}sin\varphi_4)(\dot{B}cos\varphi_4 - \dot{A}sin\varphi_4)}{(Bcos\varphi_4 - Asin\varphi_4)^2} + \frac{(\dot{C} - \dot{A}cos\varphi_4 - \dot{B}sin\varphi_4)(Bsin\varphi_4 + Acos\varphi_4)\dot{\varphi}_4}{(Bcos\varphi_4 - Asin\varphi_4)^2} \quad (13)$$

In the formula(13),

$$\dot{A} = \omega r sin\varphi \qquad \ddot{A} = \omega^2 r cos\varphi$$

$$\dot{B} = -\omega r cos\varphi \qquad \ddot{B} = \omega^2 r sin\varphi$$

$$\dot{C} = \frac{\omega r(y_{41}cos\varphi - x_{41}sin\varphi)}{l_{41}}$$

$$\ddot{C} = \frac{-\omega^2 r(y_{41}sin\varphi + x_{41}cos\varphi)}{l_{41}}$$

The angular velocity and angular acceleration for $\varphi_3$were:

$$\dot{\varphi}_3 = -\frac{\omega r sin\varphi + l_{41}sin\varphi_4\dot{\varphi}_4}{l_3 sin\varphi_3} \quad (14)$$

$$\ddot{\varphi}_3 = \frac{(\omega r sin\varphi + l_{41}sin\varphi_4\dot{\varphi}_4)cos\varphi_3\dot{\varphi}_3}{l_3(sin\varphi_3)^2} - \frac{\omega^2 r cos\varphi + l_{41}(cos\varphi_4\dot{\varphi}_4^2 + sin\varphi_4\ddot{\varphi}_4)}{l_3 sin\varphi_3} \quad (15)$$

The displacement equation of the hinge point P_{45} between the connecting rod 4 and the connecting rod 5 was:

$$\left.\begin{aligned} x_{45} &= x_{34} + l_4 cos\varphi_4 \\ y_{45} &= y_{34} + l_4 sin\varphi_4 \end{aligned}\right\} \quad (16)$$

The speed and acceleration of the hinge point P_{45} were obtained by calculating the first and second derivatives of equation (16) over time.

$$\left.\begin{aligned} \dot{x}_{45} &= \dot{x}_{34} - l_4 sin\varphi_4\dot{\varphi}_4 \\ \dot{y}_{45} &= \dot{y}_{34} + l_4 cos\varphi_4\dot{\varphi}_4 \end{aligned}\right\} \quad (17)$$

$$\left.\begin{aligned} \ddot{x}_{45} &= \ddot{x}_{34} - l_4(sin\varphi_4\ddot{\varphi}_4 + cos\varphi_4\dot{\varphi}_4^2) \\ \ddot{y}_{45} &= \ddot{y}_{34} + l_4(cos\varphi_4\ddot{\varphi}_4 - sin\varphi_4\dot{\varphi}_4^2) \end{aligned}\right\} \quad (18)$$

3.3 Grade rod group 2

Rod group 2 was shown in Figure 5. The length of the connecting rod 5 was set to be l_5, and the x coordinate x_{56} of the hinge point P_{56} was a known quantity. The$y$ coordinate(y_{56}) of the hinge point P56 was the displacement s for the slider 6. The displacement equation of hinge point P_{56} was:

$$\left.\begin{aligned} x_{56} &= x_{45} + l_5 cos\varphi_5 \\ y_{56} &= y_{45} + l_5 sin\varphi_5 \end{aligned}\right\} \quad (19)$$

The result of solving was:

$$\varphi_5 = 2\pi - 2arctg\sqrt{\frac{1-D}{1+D}} \quad (20)$$

In the formula(20),

$$D = \frac{x_{56} - x_{45}}{l_5}$$

The angular velocity and angular acceleration for $\varphi_5$were:

$$\dot{\varphi}_5 = \frac{\dot{x}_{45}}{l_5 sin\varphi_5} \quad (21)$$

$$\ddot{\varphi}_5 = \frac{\ddot{x}_{45}}{l_5 sin\varphi_5} - \frac{\dot{x}_{45}cos\varphi_5\dot{\varphi}_5}{l_5(sin\varphi_5)^2} \quad (22)$$

Therefore, the speed and acceleration of the slider 6 were respectively:

$$v = \dot{y}_{56} = \dot{y}_{45} + l_5 cos\varphi_5\dot{\varphi}_5 \quad (23)$$

$$\alpha = \ddot{y}_{56} = \ddot{y}_{45} + l_5(cos\varphi_5\ddot{\varphi}_5 - sin\varphi_5\dot{\varphi}_5^2) \quad (24)$$

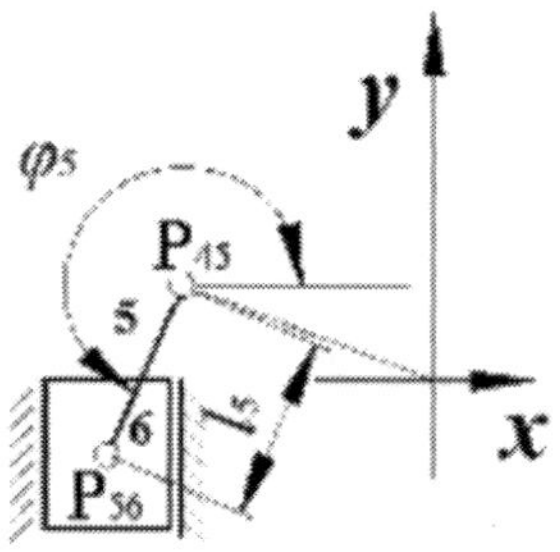

Figure 5. II grade rod group 2

4 Determination of rod design parameters

The main design parameters of the 50kN mini servo press were shown in Table 1:

Table 1. The main design parameters.

item	*value*	*unit*
Nominal force	50	kN
Stroke length(max)	40	mm
Activating point	1.4	mm
No. of strokes(max)	150	
Open height	300	mm
Overall Dimensions (L.R. xH x F.B.)	700x1000x1000	mm

4.1 Design guidelines for link system

1. Taking into account the design and manufacturing factors the design value of the crank radius r is within 50mm;

2. According to my past experience, the pin diameter of the rod system is about 30mm. Considering the wall thickness of the pin hole and the movement space of each rod, the length of each rod or part is about 80mm;

3. The connecting rod coefficient(r/l_2) should be designed within 0.1 ~ 0.2 (7),for high-speed and fast presses will be less than 0.1;

4. The crank-rocker mechanism should meet the conditions of the rod length and the existence of the crank (8);

5. In order to avoid the dead point of the mechanism due to friction, the angle between the connecting rod 5 and the vertical line should be designed within 30 ° ;

6. The overall length of the rod system should be less than the overall dimensions of the front and back directions.

4.2 Rod parameters

In the SolidWorks software, the origin was taken as the crank center point O of the mechanism, the connecting rods were replaced by line segments, the point drawn in the line segment was considered as the hinge point in the connecting rod, and the multi-link mechanism can be obtained. The length of the rods and the coordinates of the hinges were given appropriate values to ensure that the mechanism can operate without dead point when the mouse dragging the mechanism's crank. Based on MATLAB, the kinematics analysis program of the mechanism was written in the form of M file, and then the parameters of the multi-link system were adjusted slightly. At the same time, the kinematics equations were corrected (confirming the quadrant where the angle was), and the rod parameters that satisfy the design requirements were shown in Table II

Table 2. Rod length and position parameters(mm).

Item	r	l_3	l_{41}	l_4	l_5	x_{14}	y_{14}	x_{56}
Value	14.58	330	80	184	100	-380	60	-465

5 Calculation results

By moving the axes, it was guaranteed that the curve changes from top dead center (0 °) as the crank angle changes. For comparison, the slider displacement, velocity and acceleration of the crank-slider mechanism (slider stroke： 40 mm, crank radius ： 20 mm, link length ： 300 mm) were given in the figures below.

Slider displacement curves were shown in Figure 6. Compared with the crank-slider mechanism, the position of the B.D.C of the slider for the multi-link mechanism was slightly shifted to the left by about 3 ° due to the connected crank-rocker'quick-return characteristics. And the mechanismwas able to reach capacity early.

Figure 6. Slider displacement curves

Slider velocity curves were shown in Figure 7. Compared with the crank-slider mechanism, The multi-link mechanism's return speed of 292.85mm/s was slightly less than the crank-slider mechanism's 314.85mm/s. Near B.D.C (left side of the curves), the speed of the multi-link mechanism was slightly less than the crank-slider mechanism, and it was help to precision punching. When leaving B.D.C(right side of the curves), the multi-link mechanism's speed was larger than crank-slider mechanism's.

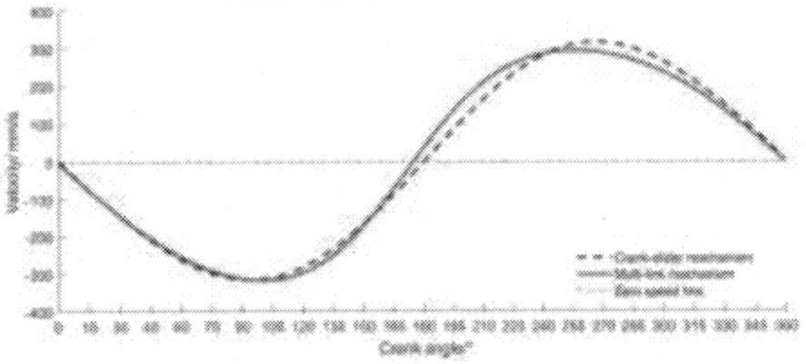

Figure 7. Slider velocity curves

Slider velocity curves were shown in Figure VIII. Compared with the crank-slider mechanism, the acceleration of the multi-link mechanism near B.D.C reaches 6650.1mm/s^2, which is 1.26 times the 5263.8mm/ s^2 of the crank-slider mechanism.

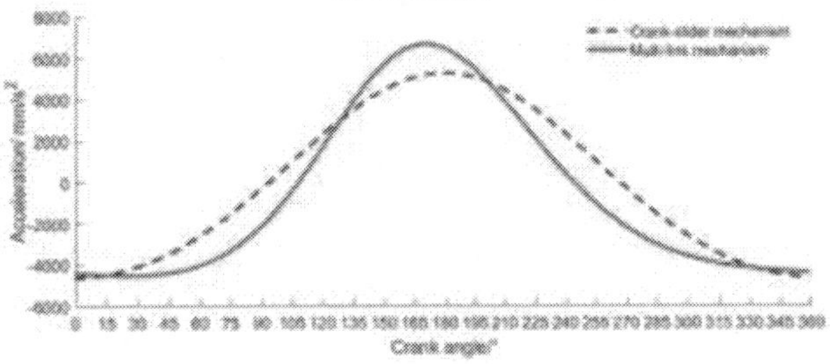

Figure 8. Slider acceleration curves

6 Conclusions

The kinematic equations of a multi-link drive mechanism for the 50kN servo press were established, and the design principles of the rod-system parameters were put forward. A set of parameters were given to meet the design requirements. The crank radius of this mechanism was only 72.9% of the crank radius in the crank-slider mechanism at the same slider stroke. Kinematics equations were established in Matlab and calculated. Compared with the crank-slider mechanism, the mechanism has the features of quick return, small return speed, slightly smaller speed and larger acceleration near B.D.C. Therefore, the speed and acceleration characteristics of the mechanism were inferior to the crank-slider mechanism in the case of reverse rotation and higher speed.

7 Acknowledgements

The author wishes to thank the Opening Project of Research Department of Intelligent Manufacturing Equipment of Nanjing Institute of Technology(project number: PTKJ201608) and The introduction of talent research fund project of Nanjing Institute of Technology(project number:YKJ201503).

References

1. Lian Xia,Jin Zhang, Jiang Han, Yunshan Wang and Liang HE. Kinematics analysis of eight-link mechanical

press[J],Forging&Stamping Technology,2012,37(4):158-161(in Chinese).

2. Guiwei Dong,Yiguo Luan,Linlin Sun and Yanjun Yu. Kinematic analysis and Optimum Design for Multi-link Mechanism in Drawing Press[J]. China Metalforming Equipment & Manufacturing Technology,2010,41(4):27-30.(in Chinese).
3. Yujian Xiang,Shisong Tang,Zhiwei Zhou,Kuan She and Zhi Liu.Optimal Design of Rod System for Double Crank Servo Press on the Basis of Genetic Algorithm[J]. China Metalforming Equipment & Manufacturing Technology,2017,52(3):17-19.(in Chinese).
4. Cankun He. Kinematic Characteristics Analysis of Multi-link High-speed Press[J]. China Metalforming Equipment&ManufacturingTechnology,2013,48(5):19-21.(in Chinese).
5. Xinjian Lu,Zunmang Ke, Sihong Zhu and Jun Zhu. Research on Bottom Dead Center (B.D.C) of High Speed Precision Press[J]. Forging&Stamping Technology,2010,35(1):87-92(in Chinese).
6. Meilin Zhang, Youchen Zhang and Liyan Zhang. Mechanical Innovation Design (2nd Edition)[M]. Beijing: Chemical Industry Press, 2010.(in Chinese)
7. Deyu He. Crank Press[M]. Beijing: China Machine Press, 1981.(in Chinese)
8. Yu Zhu. Fundamentals of Mechanical Design[M]. Beijing: Peking University Press, 2013.(in Chinese)

Research on Several Tool Chuck Technology in High Speed Milling

Zhao Zhenyu[1, a*], Xu Xiaomei[1, b], Zhou Yongquan[1, c], Xiao Haibin[1,d]

[1] *Shenzhen Institute of Information Technology, Shenzhen, China*

Abstract. In order to meet the requirements of high-speed precision machining technology, the paper elaborates the tool clamping technology commonly used in high-precision machining, such as the static pressure expansion chuck, CoroGrip chuck, and stress locking chuck and so on. How to correctly select the tool clamping system, the characteristics of the tool clamping system and the existing problems are analyzed.

1 Introduction

In recent years, due to the continuous development of CNC cutting machine tools and tool manufacturing made great achievements in the field, it makes the level of metal cutting to high-power, high-speed, high-precision continuously improved. It also sets higher and higher requirements on the important part of the tool clamping system that connects the tool and the machine tool. The application of high-speed cutting technology and precision machining technology requirements need high-performance CNC machine tool system tool chuck higher and higher requirements. As long as the tool system to improve the clamping accuracy, we must try to make the tool to be precise and reliable positioning, to ensure adequate clamping force, strict control and improve tooling system with precision, and increase the clamping length, and optimize the structural design and reasonable selection[1].

The paper focuses on several kinds of tool clips suitable for high-speed machining, such as static pressure expansion tool holder, CoroGrip chuck, and stress locking chuck and so on.

2 Technical analysis

2.1 Static pressure expansion chuck

The static pressure expansion chuck is a high precision tool holder with static pressure expansion principle. The structure is shown in Fig.1. It has excellent technical performance, which is suitable for holding drills, reamer, milling cutter in machining center, high precision boring milling machine and flexible production line. Currently in Germany, static pressure expansion chuck has been widely used in manufacturing industries such as automobiles and machinery. In China, many machining enterprises have successfully used the tool holder.

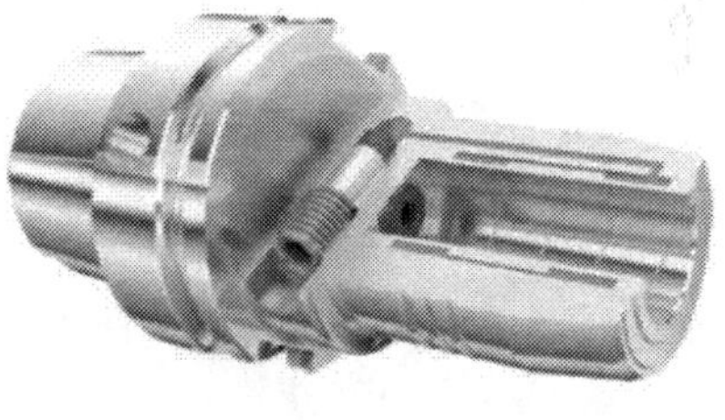

Fig.1 static pressure expansion chuck

The structural principle of static expansion chuck is shown in Fig.2.The holder main body and the expansion of the clamping hole wall have a circular closed oil cavity; the oil cavity fills with dedicated hydraulic oil. The oil pressure can be evenly delivered to every part of the oil sealed cavity, at the same time also to increase structural damping, to improve the dynamic characteristics of chuck, and to reduce the vibration and improve the quality of machining. The expansion wall of the clamping hole is set by precise calculation. It has a good elasticity and can produce the desired amount of expansion when the oil pressure reaches the set value. The clamping hole wall has ring slot, and can accommodate the cutting tool residual lubricant, and keep the clamping hole wall clean and ensure reliable transmission of torque [2-3].

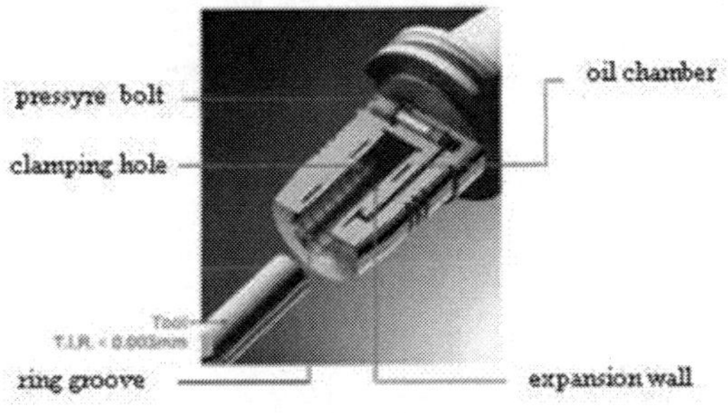

Fig.2 Static pressure expansion chuck working principle

[a] Corresponding author: yxpzzy01@163.com

© The Authors, published by EDP Sciences. This is an open access article distributed under the terms of the Creative Commons Attribution License 4.0 (http://creativecommons.org/licenses/by/4.0/).

The static pressure expansion tool holder produced by the German company has the following excellent properties. The clamping rotary precision is less than 3um. Repeated clamping precision is less than 2um. It has stable and reliable clamping force. It can deliver high torque. It has excellent damping performance. The clamping system has the whole sealing structure, no abrasion and durability. When the outer diameter of the cutter is 6-8mm, the precision level of the cutter bar is h6. When the outer diameter of the tool is greater than 8mm, the precision level of the pole can be relaxed to h7.

2.2 CoroGrip chuck

Sandvik has developed the CoroGrip chuck, as shown in Fig.3. The product uses a hydraulic device to push the cone, which is measured in 3D with a radial beat of 2-6um. The chuck is more reliable. Its rigidity is higher than the hydraulic clamping head, and the clamping time is shorter than the heat shrinkable head [4].

Fig.3 CoroGrip chuck

CoroGrip chuck is acted on the wedge type machine tool clamping jaw of the hydraulic pressure to realize the clamping function at the same level products. CoroGrip chuck clamping force is unusually high and has a very high precision, suitable for drilling and from super-finishing to reload all rough machining milling process. CoroGrip chuck has all the pins on its own balance. During the long process of processing, the pulsation amount is always consistent, thus effectively improving the service life of the cutter and ensuring the quality of the parts.

The high clamping force required for CoroGrip chuck is provided by the external hydraulic pump. The clamp has two different hydraulic pumps. They are the manual pump and the pneumatic motor drive pump. The pneumatic pump uses the general pneumatic system in the workshop. With these pumps, it is easy to change the knife. It can be changed for less than 20 seconds, and the same clamping force is always on the handle.

The HydroGrip chuck is used to achieve the tool clamping by the pressure of the membrane on the handle of the handle. The clamping principle is shown in Fig.4. The clamping force is more than three times that of the traditional chuck.

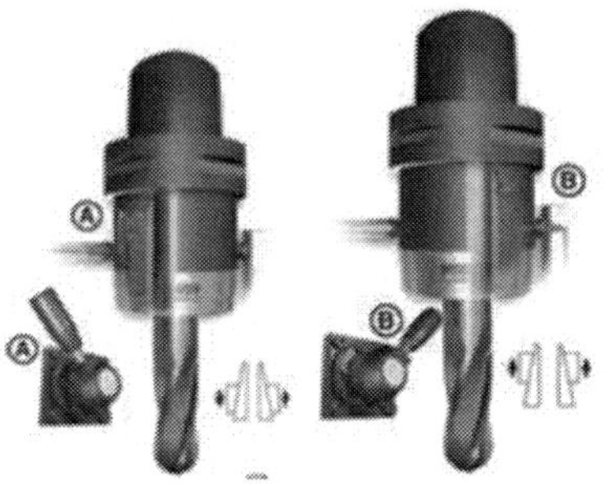

Fig.4 CoroGrip chuck working principle

CoroGrip high-precision power chuck has the following several advantages. It can improve the surface quality. At three times in diameter, its beating quantity is 6 um or less. It can prolong tool life. According to the rule of thumb, According to rule of thumb, every 10μm increase in runout results in a 50% reduction in tool life. It is almost no vibration and low noise level. The repeatability is high and the clamping force remains unchanged after using 15,000 cycles. Heavy-duty heavy processing has excellent performance, and unusually high clamping force and rigid structure prevent tool slipping.

2.3 Stress-locking chuck

There are two main types of stress locking chuck. They are TRIBOS-R and TRIBOS-S, as shown in Fig.5. The shape of the TRIBOS-R chuck is robust and has a specially designed vibration reducing cavity structure which can be used for rough cutting and heavy cutting. The TRIBOS-S chuck has a very fine external diameter, which can be used to solve the problem of interference in a very restricted area of processing space.

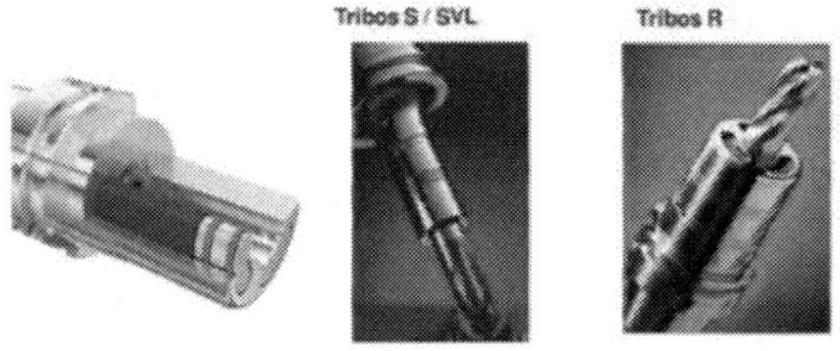

Fig.5 TRIBOS chuck

Schunk Company has two basic TRIBOS chuck. One chuck is tool extension rod, and it can have the effect of prolonging tool, but it must be associated with a cutting tool holder which is used. Another chuck is for a complete tool holder, and it can be directly as well as clamping tool with the machine tool spindle.

The clamping slewing accuracy and repeat clamping accuracy is less than 3um in Stress-locking clamping system. Compared with chuck TRIBOS-S, TRIBOS-R chuck increases the clamping piece of overall diameter, and designs the special cavity structure of vibration reduction. The cavity is filled with thermosetting plastics, make the cutter holder rigidity and clamping force have been greatly improved, can be used for heavy cutting processing. The two kinds of chuck can reduce the vibration of the tool and improve the service life of the cutter.

The holding principle of TRIBOS-R stress locking chuck is shown in Figure.6. The clamp hole has a special geometrical shape with axial-symmetric shape, as shown in Figure.6a. In the original state, the tool cannot be inserted into the holing hole. When installing a tool, a special loader is used to pressurize the holding section from outside, forcing the clamping hole to become a round hole in the range of elastic deformation, as shown in Figure.6b. Insert the cutter into the clamp hole smoothly, as shown in Figure.6c.Then, loosen the loader and remove the external load, and the cutter is firmly clamped with deformation recovery force, as shown in Figure.6d. Also, the tool can be easily removed or replaced with a loader. Because the clamping motion of TRIBOS has the characteristics of axial-symmetric, it has extremely high clamping accuracy. This tool holder is also very suitable for high speed machining [5].

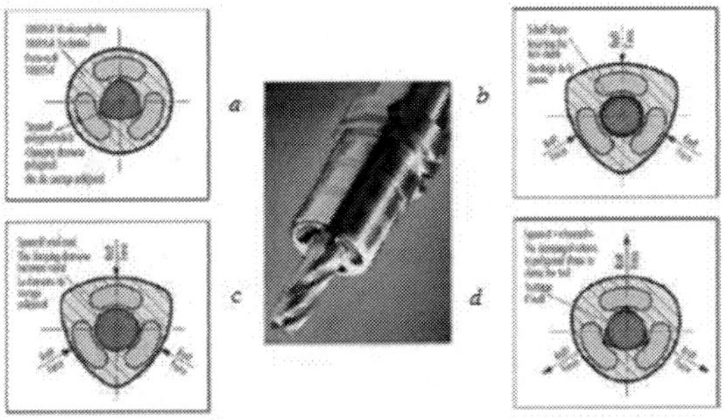

Figure.6 TRIBOS-R chuck working principle

The TRIBOS-R chuck and the TRIBOS-S chuck, the two handles have the same principle. The holding principle of TRIBOS-S is shown in Fig.7 [6]. The inside of TRIBOS-S chuck is very small. It can be used to make a tool extension rod for the occasion of needing lengthening knives, such as the processing of deep cavity mould. It can also be directly used as a tool holder for the machining of parts with geometric interference. Because of its thin outer diameter and relatively poor rigidity, it can only be used for light cutting [6].

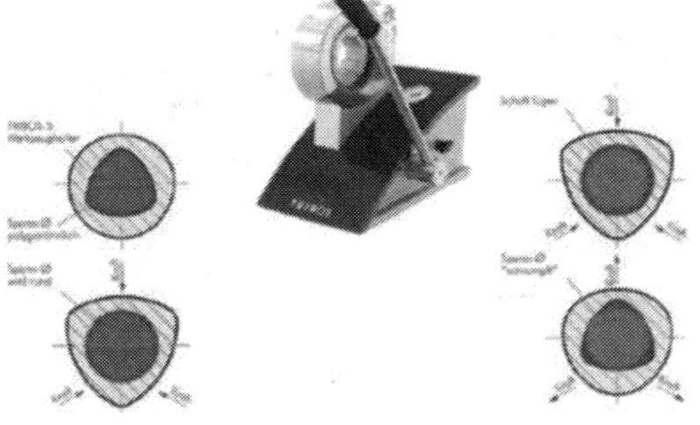

Fig.7 TRIBOS-S chuck working principle

TRIBOS-SVL is a tool extension bar, as shown in Fig.8. Its working principle and performance characteristics are the same as the TRIBOS-S chuck. Its clamping accuracy and repeat clamping precision are less than 3um.The standard series of TRIBOS-SVL cutters have a length of 100mm and 150mm, with a cylindrical handle of Φ12mm, Φ20mm and Φ 32mm.

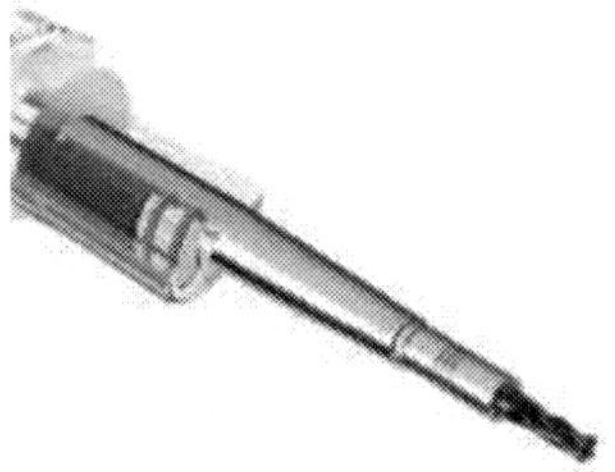

Fig.8 TRBIOS-SVL extension chucks

Static pressure expansion chuck and the stress locking chuck, their performance is shown in Table.1.

Table.1 The performance contrast of the chuck

Chuck type	Static pressure expansion chuck	stress locking chuck	
		TRIBOS -S	TRIBOS -R
Advantages	Damping, loading and unloading tools without special equipment	Super fine appearance	Damping, high radial stiffness
Clamping range	3-32mm	0.3-32mm	3-32mm
Radial runout	<0.003mm	<0.003mm	<0.003mm
Work speed	<50000min^{-1}	<85000min^{-1}	<55000min^{-1}
Change the diameter	Use a smaller diameter sleeve	Use a smaller diameter sleeve	Use a smaller diameter sleeve
Loading and unloading tools	Allen wrench	loader	loader

3 Summary

The paper introduces several common tool clamping system, the static pressure expansion type chuck, CoroGrip chuck, stress locking chuck. These clamping systems reflect the advanced level of cutting tools in international cutting processing. Their successful application in mold manufacturing industry has made important contribution to the modernization of mould manufacturing.

https://doi.org/10.1051/matecconf/201820702007

4 Acknowledgement

The authors would like to thank "Science plan project of Shenzhen" (JCYJ20170817112440533, GRCK2017042409562903, JCYJ20160415114050831) and for their financial support.

References

1. Z.Y. Zhao, C. Y. Wang: Manufacturing technology and machine tools, Vol.12, p.65 (2004)
2. J. Shen, H.Y. Zhang: Tool technology, Vol.39, p.39 (2005)
3. Z.X Yang: Modern manufacturing engineering, Vol.12, p.5 (2002)
4. T. Tan: World manufacturing technology and equipment market, Vol.5, p.100(2009)
5. Q. Y. Zhou, Z.X. Zhou: Machine tool & hydraulics, Vol.38, p.1 (2010)
6. H. Yuan: Mould industry, Vol.219, p.42(1999)

MATEC Web of Conferences 207, 02008 (2018) https://doi.org/10.1051/matecconf/201820702008
ICMMPM 2018

Finite Element Analysis of Cylindrical Rubber Fender

Chung-Ming Tan[1, a], Mau-Yiu Chang[1]

[1]Graduate School of Opto-Mechatronics and Materials, WuFeng University, ChiaYi, Republic of China

Abstract. This research is to develop a methodology to optimize the performance of fenders using finite element analysis. The design of fender geometry is done using software SolidWorks. Simulation is one of several add-in tools in SolidWorks. The finite element analysis of the performance of the fender designed is then evaluated by simulation tool in SolidWorks. This study also shows the potential applications of computer aided engineering and its benefits in verifying and reinventing various fenders design.

1 Introduction

The berthing of ship to the jetty is the last stage of navigation process in marine shipping. During the ships berthing the impact in berth can occur and in result the appearing of energy in ship - fender device - berth system [1-6]. The exceeding of admissible value of this energy leads to damage of one of these elements and in effect causes the accident. The accident can happen as unwanted event with negative consequences such as a damage of ship and port structure. The fenders are devices to protect the port structures and vessel from damage by absorbing the kinetic energy of vessel while vessel berthing to the jetty or quay [7-9]. For this study, the performances of fender system are analyzed using finite element analysis model. The finite element method is used to analyze the performance of a cylindrical rubber fender system when subjected to the impact load caused by vessel collision. The cylindrical fender system is one of the buckling type fenders which exhibit good energy absorption capacity during post-buckling deformation. The inner diameter of a hollow cylindrical fender is chosen to be the parameter which will be optimized with the aid of finite element analysis.

2 Material and Methodology

Two phenomenological models for hyperelastic materials—the Neo-Hookean, Mooney-Rivlin—are predefined in the SolidWorks Material models. One can use both the Neo-Hookean and the Mooney-Rivlin material models for modeling nearly incompressible hyperelastic materials such as rubber. Extracting the parameters for the material models from experimental stress strain curves involves curve fitting with the appropriate equations for the nominal (Piola-Kirchhoff) stress. For a thin sheet of incompressible hyperelastic material under uniaxial tension the stretch along the axis of the loading is $\lambda_1 = \lambda = 1 + \varepsilon$, where ε is the strain. Due to symmetry and incompressibility the stretch ratios in the transverse directions are $\lambda_2 = \lambda_3 = \lambda{-}1/2$. The stress state corresponding to this state of deformation is reduced to $P_1 = P$, $P_2 = P_3 = 0$, where

$$P = 2(\lambda - \lambda^{-2})\left(\frac{\partial W_{hyp}}{\partial I_1} + \frac{1}{\lambda}\frac{\partial W_{hyp}}{\partial I_2}\right) \qquad (1)$$

Assume a strain energy function, W_{hyp}, according to the Mooney-Rivlin material model, for which Equation (1) becomes

$$P = 2(\lambda - \lambda^{-2})C_{10} + 2(1 - \lambda^{-3})C_{01} \qquad (2)$$

Then use Equation (2) to curve fit the parameters C_{10} and C_{01} from a uniaxial stress-strain curve.

2.1 Computer Aided Engineering Methodology

In this study, we have designed a three-phase structure which consists of initial CAD model, FEA performance evaluation and CAD model modification as shown in figure 1. In this study, we optimize the inner diameter of a hollow cylindrical rubber fender.

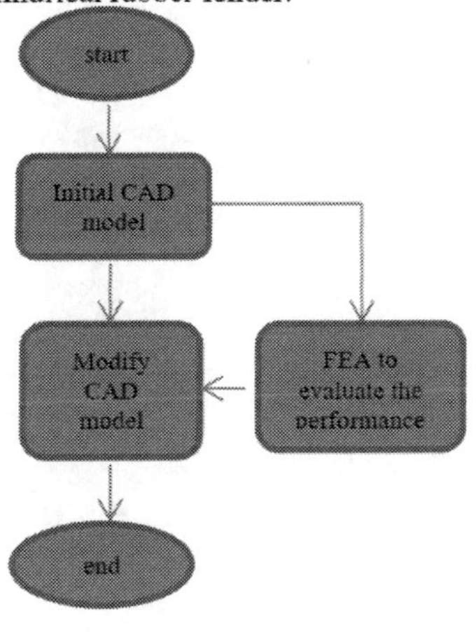

[a] Corresponding author: cmtan@wfu.edu.tw

© The Authors, published by EDP Sciences. This is an open access article distributed under the terms of the Creative Commons Attribution License 4.0 (http://creativecommons.org/licenses/by/4.0/).

Figure 1. The flowchart of the parameter study process in this study.

2.2 Finite Element Analysis

The software uses the Finite Element Method (FEM). FEM is a numerical technique for analyzing engineering designs. FEM is accepted as the standard analysis method due to its generality and suitability for computer implementation. FEM divides the model into many small pieces of simple shapes called elements effectively replacing a complex problem by many simple problems that need to be solved simultaneously.

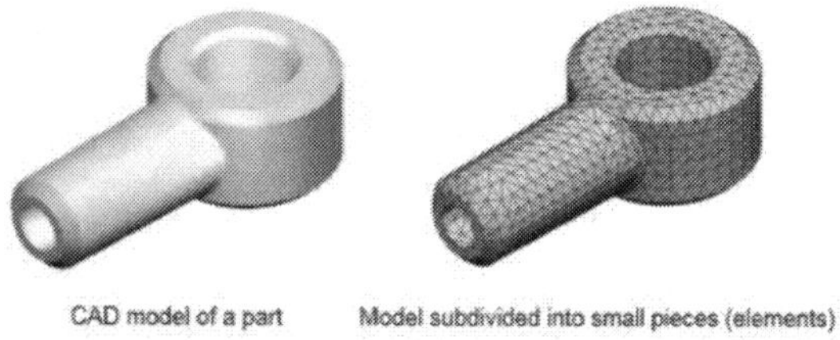

Figure 2. Illustration of meshing in finite element analysis.

Elements share common points called nodes. The process of dividing the model into small pieces is called meshing as shown in the figure 2. The behavior of each element is well-known under all possible support and load scenarios. The finite element method uses elements with different shapes. The response at any point in an element is interpolated from the response at the element nodes. Each node is fully described by a number of parameters depending on the analysis type and the element used. For example, the temperature of a node fully describes its response in thermal analysis. For structural analyses, the response of a node is described, in general, by three translations and three rotations. These are called degrees of freedom (DOFs). Analysis using FEM is called Finite Element Analysis (FEA).

In nonlinear finite element analysis, a major source of nonlinearities is due to the effect of large displacements on the overall geometric configuration of structures. Structures undergoing large displacements can have significant changes in their geometry due to load-induced deformations which can cause the structure to respond nonlinearily in a stiffening and/or a softening manner. There are different numerical procedures that can be incorporated in the solution of nonlinear problems using the finite element method. A successful procedure must include the following:

•A control technique capable of controlling the progress of the computations along the equilibrium path (s) of the system.

• An iterative method to solve a set of simultaneous nonlinear equations governing the equilibrium state along the path (s).

•Termination schemes to end the solution process.

Additional schemes such as line search, acceleration, and/or preconditioning may be augmented to enhance the solution procedure. In this study, we adopted the add-in tool called "simulation" in SolidWorks to implement the nonlinear finite element analysis described roughly in this section.

3 Results and Discussion

The initial CAD design for the cylindrical fender is represented by a hollow cylinder which has the height of 100 mm, outer diameter of 100 mm, and inner diameter of 50mm. Then we started using SolidWorks to create a hollow cylinder and employed the simulation tool to undergo the finite element analysis. One end of the cylinder is set to be fixed and the other end is set to have a prescribed displacement of 60 mm to compress the cylinder. From the experimental data of an actual tension test, the Mooney-Rivlin coefficient c01 and c10 can be found using the above mentioned equation (2). By fitting these data to the equation, the coefficients C01 and C10 of Mooney-Rivlin model used in the simulation are 1.7 Mpa and 0.55 Mpa respectively. The stress flooded plot at maximum compression displacement and the reaction forces versus compression displacement curves for the initial CAD model are presented in the figure 3.

From figure 3, it is obvious that buckling of the whole fender occurred around the pseudo time of 0.68 second. The pseudo time 1 second here stands for how long it would take when the whole prescribed displacement at one end is completed in the simulation. The simulation is terminated at pseudo time of 0.86 second. This is due to the structural instability induced by the local buckling at both ends. Another reason for termination is that self-contact condition at both ends also failed.

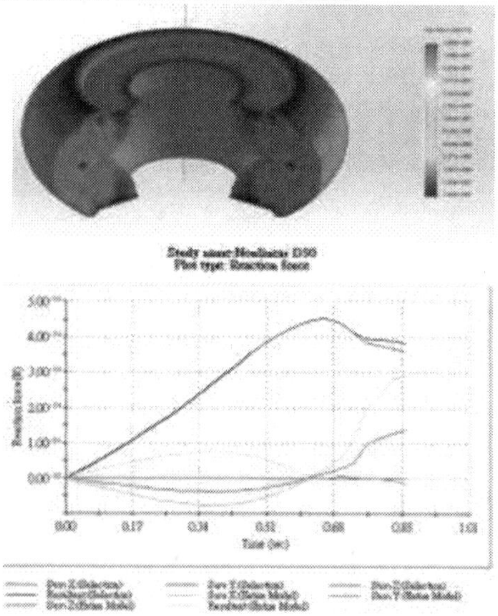

Figure 3. The stress flooded plot at maximum compression displacement and the reaction forces versus compression displacement curves for inner diameter of 50 mm.

In order to improve the result of the initial CAD model, we varied the inner diameter by increment of 5 mm at one time and repeated the simulation by 4 times till the inner diameter reaches 70 mm. Simulation results for different inner diameters are presented in 4 following figures. First, we compared the results in figures 3 and 4. Global buckling behavior and local buckling at both ends are similar. Therefore, no obvious better performance can be seen except the axial stiffness decreased by 20 % or so.

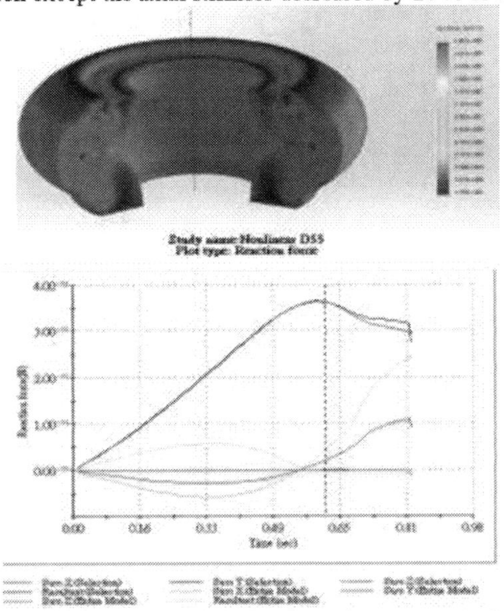

Figure 4. The stress flooded plot at maximum compression displacement and the reaction forces versus compression displacement curves for inner diameter of 55 mm.

From figure 5, one can easily tell that the postbuckling behavior observed can be improved the energy absorption capacity a lot. The reaction force versus displacement curve revealed favorable s-shape. This means that post-buckling has a large region maintains high reaction force that improves the energy absorption capacity. Fenders consisting of buckling elements are connected in a manner assuring its desired characteristic. This generally features rapid growth of reaction force at small deflection, and then the reaction remains more or less the same although deflection increases. Examining strength characteristics of fender materials, engineers attempt to obtain possibly small elasticity constant with possibly large relative elongation. Such material should have high level of maximum admissible stresses. The vessel loses its speed and kinetic energy by deforming a fender along a certain distance. The vessel and accompanying water have a certain mass when berthing at a certain speed and hitting the fender. Therefore, it can be said that the measure of fender elasticity is the length of elastic deflection caused by a certain force. In practice, relative deflection (percentage) is used.

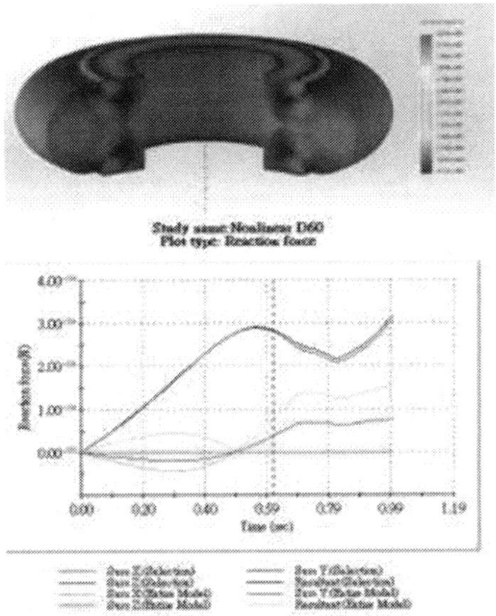

Figure 5. The stress flooded plot at maximum compression displacement and the reaction forces versus compression displacement curves for inner diameter of 60 mm.

Further increase of the inner diameter, figure 6 revealed the further improvement but figure 7 revealed no improvement can be seen. Comparing the figures 6 and 7, we noticed that there is a further force drop happened during the post-buckling for inner diameter of 70 mm. The phenomenon stops the improvement of the energy absorption.

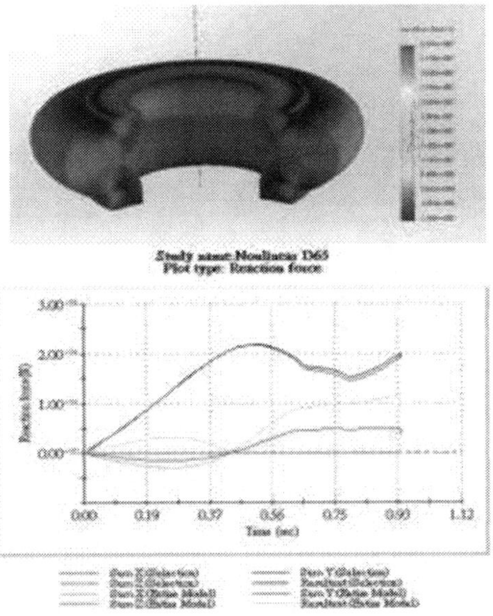

Figure 6. The stress flooded plot at maximum compression displacement and the reaction forces versus compression displacement curves for inner diameter of 65 mm.

https://doi.org/10.1051/matecconf/201820702008

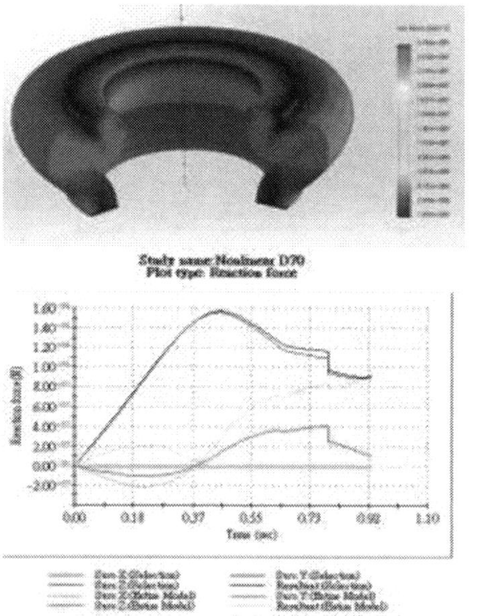

Figure 7. The stress flooded plot at maximum compression displacement and the reaction forces versus compression displacement curves for inner diameter of 70 mm.

4 Conclusion

This research showed the potential applications of computer aided engineering and its benefits by optimizing the performance of a hollow cylindrical rubber fender. The initial CAD design was drawn in SolidWorks. Using the add-in simulation tool in SolidWorks, one can implement the nonlinear finite element analysis to evaluate the postbuckling behavior of a hollow cylindrical rubber fender. Choosing the inner diameter to be the parameter, we can apply the iterative methodology to maximize the energy absorption capacity of the fender. Finally an optimal inner diameter was accomplished which is between 65 mm and 70 mm. The research specially emphasizes the implementation of computerized simulation which ultimately saved a lot of time and loss of material and gave an excellent result.

References

1. M. N. Hamza, H. M. Alwan, Eng. &Tech. J. 23 2560-2575 (2010).
2. A. Chegenizadehi, B. Ghadimil, H. Nikrazi, E. Heshmati, J. Eng. Sci. Tech. **10** 1054 – 1064 (2015).
3. W. Galor, J. JONES powertrain transp. **14** 169-175 (2007).
4. G. Neser, D. Unsalan, Ocean Eng. **33** 1919–1934 (2006).
5. M. Zikra1, K. Sambodho1, T. A. Manan, and Y. Mulyadi1, J. Eng. Appl. Sci. **11** 896-900 (2016).
6. M. S. Zala, K. J. Motwani, N. J. Patel, Int. J. Adv. Eng. Res. Dev. **2** 1144-1149 (2015).
7. S. Sakakibaraa, M. Kubob, Ocean Eng. **34** 1174–1181 (2006).
8. Z. Jiang, M. Gu, Mat. Des. **31** 1085–1095 (2009).
9. M. Gohil, U.V. Shah, Int. J. Sci. Res. **5** 2277 – 8179 (2013).

Design and analysis of uptilted aerial ladder for fire truck format

Zhi-Jian TIAN[1,2], Xiang-Qin HAN[1], Lei XU[1], Li-Juan YAN[3], Chong-Yi WEI[4], Jie LI[4], Zhi-Song HOU[4] and Wen-Guang JIANG[4,a]

[1]*XCMG Construction Machinery Co., Ltd, XCMG fire-fighting safety equipment Co., Ltd, Xuzhou, China*
[2]*School of mechatronic engineering, China university of mining and technology, Xuzhou, China*
[3]*XCMG state key laboratory of intelligent manufacturing of advanced construction machinery, Xuzhou, China*
[4]*School of mechanical engineering, Yanshan university, Qinhuangdao, China*

Abstract. Aerial ladder fire truck is a kind of fire-fighting truck which is used for rescue of people and goods from high altitude locations. The aerial ladders of the fire truck are typically nested U-shaped truss structures manufactured with thin-walled steel by welding. Compared with close-shaped truss structures, their stiffnesses are relatively low. This means that they are quite slender structures, and they will deflect quite significantly due to their own weight and the working loads applied. If the original design of the ladder is straight, then the ladder will deform to a 'fishing-rod' shape in the air. The consequent 'bent-down' shape of the deformed ladder causes significant inconvenience during rescuing operations. To deal with this issue, a practical solution to the problem is that the ladder is manufactured with an appropriate uptilted curved shape, and it could deform to an approximately straight ladder when it is under certain desired working condition. In this paper, the effectiveness of the curved ladder design proposed in this paper has been validated by both finite element analysis and experimental results.

1 Introduction

Aerial ladder fire truck is a special fire-fighting truck equipped with extendable ladder system used for fire-fighting and rescuing from high altitude locations such as skyscrapers. A typical structure of an aerial ladder fire truck is shown in Fig. 1. Its structure can be considered to be composed of two major parts: the undercarriage and the boarding system. The undercarriage mainly consists of car chassis and an outrigger system which is used to enhance the against-overturning ability of the whole truck during working. The boarding system consists of turntable and ladder system. The turntable is mounted onto the chassis using a slewing bearing. The extendable ladder system consists of a base ladder mounted on the turntable and five telescopic sub-ladders. They are U-shaped truss structures nested together (see Fig. 1). There are sliding tracks designed on the sub-ladders of the ladder system which allow the adjacent sub-ladders to slide smoothly on each other during extension/retraction operations. The ladder system is powered by two sets of separate wire rope driving systems in parallel to realize synchronous extension or retraction operations. There are working platform and fire-fighting equipment installed on the tip of the ladder system, i.e. the end of sixth ladder. The inclined angle of the ladder system is controlled by the variable amplitude hydraulic cylinder. The slewing bearing allows the turntable together with the ladder system mounted on it to rotate full 360° horizontally and this is controlled by a hydraulic motor.

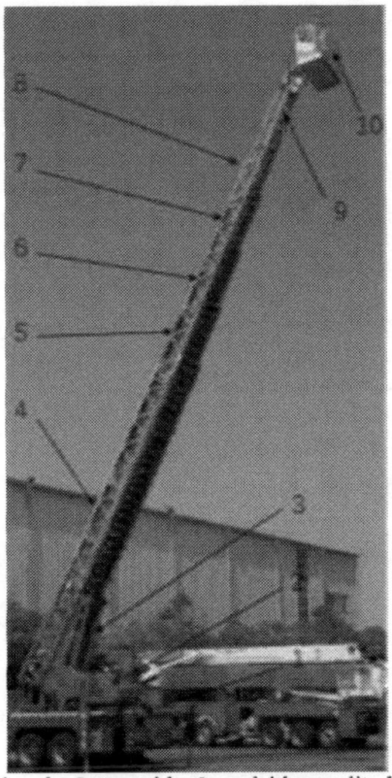

1.Car chassis; 2. turntable; 3. variable amplitude hydraulic cylinder; 4. base ladder; 5. second ladder; 6. third ladder; 7. fourth ladder; 8. fifth ladder; 9. sixth ladder; 10. working platform

Figure 1. Structure diagram of the aerial ladder fire-fighting truck

[a] Corresponding author: wgj@ysu.edu.cn

© The Authors, published by EDP Sciences. This is an open access article distributed under the terms of the Creative Commons Attribution License 4.0 (http://creativecommons.org/licenses/by/4.0/).

https://doi.org/10.1051/matecconf/201820702009

In order to realize lightweight, the ladder system usually adopt the design of using U-shaped welded thin-walled truss structure [1]. When the ladder system is fully extended it is quite slender just like a 'fishing-rod'. From the point of view of both smooth operation of the machine and achieving clear vision for the operators, it is highly desirable that the fully extended ladder could stay in a straight-line configuration during working. A practical engineering solution to this problem is that the ladder system is designed with an uptilted curved original shape, the magnitude of the uptilt could just compensate the deflection caused by its self-weight and working loads, and as a consequence the ladder will show the desired straight configuration when loaded. To consider the importance of the smoothness of the extending-retracting operation process of the ladder system and easy-to-fabricate, the pre-bending ladder is preferably designed to have a single curvature circular arc shape. The aim of this paper is to develop numerical model to help decide the optimal radius for the pre-bending arc of the ladder system for a newly designed fire truck.

2 Finite Element Model

Due to the structural complexity of the aerial ladder, analytical solution to this mechanical system is very difficult to develop. In order to accurately predict the deformed shape of the complex aerial ladder system, finite element analysis has been employed [2]. To reduce the complexity of the establishment of the finite element mesh of the aerial ladder system with pre-bending curvature, we first build a finite element mesh of the straight ladder system, and then we use coordinate transformation to transform the finite element mesh of the straight ladder to the finite element mesh of the ladder system with a predefined pre-bending curvature. The transformation relation between the straight ladder and the pre-bending ladder is given by the following nodal coordinate transformation equations:

$$x_1 = (R - y_0)\sin(x_0/R) \tag{1}$$

$$y_1 = R - (R - y_0)\cos(x_0/R) \tag{2}$$

$$z_1 = z_0 \tag{3}$$

where R is the predefined radius for the uptitlted pre-bending arc shape of the ladder system, (x_0, y_0, z_0) are the nodal coordinates of the straight ladder, (x_1, y_1, z_1) are the nodal coordinates of the pre-bending ladder.

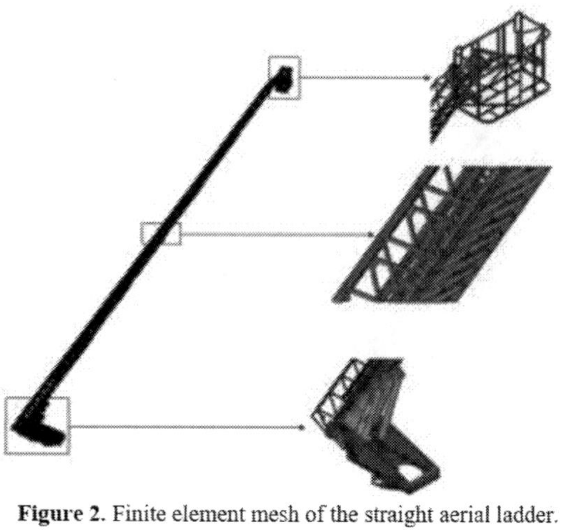

Figure 2. Finite element mesh of the straight aerial ladder.

Fig. 2 shows the finite element mesh of the straight ladder established. 8-noded solid brick elements have been used for the structural discretization. The total number of elements is 1029043 and the total number of nodes is 1718823. Contact elements have been used to simulate the contact relationships between the neighbouring contacting sub-ladders. 2-noded link elements have used to simulate the rope systems driving the extension/extraction of the ladder systems. Fig. 3 illustrates the building process of the finite element mesh of the ladders with pre-bending from the already built finite element mesh of the straight ladder (see Fig. 2) using the transformation relationships of Eq. 1, Eq. 2, Eq. 3.

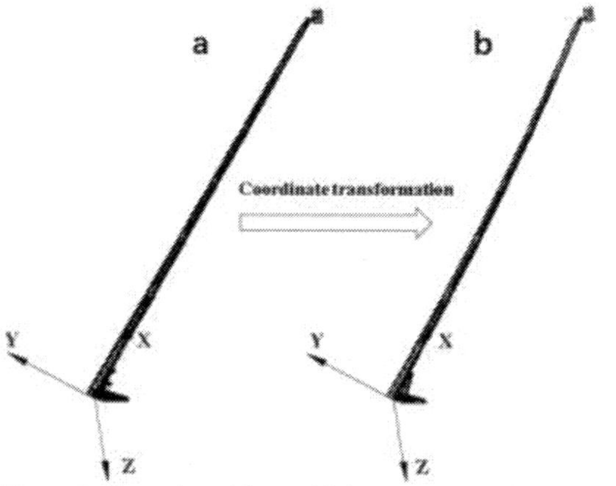

Figure 3. Illustration of the establishment process of the finite element mesh of the pre-bending ladder using coordinate transformation technique, (a) finite element mesh of an initially straight ladder, and (b) finite element mesh of the pre-bending ladder.

3 Determination of the Optimal Pre-bending Radius and Experimental Verification

The false position searching method [3] has been employed to find the optimal pre-bending radius of the ladder. It is an iterative numerical process. In each iterative step, finite element analysis has to be re-performed again, and from the nodal degree- of-freedom analysis result to obtain the extent of deviation from the ideal straight deformed shape. The iterative process continues until the deviation becomes negligibly small, i.e. the deviation of the inclined tip angle of the ladder is less than 0.1°. Due to the nature of large deformation of the structure, geometric nonlinear effect has been considered in the finite element analyses. Fig. 4 shows the deformed shape of the fully extended ladder superimposed with the undeformed configuration from the finite element analysis results from the ladder with the optimal radius. It can be seen from this figure that the pre-bending structure (shown as the configuration free from load in the figure) is obviously a curved shape, and it becomes a straight ladder (shown as the configuration after loading in the figure).

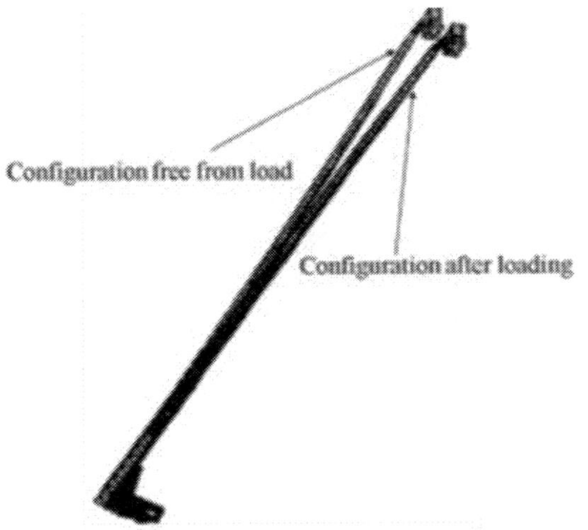

Figure 4. Deformed shape of the fully extended ladder superimposed with undeformed configuration

A real ladder using the predicted optimal radius obtained from this research has been manufactured, and a verification experimental test has been conducted on it. A specialized instrument called Focus X330 3D laser scanner (shown in Fig. 5) has been used to measure the deformed shape of the circular arc shaped aerial ladder with the predicted optimal bending radius. The measuring range of the laser scanner is 0.6m-330m and the measuring accuracy is ±1mm, which is very accurate. The measurement process scanned and recorded spatial positions of all sample points which densely arranged along the whole length of the ladder system. Figure the location distribution of the sample points recorded which could represent the spatial shape of the extended aerial ladder under working load. In this figure, a reference straight line is also plotted as a baseline to show the effect of the uptilted arc shape design of the ladder system. The deformed shape of the ladder system can be seen to be very closely overlapped with the reference baseline, which indicates that Fig. 6 shows the pre-bending design is perfectly validated.

4 Conclusion

Finite element model for a pre-bending aerial ladder system for a newly designed fire truck has been developed. A numerical search process has been implemented to find the optimal radius of the pre-bending arc shape of the ladder system using the finite element model developed. The predicted optimum curved shaped aerial ladder has been manufactured. Verification experiment has been carried out on it. The optimal radius predicted in this paper has been validated by the experimental results.

5 Acknowledgement

The authors would like to acknowledge the financial support from National Key R&D Program in China, Project title: Key Technology Research and Application Demonstration for High Mobility and Multi-Function Emergency Rescue Vehicle (funding number: 2016YFC0802900)

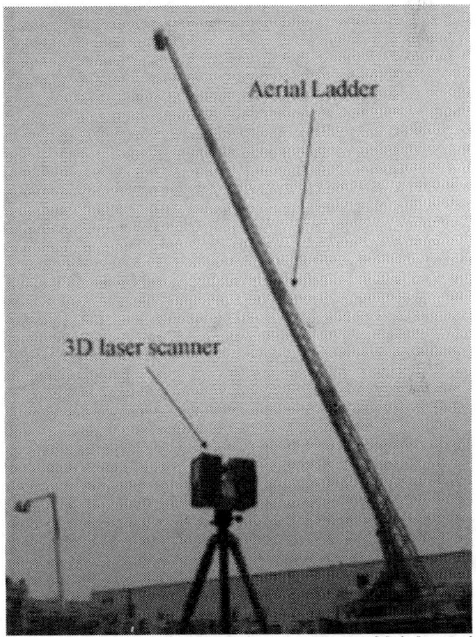

Figure 5. Photo showing a 3D laser scanner which is used to measure the deformed shape of the aerial ladder

 https://doi.org/10.1051/matecconf/201820702009

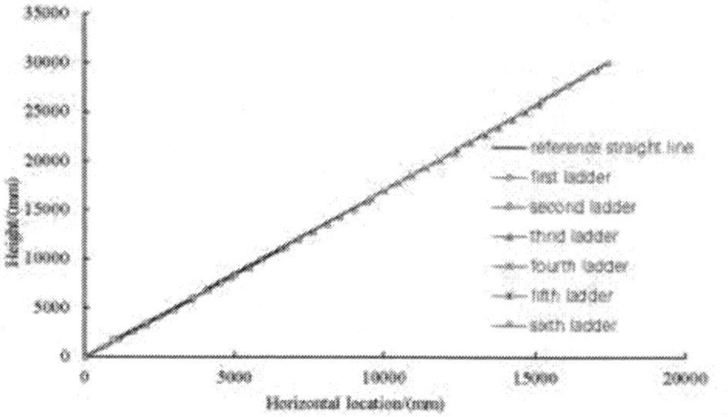

Figure 6. Experiment result showing that the deformed shape of the pre-bending ladder system is very close to an ideal straight baseline

References

1. Z. J. Tian et al. Design manual of aerial ladder truck, XCMG group, *Internal document*, (2017)
2. K. J. Bathe. Finite element procedures. *Prentice Hall*, (2006)
3. D. V. Griffiths and I. M. Smith. *Numerical Method for Engineers*. 2nd edition, Chapman &. Hall/CRC press, (1991)

A Study on Laser Micromachining

Muhammad Sajjad[1], Jithin Ambarayil Joy[1] and Dong Won Jung[1,a]

[1]*Department of Mechanical Engineering, Jeju National University, Jeju, South Korea*

Abstract. Ultra Violet (UV) pulse are widely used for micromachining in the research and development of Information Technology (IT), Nano Technology (NT), and Bio Technology (BT) products because due to short wavelength of the laser it not only allows micro drilling, micro cutting and micro grooving with high accuracy, but acts also as high penetrating coefficient, which give opportunity for different materials to machined properly. A finite element model considering photo-thermal and photomechanical properties was recommended to study the micro phenomena subject by mechanical pressure impacts and thermal heat transfers during a very short processing time. From the results of this research work, few of dynamic deformation behaviours such as deformation shape, strain, and stress distributions were examine and compare it with the experimental results of previous research work. The results from this comparison will help to know better knowledge about interference of UV laser beam and metal materials.

Keywords: UV pulse, Photothermal, Photomechanical, Deformation shape, Stress distributions.

1 INTRODUCTION

It is established that nanosecond pulsed laser processing of metals, ceramics, and polymer materials is a complex phenomenon associated with photochemical, photothermal and photomechanical effects. Photothermal effects are the dominant phenomena in processing metals and photochemical effects are dominant in processing polymers, especially under high-density lasers. During laser irradiation on the surface of the material, some parts of laser light are reflected to the air or transmitted and the remainder is absorbed inside the material. Absorbed energy is transferred to the inside of the material by thermal conduction, and then the material melts in a very short time. When the amount of energy increases, some of material is rapidly removed by explosive vaporization with chemical and physical decomposition, this process is called the photothermal ablation mechanism [1].

Recently the study of LSP(Laser Shock Processing) on micro copper and aluminum were presented by Wenwu Zhang and others. A comparative study on experimental and computational analysis about micromachining using micro beam laser was reported using LSP. As soon as the strong-pulsed laser beam arrives and focuses on the surface of materials, the surface layer evaporates immediately and the plasma is generated. This plasma induces shock waves from the irradiated surface during the air expansion, and the mechanical impact pressures are transferred to the target. If it is confined by water, or another type of medium, the shock pressure of being under-water can be over five times that of open-air. The medium, water, also protects the target materials from thermal energy so that only pure mechanical energy effects the processing [2, 3]. When lasers irradiate the surface of metals, the impact processing caused by shockwave pressure is called the photomechanical ablation mechanism. When laser micromachining is processed in nanoseconds, the average strain rate was approximately 104~105(sec-1), which is very high [5].

Therefore, analysis results which do not considered the strain rate effect show a lot of difference with the experimental results. So far, the theoretical analyses for machining metals using UV lasers with wavelengths between 200 and 400nm have studied the photothermal effect to predict ablation depth and diameter. However, the behavior of molten metal and the analysis of temperature distribution are very complex problems because there are a lot of process parameters such as convection, radiation and conduction and it occurs for a very short time. So, the behavior is difficult to predict accurately because of the wide variance between model and experiment.

In this paper, a computational analysis of photothermal ablation of copper using UV pulse laser was performed in two parts for thermal and mechanical effects. Then, a new finite element model that combined photothermal and photomechanical effects was suggested. Finally, the analyzed results of this new model were compared and verified with previous experiment results.

2 FINITE ELEMENT MODEL

A 355nm UV pulse laser was used to perform this experiment and model analysis the layout of this system is shown in Fig.1. Copper material was used in this analysis, the model has a 40 μm diameter and thickness of 20 μm. Each analysis termination time in this analysis was 50 nanoseconds, and the pulse duration time was 25 nanoseconds. The model was composed of total 16,544 elements and mesh relevance was chosen at the value of +100 in order to improve the accuracy of analysis results. Only 1/4 of the copper thin film was modeled due to its symmetric geometry. The metric units used μm-kg-μN-s-system and the mm-tone-N-s-K units for ANSYS and LS-DYNA, respectively. Table 1 shows the mechanical and Thermo-physical properties of copper material [6].

[a]Professor Dong-Won Jung, Department of Mechanical Engineering, Jeju National University, Jeju, Republic of Korea : jdwcheju@jejunu.ac.kr

© The Authors, published by EDP Sciences. This is an open access article distributed under the terms of the Creative Commons Attribution License 4.0 (http://creativecommons.org/licenses/by/4.0/).

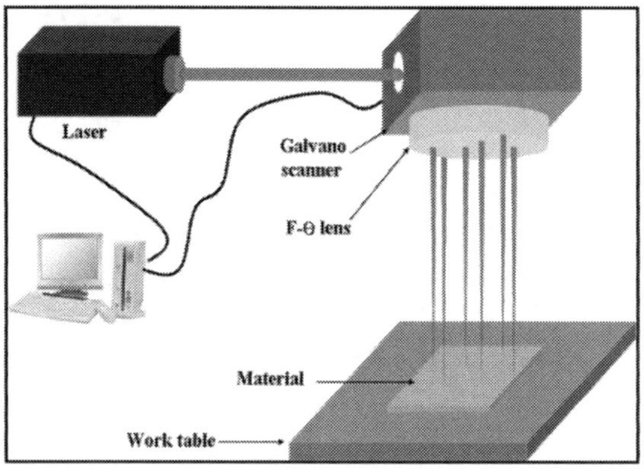

Figure 1. Schematics of Galvano scanner and work table.

At first, transient thermal analysis and static structure analysis of thermal stress was performed in one coupled environment of ANSYS Workbench, and then explicit dynamic analysis was introduced from the result of ANSYS, and finally executed as finite element analysis tool using LS-DYNA program. The FEM analysis procedure is shown in Fig. 2

Table 1. Mechanical and Thermo-Physical Properties of Copper Material.

Density (kg/m3)	8,300
Elastic Modulus (Pa)	1.1×10^{11}
ʋ (Poisson's ratio)	0.34
Yield Strength (Pa)	2.8×10^{8}
Isotropic Thermal Conductivity(W/m/ °C)	401
Specific Heat(J/kg/ °C)	385
Coefficient of Thermal Expansion(°C $^{-1}$)	1.85×10^{-5}
Reference Temperature(°C)	22

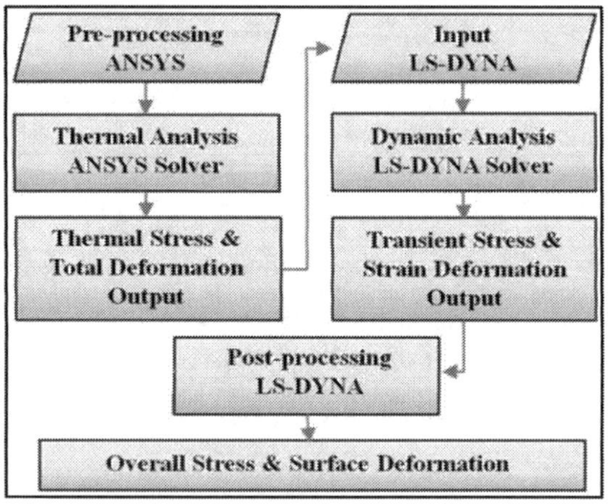

Figure 2. FEM analysis procedures

2.1 Photomechanical Analysis

The photomechanical analysis model was considered under the following assumptions:
- The impact pressure wave is uniaxial and planar.
- The pressure produced due to the laser pulse is uniform.
- The material obeys the Von-Misses yielding criterion.
- The viscous effect is ignored.

The Cowper-Symonds relation was used in this analysis model when considering the strain rate effect in pulse laser processing. The relation equation was the constitutive equation of the copper material, represented by multiplying the strain hardening by the strain rate hardening. The peak pressure, 3,350 MPa, generated by the expansion of plasma was referenced by previous research.

2.2 Photo-thermal Analysis

From the results of transient thermal analyses, the distribution of temperature, heat flux, and other thermal quantities were determined over the total analysis time. In order to obtain more accurate transient analysis simulating real phenomena, it was very important to input the quantity of heat flux by introducing the graph of the relationship between heat flux and time for each time period. Initial temperature conditions could be adapted as uniform or non-uniform thermal condition, but the temperature of the copper surface was assumed as room temperature of 22 °C in this analysis model because only single pulsed ablation was considered in this analysis.

The absorption of copper is assumed at approximately 70 percent when the laser wavelength (λ) was 355 nm. Heat flux was applied and represented as distributed thermal load and transfer heat flow per area and time. The positive value means that energy will come into the model, while the negative means that energy will come out of the model. The values of thermal conductivity, density, and specific heat of copper material had to be defined for a transient thermal analysis. Thermal conductivity was assumed as isotropic in this model. The thermo-physical properties of copper are shown in Table 2.

3 EXPERIMENT AND ANALYSIS RESULTS

3.1. Photo-thermal and Photomechanical Analysis Results

Thermal stress distributions and dilated deformation were obtained through photo-thermal analysis as shown in Fig. 3. The surrounding of the dent was raised up about 0.09 μm because of thermal expansion. The scale was magnified to easily identify the raised part as shown in Figure 3. The impact pressure analysis was accomplished after introducing the results of thermal analysis, including the raised surface geometry data, and the analysis procedure was coupled with ANSYS and LS-DYNA programs. With laser shock waves analysis, the pressure wave generated during the pulse time of 25 nanoseconds strongly impacted the surface and then a dent was formed with a very small area.

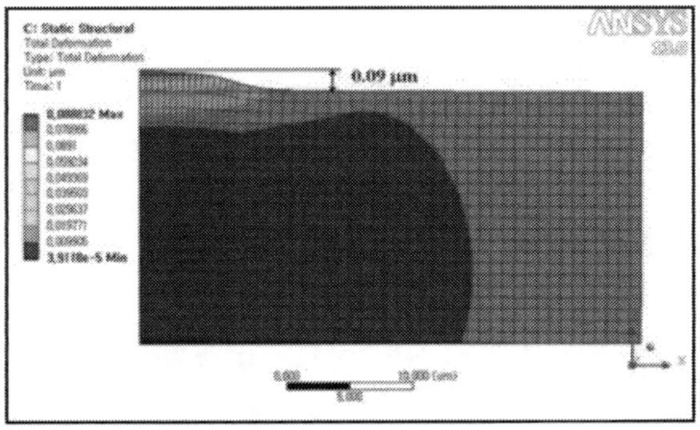

Figure 3. Total deformation of dilatational surface.

3.2 Photomechanical Analysis Results without the strain rate effect

The effective von-Mises stress and plastic strain distribution, ignoring the strain rate effect, are shown in Figs.4. After 50 nanoseconds, pressure of 500 MPa was calculated in the model. It was also found that the dent diameter and depth was about 16 μm and 7.24 μm, respectively. The burr height around the dent was measured at about 1.5 μm. The analysis model without strain rate effect was more locally deformed than the experimental results on the surface of the laser beam focused area.

Figure 4 Effective stress (v-m) distributions for copper without the strain rate effect

3.2 Photomechanical Analysis Results with the strain rate effect

Figure 5 show the effective von-Mises stress and plastic strain distribution with the strain rate effect. After 50 nanoseconds, 3,500 MPa of pressure was calculated in the model. In this case the diameter and depth of the dent was about 11.9 μm and 0.64 μm, respectively. If the strain rate increases, the fracture time and the strength of material increase because of the strain-rate hardening effect. The maximum stress in the model considering the strain rate effect was higher than non-considering model, and the burr around the dent was barely observed, despite strong laser pressure. Deformation energy from the laser shock wave was absorbed over the whole material. The SEM picture of the real dent processed by LSP was also compared with two FE analysis models, with and without strain-rate effect. When compared and measured between the analysis and experimental results, the proposed model considering strain rate effect was more effective in predicting micro dent shape processed by laser shock wave.

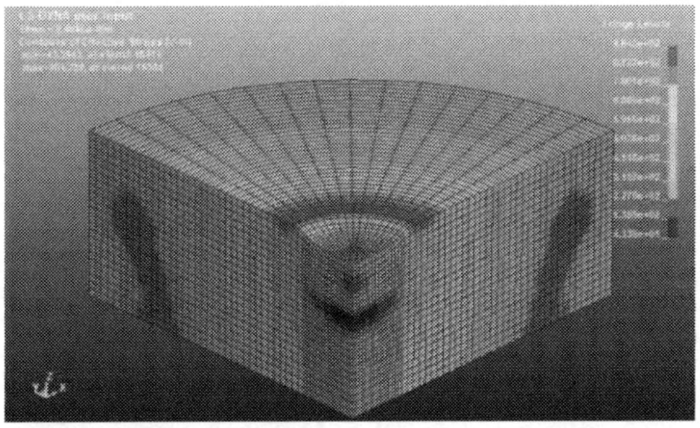

Figure 5. Effective stress (v-m) distributions for copper with the high strain rate effect

3.3 Experiment Results

SEM pictures of experimental results of thin copper sheets with respect to the change in fluence of the 355nm UV laser are shown in Figure 6. When processed at a fluence of 0.23 J/cm^2, the middle part irradiated and melted by the laser beam was raised up a little as shown in Fig. 6(a). The small crater in the middle of the raised part, as shown in Fig. 6 (b), was created when processed with the fluence of 0.8 J/cm^2. That is, the threshold energy capable of removing the metal can be determined as a fluence of 0.8 J/cm^2. Small crown-like craters formed by copper melted and pushed out were shown with the fluence of 13.56 J/cm^2 as shown in Fig. 6 (c). Generally, photo-thermal ablation provided the thermal expansion of material at low laser fluence, which was almost near to threshold energy. On the other hand, the burr consisted of particles oxidized by both re-solidification of the melt and re-condensation of evaporated material, which were remained near the crater hole at the high fluence. In this paper, the coupled analysis model was introduced with a fluence of 0.8 J/cm^2 to compare with simulated and experimental dent shapes.

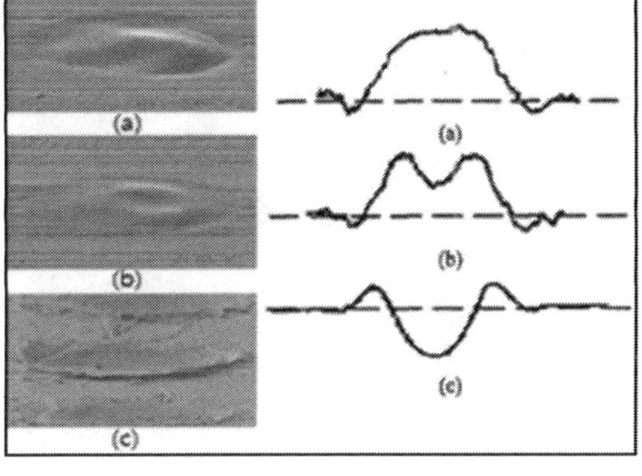

Figure 6. SEM photos of copper processed by UV laser under the fluence of (a) 0.23 J/cm2 , (b) 0.8 J/cm2 , and (c) 13.56 J/cm2 with 30 shots, respectively

4 CONCLUSION

From these experiments and analyses, we reached the following conclusions. The photomechanical analysis model that considers the effect of high strain-rate was very useful in predicting the LSP ablation phenomena and confirmed by comparison with experimental results. Photo-thermal ablation problem was performed using transient thermal analysis so that the maximum temperature was observed on the surface of copper sheet, which was a little above of the melting point of work piece material. Threshold energy condition of laser ablation was therefore found, which meant the starting point of material removal for laser micromachining process. Finally a new simulation model for photomechanical analysis coupled with photo-thermal effects was suggested and the results were verified with excellent agreement with the real ablation.

References

1. A. Mecke, I. Lee, J.R. Baker jr., M.M. Banaszak Holl, B.G. Orr, Eur. Phys. J. E **14**, 7 (2004)
2. A M. Ben Rabha, M.F. Boujmil, M. Saadoun, B. Bessaïs, Eur. Phys. J. Appl. Phys. (to be published
3. A J. Y. Oh, B. S. Shin, J. of KSPE **24**, 4 (2007)
4. A W. Zhang, Y. L. Yao, J. of Manuf. Process **3**, 2 (2001),
5. A T. Hu, Z. Yao, J. Hu, Surf. & Coat. Technol. **201**, (2006)
6. A B. S. Shin, T. J. J, J. of KSPE **5**, 2 (2006

T-phase formation and its effect on mechanical properties of Al-Cu-Mn alloys

SobanMuddassir Dar[1], HengCheng Liao[1,a] and GuangJing Li[1]

[1]*School of Materials Science & Engineering, Southeast University, Jiangsu Key Laboratory for Advanced Metallic Materials, Nanjing 211189, China*

Abstract. Al-Cu-Mn alloys are used in components in automotive and aircraft industry which work at high temperature. Recent research is focused on Al-Cu-Mn alloys to enhance their performance by promoting the formation of thermally stable phase particles. In this study, the effect of Cu/Mn content on microstructure and mechanical properties of two Al-Cu-Mn alloys (Cu/Mn wt.% ratio = 5.56, 2.75) is investigated by optical microscopy, scanning electron microscopy, energy dispersive spectroscopy and tensile testing. Results reveal that as-cast microstructure of both the alloys consists of trunks of α-Al dendrites surrounded by Cu-rich interdendritic region. After solution treatment, a great amount of Al-Cu-Mn intermetallic compound particles are found to precipitate adjacent to this region. Alloy with low Cu/Mn content ratio exhibits high strength due to relatively high amount of Al-Cu-Mn (T-phase) particles.

1 Introduction

Al-Cu-Mn alloys are widely used in automotive and aircraft industry due to their high strength and light weight[1]. These alloys possess excellent strength because of high density of meta-stable θ″/θ′ precipitates after solutionizing and aging treatment. However, the service temperature of these alloys is usually limited to 250°-300°C because of coarsening of these metastable precipitates[2, 3]. To improve the properties of these alloys, the effect of rare earth addition on the properties of these alloys is studied[4-6]. Recently, scientists have focused their attention in these alloys to completely avoid or suppress the formation of Al_2Cu (θ-phase) and to maximize the formation of Mn-rich phase (T-phase) [7, 8]. Some of these studies have been proved to be fruitful in terms of improved properties. However, no attention is paid to microstructure evolution during/safter solidification (as-cast) and its effect on the properties. Furthermore, the solidification reactions proposed by Mondolfo[9] long ago in Al-rich portion of Al-Cu-Mn ternary phase diagram are largely of hypothetical nature and have no physical significance[10]. So, in this study, the effect of Cu/Mn content on microstructure and mechanical property of two Al-Cu-Mn alloys is investigated.

2 Experimental procedures

Two alloys of varying Cu and Mn contents, as listed in Table 1, are produced from commercially pure aluminum, Al-20%Cu and Al-10%Mn master alloys. Cu/Mn content ratio in these alloys corresponds to 5.56(hereafter Alloy-1) and 2.75(hereafter Alloy-2), respectively. The alloys are prepared by melting at 760°C for 3 hrs in an electrical resistance heating furnace. After degassing by C_2Cl_6 at 720°C, the molten alloys are cast into a plate like cast-iron mold with a cavity of 170×150×20mm3 which was preheated at 250°C for at least 5h. The composition of the alloys is measured by MAXx LMF15 spark emission spectrometer. The as-cast alloys are solution treated at 530°C for 16 hrs, then water quenched and finally aged at 170°C for 6 hrs. Microstructure of alloys is characterized by using optical microscopy (OM) and scanning electron microscopy (SEM). For this purpose, samples are prepared by mechanical grinding, polishing and final etching with Keller's reagent. Phase constituents are determined by energy dispersive spectroscopy (EDS). Samples for tension test are cut from the central part of casting in accordance with GB/T228-2002 (a Chinese standard). Mechanical properties are determined at 25°C (room temperature) and 300°C on a universal testing machine (CMT4503) with a strain rate of 2 mm min-1. For each alloy, three samples are tested at every temperature and mechanical properties are reported as an average.

Table 1: Composition of the experimental alloy by MAXx LMF15 spark emission spectrometer

Alloy Name	Composition, wt.%			Cu/Mn, wt.% ratio
	Al	Cu	Mn	
Alloy-1	Bal.	1.67	0.29	5.57
Alloy-2	Bal.	2.27	0.77	2.75

[a] Corresponding author: hengchengliao@seu.edu.cn

© The Authors, published by EDP Sciences. This is an open access article distributed under the terms of the Creative Commons Attribution License 4.0 (http://creativecommons.org/licenses/by/4.0/).

Fe<0.15%, Si<=0.08%

3 Results and discussion

3.1 Microstructure

Optical micrographs of as-cast alloys (Alloy-1 and Alloy-2) are shown in Figure 1. There is no obvious difference between the microstructure of two alloys. The as-cast microstructure consists of trunks of α-Al dendrites surrounded by Cu-rich interdendritic region, as demonstrated in Figure 2(a), (b) for Alloy-1 and Alloy-2, respectively. Moreover, some black particles are also observed within the interdendritic region. SEM images and EDS analysis shows that these are particles of θ-Al_2Cu phase formed as a result of divorced eutectic reaction. In Alloy-1, the interdendritic region consists of α-Al + globular particles of Al_2Cu phase, as shown in Figure 2(a) and in Alloy-2, the interdendritic region consists of α-Al + broken network of Al_2Cu phase. Two morphologies of Al_2Cu phase i.e. globular particles and elongated particles are observed in Alloy-2, as shown in Figure 3(a). It is because of relatively high amount of Cu present in this alloy. SEM-EDS analysis shows that in addition to non-equilibrium θ-Al_2Cu phase, some Fe-rich and Si-rich intermetallics, formed during solidification, are also observed in the microstructure because of minor amounts of these elements present in these alloys. Distribution of alloying elements within the microstructure is studied by EDS analysis. Results reveal that most of Cu is present in the form of θ-Al_2Cu phase. No Mn-rich binary or ternary intermetallics are observed. It is worth noting that, in both the alloys, amount of Cu is low in Al matrix as compared to nominal composition of the alloy (Fig. 2(b) and Fig.3(b)). On the other hand, amount of Cu is higher in interdendritic regions (Fig.2(c) and Fig.3(c)). However, Mn is completely dissolved into the α-Al matrix.

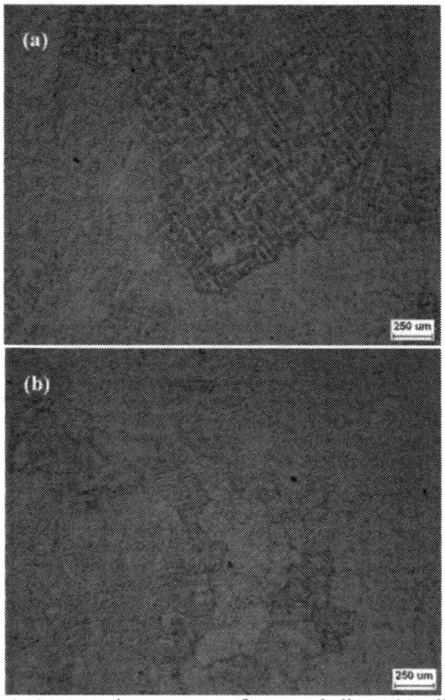

Figure 1: As-cast microstructure of prepared alloys, (a) Alloy-1and (b) Alloy-2

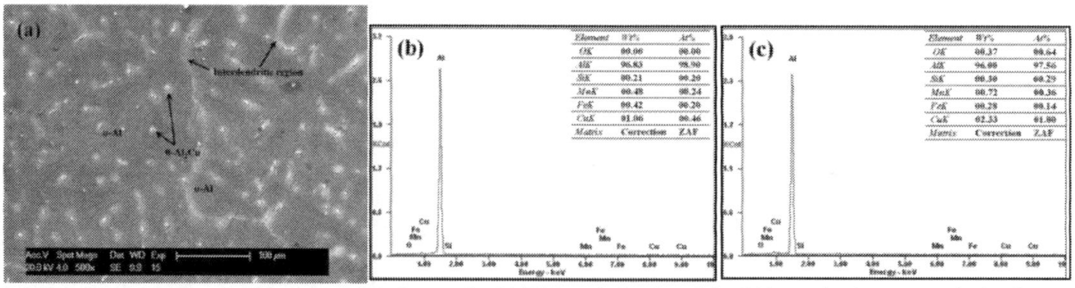

Figure 2: SEM and EDS results of Alloy-1: (a) SEM micrograph, (b) EDS analysis of α-Al dendritic trunks, (c) EDS analysis of interdendritic Cu-rich zone

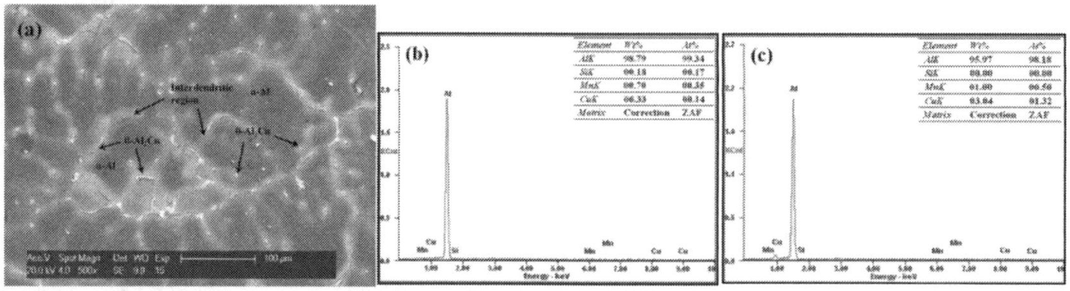

Figure 3: SEM and EDS results of Alloy-2: (a) SEM micrograph, (b) EDS analysis of α-Al dendritic trunks, (c) EDS analysis of interdendritic Cu-rich zone

https://doi.org/10.1051/matecconf/201820703001

Optical micrographs (Figure 4) of two alloys after solution treatment (530°C for 16 hrs) show that there are a great amount of intermetallic particles precipitated adjacent to the interdendritic region(described in the as-cast microstructure). SEM micrographs more clearly reveal it. The amount of precipitates in Alloy 2 is much greater than that in Alloy 1. In Alloy 1, most of the precipitates are thought to be $CuAl_2$ phase due to the EDS results (Fig.5(b,c)). In Alloy 2, however, by the EDS results (Fig.6(b,c)), most of the precipitates are rod/plate-like and are thought to be Mn-rich phase (T-phase). It is interestingly noted that the precipitation behavior during solutionizing occurs mostly adjacent to the interdendritic regions where it is Cu-rich due to non-equilibrium solidification. In these regions, the Cu and Mn solutes are locally supersaturated, thus $CuAl_2$ and/or T-Mn phases are precipitated during solutionizing.

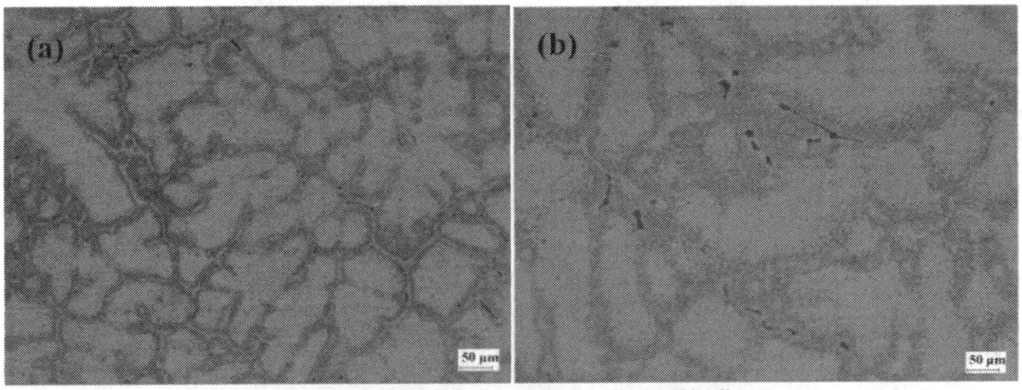

Figure 4: Optical micrographs after solution treatment (530°C for 16 hrs): (a) Alloy-1, (b) Alloy-2

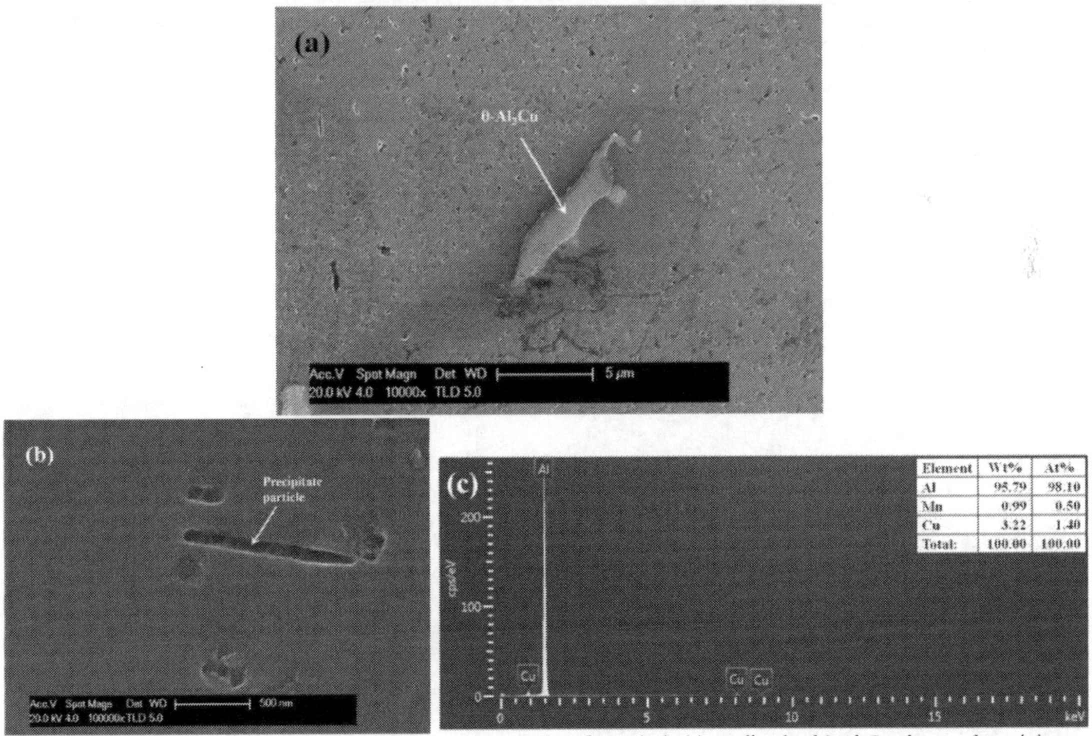

Figure 5: SEM micrographs of Alloy-1 after solution treatment (530°C for 16 hrs), (a) un-dissolved θ-Al_2Cu phase and precipitate particles adjacent to interdendritic region, (b) morphology of fine precipitatates, (c) EDS results of the precipitate in (b)

[a] Corresponding author: hengchengliao@seu.edu.cn

 https://doi.org/10.1051/matecconf/201820703001

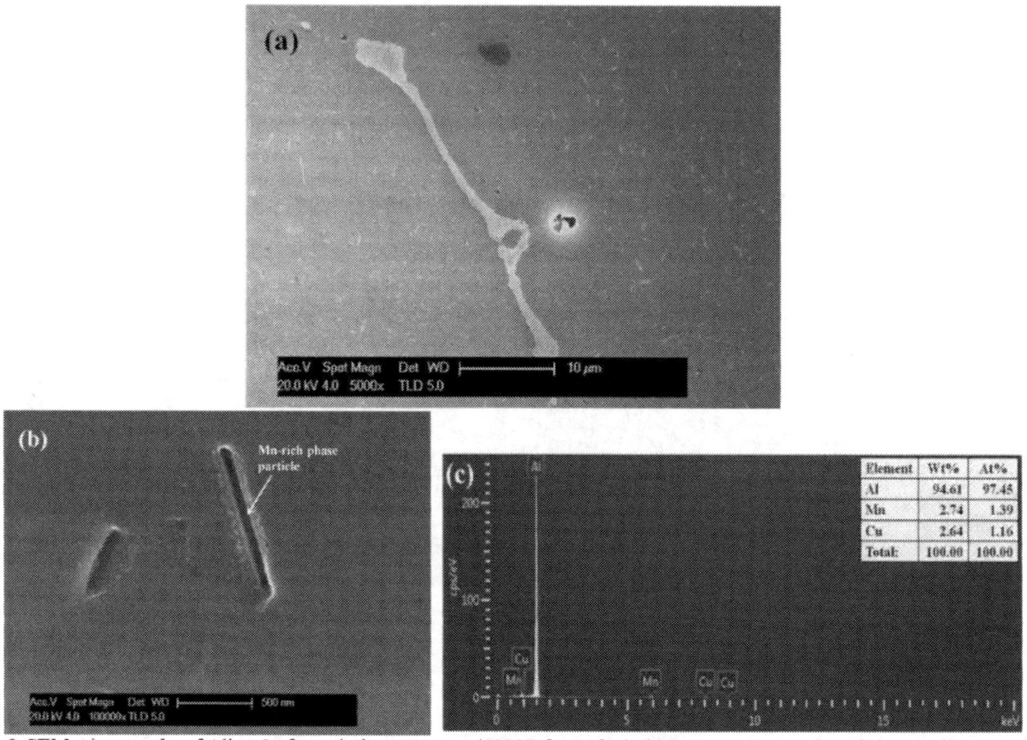

Figure 6: SEM micrographs of Alloy-2 after solution treatment (530°C for 16 hrs), (a) a great amount of precipitates adjacent to interdendritic region, (b) morphology of fine precipitatates, (c) EDS results of the precipitates in (b)

3.2 Mechanical properties

Results of tensile properties of Alloy-1 and Alloy-2, at different temperature, are summarized in Table 2. It shows that Alloy-1 has low value of yield strength i.e. 44MPa at room temperature when compared to that of Alloy-2 i.e. 66MPa. Similar trend is observed in ultimate tensile strength of both the alloys. However, Alloy-1 has experienced higher plastic deformation prior to failure due to low amount of Cu & Mn contents and Alloy-2 has experienced lower plastic deformation. High temperature tensile test is carried out at 300°C. Results of these tests representing the mechanical behavior of Alloy-1 and Alloy-2 are shown in Figure 7(a) and (b), respectively. It is observed that the ultimate tensile strength of Alloy-1 is dropped from 120MPa to 48MPa and the elongation is increased significantly from 37% to 56%. It is considered to be because of dissolution and/or coarsening of the precipitates at high temperature. While the ultimate tensile strength of Alloy-2 decreases to 82MPa and elongation increases from 23% at room temperature to 37% at 300°C. It is clear that the ultimate tensile strength of Alloy 2 is much higher than that of Alloy 1 and it also experiences less plastic deformation. This behavior of Alloy-2 is attributed to the presence of Mn-rich (T-phase) particles within the grain interiors hindering plastic deformation at high temperature. These results are in agreement with literature suggesting that Mn-rich (T-phase) particles can improve high temperature properties by impeding the grain boundary movement[11].

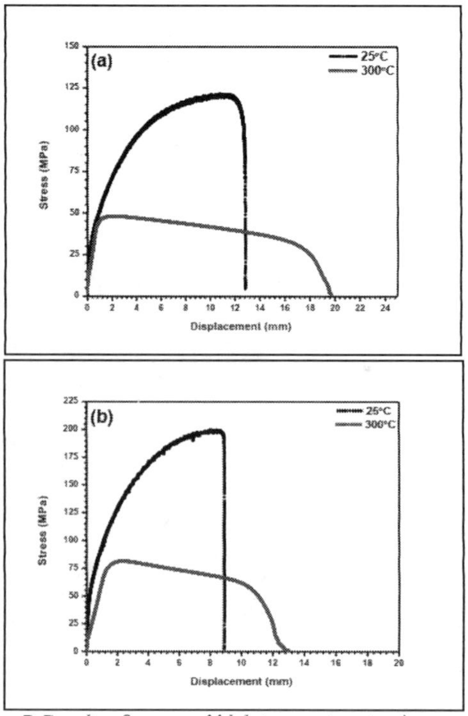

Figure 7: Results of room and high temperature tension experiment, (a) Alloy-1 and (b) Alloy-2

Table 2: Mechanical properties of experimental alloys at 25°C and 300°C

Alloy Name	Temp.	YS (MPa)	UTS (MPa)	Elong.
Alloy-1	25°C	44	120	37%
Alloy-2		66	199	23%
Alloy-1	300°C	-	48	56%
Alloy-2		-	82	37%

4 Conclusions

(1) As-cast microstructure of both the alloys (with varying Cu/Mn content ratio) consists of trunks of α-Al dendrites and interdendritic Cu-rich regions where θ-Al2Cu particles are formed. No ternary (Al-Cu-Mn) intermetallics are observed

(2) After solution treatment, a great amount of precipitates are found to form adjacent to the Cu-rich regions, most $CuAl_2$ particles in Alloy-1 and most T-phase(Al-Cu-Mn intermetallic compound) in Alloy-2

(3) Alloy-2 with low Cu/Mn content ratio exhibits high strength due to higher thermal stability and much greater amount of T-phase particles

References

1. X.W. Yang, J.C. Zhu, Z.S. Nong, M. Ye, Z.H. Lai, and Y. Liu, Modern Physics Letters B, **27**, 13410361-13410368 (2013).
2. A.R. Toleuova, N.A. Belov, D.U. Smagulov and A.N. Alabin, Metal Science and Heat Treatment, **54**, 27-31 (2012).
3. L. Fan, Q.T. Hao, and W.K. Han, Rare Metals, **34**, 308-313 (2014).
4. Z.W. Chen, M.J. Tang, and K. Zhao, International Journal of Minerals, Metallurgy, and Materials, **21**, 155-161 (2014).
5. Z.W. Chen, P. Chen, and S.S. Li, Materials Science and Engineering: A, **532**, 606-609 (2012).
6. Z.W. Chen, P. Chen, and C.Y. Ma, Rare Metals, **31**, 332-335 (2012).
7. N.A. Belov, A.N. Alabin, and I.A. Matveeva, Journal of Alloys and Compounds, **583**, 206-213 (2014).
8. A.S. Prosviryakov, K.D. Shcherbachev, and N.Y. Tabachkova, Materials Science and Engineering: A, **623**, 109-113 (2015).
9. L.F. Mondolfo, *Aluminum Alloys: Structures and Properties*, Butterworth and Co (Publishers) Ltd, 3-10 (1976).
10. N.A. Belov, D.G. Eskin, and A.A. Aksenov, *Multicomponent Phase Diagrams*, Elsevier: Oxford, 159-192 (2005).
11. Y.Q. Chen, D.Q. Yi, Y. Jiang, B. Wang, D.Z. Xu, S.C. Li, Journal of Material Science, **48**, 3225–3231(2013).

Modelling of macrosegregation with mesosegregates in a binary metallic cast by the diffuse approximate meshless method

Katarina Mramor[1,2,a], Vanja Hatić[1], Boštjan Mavrič[1,2] and Božidar Šarler[1,2]

[1]*Institute of Metals and Technology, Laboratory for Simulation of Materials and Processes, Lepi pot 11, 1000 Ljubljana, Slovenia*
[2]*University of Ljubljana, Faculty for Mechanical Engineering, Laboratory for Fluid Dynamics and Thermodynamics, Aškerčeva cesta 6, 1000 Ljubljana, Slovenia*

Abstract. The solidification of binary metallic alloy (Sn-10%Pb) is simulated in 2D with a diffuse approximate meshless method. The macrosegregation with mesosegregates is described by a coupled set of volume-averaged partial differential equations. The incompressible Newtonian fluid is described by coupled mass and momentum equations and the mushy zone is treated as a Darcy porous media. The energy transport is described with enthalpy formulation and the species transfer is incorporated by considering the Lever rule approximation. The thermo-solutal Boussinesq hypothesis is applied in order to account for the double-diffusive effects in the melt. The mathematical model is solved by a diffusive approximate method on local subdomains with Euler time stepping. Pressure-velocity coupling is incorporated by applying the fractional step method. The instabilities due to the convective terms are smoothed out by applying the upwinding technique. The results are presented on a regular node arrangement and are compared to other benchmark results. The present paper demonstrates that the results calculated with the diffuse approximate meshless method are in good agreement with reference results.

1 Introduction

Solidification is a phase change process that combines many different research areas ranging from continuum mechanics to thermodynamics. In recent years, the demand for materials with better structural quality and advanced physical properties resulted in steady raise in the research topic interest as the number of research articles has grown from 83 in 1970 to 3383 in 2017 [1]. The growth in the popularity of this research topic is mainly due to the requirements from industry, particularly those involving casting, as the solidification process plays an important role in the quality of the final cast product. By better understanding the physical phenomena during the solidification process, many defects, such as inhomogeneities can be either avoided or diminished. As experimenting with the casting of industrial size ingots can be often very expensive and time-consuming, the simulation programs can play an important role in the adjustment of the cast parameters. The simulation programs have to be thoroughly tested before they can be used for the application on an industrial project, especially when using novel numerical methods, such as meshless methods.

Among many varieties of meshless methods developed in recent years [2–6], the diffuse approximate method (DAM) [7] is chosen for the model presented in this paper. The method applies weighted least squares functions to determine a locally smooth and differentiable approximation of the discontinuous data in each particular computational point.

The DAM has already been applied to convective diffusive problems [8], natural convection in porous media [7], macrosegregation in columnar solidification [9], heat conduction and transport [10, 11] problems. In the present paper, a benchmark test case devised for macrosegregation with mesosegragtes in binary metallic alloy by [12] was chosen and tested for Sn-10%Pb.

The problem describes the solidification of binary alloy in a rectangular cavity. Numerically, it is described by a set of tightly coupled momentum, mass, energy and solute transport equations. The strong coupling and the transition from fully liquid to fully solidified state results in the formation of mesoscale channels– mesosegregates. The formation of mesosegergates has been previously observed for a similar alloy (Pb-18%Sn), both experimentally [13] and numerically [9, 14]. The present paper provides the results for Sn-10%Pb alloy and compares them to reference results calculated in [14].

2 Model description and numerical solution

2.1 Problem description

The present paper addresses a simplified solidification model previously described in [12, 14]. The referenced

[a] Corresponding author: katarina.mramor@imt.si

© The Authors, published by EDP Sciences. This is an open access article distributed under the terms of the Creative Commons Attribution License 4.0 (http://creativecommons.org/licenses/by/4.0/).

SMACS (Simulation of MACrosegregation and Structures project founded by the French Research Agency) benchmark solidification example is here presented for Sn-10%Pb alloy for thermo-physical properties listed in Table 1. The test case is defined for a rectangular domain which is schematically represented in Figure 1. The size of the domain is 10 cm x 6 cm. Due to the symmetry, the computational domain is reduced to 5 cm x 6 cm.

Table 1. Thermo-physical properties of Sn-10%Pb and process parameters.

Property	Symbol	Value	Unit
Density	ρ	7.00e+03	kg/m^3
Specific heat	c_p	2.60e+02	J/kgK
Thermal conductivity	λ	5.50e+01	W/mK
Latent heat of fusion	L	6.10e+04	J/kg
Liquid dynamic viscosity	μ	1.00e-03	Pa s
Thermal expansion coefficient	β_T	6.00e-05	K^{-1}
Solutal expansion coefficient	β_C	-5.30e-03	%$^{-1}$
Reference temperature	T_{ref}	2.19e+02	°C
Reference concentration	C_{ref}	1.00e+01	%
Reference density	ρ_{ref}	7.00e+03	kg/m^3
Permeability constant	K_0	2.34e-11	m^2
Melting temperature	T_f	2.32e+02	°C
Eutectic temperature	T_e	1.83e+02	°C
Eutectic concentration	C_e	3.81e+01	%
Solubility at eutectic temperature	C_{eS}	2.49e+00	%
Liquidus slope	m_L	-1.28e+00	°C/%
Gravity acceleration	g_y	-9.80e+00	m/s^2
External temperature	T_{ext}	25	°C
Heat transfer coefficient	q	400	W/m^{2}K
Equilibrium partition coefficient	k_p	0.0656	/

2.2 Mathematical model

2.2.1 Governing equations

The problem is described by a set of coupled partial differential equations describing energy and solute transport, and incompressible Newtonian fluid flow with Darcy's approximation for porous media. The microscopic transport is considered through coupling of the solute transport equation and Eutectic phase diagram. Kozeny-Carman relation is used to account for the porosity of solidifying fluid and Boussinesq approximation is applied to account for the buoyancy forces.

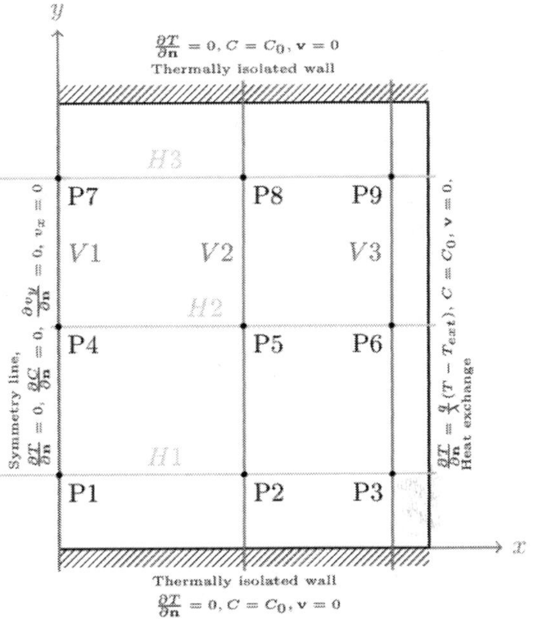

Figure 1. Schematic representation of the computational domain.

$$\nabla \cdot \mathbf{v} = 0 \tag{1}$$

$$\rho \frac{\partial \mathbf{v}}{\partial t} + \frac{1}{f_L} \rho \mathbf{v} \nabla \mathbf{v} = -f_L \nabla P + \mu \nabla^2 \mathbf{v} - \frac{f_L \mu}{K} \mathbf{v} + f_L \mathbf{b} \tag{2}$$

$$\rho \frac{\partial h}{\partial t} + \rho \mathbf{v} \cdot \nabla h_L = \lambda \nabla^2 T \tag{3}$$

$$\frac{\partial C}{\partial t} + \mathbf{v} \cdot \nabla C_L = 0 \tag{4}$$

The liquid velocity $\mathbf{v} = f_L \mathbf{v}_L$ depends on liquid fraction f_L and intrinsic liquid velocity $\mathbf{v}_L$. ρ represents density, P stands for pressure, μ for viscosity and $K = K_0 f_L^3 / (1 - f_L)^2$ for permeability, where K_0 is permeability constant. The buoyancy term $\mathbf{b} = \rho_{ref}[1 - \beta_T (T - T_{ref}) - \beta_C (C - C_{ref})]\mathbf{g}$ depends on thermal expansion coefficient β_T, solutal expansion coefficient β_C, temperature T, concentration C, reference values of temperature T_{ref}, density ρ_{ref} and concentration C_{ref}, and gravitational acceleration $\mathbf{g}$. Liquidus temperature $T_L = T_F + m_L C_L$ is defined as a sum of fusion temperature of the pure solvent T_F and a product of liquidus slope m_L and liquid concentration C_L, whereas concentration is defined as

$C=[f_L+(1-f_L)k_p]C_L$, where k_p is equilibrium partition coefficient. The enthalpy $h=c_pT+f_LL$ depends on specific heat at constant pressure c_p, T, f_L, and latent heat of fusion L. λ is thermal conductivity and t is time.

The pressure-velocity coupling is performed by a fractional step method, where the intermediate velocity is calculated by omitting the pressure correction term. The pressure is then calculated from the Poisson's equation and the velocity is then incremented by a pressure gradient term in order to obtain the divergence-free velocity in the next time step.

2.2.2 Boundary and initial conditions

The boundary conditions (Figure 1) at the symmetry line are

$$\frac{\partial T}{\partial \mathbf{n}}=0,\ \frac{\partial C}{\partial \mathbf{n}}=0,\ \frac{\partial v_y}{\partial \mathbf{n}}=0,\ v_x=0. \quad (5)$$

At the thermally isolated walls, the boundary conditions are

$$\frac{\partial T}{\partial \mathbf{n}}=0,\ C=C_0,\ \mathbf{v}=0 \quad (6)$$

and at the heat exchange wall, the boundary conditions are

$$\frac{\partial T}{\partial \mathbf{n}}=\frac{q}{\lambda}(T-T_{ext}),\ C=C_0,\ \mathbf{v}=0. \quad (6)$$

The initial conditions in the whole domain and on the boundaries are

$$\mathbf{v}(t=0)=0,\ T(t=0)=T_{ref},\ C(t=0)=C_0. \quad (7)$$

2.3 Meshless solution procedure

The main challenge addressed in this paper is the application of a meshless approach for the solution of macrosegregation modelling of mesosegregates. The DAM is used for spatial approximation and Euler time stepping scheme is used for time integration.

The global computational domain is divided into local subdomains in such a way that each of the computational nodes k has its own subdomain (k). Each subdomain has n computational nodes. The number of nodes in the subdomain must be at least equal to the size of a polynomial basis m. In the present paper, the second order polynomial basis is used ($m=6$) and each of the subdomains has 13 nodes ($n=13$). For each of the subdomains, a smooth and differentiable approximation function

$$\hat{f}^k(\mathbf{x})=\mathbf{p}(\mathbf{x},\mathbf{x}_k)\boldsymbol{\alpha}^k=\sum_{j=1}^{m}p_j(\mathbf{x},\mathbf{x}_k)\alpha_j^k \quad (8)$$

is obtained. The $\mathbf{p}(\mathbf{x},\mathbf{x}_k)$ is the second order polynomial basis vector for subdomain k with central computational point $\mathbf{x}_k$ and $\boldsymbol{\alpha}^k$ vector of coefficients. In order to obtain the approximation coefficient, the following expression has to be minimized

$$J^k=\sum_{i=1}^{m}\theta(\mathbf{x}_i,\mathbf{x}_k)[f(\mathbf{x}_i)-\hat{f}^k(\mathbf{x}_i)]^2. \quad (9)$$

The $f(\mathbf{x}_i)$ are the exact values in point $\mathbf{x}_i$ and the $\theta(\mathbf{x}_i,\mathbf{x}_k)=exp(-cr_i^2/h^2)$ is Gaussian weight function, with r_i representing the Euclidian distance between points $\mathbf{x}_i$ and $\mathbf{x}_k$, $c=5$ standing for free parameter of the weight function and $h=\max(r_i)$ is a scaling factor. Due to the dominance of the convection term, an upwind scheme is implemented in the DAM weight function [15].

3 Numerical example

The code for the present example is implemented in FORTRAN 2008. Intel Visual Fortran compiler (XE14) is used to compile a 64-bit executable. The examples are run on 6 parallel threads on an Intel Xenon at 2GHz. It takes about 17 h to calculate the present example. The calculations are performed on 151 x 181 node arrangement with time step $\Delta t=1\cdot 10^{-3}$ s.

4 Results and discussions

The present paper solves the solidification of Sn-10%Pb alloy defined by process parameters presented in Table 1. The results are compared to the results obtained by local radial basis functions colocation method (LRBFCM) [14] and are presented for representative evaluation points P1 to P9 listed in Table 2. The horizontal and vertical cross-sections were observed for three different horizontal (H1 to H3) and vertical (V1 to V3) lines, the details of which are listed in Table 3. The upwinding factor is set to 1, which means that the calculation point is moved at most halfway in the upwind direction. At the end of each time iteration, the smoothing stage described in [16] is performed and the smoothing factors, which are one of the free parameters of the model are set to $f_c=1.0$ for velocity, and $f_c=0.025(1-f(t'))$ for concentration. The smoothing factor for concentration is applied with a delay

$$f(t')=\begin{cases}1; & t'>1\\ t'^2(6-t'(8-3t')); & t'<0;\ t'=\dfrac{t}{\tau}\\ 0; & 1>t'>0\end{cases} \quad (10)$$

where ramp factor is $\tau=0.25$.

Figures 2a-f represent time development of concentration and velocity throughout the solidification process. The thermo-solutal convection starts as a consequence of the cooling of the liquid due to the heat exchange at the outer wall. Thus the flow in the clockwise direction is formed. As the temperature drops below the liquidus temperature, the solidification starts. The solute-rich liquid is carried by the flow and thus the solute-poor region is formed at the top of the cavity (Figures 2b-f) and the solute-rich area is gradually formed at the bottom (Figures 2b-f) and at the centre of the cavity (Figures 2e-f). The mesosegregates form in the shape of channel-like anomalies in the mushy zone region (Figures 2d-f). The instabilities that are the driving force behind the formation of these channels propagate the flow and consequently the segregation, thus forming the patches with alternating areas of positive and negative segregation (Figures 2d-f). Although the mechanism behind the formation of these channels is not well known, the numerical instabilities are not the reason for their appearance, as these mesosegragate formations have also been observed experimentally [13].

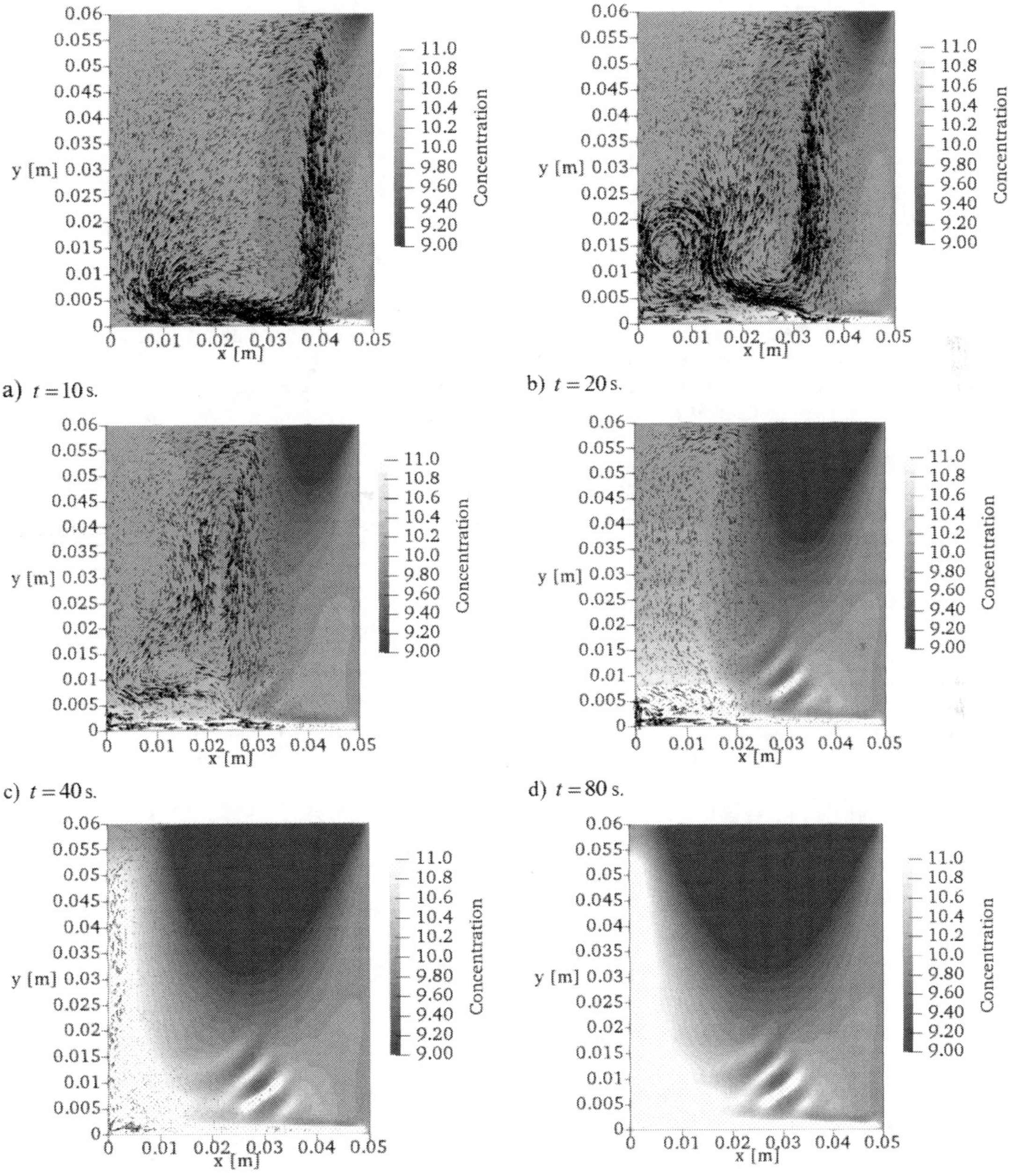

a) $t = 10$ s.

b) $t = 20$ s.

c) $t = 40$ s.

d) $t = 80$ s.

e) $t = 120$ s.

f) $t = 600$ s

Figure 2. Concentration [%] and velocity vector at different times.

The results have also been analysed for reference points defined in Table 2 and cross-sections defined in Table 3. Figure 3 presents concentration profiles for horizontal cross-sections. The profiles correspond to the final time step of the computational example (Figure 2f). The H3 line shows a smooth distribution of concentration in the top part of the cavity. The Sn concentration is the highest at the centre of the cavity and at the outer (heat exchange) wall and the lowest in the middle of the computational domain. The H1 line shows horizontal cross-section close to the bottom of the cavity and shows a perturbation at the middle of the computational domain. The perturbation corresponds to the segregation channels (mesosegregates). Figure 4 presents the concentration of Sn for vertical cross-sections at V1 to V3. The peak in blue line (V3) corresponds to the increased concentration at the bottom of the cavity close to the outer wall. The concentration of Sn then gradually drops toward the top of the cavity. The V1 line presents the concentration at the centre of the cavity where the area with the positive Sn concentration is formed. The slight decrease in the middle is due to the velocity recirculation pattern in the final stages of solidification as can be seen in Figure 2e. The pattern is formed due to the temperature, which in term causes the change in liquid fraction. The liquid consequently flows around this part and thus the velocity in the totally liquid part of the fluid is increased in contrast to the velocity in the partially solidified fluid, which is decreased. In both cases (horizontal and vertical), the H2 and V2 lines present the cross-sections in the middle of the horizontal and vertical domains.

Table 2. Definition of benchmark control points defined for SMACS benchmark test case.

Point	Coordinates (x [m], y[m])
P1	(0,0.01)
P2	(0.025,0.01)
P3	(0.045, 0.01)
P4	(0.0,0.03)
P5	(0.025, 0.03)
P6	(0.045,0.03)
P7	(0.0, 0.05)
P8	(0.025,0.05)
P9	(0.045, 0.05)

The results for temperature, liquid fraction and concentration are listed in Table 4 for all representative points P1 to P9. The results are listed at $t = 570$ s, when the whole computational domain is solidified. It should be noted here, that in comparison with [14], the results of the present study show lower positive segregation in the central area than the results calculated with LRBFCM [14]. The results for V2 and V3 are in good agreement with reference results as are points P2, P3, P5, P6, P8 and P9 that do not lie on the centre line.

Time developments of temperature, liquid fraction and concentration fields are presented in Figures 5 to 7. As already mentioned, the solidification starts once the temperature drops below the liquidus temperature and lasts until it reaches the eutectic temperature ($T_{eut} = 183$ °C). Figure 5 compares temperature for calculated results (lines) with reference results [14] (crosses) in representative points (P1 to P9). The results are in good agreement.

Table 3. Definition of benchmark control lines defined for SMACS benchmark test case.

Line	Location ([m])
H1	y = 0.01
H2	y = 0.03
H3	y = 0.05
V1	x = 0.00
V2	x = 0.025
V3	x = 0.045

Table 4. Representation of results in representative points at $t = 570$ s.

Point	T [°C]	f_L	C [%]
P1	109.30	0.00	12.76
P2	105.91	0.00	10.81
P3	98.49	0.00	10.43
P4	109.30	0.00	10.72
P5	105.90	0.00	9.17
P6	98.49	0.00	9.97
P7	110.29	0.00	12.29
P8	105.91	0.00	7.77
P9	98.49	0.00	9.28

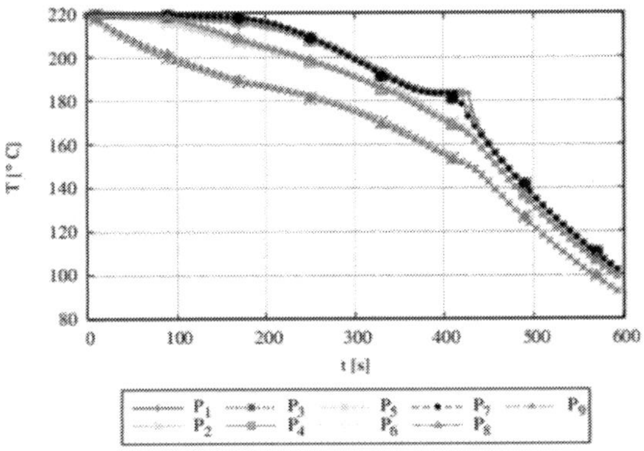

Figure 5. Time development of temperature in representative points P1 to P9. Lines and smaller nodes depict results calculated with DAM and bigger nodes depict results calculated with LRBFCM [14].

The time development of liquid fraction is presented in Figure 6 for calculated results. In the present case, the solidification stops, when the temperature falls below eutectic temperature, which can be clearly seen in Figure 6, where the liquid fraction drops to zero after reaching the eutectic temperature. The effect of the described model is best observed for points at the centre line (P1, P4 and P7) and can be also seen on the temperature graph (Figure 5).

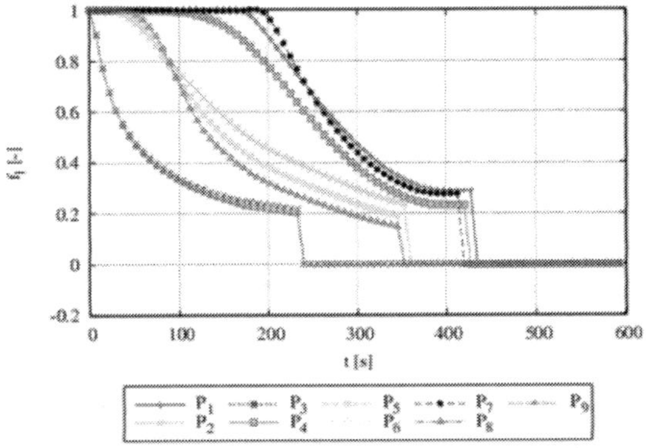

Figure 6. Time development of liquid fraction in representative points P1 to P9.

Figure 7 presents the time development of Sn concentration in representative points. The solidification begins when the temperature drops below the liquidus temperature and stops when the eutectic temperature is reached.

Finally, a comparison is made for a uniform and non-uniform node arrangement with randomly positioned nodes [17] and is presented in Figures 2f (uniform) and 8 (non-uniform). Although the final macrosegregation pattern is similar for both of the cases, there are observable differences in the mesosegregates. The formed channels are much more pronounced in non-uniform grid example, which can be clearly seen in Figures 3 and 4. The difference might be due to the upwind and smoothing parameters, which might require different values for a non-uniform node arrangement. However, this was not observed in previous simulations of macrosegregation [15] without mesosegregates. The results show that formulation of channels strongly depends on the node density and arrangement. Similar observations were made in [9]. The simulation with various models and methods showed that it is not possible to obtain mesh-converged results.

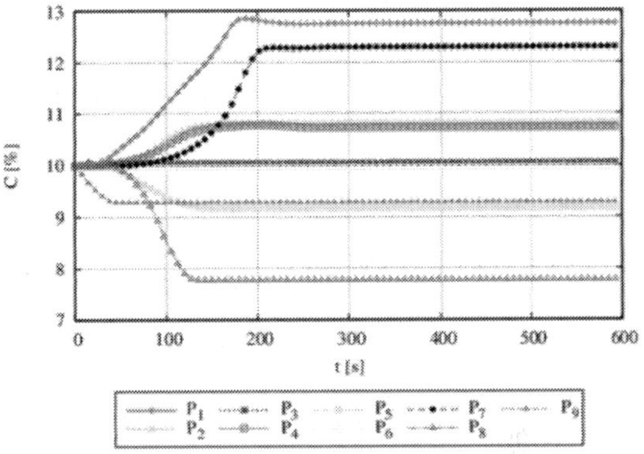

Figure 7. Time development of concentration in representative points P1 to P9.

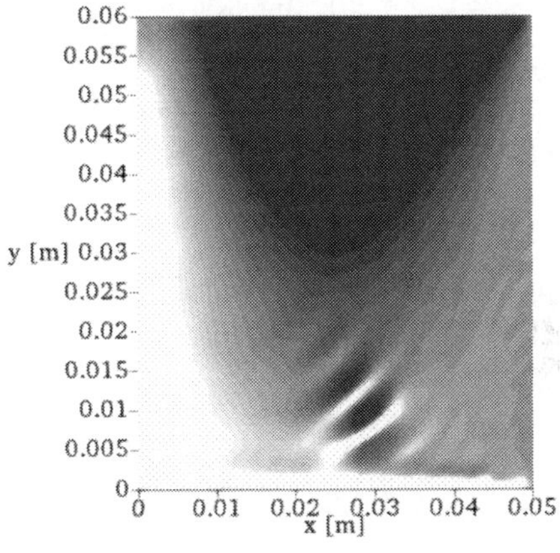

Figure 8. Concentration [%] and velocity vector for a fully solidified case for a non-uniform node arrangement ($t = 600$ s).

5 Conclusions

The present paper presents a solution of a mathematical model for macrosegregation with mesosegregates in a binary substrate solved by DAM. The problem corresponds to the benchmark example for Sn-10%Pb alloy proposed in [12] and the results are compared to those obtained with LRBFCM presented by [14]. The results are in good agreement with reference results.

Acknowledgement

This work was founded by Slovenian research agency (ARRS) in the framework of basic research project J2-7384 and program group P2-0162.

References

1. "Web of Science [v.5.28] - Web of Science Core Collection Result Analysis." [Online]. Available: http://wcs.webofknowledge.com/RA/analyze.do. [Accessed: 10-Apr-2018].
2. E. J. Kansa, CMA, **19**, 127 (1990)
3. W. Chen, *Meshfree methods for partial differential equations*, (Springer, 2003)
4. S. N. Atluri, T. Zhu, Comput. Mech. **22**, 117 (1998)
5. G.-R. Liu, *Mesh free methods: moving beyond the finite element method*. (CRC Press, 2002)
6. B. Šarler and R. Vertnik, CMA, **51**, 1269 (2006)
7. C. Prax, H. Sadat, P. Salagnac, Transp. Porous Media, **22**, 215 (1996)
8. B. Šarler, R. Vertnik, J. Perko, CMA, **2**, 77 (2005)
9. H. Combeau *et al.*, Mater. Sci. Eng., **33** (2012)
10. H. Sadat, N. Dubus, L. Gbahoue, T. Sophy, Numer. Heat Transfer, Part B, **50**, 491 (2006)
11. H. Sadat and C. Prax, Int. J. Heat Mass Transfer, **39**, 214 (1996)
12. M. Bellet *et al.*, Int. J. Therm. Sci. **48**, 2013, (2009)
13. J. Sarazin, A. Hellawell, Metall. Trans., **19**, 1861 (1988)
14. G. Kosec, B. Šarler, EABE, **45**, 36 (2014)
15. V. Hatić, B. Mavrič, B. Šarler, Int. J. Numer. Methods Heat Fluid Flow, **28**, 361 (2018)
16. K. Reuther, B. Sarler, M. Rettenmayr, Int. J. Therm. Sci., **51**, 16 (2012)
17. V. Hatić, B. Mavrič, N. Košnik, B. Šarler, Appl.Math. Model., **54**, 170 (2018)

A Rolling Simulation System for Reversing Rolling Mills

Umut Hanoglu[1,2,a] and Božidar Šarler[1,2]

[1]*Laboratory for Simulation of Materials and Processes, Institute of Metals and Technology, Lepi pot 11, 1000 Ljubljana, Slovenia*
[2]*Laboratory for Fluid Dynamics and Thermodynamics, Faculty of Mechanical Engineering, University of Ljubljana, Aškerčeva 6, 1000 Ljubljana, Slovenia*

Abstract. In this work a rolling simulation system has been developed for rolling schedules which consists of multiple reversing rolling mills. A slice model approach is applied where the position of a slice can only be determined by considering total deformation. Each slice is parallel to each other and perpendicular to the rolling direction. The solution of coupled thermal and mechanical models over each slice, at a given time and position, are achieved by a novel meshless Local Radial Basis Function Collocation Method (LRBFCM). Mechanical material model obeys ideal plastic flow rule defined by Von Mises. Unknown fields over the slices are interpolated by a certain number of collocation points distributed over the physical domain and its boundary. A system of equations is solved for each collocation point considering its local neighbouring points in the range between 5 and 7. A non-linear system of equations is solved by direct iteration. Groove geometries of each roll are implemented in a compatible way with the slice model and every roll has a horizontal orientation. In between each rolling pass the billet is rotated either 90 or 45 degrees clockwise or counter clockwise. Reduction at each of the passes can be very high, and in such cases, the material completely fills up the groove. This requires a special attention regarding the contact boundary conditions and the collocation node distribution due to numerical instability issues. Coulomb model of friction or sticking boundary conditions are used at the contact boundaries and Gauss-Seidel iterative elliptic node generation algorithm is used for redistributing collocation nodes over the physical domain, when necessary. The simulation results for arbitrary initial position of the slice in the billet are shown in terms of temperature, displacement, strain and stress fields as well as roll forces and torques. A user friendly computer application is created for industrial use based on C# and .NET.

1 Introduction

In this paper a rolling simulation model is explained for reversing rolling mills. A steel billet is passed through a rolling stand multiple times and at each time the direction is reversed. At each pass, a different groove is used for shaping the material. In between the passes the billet is rotated either 90 or 45 degrees. The rotation is necessary since the mill has exclusively horizontal orientation. At each pass the billet becomes longer and takes more time to go through next rolling. The model expressed here is capable of simulating user defined reversing rolling mill uninterruptedly from the beginning until the end.

Simulation of rolling is in general very hard task of computational solid mechanics [1,2]. The complex groove geometries together with the deformed billet are reduced to 2D computational domains during present simulations. A meshless solution procedure is used together with the slice model assumption explained below. The Local Radial Basis function Collocation Method (LRBFCM) is used for interpolation of all the unknown fields over each slice at given time and position. LRBFCM has been previously described [3], applied to simulation of fluid flow problems [4], thermo-mechanics [5], dendritic growth [6], finishing rolling mills [7]. The rest of the paper is organized as follows. The physical model and the meshless solution procedure of the rolling simulation system based on both thermal and mechanical models are explained. Later the preferred node generation algorithm is elaborated. At the end, numerical simulation results of a reversing rolling mill with ten passes are shown and concluding remarks are given.

2 Physical model

In this research two major models are solved in a coupled way. Both thermal and mechanical models are defined in their strong form. Temperature field is obtained from the thermal model by solving it by explicit time stepping. The mechanical model is solved to get displacement, stress and strain fields.

2.1 Slice model assumption

The thermo-mechanical simulation of reversing rolling mill is a complex 3D process however, a simplified 2D

[a] Corresponding author: umut.hanoglu@imt.si

© The Authors, published by EDP Sciences. This is an open access article distributed under the terms of the Creative Commons Attribution License 4.0 (http://creativecommons.org/licenses/by/4.0/).

modelling assumption may be introduced here based on slices aligned in parallel and towards the rolling direction as shown in Fig. 1. Each slice becomes a computational domain at given position and time. Slice is rolled multiple times until the end of the simulation and rotates in between rolls.

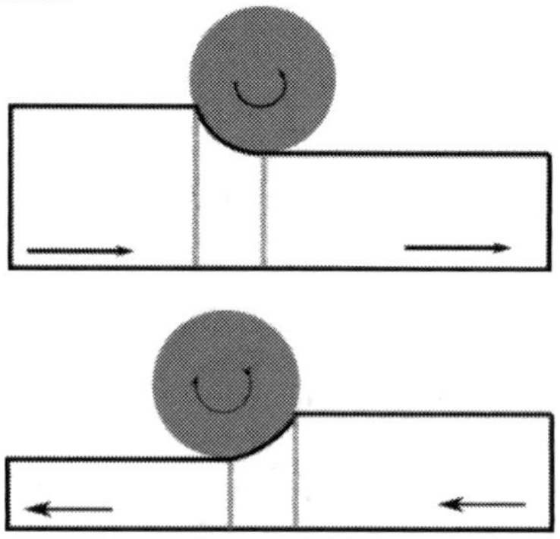

Figure 1. Computational slice positions and traveling of a slice through reversing rolling mill.

2.2 Thermal model

The temperature field is obtained from the following heat transfer equation applicable for slice model assumption

$$\rho c_p \frac{\partial T}{\partial t} = \nabla \cdot (k \nabla T) + \dot{Q}, \tag{1}$$

where $\rho, c_p, T, t, k, \dot{Q}$ stand for density, specific heat, temperature, time, thermal conductivity and internal heat generation rate due to deformation which is defined as

$$\dot{Q} = \eta \frac{\partial}{\partial t} \int_{\bar{\varepsilon}} \bar{\sigma} d\bar{\varepsilon} . \tag{2}$$

where η is Taylor-Quinney parameter, $\bar{\sigma}$ is effective stress and is $\bar{\varepsilon}$ effective strain. Robin type of boundary condition is applied over all of the boundaries.

$$-k \frac{\partial T}{\partial \mathbf{n}_\Gamma} = h\left(T - T^{ref}\right). \tag{3}$$

When there is a contact with the roll, the heat transfer coefficient h is chosen much larger than to the air.

2.3 Mechanical model

For each slice in contact with the roll, the displacement field is solved directly from the governing equation used for the mechanical model.

$$\mathbf{L}^T \boldsymbol{\sigma} + \mathbf{b} = \mathbf{0}, \tag{4}$$

where $\mathbf{L}$ is the derivative operator matrix, $\boldsymbol{\sigma} = \left[\sigma_{xx}, \sigma_{yy}, \sigma_{xy}\right]^T$ is the stress vector, and $\mathbf{b} = \left[b_x, b_y\right]^T$ is the body force vector, however it is neglected in the current model. Essential and natural boundary conditions, used over the boundaries depending on whether the prescribed displacement $\bar{\mathbf{u}}$ or traction $\bar{\boldsymbol{\tau}}$ is known

$$\bar{\mathbf{u}} = \mathbf{u}, \; \bar{\boldsymbol{\tau}} = \mathbf{n}\boldsymbol{\sigma} \tag{5}$$

3 Meshless solution procedure

Instead of grid generation over the physical domain such as in FEM, collocations nodes without any geometrical shapes between them are used over the domain and boundaries. An example of an unknown field $\theta(\mathbf{p})$ at position $\mathbf{p}$ is interpolated with multi-quadric RBF $\psi(\mathbf{p})$ over neighbouring nodes ($N = 5,7$) together with the augmented first order ($N_p = 3$) polynomial functions $\vartheta(\mathbf{p})$,

$$\theta(\mathbf{p}) = \sum_{i=1}^{N} \psi_i(\mathbf{p}) \alpha_i + \sum_{i=1}^{N_p} \vartheta_i(\mathbf{p}) \alpha_{N+i} . \tag{6}$$

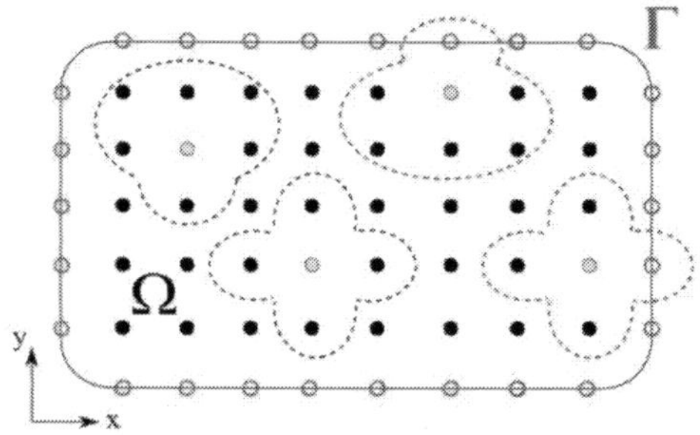

Figure 2. Examples of 5 or 7 noded local influence domians considered in the meshless solution.

3.1. Solution of the thermal and mechanical models

For the solution of the temperature field, the heat transfer equation is considered in the following discrete form

$$\rho c_p \frac{T_i - T_{i0}}{\Delta t} = \nabla \cdot (k \nabla T) + \frac{\Delta Q}{\Delta t}. \tag{7}$$

Then the temperature field is interpolated locally for each collocation node considering only its neighbours.

$$T(\mathbf{p}) = \sum_{i=1}^{N} \psi_i(\mathbf{p})_T \alpha_i + \sum_{i=1}^{N_p} \vartheta_i(\mathbf{p})_T \alpha_{N+i} . \tag{8}$$

Temperature field in the governing equation and boundary conditions is discretized in this way and same collocation constants are used to obtain the temperature field.

The governing equation used in the mechanical model may be written in terms of displacement vector $\mathbf{u}$ with using stiffness matrix $\mathbf{C}$, written in terms of effective stress-effective strain rate $\left(\bar{\sigma}/\bar{\varepsilon}\right)$, explained in [8],

$$\mathbf{L}^{\mathrm{T}}\mathbf{C}\mathbf{L}\mathbf{u}+\mathbf{b}=\mathbf{0}. \tag{9}$$

The displacement vector $\mathbf{u}=\left[u_x,u_y\right]^{\mathrm{T}}$ components are interpolated in a similar way as the temperature field

$$u_x(\mathbf{p})=\sum_{i=1}^{N}\psi_i(\mathbf{p})_{u_x}\alpha_i+\sum_{i=1}^{N_p}\vartheta_i(\mathbf{p})_{u_x}\alpha_{N+i}, \tag{10}$$

$$u_y(\mathbf{p})=\sum_{i=1}^{N}\psi_i(\mathbf{p})_{u_y}\alpha_i+\sum_{i=1}^{N_p}\vartheta_i(\mathbf{p})_{u_y}\alpha_{N+i}. \tag{11}$$

The system of equations are written in global form for each node to obtain the solution of displacements $(\mathbf{U})$ by direct iteration,

$$\mathbf{K}\left(\mathbf{U}^{j-1}\right)\mathbf{U}^{j}=\mathbf{f}, \tag{12}$$

where $\mathbf{K}(\mathbf{U})$ is the global sparse matrix and $\mathbf{f}$ is the adjacent vector.

3.2 Node generation algorithm

The initial node configuration has an important role during the rolling simulation. If the initial size is a rectangle then a uniform node arrangement is applied at the beginning. If the initial slice has a unique non-rectangular shape then the nodes are distributed depending on iterative Gauss-Seidel elliptic node generation algorithm [9] as shown in Fig. 3.b.

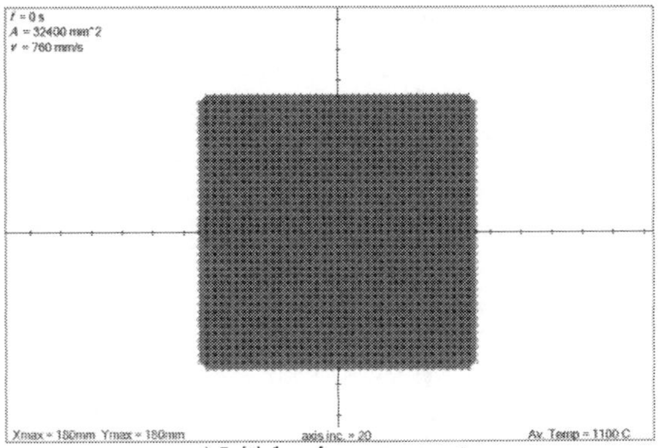

a) Initial node arrangement

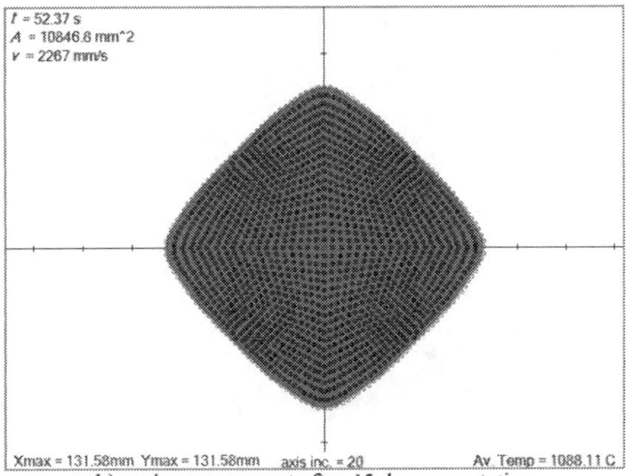

b) node arrangement after 45 degrees rotation

Figure 3. 1517 collocation nodes are shown in each figure. First figure shows the initial configuration, second figure shows the configuration after 45 degrees rotation before the second pass.

3.3 Development of the simulation code

A simulation code has been written from scratch using Microsoft Visual Studio C# 2015 Express Edition [10]. A Windows desktop application is created with user friendly interface by using features of the Microsoft .NET framework. The only additional library used in this simulation code is Math.Net for solving the system of equations. The code is able to read multiple rolling schedules written in an excel file with all the involved groove types. Each groove is described based on 2D coordinates of multiple groove surface points written in a .txt file. The simulation then calculates and redraws the corresponding groove line at each slice position and updates the boundary conditions accordingly. The user is able to change the slice position to any desired value and view all the results belonging to that slice.

4 Simulation results

Two different reversing rolling mills are used in the numerical simulations. During the first 5 passes the roll diameter is 800 mm and later for the subsequent 5 passes it is used as 650 mm. The material model used is $\bar{\sigma}=589\bar{\varepsilon}^{0.214}\exp\left(R\left(1200-T\right)\right)$ MPa and the coefficient of friction is taken as 0.07.

4.1 Reversing rolling mill simulation with 800 mm roll diameter

The initial shape and the node arrangement is used as shown in Fig. 3.a results of the first 5 passes are shown with groove slice lines drawn from the first contact until the last in Fig. 4.

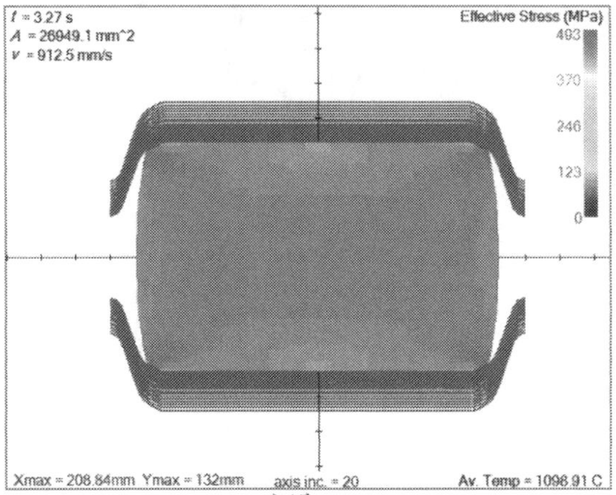

a) 1st pass

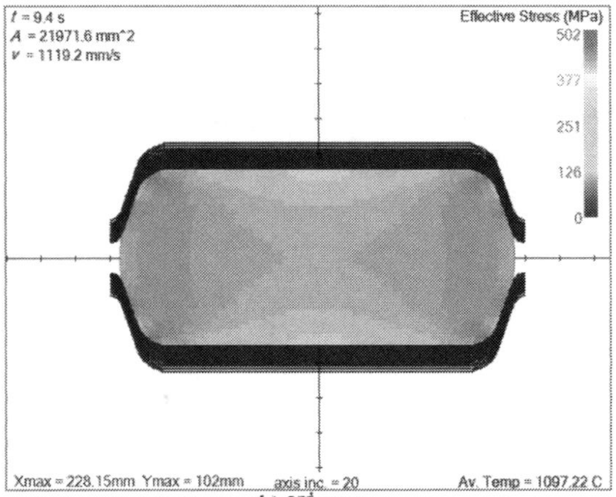

b) 2nd pass

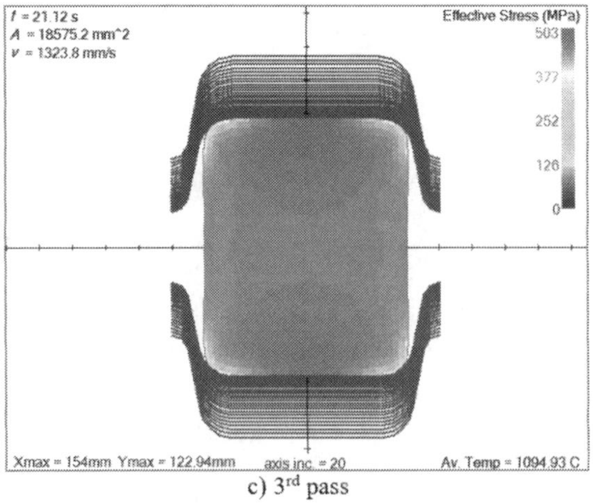

c) 3rd pass

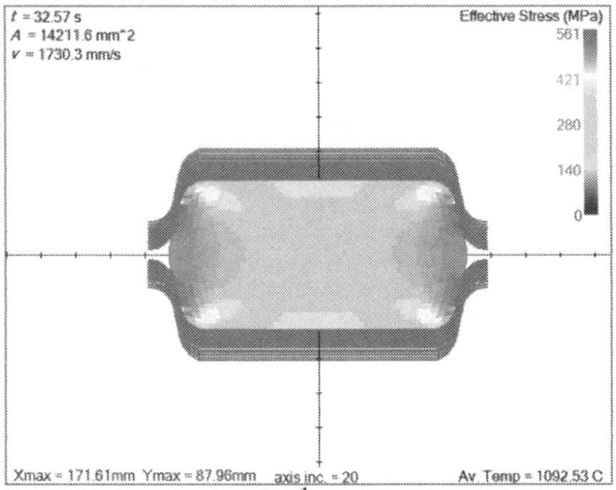

d) 4th pass

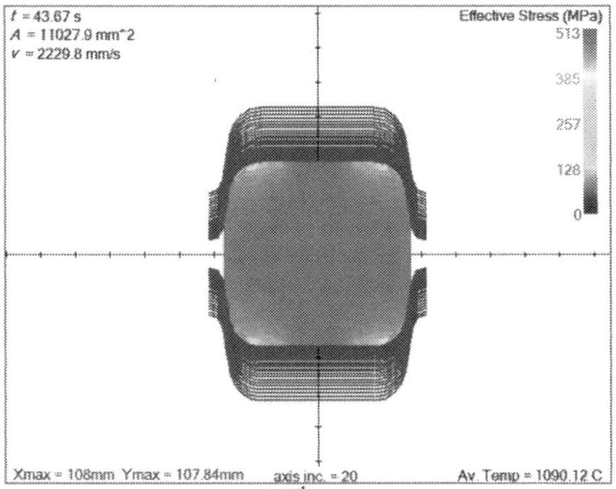

e) 5th pass

Figure 4. Effective stress filed resulting after the first 5 passes when the billet is rotated 90 degrees after each pass starting from the second pass.

4.2 Reversing rolling mill simulation with 650 mm roll diameter

This part of the simulation is a continuation of the previous simulation shown in Fig. 4. The slice in Fig 3.b is used for the initial slice shape and node arrangement. It has the same size as the slice shown in Fig. 4.e, but it is rotated 45 degrees.

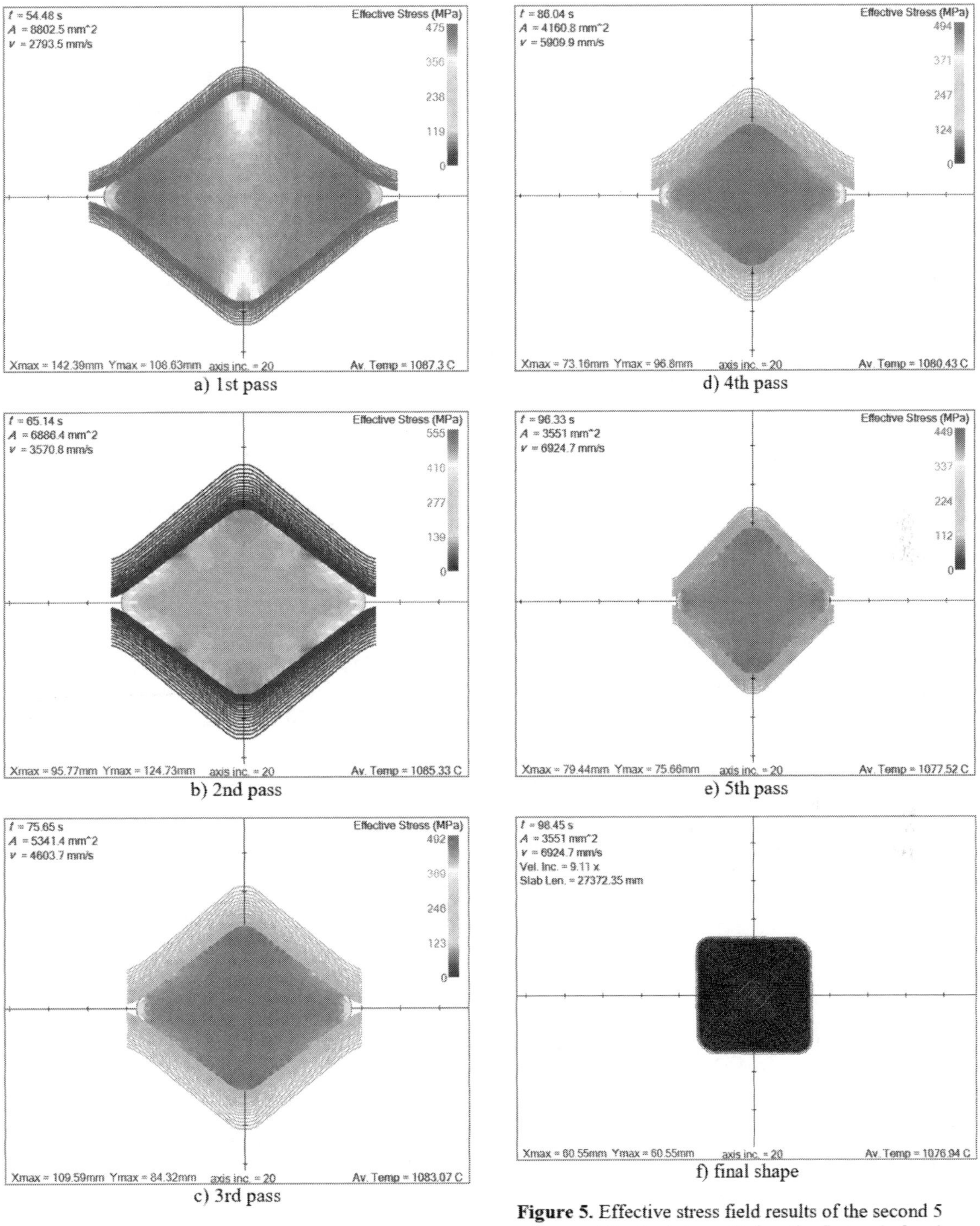

Figure 5. Effective stress field results of the second 5 passes and the final shape. Before the first and after the last pass the billet is rotated 45 degrees and in-between other passes it is rotated 90 degrees.

The simulation results shown in Fig. 4 and Fig. 5 are directly obtained from the user interface of the simulation code after simulation. Results may be seen over one slice at a time, based on its position, until the end of the rolling process. Results in this chapter are shown exactly at positions where the slice leaves each rolling stand and all the previously contacting groove lines from the same rolling stand are drawn as well. Except for the first rolling stand, after each pass the slice is rotated and a 6 seconds of rotation time is considered. During the first 5 stands (800 mm) the billet is rotated 90 degrees, however for the subsequent 5 stands (650 mm) the rotation is only 45 degrees. In this way, it is possible to consider only the horizontally oriented rolling stands. The increment between the slice positions is 5 mm when there is a contact. If there is no contact, the incremental distance is increased to 20 mm. Relatively large reduction rates are applied and the maximum effective stress at each pass is calculated to be around 500 MPa. At the end of the simulation a 60 mm square cross section is obtained just like desired. Initially 3 meter long billet at the end becomes 27.372 meters. The slice is also constantly loosing heat either to the air or to the roll during rolling. Therefore, the average temperature is always decreasing.

5 Conclusions and future work

In this work a rolling simulation system is explained and applied to reversing rolling mills. A coupled thermo-mechanical solution procedure is used and the solution is achieved by using a local meshless collocation method with radial basis functions. Simulation results are shown in two sets where in each set billet goes through 5 passes. In between the passes billet is rotated 90 degrees, in between the sets it is rotated 45 degrees. Corresponding grooves are drawn and the results are shown in terms of effective stress.

A macro simulation model is shown here where the shape of the billet may be predicted for each rolling pass. However, this model should also be coupled with the micro scale simulations to get a complete overview of the deformation and material properties of the rolled material.

6 Acknowledgements

The authors of this paper are supported by the program group P2-0162, funded by Slovenian Research Agency, and Štore Steel (www.store-steel.si) company.

References

1. J.G. Lenards, *Primer on flat rolling* (Elsevier, Oxford, 2007)
2. J.G. Lenards, M. Pietrzyk, L. Ceser, *Mathematical and physical simulation of the properties of hot rolled products* (Elsevier, Oxford, 1999)
3. R. Vertnik, B. Šarler, Comput Math Appl, **51** (2006)
4. R. Vertnik, B. Šarler, Eng Anal Bound Elem, **45** (2014)
5. B. Mavrič, B. Šarler, Inter J Numer Method H, **25** (2015)
6. T. Dobravec, B. Mavrič, B. Šarler, J Comput Phys, **349** (2017)
7. U. Hanoglu, B. Šarler, Comput Struct **194** (2018)
8. S. Kobayashi, S. I. Oh, T. Altan, *Metal forming and finite element method* (Oxford University Press, New York, 1989).
9. A J. F. Thompson, B. K. Soni, N. P. Weatherill, *Handbook of grid generation* (CRC, Florida, 1999)
10. Microsoft Corporation, Microsoft Visual Studio Express 2015 for Windows Desktop, version 14.0.23107.0, (2015)

Use of Web technologies to create an information processing system of pressure metal forming processes

Valentin A. Osadchy[1], Alexander V. Osadchy[2], Valeriy V. Savin[3], Lyudmila A. Savina[3], Nikolai V. Novoseltcev[3,a]

[1]*National University of Science and Technology MISIS, Leninskiy ave. 4, Moscow, 119049 Russia*
[2]*Prokhorov General Physics Institute, Russian Academy of Sciences, ul. Vavilova 38, Moscow, 119991 Russia*
[3]*Immanuel Kant Baltic Federal University, ul. Nevskogo 14, Kaliningrad, 236041 Russia*

Abstract. The problem of obtaining reliable information when performing calculations, can be solved, thanks to the intensive development of computer facilities, using specially created databases and information systems, as well as improving data processing techniques and optimizing the characteristics of materials and products, parameters of technology and equipment. Following tasks have been solved: a) an information system in the field of metal forming and related science fields has been developed; b) a database on metal materials, metal products, technology and equipment of rolling production has been created; c) a mathematical apparatus and model of the rolling process for material and process parameters selection have been developed. Obtained as a result framework can be applied to solve a broad aspect of practical problems.

1 Introduction

Modern system approach to the technological processes design is based on the development of mathematical models of objects and automated search for optimal solutions. This puts on the need to change approaches to the creation and implementation of mathematical models, requires the development of complex computer systems that make modelling tasks as easy as possible. Such systems should contain information and reference materials, databases, the corresponding mathematical apparatus both general character, and for describing specific processes. In this case, they can be used not only for research purposes, but also as independent information-reference, as well as training systems.

2 Results

The achieved level of development and dissemination of computer equipment, as well as the accumulated baggage of mathematical methods for calculating the parameters of metal smelting and processing, allow us to begin creating a software environment for designing simulation models of processes and optimizing technological parameters, considering specific conditions and requirements with a wide use of computer information technologies in the Internet.

The use of information technologies in the development of new and improvement of existing technological processes makes it possible to fully benefit from the following opportunities:

1. Fast and complete provision of information on the problem being developed.
2. Selection of suitable calculation methods using computer analysis.
3. Search in the database for the required characteristics of materials and process parameters.
4. Construction of empirical dependencies for solving technological and research problems by statistical processing of information from the database.
5. Performing an effective comparative analysis of the results obtained at both intermediate and final stages.
6. Search for optimal process flow parameters.

The development and improvement of technological processes based on theoretical research using mathematical models is indisputably an effective way of developing production. At the same time, the successful application of mathematical models is largely determined both by the completeness of the process description and by the quality of the information used as the initial data. Further development of computational methods is required for rolling production, especially in the direction of taking account of the rolling peculiarities in calibers in determining the shape and temperature field, as well as the rheological characteristics and properties of the metal.

The problem of obtaining reliable information when performing calculations, in turn, can be solved now, thanks to the intensive development of frameworks, using specially created databases and information systems, as well as improving data processing techniques and optimizing the characteristics of materials and products, parameters of technology and equipment. Therefore, for the successful use of information technology in the

[a] Corresponding author: niknovo@me.com

© The Authors, published by EDP Sciences. This is an open access article distributed under the terms of the Creative Commons Attribution License 4.0 (http://creativecommons.org/licenses/by/4.0/).

development and improvement of rolling production processes, the following tasks must be solved:

- Develop an information system in the field of metal forming with pressure and related fields of science;
- Create a database on metal materials, metal products, technology and equipment of rolling production;
- Develop a mathematical apparatus and model of the rolling process to select materials and process parameters that ensure the required level of properties.

The framework thus obtained can be applied to solve a broad aspect of practical problems.

In this paper, we present an information framework in the form of a Web application, including an information and training system, a database, programs for processing data, calculating process parameters, equipment parameters, and the characteristics of metal products.

The software package is designed for use on the Internet and designed in accordance with the currently accepted rules for organizing the interface in application systems on personal computers and any researcher or designer who does not have special skills in working with a personal computer can use it.

2.1 Informational and training system for metallurgical production

To facilitate the search for the required information of a descriptive nature in the field of metal working with pressure and related parts of science, an information and training computer system has been created that represents a considerable amount of information equivalent to dozens of textbooks broken down into small fragments (articles) combined with a cross-reference system (hyperlinks) and table of contents.

The structure of information storage and data access system has been developed, information on the theory, technology and equipment of metal processing processes in the press and related fields of science has been collected, processed and included in the database, which can be used as a reference in the development and improvement of technological processes, and as a computer textbook for 19 courses.

Depending on the chosen method of information access, the computer system can be used as a reference book, encyclopaedia, computer (electronic) textbook. The system is multilevel and allows you to study the problem with varying degrees of detail.

2.2 Database on metal materials and metal products

The database contains information on the production of ferrous and non-ferrous metals in the CIS countries. The developed structure of the database, as well as its filling and software shell, allows a quick search for information about chemical compounds and analogues of domestic and foreign non-ferrous and ferrous metals and alloys, the purpose and conditions of application, heat treatment regimes, mechanical and physical properties, the requirements of standards, technology and the characteristics of proprietary equipment available at the enterprises, etc. It is possible to obtain information about manufacturers of various types of metal products.

The main components of the database are tables, which allows you to modify the information without changing the program itself.

The software shell includes algorithms for preliminary data processing calculation of predicted values of material characteristics in the absence of required data in the database.

The database "Ferrous and non-ferrous metals" has been registered in the State Register of Databases, awarded with certificates at exhibitions, purchased and used by many Russian and foreign organizations and private individuals.

2.3 Information and calculation system for rolling fabrication

The correctness of the comparative evaluation of technology options is largely determined by the reliability of the data used, including technological parameters of the processes. In the CAD of rolling shops, the main component is the subsystem for the implementation of the technological part of the project, which determines the main technical solutions and the level of technical and economic indicators of the facility. It is calculations based on technological models that allow us to give an economic assessment and make a choice of technology in the future. In technological models, natural indicators of resource consumption and output are combined simultaneously. Based on such indicators, the calculation of the overall project efficiency is carried out.

Taking this into account, a mathematical model of the longitudinal rolling process has been developed for their determination, which makes it possible to determine deformation, temperature, velocity and energy parameters, and predict mechanical characteristics of the metal.

The developed mathematical model is based on engineering methods of calculation of technological parameters, provides for the possibility of process optimization, is connected to the database and information system.

Since the existing dependencies for the calculation of the most important parameters of the rolling process do not always reflect the influence of various process parameters quite adequately, and in all cases, provide sufficient reliability of the results obtained, the mathematical model provides for the choice of alternative methods.

Based on statistical processing of the results of tests of domestic and foreign ferrous and non-ferrous metals presented in the database, regression dependences of thermophysical characteristics on the temperature and the percentage content of chemical elements in steel or alloy were obtained with the help of the statistical processing program. Black and non-ferrous metals and alloys are divided into groups and equations are constructed for each of them.

To determine the deformation resistance of various deformable materials - ferrous and non-ferrous metals and alloys, they are divided into 63 groups. An algorithm for determining the material group and the necessary parameters for a given chemical composition of the metal has been developed. The construction of the equations considered the influence of 66 basic chemical elements in the composition of the material.

When predicting the mechanical characteristics of rolled products, an analytical approach is used in combination with the experimental one.

When entering the chemical composition of materials, determining the mechanical and thermal properties of materials, equipment parameters, databases are used.

For efficient processing of discrete data and obtaining regression dependencies in the research and modelling of technological processes of rolling, an algorithm is developed, and a program is developed that performs statistical processing of significant volumes of data, determination of correlation coefficients, and search for the best kind of functional connection.

In constructing the regression equation, all possible combinations of power terms, including negative ones, are used (logarithm, exponent) applied to individual arguments and to the whole expression.

The efficiency of solving most of problems in the development and improvement of technological processes depends to a large extent on the correctness of the postulate of the optimization problem and the quality of the applied methods of solution. In this connection, an approach and program were developed to find optimal conditions in the case of many variable parameters and a complex process model.

The search for unknown parameters is carried out by combining three numerical methods of searching for an extremum. First, a search of possible combinations of values of unknown parameters in each interval with a certain step in the parameter (scanning method) is performed. This allows in most cases to find the region of the global optimum. Then the best set of values of the required parameters is used to find the local optimum jointly by the methods of configurations and steepest descent with some additions.

2.4 Software for assemblies and technological tools design

In a market economy, one who survives can respond quickly to changing market needs. Recently, demand for bent rolled sections has increased substantially.

The equipment development and rolls calibration to produce bent roll profiles is a rather complicated and time-consuming task. This is due to the need to perform a large amount of calculations, from the preparation of drawings, as well as the need for optimization of calibration.

The adopted method for describing the shape of the strip and the developed method for determining the profiling route made it possible to compile a general algorithm for calculating roll calibrations in the production of bent symmetrical and asymmetric profiles of practically any real configurations such as a corner, channel, snake, trough, closed profile, etc. The program is based on a mathematical model of the strip profiling process, which allows calculating geometric and energy parameters. The developed algorithm allows to calculate the calibration of forming stands, deduce the drawings of rings and banded rolls in assembled form, includes the investigation of rational ways of profiling, in-depth analysis of situations, possibilities, numerous checks and directions of the process of flexibility at each stage production.

The main result of the program is working drawings of rollers and bandages, which can be transferred directly to production.

The program has been registered in the State Register of Databases. It is acquired and used by many Russian enterprises. With its help, calibrations of rolls have been developed and successfully used to produce all kinds of bent profiles.

Currently, work is underway to create a software for designing equipment for a roll forming mill.

Thus, a software package was created that is designed to work on the Internet and implemented in the Visual Studio environment, which made it possible to organize a convenient dialogue between a person and a computer in a dialog mode in accordance with the currently accepted rules for creating an interface in software with the use of automatically appearing forms (windows) containing input fields with explanations, buttons, menus (drop-down lists), graphs, drawings, etc. It is included in the distance learning system.

Description of the framework is available on the website [13]

Programs are provided with prompts with a tree structure with a detailed description of the methods used, explanations and recommendations for any program fragment that appear after clicking or the "Help" button on each form. The descriptions are in the form of files in HTML format using hyperlinks.

Evaluation programs include extensive diagnostics of the information to be entered to detect unrealistic values and to warn about the occurrence of errors.

3 Summary

The developed apparatus of computer software for the use of information technologies allows the most rational approach to improving the production processes of rolling products. The general scheme for solving practical problems with the use of the created complex includes the following main stages: search and study of information on the problem under consideration with the help of the information-learning system and literature; search for data characterizing the material and technology of its production in the database and in production; constructing regression equations from these data using a statistical processing program; execution of analysis and preliminary calculations using the mathematical model of the rolling process; search for the chemical composition and technological parameters that ensure the fulfilment of the required conditions, with the help of an optimization program, the design of tools and equipment elements.

Previously, this approach was tested in the design of technological processes at all stages of production - from the search for the chemical composition of the metal up to the design of the technology that provides the required characteristics of the finished products.

Acknowledgments

Research was supported by Ministry of Education and Science of the Russian Federation, agreement no. 14.578.21.0252, unique project ID RFMEFI57817X0252.

References

Online references will be linked to their original source, only if possible. To enable this linking extra care should be taken when preparing reference lists.

References should be cited in the text by placing sequential numbers in brackets (for example, [1], [2, 5, 7], [8-10]). They should be numbered in the order in which they are cited. A complete reference should provide enough information to locate the article. References to printed journal articles should typically contain:

• The authors, in the form: initials of the first names followed by last name (only the first letter capitalized with full stops after the initials),
• The journal title (abbreviated),
• The volume number (bold type),
• The article number or the page numbers,
• The year of publication (in brackets).

Authors should use the forms shown in Table 3 in the final reference list.

Table 3. Font styles for a reference to a journal article.

Element	Style
Authors	Normal Initials followed by last name
Journal title	Normal Abbreviated
Book title, Proceedings title	Italic
Volume number	Bold
Page number	Normal
Year	Normal In brackets

Here are some examples:

1. O.V. Sokolova, A.E. Lepestov, Forecasting the quality of welded pipe., Science and education. **6** (2011)
2. V.A. Rymov, P.I. Polukhin, I.N. Potapov, *Improving the production of welded pipes* (Metallurgy, Moscow, 1983)
3. B.A. Romantsev, A.V. Goncharuk, N.M. Vavilkin, S.V. Samusev, *Pipe production* (MISIS, Moscow, 2011)
4. V.Ya. Osadchiy, A.S. Vavilkin, V.G. Zimovets, A.P. Kolikov, *Technology and equipment of pipe production* (Intermet Engineering, Moscow, 2011)
5. V.N. Danchenko, A.P. Kolikov, B.A. Romantsev, S.V. Samusev, *Technologies of pipe production* (Intermet Engineering, Moscow, 2002)
6. S.V. Samusev, V.K. Mikhailov, E.Yu. Ivanova, V.V. Terentyev, *Modern methods of calculating the calibrations of the rolls of electric-welding rigs* (MISIS, Moscow, 2004)
7. Yu.M. Matveev, *Theoretical foundations for the production of welded pipes* (Metallurgy, Moscow, 1967)
8. S.V. Samusev, A.N. Fortunatov, *Methods for calculating the stress-strain state during the production of welded pipes in the TESA line* (VF MISIS, Vyksa, 2008)
9. S.V. Samusev, A.N. Fortunatov, N.A. Frolova, N.G. Pashkov, *Methods for calculating instrument calibrations and energy-force parameters of the process of production of welded pipes in the press line and TESA* (VF MISIS, Vyksa, 2006)
10. Yu.M. Matveev, Ya.L. Vatkin, *Calibrating the tool of pipe mills* (Metallurgy, Moscow, 1970)
11. A.A. Ilyushin, *Continuum mechanics* (Moscow University Publishing, Moscow, 1978)
12. V.I. Feodosiev, *Resistance of materials: textbook for technical colleges (9th ed.)* (Science, Moscow, 1986)
13. Information on http://econom.misis.ru

Novel Materials Synthesized from Red Mud, Bagasse, and Bentonite for Gas Treatment by CO_2 Absorption

Nguyen Hoc Thang[1,a], Nguyen Hoang Luong Ngoc[1], Vo Thi Nha Uyen[1] and Pham Trung Kien[2]

[1]*Faculty of Chemical Technology, Ho Chi Minh City University of Food Industry, Ho Chi Minh City, Viet Nam*
[2]*Faculty of Materials Technology, Ho Chi Minh City University of Technology, Ho Chi Minh City, Viet Nam*

Abstract. Carbon dioxide (CO_2) is a gas which causes both impact to atmosphere (one of greenhouse gases) and decrease heating value of gaseous fuel (such as natural gas, biogas, landfill and sewage gas). Hence, there are many investigations to find solutions for gas treatment and carbon dioxide absorption from researchers. Catalyst or synthesized materials is to optimize processes of CO_2 treatment and absorption to obtain the best benefit for factories and community. This study utilized industrial wastes of red mud and bagasse in combination with bentonite to synthesis the novel material (absorbent) responding requirements for the process of gas treatment. More specially, raw materials are impacting negatively on the environment. In which, red mud is solid waste of Bayer process from bauxite mining which is being the hard problem to have solutions for its management and utilization and bagasse is industrial waste of sugar factories. Wet pressing method was applied to form the absorbent samples by mixing red mud, bagasse, bentonite, and water in mixtures. The samples were put in furnace for heating process at 300°C for 2 hours. The final samples were characterized for microstructure using X-ray diffraction (XRD) and fourier transform infrared spectroscopy (FTIR). And then the absorbents were tested for gas absorption capacity of carbon dioxide. The results showed a potential application of the novel absorbent materials for gas treatment.

1 Introduction

Development of industries helped human control everything and produce many convenient products for life. However, the process also caused many problems impact negatively to environment, ecological system, and community. For examples, Bauxite mining industry produced aluminum (Al), aluminum oxide (Al_2O_3), aluminum hydroxide ($Al(OH_3)$) to supply for other fields but the Bayer process discharged a large of red mud (RM) amount. RM or red sludge can cause risks such as impacts to the environment of high alkaline (pH 10-13), RM particles are fine which caused silicosis disease, potential dangers from RM dams, e.g... The annual amount of RM is estimated of 2.7 billion tonnes all over the world and around one million tonnes in Viet Nam (these amounts are continuing to grow up on the project of bauxite exploitation and processing of alumina production). Therefore, the management and utilization of RM waste are necessary requirements for researchers and managers [1]. Several studies have reported the reuse of red mud in another industry such as for pigment [2] and fillers in polymer materials, among others [3-8]. Recently, there also have been growing interests of using red mud as a raw material for building materials [8-11].

Sugarcane is widely grown in tropical countries to supply raw material resource for sugar industry and others. Sugarcane is a grass group which has height from 3 to 4 meters, diameter about 5 cm. The mature cane at harvest stage contains 12–16% soluble sugars, 2–3% nonsugars, 63–73% water, and 11–16% fiber [12]. According to the Viet Nam Sugar Association (VSSA), anually, the sugar factories use more than 15 million tons of sugarcane, discharge around 4.5 million tonnes of bagasse. Therefore, the bagasse must be managed and reused to avoid effected to environment and other production activities. The bagasse can be used as organic fuel for biomass process, furnace, and boiler power generation. However, these processes discharged cacbon dioxide which caused impacts to atmosphere and decrease calorific value of gaseous fuels.

There have been investigations for absorption of CO_2 using zeolites or activated carbon. At the levels discussed here, CO_2 is not toxic and one should not confuse it with the highly toxic carbon monoxide. On the contrary, CO_2 increases plant growth as it provides, together with water, the basic materials needed for photosynthesis. The principal risk of an increase in atmospheric CO_2-concentration IS its impact on the radiation balance of the atmosphere, the so-called greenhouse effect. For biogas or natural gas, carbon dioxide is chemically inert in nature and slightly heavier than air. It has high specific heat capacity and does not support combustion. Methane is known to be the main constituent gaseous fuel (such as natural gas, biogas, landfill and sewage gas) that contributes to the heating value of fuels. The presence of

[a] Corresponding author: thangnh@cntp.edu.vn

© The Authors, published by EDP Sciences. This is an open access article distributed under the terms of the Creative Commons Attribution License 4.0 (http://creativecommons.org/licenses/by/4.0/).

high amounts of carbon dioxide in a mixture of natural gas/carbon dioxide can cause a lower heating value of the mixture when compared with pure natural gas. This, in turn, leads to a reduction in the burning velocity which ultimately affects the performance of the engine [13].

2 Methodology

2.1 Synthesis of Absorbent Materials from Red Mud, Bagasse, and Bentonite.

Raw materials (red mud, bagasse, and bentonite) were prepared in particle size distribution less than 90μm using ball miller and 90μm-siever, and then dried at 110°C for 24 hours. The materials were tested physico-chemical and microstructural properties before mixed together to synthesis the absorbent. The mixtures had the ratio of RM, BG, and Bentonite as listed in the following table:

Table 1. Mix propotions (%, in weight) of the gas absorption materials only.

Mix	RM	BG	Bentonite
Mix 613	60	10	30
Mix 523	50	20	30
Mix 433	40	30	30

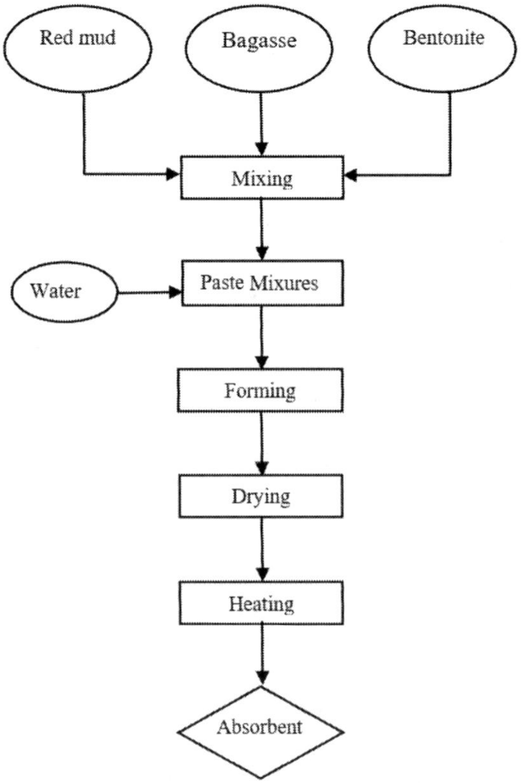

Figure 1. Experimental diagram for production of absorbents

Water was added the mixtures with 40% (liquid per solid phase in weight) to form the samples by wet-press process. All samples were dried at room condition (30°C, 80% humidity) and then heated at 300°C for 2 hours. The experimental processes are shown in Fig.1. Finally, the absorbent materials were tested for phase structure using X-ray diffraction and absorption capacity of CO_2 gas.

2.2 Testing Technique of CO_2 Absorption.

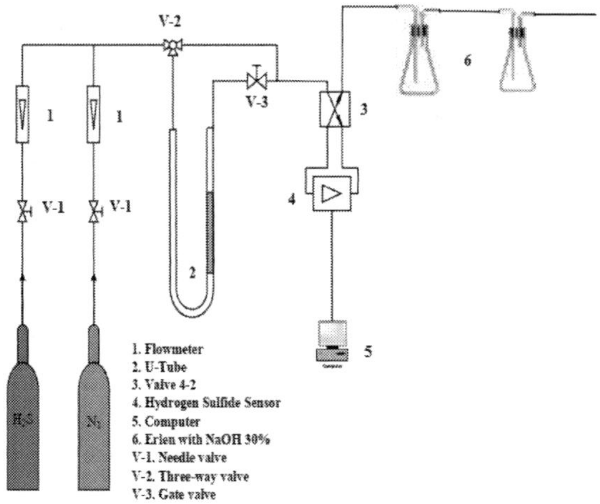

Figure 2. Diagram of CO_2 adsorption system [14].

Fig. 2 shows a system which was used to test for carbon dioxide absorption of the materials synthesized from red mud, bagasse, and bentonite. In this diagram, the main system included two tanks of N_2 and CO_2. The adsorption process was conducted at room condition (30°C, 1 atm). Flowmeter (No. 1) is used to adjust with fixed flow at 3 L/h. The initial concentration of CO_2 gas in the mixture was fixed at 1000 ppm. The synthesized material (absorbent) from red mud and bagasse/ bentonite was contained in the reaction tube of U-shaped Pyrex (0.8 cm in diameter and 25 cm in height) with the mass was a range of 0.18 - 0.19 gram. At input, flow of gas mixture was led into the reaction tube absorbed by the absorbent, and then at output, the concentration of CO_2 gas was measured and recorded continuously by the probe sensor [14].

Based on the adsorption graphs obtained from the computer, the degree of absorbent for the CO_2 gas can be determined by the equation:

$$C_{cap} = \frac{M_C \times 10^{-3}}{m_{absorbent}} \times \frac{P}{RT} \times F\int (C_{in} - C_{out})dt \qquad \text{(mgC/g)} \qquad (1)$$

In which:

M_C = 12 (g/mole) molecular weight of carbon.

$m_{absorbent}$ (g) weight of absorbent used to absorb CO_2 gas.

P = 1 (atm) pressure during CO_2 adsorption.

T = 303 (K) temperature at the time for testing CO_2 adsorption.

R = 0.082 (L.atm/mol.K) the ideal gas constant.

C_{in} (ppm): input concentration of CO_2 in the gas mixture.

C_{out} (ppm): concentration of CO_2 at the point of breakthrough.

F (L/h): flow of input gas.

t (h): time to reach breakthrough levels.

3 Results and Discussion

The results of XRD analysis showed the absorbents have the same structure with high amorphous phases because of broad and noise of XRD backgrounds. There are also peaks with high intensity were detected. They are crystal structures of alumina (Al_2O_3), hematite (Fe_2O_3), and quartz (SiO_2) as shown in Fig. 3.

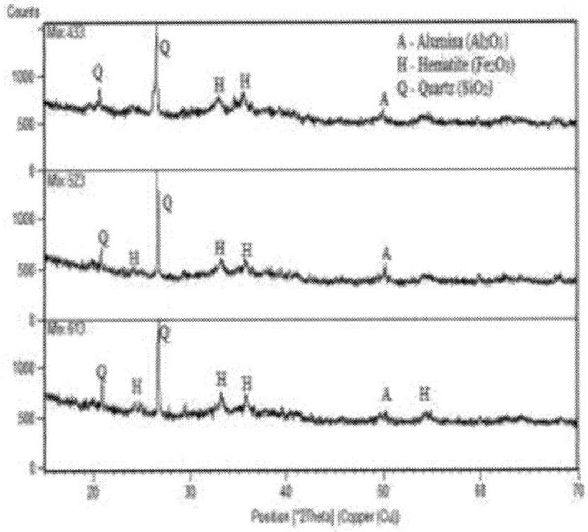

Figure 3. XRD patterns of absorbents (mix 433, 523, and 613).

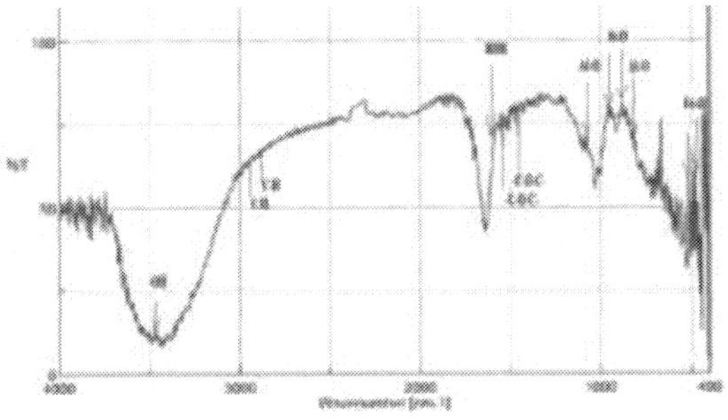

Figure 4. FTIR spectrum of absorbent (mix 433).

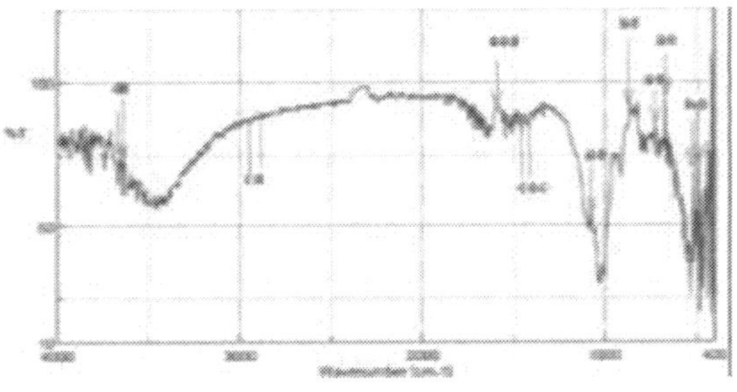

Figure 5. FTIR spectrum of absorbents (mix 523).

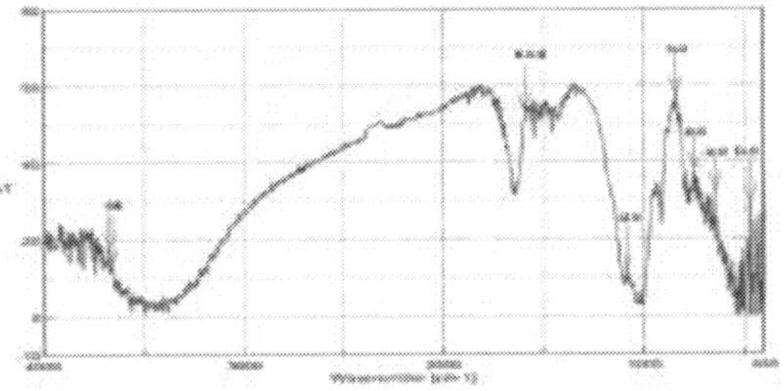

Figure 6. FTIR spectrum of absorbents (mix 613).

The results of FTIR analysis showed the absorbents have the structures in bondings of Si-O, Al-O, Fe-O, C-H, O-H, and others. These groups are characterized by vibrational wavenumber such as 3000, 3800, 1700, 1650, 1030, 1020, 1000, 950, 820, 780, 680, 450 (cm^{-1}) as shown in Fig. 4,5,6.

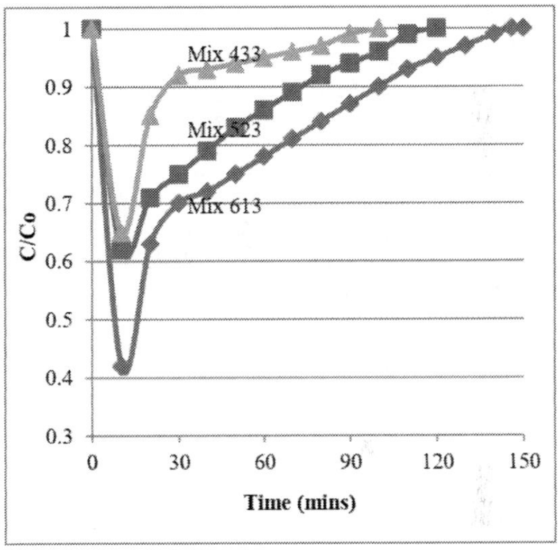

Figure 7. Absorption capacity of absorbents with CO_2 gas.

Table 2. Input and output parameters for CO_2 absorption processes of absorbents.

Parameters	Samples		
	Mix 613	**Mix 523**	**Mix 433**
Capacity [mgC/g]	157.27	180.64	125.37
Max [ppm]	1021.00	1208.00	1023.00
Min [ppm]	301.80	341.60	206.60
Flow of input gas [L/h]	3.00	3.00	3.00
Time of absorption [mins]	125	146	96
Weight of absorbent [g]	0.1827	0.1826	0.1831

The experimental data show that the values of CO_2 absorption capacity are in range of 125.37-180.64 (mgC/g). In which, the sample with 20% BG and 50% RM (Mix 523) has the highest value of the capacity at 180.64 (mgC/g). This sample also has the longest time of absorption at 146 minutes in comparison with the others. Because the process of combustion burned the BG caused increase porosity and concentration of charcoal in the absorbents. Both two factors (porosity and concentration of charcoal) are as main role effect to the absorption capacity of CO_2 gas. In comparison with other studies using red mud and/or bagasse to syntheses absorbents, the values obtained in this research are better when evaluated in the same conditions in recent years.

4 Conclusions

This study proposes a method to utilize red mud (Bauxite mining industry) and bagasse (sugar production process from sugarcane) as raw materials for synthesizing the absorbents in combination with bentonite. The results showed that the materials have amorphous and crystal phases in the structures. Using the CO_2 adsorption system and the equation for determination of absorption capacity, the absorbents have high capacity at 180.64mgC/g for 146 minutes. The values are better than other research using red mud and bagasse. Further researches will be carried out for optimization of mix design among raw materials as well as evaluation effect of heat treatment to absorption capacity of CO_2 and others gas. Analysis of microstructures and surface area of absorbents will be conducted using BET, FTIR, and SEM for explanation of the experimental data.

References

1. H.T. Nguyen, T.K.A. Le, N.H. Nguyen, Q.L. Nguyen, T.K. Pham, M.A.B. Promentilla, 23rd Regional Symposium on Chemical Engineering (RSCE2016) (2016)
2. S. Chandra, *Waste materials used in concrete manufacturing*, Noyes, USA, (1997)
3. S.S. Amritphale, A. Anshul, N. Chandra, N. Ramakrishnan, Journal of the European Ceramic Society **27** (2007)
4. S. Ordonez, H. Sastre, F.V. Diez, Journal of Hazard Materials **B81** (2001)
5. S. Shushil and V.S. Batra, Applied Catalysis B: Environmental **81** (2008)
6. H.T. Nguyen, V.T.H.Q. Pham, T.P. Dang, and T.K. Dao, AIP Conference Proceedings **1954:1** (2018)
7. A.K.R. Sumabat, A.J. Manalac, H.T. Nguyen, M.E. Kalaw, R.R. Tan, M.A.B. Promentilla, Chemical Engineering Transaction **45** (2015)
8. H.T. Nguyen, T.N. Le, V.T.H.Q. Pham, T.P. Dang, T.K. Dao, and V.P. Nguyen, AIP Conference Proceedings **1954:1** (2018)
9. M.A.B. Promentilla, H.T. Nguyen, T.K. Pham, H.Hinode, F.T. Bacani, S.M. Gallardo, Waste biomass valorize **7** (2016)
10. H.T. Nguyen, T.K. Pham, M.M.A.B. Abdullah, MATEC Web Conference **97** (2017)
11. M.A.B. Promentilla, M.E. Kalaw, H.T. Nguyen, K.B. Aviso, R.R. Tan, Computer Aided Chemical Engineering **40** (2017)
12. R. Perez, *Feeding pigs in the tropics, Food and Agriculture Organization of the United Nations*, (1997)
13. S. Bari, Renewab. Energy **9** (1996)
14. H.T. Nguyen, V.T.H.Q. Pham, T.N. Le, T.P. Dang, and N.K.T. Nguyen, AIP Conference Proceedings **1954:1** (2018)

Performance analysis of the bump beams made of High Strength Steel and Carbon Fiber reinforce plastics

Yi Yao[1], Quan Yuan[2,3,a*], Sihuan Fu[2, 4]

[1] *Pan Asia Technical Automotive Center Co., Ltd. Shanghai 201206, China*

[2] *School of Automotive Engineering, Wuhan University of Technology, Wuhan 430070, China*

[3] *Hubei Key Laboratory of Advanced Technology for Automotive Components, Wuhan 430070, China*

[4] *SAIC GM Wuling Automobile Co., Ltd, Liuzhou 545007, China*

[a*] *Corresponding author, email:864177940@qq.com*

Abstract: Lightweight and safe are the key research directions of the automobile industry. Bump beam is an important part of the car safety protection and it's lightweight design has drawn much attention. In this paper, the static strength and the low-speed impact performances of two types of bump beam with the same structure, which made of carbon fiber reinforced plastics (CFRP) and high strength steel (HSS), were investigated. Under the condition of static press, frontal and 40% offset impact, part structural stiffness, component crush intrusion, energy absorption were comparison analyzed. In order to guide the product design, the deformation and failure of CFRP beams with five different laminate structures were simulated under frontal and 40% offset impact. The results indicate that, with the same structure, the properties of CFRP bump beam are generally superior to HSS product, more conducive to lightweight and the optimum design of the laminate is [0 ° / 60 ° / 120 °]

1 Introduction

Body lightweight and safety is an important aspect of automotive technology [1]. Bump beam is one of the main crash safety structures in the front of the car, and its lightweight technology has drawn wide attention [2].

In recent years, a large number of researchers have carried out research work on the crashworthiness and lightweight design of vehicle bump beam. M.M. Davoodi Ramin Hosseinzadeh and Zhang Haiyang etc[3,4,5] studied the structural design and composite design of lightweight bump beam. The results show that, for satisfying the requirements of the crashworthiness performance, lightweight effect is more obvious when use Composites; Dong-Kyou Park and Zhang Jing wen etc[6,7] established a numerical analysis model to analyze the performance of HSS bump beam; Sun Huiwei etc[8] conducted simulation analysis to the structure of the bump beam, put forward a structural improvement program to optimize the product performance; Hong Qiucai etc [9]with the acceleration wave form, front bulkhead intrusion and the deformation pattern of front side rail as the development objective of structure, the optimization design of bump beam and body parts is carried out.

Due to carbon fiber composite materials in the obvious advantages of lightweight, it has broad application prospects in the automotive industry [10]. In order to explore the design and development of

© The Authors, published by EDP Sciences. This is an open access article distributed under the terms of the Creative Commons Attribution License 4.0 (http://creativecommons.org/licenses/by/4.0/).

https://doi.org/10.1051/matecconf/201820703006

lightweight bumper, this paper comparatively studied the crashworthiness of bump beam which made of CFRP and HSS, and simulated the performance of CFRP bump beam with different plies design, for reference in engineering design.

2 Set the constraints conditions

The main analysis of this paper is the static load and the regular pattern of the low-speed crash of the same type of bump beam in different materials. Thus using a simplified model of the bump beam, the thickness is 3.34 mm. Based on experience and calculation, the mesh size is initially determined to be 10mm, while some key sections is controlled to 5mm.

The high strength steel is selected No. 20 material model from the LS-DYNA material library[11], the composite material selected carbon fiber material No. 54 model.

Collider intrusion can not exceed the allowable intrusion. In this paper, the size of the bump beam sets the maximum intrusion amount of the impact beam to be 50mm when the low-speed frontal impact, and the low-speed offset impact value should be 25mm.

In the static analysis, the loading on the red part is given a total of 8 points with a load of 200N at each point, while the yellow part shows the constraint. The point of the 6 degrees of freedom all constraints (as shown in Fig. 1 (a))and using the same constraints and loads for static analysis of two types of bump beams ,which made of high strength steel and carbon fiber composite. In low-speed frontal and 40% offset impact, in order to simplify the process of solving the problem, the following only select the bump beam system for analysis, extraction of bump beam and energy-absorbing box as the crash analysis object, set the boundary conditions of the bump beam 6 degrees of freedom constraints. And giving the collider a mass of 1200 kg and constraint the degree of freedom of rotation of the collider about these three axes (as shown in Fig. 1(b) and(c)). Set the crash speed 4km / h, when occurred frontal impact; Under the condition of 40% offset impact, the collider crashed into the bump beam at a speed of 2.5 km / h.

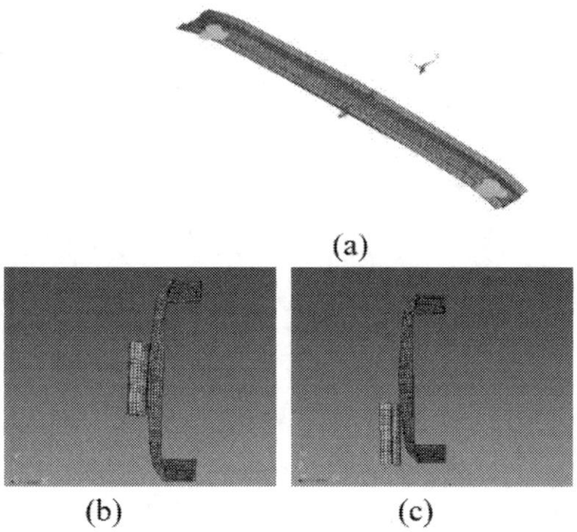

Figure 1 Constraints and Loads in three Conditions

(a): static load; (b) frontal impact; (c) 40% offset impact

3 Results and analysis

3.1 Static Load Conditions

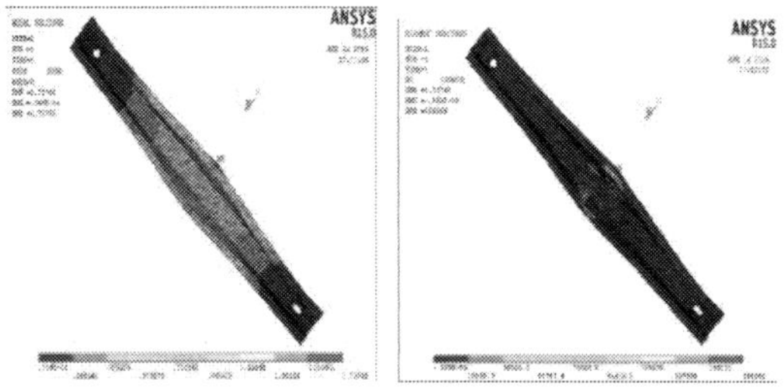

Figure 2 Maximum displacement cloud and stress distribution of CFRP bump beam

Due to the two materials are different, the stiffness is usually stress analysis, the results of larger errors. The general principle of equal stiffness to be analyzed, the maximum displacement as the research object for analysis. It can be seen from the Fig.2 that the maximum displacement of the HSS bump beam is 1.84 mm and the maximum displacement of the CFRP bump beam is 1.74 mm. Under the same constraint and load conditions, the CFRP bump beam not only reduces the weight by 64.12%, but also reduces the displacement.

3.2 Frontal Impact

Displacement comparision chart of high strength

https://doi.org/10.1051/matecconf/201820703006

steel bump beam and carbon fiber composite beam under frontal impact are shown in Fig.3. The maximum deformation of two kinds of bump beam are 44.35mm and 41.12mm respectively, which are less than the allowable displacement of 50mm and meet the requirements. However, the displacement of carbon fiber composite material smaller, more superior performance.

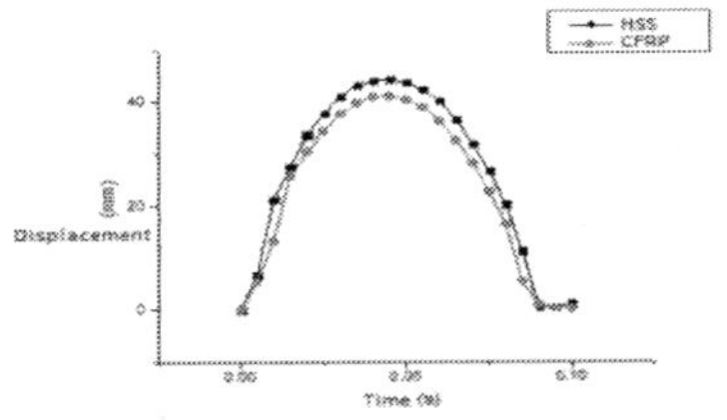

Figure 3 Frontal Impact Displacement Comparison Chart

Fig.4 shows the variation of system energy under frontal impact conditions. The total energy of the system is equal to the initial kinetic energy of the collider being 550J and remains roughly the initial value during the collision. During the collision, the kinetic energy was continuously transformed into the internal energy of the system, about 42 ms, the internal energy of the two materials reached the peak values, respectively 548 J and 547 J. The proportion of total energy is roughly the same and with a small difference. Subsequently, the internal energy continued to decrease. At 90 ms , the elastic deformation of bump beam has basically recovered. The internal energy and kinetic energy were maintained at a stable value.

Although the energy curves of the two materials are almost the same, the numerical difference is small, but the quality of the carbon fiber composite material bump beam is far less than the high strength steel bump beam, resulting in large difference between specific energy absorption, as shown in Fig.5. The specific energy absorption of composite material is 109J / kg, while the high strength steel's specific energy absorption is only 50J / kg.

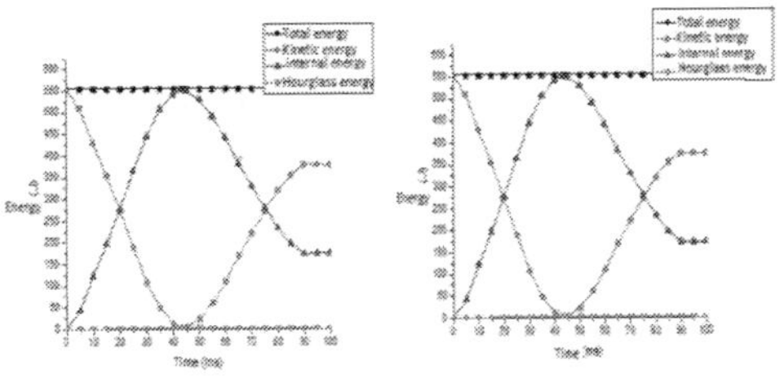

(a) (b)

Figure 4 Frontal Impact Energy Change Diagram(a)HSS; (b) CFRP

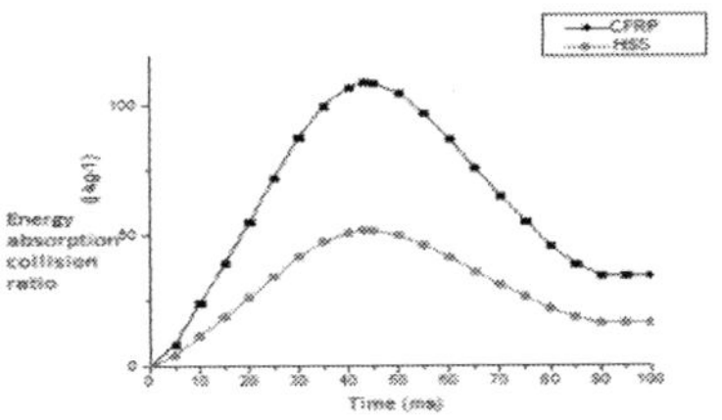

Figure 5 Frontal Impact of Two Materials Specific Energy Absorption Comparison Chart

3.3 40% Offset Impact

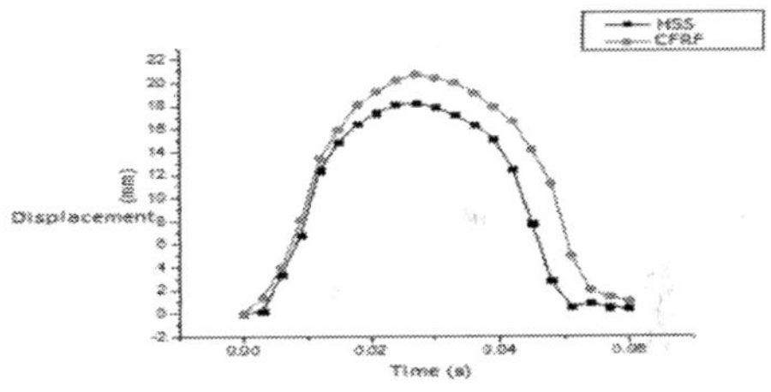

Figure 6 40% Offset Impact Displacement Comparison Chart

As shown in Fig.6, the maximum deformation of high strength steel bump beam and carbon fiber composite beam are 18.35mm and 20.92mm, respectively, which are less than the allowable displacement of 25mm, meet the requirements. But the deformation of carbon fiber composite material is slightly larger.

In 40% offset impact condition, the energy curves of various parts of the system over time are roughly the same as the frontal impact. The total energy of the system is equal to the initial kinetic energy of the collider is about 221J, and the value is basically maintained during the collision. During the collision, the kinetic energy was continuously transformed into the internal energy of the system, about 23 ms, the internal energy of the two materials reached the peak values, respectively 219 J and

https://doi.org/10.1051/matecconf/201820703006

218 J. The proportion of total energy is roughly the same, simultaneously, the energy of structure difference is minimal; Then, the bump beam system continuously changes back to the initial shape, and the internal energy decreases. At 43 ms of collision, the deformation of the bump beam disappears, and the internal energy and kinetic energy remain at a certain value. However, the specific energy absorption of carbon fiber composites is 42J / kg, higher than the specific energy absorption of high strength steel 21J / kg.

4 Stacking Design of CFRP Bump Beam

Before analyzing the bump beam layer structure, the multi-structural analysis of static load condition determines a fixed laminate thickness to simplify the analysis process and avoid the complexity of the analysis process and increase the analysis time due to the change of the number of layers. In general, ply orientation of 0 ° carry axial load, ± 45 °carry shear loads, and 90 °carry lateral loads and control Poisson's ratio[12]. Therefore, according to the comparative analysis of the above two materials, under the condition that the number of layers of carbon fiber composite material is 20, applying constraint and load according to the different working conditions of the beam in accordance with the comparative analysis of the beam with high strength steel and carbon fiber composite material by changing stack sequence.

In this paper, selecting stack sequence are layer1 [0°/90°]nt, layer2 [0°/60°/120°]nt, layer 3 [0°/90°/45°/-45°]nt, layer4 [45°/-45°]nt, layer 5 [0°/45°/-45°]nt.

In the frontal and 40% offset impact conditions, the maximum displacement and stress curve under five kinds of layer design, is shown in Fig.7 and Fig.8:

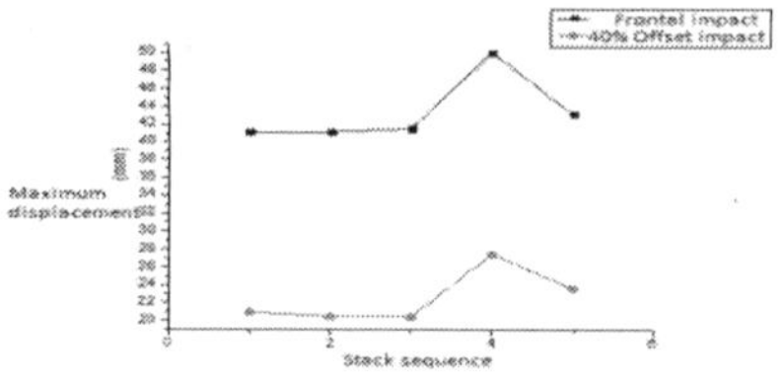

Figure 7 Frontal and 40%Offset Impact Maximum Displacement Comparison Chart

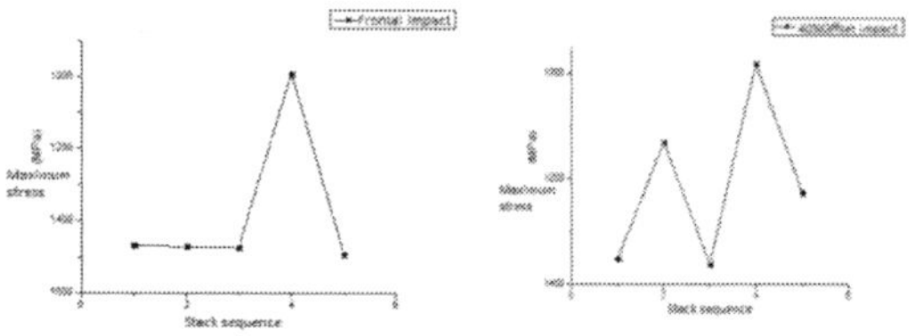

Figure 8 Frontal and 40%Offset Impact Maximum Stress Comparison Chart

From the comparison of the above graphs, we can see that for the maximum displacement, under the conditions of low-speed frontal and 40% offset impact, the maximum displacement changes little with stack sequence, except for the layer order 4 and the maximum displacement is smaller at [0 ° / 90 °], [0 ° / 60 ° / 120 °] and [0 ° / 45 ° / -45 ° / 90 °]. For the maximum stress, under the low-speed frontal impact, the maximum stress of layer 4 hardly changes with stack sequence, and the maximum stress reaches the minimum under the layer of [45 ° / -45 °]. Under the condition of 40% offset impact, from the analysis results, it can be seen that the maximum stress of other lay-up sequences is basically the same except that the maximum stress of the layer of [0 ° / 60 ° / 120 °] and [45 ° / -45 °] are smaller. Therefore, the optimal lay-up order is [0 ° / 60 ° / 120 °].

5 Summary

(1) In the same structure, CFRP bump beam is superior to HSS bump beam in terms of static stiffness, crash intrusion, energy absorption and specific energy absorption to meet the performance requirements of the beam, and the quality is lighter.

(2) Among the various layer design schemes that meet the static stiffness and low-speed collision conditions, the optimal lay-up order is [0 ° / 60 ° / 120 °].

References

1. Society of Automotive Engineers of China. Trace research on world automotive technology development (lightweight) [M]. Beijing: Beijing Institute of Technology Press, (2013).
2. Z.G. An, Y.Y. Zhao, The structural strength design of the light weight of the car crash beam of

anti-collision beam [J]. Computer Simulation, **33**, 7 (2016),

3. M.M. Davoodi, S.M. Sapuan, D. Ahmad. Concept election of car bumper beam with developed hybrid bio-composite material [J]. MATER DESIGN, **32**, 9 (2011)
4. Ramin, Hosseinzadeh, Mahmood M. Ramin Hosseinzadeh, Larry B.Lessard. Parametric study of automotive composite bumper beams subjected to low-velocity impacts [J]. COMPOS STRUCT, **68**, 9 (2005)
5. H.Y. Zhang, H. Fu, Z. Xiao, B. Li, S.Y. Duan. Lightweight research on bumper beam of fiber reinforced composite [J]. FIBER COMPOS, **10**, 7 (2017)
6. D.K. Park. A development of simple analysis model on bumper barrier impact and new IIHS bumper impact using the dynamically equivalent beam approach[J]. J MECH SCI TECHNOL, **25**, 8 (2011)
7. J.W. Zhang, T.Q Fan, Q.J. Zhao, Y. Li. Crash performances comparison and selection of bumper beam with different lightweight methods [C]. Proceedings of the 11th International Forum of Automotive Traffic Safety, (2014)
8. H.W. Sun , X.C. Zhang, Z.F, Liu, Y. Xin, Z. Yang Optimization of front anti-collision beam structure in offset crash [J]. AUTO ENG, **6**,4,(2017)
9. Q.C. Hong, W.G, Liu, D.Y. Zhou, A Study on Body Structure Lightweighting in the Transfiguration Development of a Sedan [J]. AUTOMOT ENG, **39**, 5,(2017)
10. X.P. Yang, Z.B. Huang, Z.Y, Zhang, H.P. Yang, Y.J. Zhou, Application progress of carbon fiber reinforced polymers for energy saving and emission reduction [J]. MATER REV, 5 (2010)
11. Z.Y. Hu, B.Q. Zeng, S.G. Xie. Vehicle Safety Simulation and Analysis Based on LS-DYNA and HyperWorks [M]. Beijing: Tsinghua University Press (2011)
12. J. Zhang. Structural Analyses and Layer Optimization of Large Scale Wind Turbine Blade [D].Wuxi: Jiangnan University, (2013)

A Study on Proposal of Flank Wear Criterion by Using a Built-in Current Sensor when Manufacturing the Mold Materials in a Smart Machine Tool

Seung-Yub Baek[1,a], Sung-Taek Jung[1], and Dae-Yu Park[2]

[1]*Induk University, Department of Mechanical Design Engineering, Nowon-gu, Seoul, Republic of Korea*
[2]*Hwacheon Machine Tool Co., Ltd, Gangseo-gu, Seoul, Republic of Korea*

Abstract. Recently, it has been increased with respect to the safe and reliable operations in industry of machine tools and intelligent of the machine tool has consistently been developing in term of an unmanned manufacturing. For such realization, diagnosis monitoring of machining must be carried out while being processed in real-time. When tool wear is reached to criteria of flank wear and crater wear, the tools must be changed to new tools for improving the manless rate of operation. However, time of tool change was when spark generated because of wear about 0.3 mm on a flank face during manufacturing in the field. So, built-in sensor system in a smart machine tool must be necessary for high efficiency unmanned of manufacturing. As mentioned earlier, the various technique for measuring the tool wear was already defined such as sensing of acoustic emissions, vibrations, sounds, currents, cutting force, and other. The representative one of measuring method is current signal, which is used as a representative index of tool state. In this study, we carried out the proposal of tool wear criterion by using built-in wireless current signal system when manufacturing the mold materials of KP-4M and it was investigated via smart machine tools.

1 Introduction

Nowadays, competitive business environment, companies are facing as a goal with big data and developing decision-making algorithms for improving the productivity. Many manufacturers are trying to fuse with between previous machine tools such as a milling, turning, drilling and other, and smart system, which included big data, artificial intelligence and automation. To realizing the smart machine tools, it must be monitoring in real-time and self-diagnosing with respect to the core parts and tools. Especially, tool wear monitoring is one of the most important techniques to be developed for preventing tool breakage and improving the product quality.

Tool condition monitoring is essential to obtain good quality product and core parts. Also, tool wear monitoring techniques are classified as direct and indirect methods [1-2]. Direct methods measure real wear changes of cutting edge by machine vision [3] and scanning electron microscope (SEM) [4]. So, these methods are affected by mist, manufacturing environment, reflecting condition and condition of machine tools. On the other hand, indirect methods are such as acoustic emissions, vibrations, sounds, currents, cutting force, and they are utilizing as tool to diagnose the tool wear and machine tools and representative research case is an in the following.

Sadettin Orhan developed a monitoring system and tool wear evaluation methods by using a vibration measurement when turning by end milling of AISI D3 cold work tool steel with 35HRC hardness [5]. The condition monitoring of milling tool was used to predict tool wear which is by linear relation via acoustic spectrum by C. S. Ai [6]. Also, Jun-Hong Zhou studied the measurement of tool wear using the acoustic emission from an embedded sensor for computation of features and prediction of tool wear [7]. However, previous methods are unsatisfactory because it can be checked after generating the wear or breakage.

To sensitively detect, in drilling process, Karali Patra [8] developed prediction model by regression model of flank wear through an artificial neural network based on the current signal. The measuring methods by the cutting force were developed for fault detection diagnosis based on an observer model of an uncertain linear system from [9]. But, these are not suitable for detecting both tool wear and machine tool conditions, and a built-in sensor is required to satisfy this. Among proposed sensing system, the monitoring of tool wear and machine tool is highly detected via built-in current sensors and flank wear can be determined by the current signal criterial.

In this study, we were able to propose the criterion with tool wear by using built-in current sensors and examined by using the measurement of two-dimensional image. In the future, it is necessary to develop algorithms suitable for diagnosing each case by built-in sensor system.

[a] Corresponding author: sybaek@induk.ac.kr

© The Authors, published by EDP Sciences. This is an open access article distributed under the terms of the Creative Commons Attribution License 4.0 (http://creativecommons.org/licenses/by/4.0/).

2 Experimental method

The experiment method was carried out by the conditions which are utilizing the mold industry and test material is a KP-4M (150 mm x 150mm x 150mm). Therefore, cutting condition has been mainly selected the suitable cutting speed and feed rate due to its high hardness of mold materials as shown in Table 1. Fig. 1 shows a change in the schematic illustration of the tool load during machining by getting the current signal transmit data. The experiment of tool wear was performed by using tools and workpiece and it indicated working distance and down milling method in Fig. 2. The insert tool type was decided as an experimental tool because end mill is very expensive than insert type tools. The wear characteristics were analyzed by this experiment in environment of dry condition without lubrication. The analysis was performed to show influences on tool wear characteristics by rough process. Finally, wear characteristics was investigated by using built-in current sensor and two-dimensional microscopes.

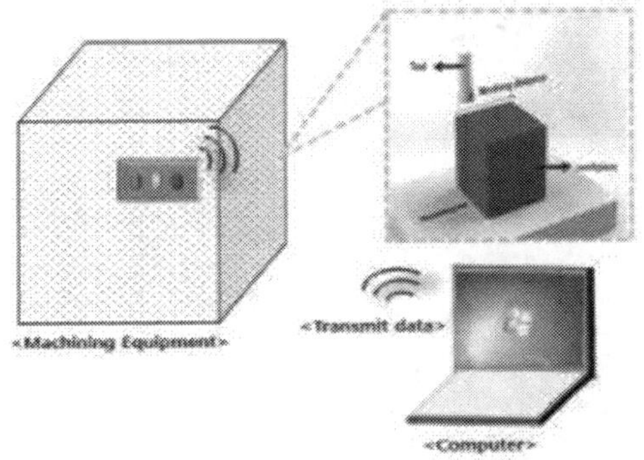

Fig. 1 Schematic illustration of gathering data of cutting load signal

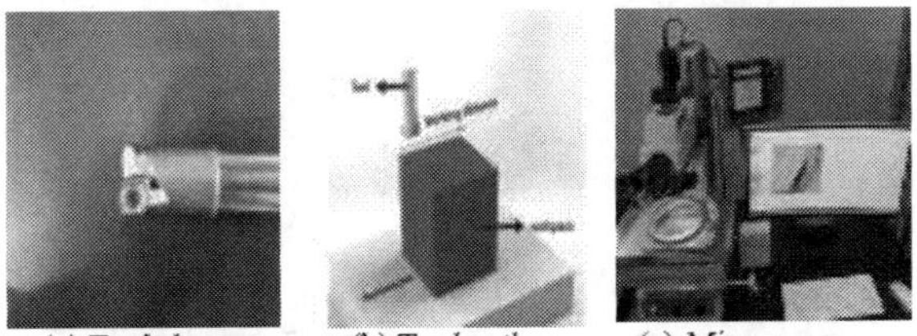

(a) Tool shape (b) Tool path: one-way (c) Microscopes

Fig. 2 Photograph of tool geometry and experiment method for analyzing wear characteristics

Table 1. Cutting conditions of KP-4M.

No	RPM (min^{-1})	Feed (mm/min)	Depth of cut (mm)
1	6,700	6,000	0.2
2	6,700	5,500	
3	5,400	6,000	
4	5,400	5,500	
5	4,800	6,000	
6	4,800	5,500	

2.1 Tool wear characteristic analysis by using current signals

We carried out analysis of the load signal with respect to its characteristics of the tool wear depending on built-in current signals as shown in Fig. 3. This paper were derived the correlation between tool wear and load signal. Tool wear normally is decided on criteria of 0.3 mm according to the international standard. So, we indicated the condition of tool wear from initial current signals to final current signals for deciding the tool life in wear zone. Because cutting speed is higher in case of 6,700 RPM, the current signals indicated high level due to increasing the coefficient of friction. Current signals continuously gathered until it reached to specific wear level. Especially, in case of condition 6, initial value of load signal is 5.5 % because feed rate is lower than other cutting condition and increasing current load lead to the tool wear caused by forced thermal convection. Experiments of Previous tool wear needed worker to consistently check the tool condition in the field. Also, it must be reduced accessing to the machine tools in terms of unmanned manufacturing and the safety. In this experiment, we proposed the assessment criteria of tool wear by current signal reference and it was investigated. Results are as in the following. The current signal is about 6.5 to 7 % in RPM 6,700 and is about 4.5 to 6 % in RPM 5,400. Finally, the current signals are about 4.5 to 5 % in RPM 4,800. These results showed that it has been experimentally verified that the higher the RPM, the higher the current signal level, which is directly related to tool life.

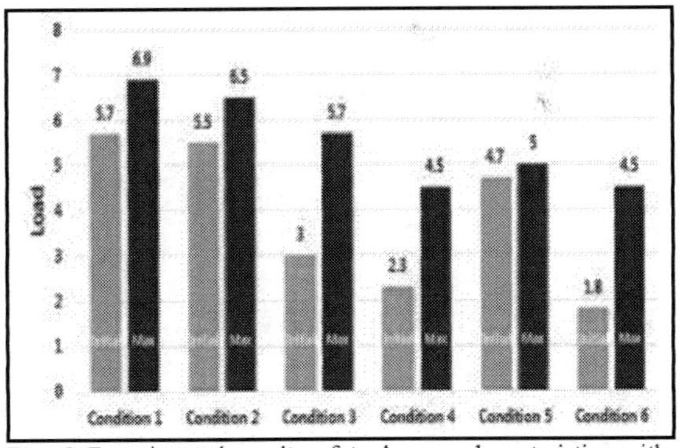

Fig. 3 Experimental results of tool wear characteristics with respect to the current signals

2.2 Tool wear characteristic analysis by using two-dimensional microscopes

The overall characteristics of tool wear were shown by using the two-dimensional microscope. The results of machining test were as in the following. As shown in Fig. 4, 6, tool wear with flank and crater and cutting time is indicated and these results by increasing the cutting speed had critical effects on the tool face because of its heat. Furthermore, the tear type chip is adsorbed onto cutting edge because of built-up edge. Their condition from 3 to

6 was sensitively detected because cutting speed is lower than other condition. Also, the cutting time is comparatively improved in term of tool cost and however, the condition from 3 to 6 is slipped in view of product productivity. Latest trends are required improving the productivity efficiency. So, we were able to find the optimized cutting condition based on the experimental results. Moreover, behavior of tool wear was changed according to the materials and geometry of tool. Such conditions used in this experiment by built-in current sensor can very sensitively be detected with respect to the generating of wear by using rough process and mold materials. In the case of the characteristic of the tool wear, it is difficult to confirm all step because the characteristic change with respect to the heat sensitively occurs. Particularly, the tool wear characteristic of the nonlinear section may be different and the tool wear was observed in the linear section. We carried out characteristic analysis in terms of tool wear by the tool wear test time as shown in Fig. 5. And, as a result, we found the tool wear point by using current signal and criterion of flank wear proposed via our experiments without quantitative measurement.

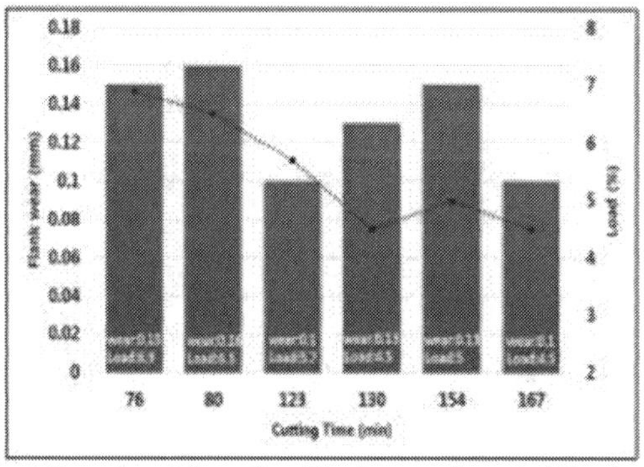

Fig. 4 Experimental results of tool wear characteristics in the time

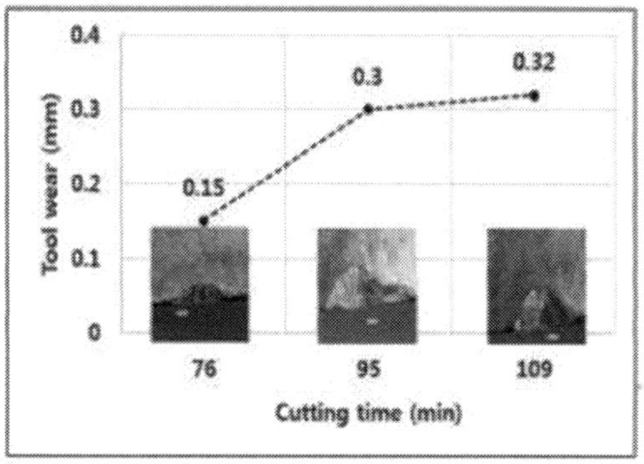

Fig. 5 Experimental results of tool wear characteristics in the time

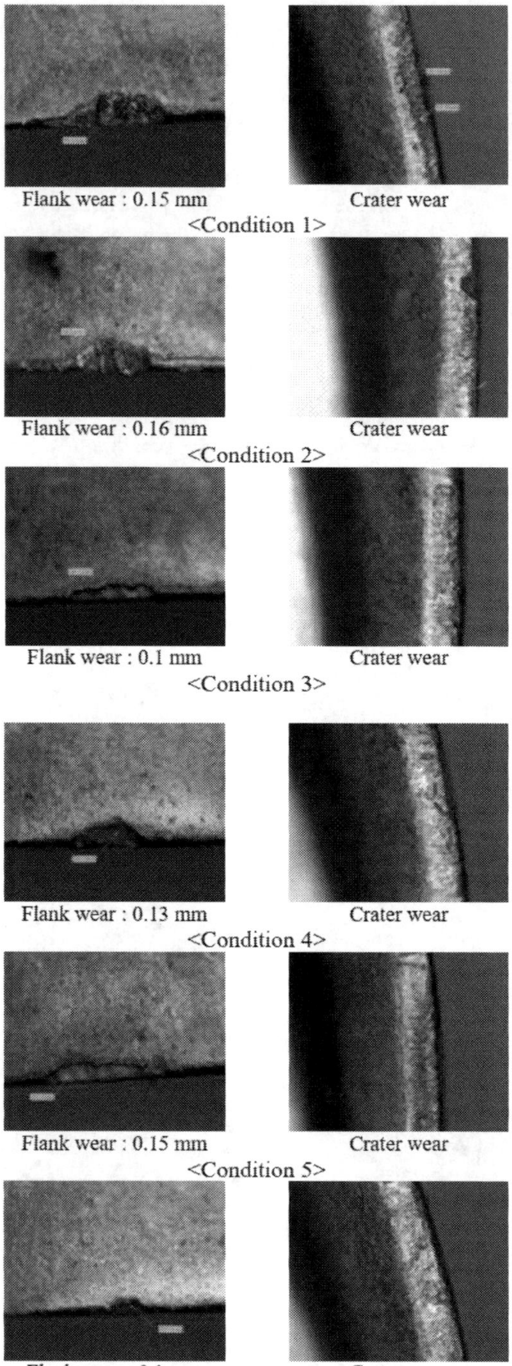

Fig. 6 Photograph of tool shape in microscopes

3 Summary

In this study, we presented a decision criteria that can predict the flank wear by using built-in current sensor, and they verified condition of flank wear due to its real-time monitoring by using two-dimensional microscopes.

These experimental results showed that as the cutting speed increases, the load of the current signal increases, and the feed rate also tends to be similar. This is analyzed as the effect of increased frictional force due to the gradual rise of tool wear due to forced thermal convection. The specified criteria of this current signal became the guide line for flank wear, and the load gradually increased as the wear progressed. If the existing wear detection was performed directly by the dimensional microscopes, the current sensor could be installed inside the smart machine tools, allowing more sensitive evaluation than the existing one. This method can be used effectively because it depends on the amount of movement of the current level even if the cutting condition changes. Furthermore, built-in sensor systems had advantage. As s first, it is possible to monitor both tool condition and machine tools, second, it can notify to worker many information with respect to manufacturing situation. Finally, these manufacturing information will be piled up at a big data server and leading to information and communications technology and internet of things.

Acknowledgements

This work is supported by the Ministry of Trade, Industry and Energy Republic of Korea. (No. 10060188)

References

1. Haber, R. E., Jiménez, J. E., Peres, C. R., & Alique, J. R, An investigation of tool-wear monitoring in a high-speed machining process. ,Sensors and Actuators A: Physical, 116**(3)** 539-545 (2004).
2. N. Ghosh, Y. B. Ravi, A. Patra, S. Mukhopadhyay, S. Paul, A. R. Mohanty, A. B. Chattopadhyay, Estimation of tool wear during CNC milling using neural network-based sensor fusion. MSSP, 21**(1)**, 466-479, (2007).
3. T. Pfeifer, L. Wiegers, Reliable tool wear monitoring by optimized image and illumination control in machine vision. Measurement, 28**(3)**, 209-218, (2000).
4. M. Cabibbo, A. Forcellese, R. Raffaeli, M. Simoncini, Reverse engineering and scanning electron microscopy applied to the characterization of tool wear in dry milling processes. Procedia CIRP, **62**, 233-238, (2017).
5. S. Orhan, A. O. Er, N. Camuşcu, E. Aslan, Tool wear evaluation by vibration analysis during end milling of AISI D3 cold work tool steel with 35 HRC hardness. NDT & E International, 40**(2)**, 121-126, (2007).
6. C. S. Ai, Y. J. Sun, G. W. He, X. B. Ze, W. Li, K. Mao, The milling tool wear monitoring using the acoustic spectrum. Int J Adv Manuf Technol, 61**(5-8)**, 457-463, (2012).
7. J. H. Zhou, C. K. Pang, Z. W. Zhong, F. L. Lewis, Tool wear monitoring using acoustic emissions by dominant-feature identification. IEEE Transactions on instrumentation and measurement, 60**(2)**, 547-559, (2011).
8. K. Patra, S. K. Pal, K. Bhattacharyya, Artificial neural network based prediction of drill flank wear from motor current signals. Applied Soft Computing, 7**(3)**, 929-935, (2007).
9. S. N. Huang, K. K. Tan, Y. S. Wong, C. W. de Silva, H. L. Goh, W. W. Tan, Tool wear detection and fault diagnosis based on cutting force monitoring. International Journal of Machine Tools and Manufacture, 47**(3-4)**, 444-451, (2007).

Study of feature extraction method of multi-information source for continuous casting process parameters

Xianqi Liang[1, a], Zhufeng Lei[1], Wenbin Su[1], Fali Zhong[1], Hang Zhao[1], Yanni Zheng[1] and Mingjie Ding[1]

Abstract. The control of the roll gap of the segment is one of the key links to ensure the quality of cast billet. In this paper, the big data in traditional continuous casting production operations is studied through in-depth experimental comparative analysis of linear and nonlinear dimension reduction method. The method is suitable for continuous casting to obtain the data of the dimension reduction. The method of principal component analysis is improved by using standardized data increment method. A faster and more efficient method of dimension reduction is obtained when the unrelated data, training time and reconstruction error are removed. Actual data simulation results show that this method is more efficient and suitable for continuous casting than any other dimension reduction method.

1. Introduction

In the modern steel industry, high-efficiency continuous casting technology, including key continuous casting equipment and new processes, has become the most internationally competitive core technology. The process of continuous casting is a complex continuous phase change process. There are many control aspects that affect the quality of slab. The key technology and core research of high quality steel continuous casting process are mainly focused on the precise control of mould, sector and composition of liquid steel [1]. Among them, the control of sector roll gap is one of the key links to ensure the quality of slab [2]. Not only the establishment of a dynamic and adaptive prediction model for the roll gap value of the sector continuous casting machine but also the real-time adaptive adjustment of the sector roll gap value according to the actual working conditions has important theoretical and practical values for improving the quality of slab.

1.1 motivation()

Due to the complexity of the continuous casting process as well as the non-intuitive and variability of the internal condition of the high-temperature slab, there are many factors affecting the roll gap parameter in the production process, resulting in many measurement points required for data collection. Therefore, the sampling frequency is higher and the sampling time is longer, so that massive data can be obtained [3.4.5]. For example, Handan Iron and Steel Group installed 6000 measuring points in a continuous casting production line to collect data every 3 seconds, so the amount of monitoring data is extremely large. Because of a large amount of data, how to refine the data related to the roll gap parameter and how to establish the roll gap prediction model service are the key to handle these data. Therefore, how to research and use the new method for data mining process parameters of continuous casting is a problem faced by this research.

Traditional data processing is usually based on the traditional method of dimension reduction, such as principal component analysis (PCA) and linear discriminant analysis (LDA). The process parameters of continuous casting have a lot of problems; for instance, multi-information source, real-time and large amount of data. Hence the traditional method of dimension reduction obviously cannot meet the requirements.

The third literature introduces an incremental principal component analysis (Candid Covariance-free Incremental PCA, CCIPCA) that does not compute covariance, which converges quickly and reduces computational cost when dealing with high-dimensional data. The fourth article introduces an incremental principal component analysis (Chunk IKPCA, CIKPCA), which decomposes bulk data. This method performs an eigenvalue calculation on a large data block and reduces the computational cost. MKaur[26] used LDA as the classification method from accuracy, sensitivity and prediction accuracy in the process of ECG signal analysis, and finally found that the isometric rotation method has the highest accuracy.

1.2Contribution

Many predecessors did a lot about the method of feature reduction which is better to solve the problem of real-time data and large amount of data, but some of them considered less about the accuracy and effectiveness of the reduction. The traditional linear dimension reduction method (PCA) and the emerging nonlinear dimension reduction methods (t-SNE, LLE, LTSA) were compared on the requirement of multi-information source, real-time and large amount of data of continuous casting production process parameters. Based on that, the dimension reduction methods which are suitable for predicting the roll gap parameters in the continuous casting process are analyzed and improved. Finally, a standardized robust incremental principal component analysis (Normalization Robust Incremental Principal Component Analysis, NRIPCA) was found. Before the reduction of the data of continuous casting, this method

© The Authors, published by EDP Sciences. This is an open access article distributed under the terms of the Creative Commons Attribution License 4.0 (http://creativecommons.org/licenses/by/4.0/).

firstly performs outlier monitoring and elimination and normalizes the processed data. This method considers not only the accuracy of feature reduction but also the validity of feature reduction.

2 Methodology(MATEC Web of Conferences)

Feature reduction removes irrelevant and redundant information from features based on guidelines. It maintains the important features and it can effectively reduce the data dimension for overcoming the 'dimension disaster' problem. This section compares the traditional linear dimension reduction method with the emerging nonlinear dimension reduction method. Thus, it finds out the dimension reduction method which is suitable for the prediction of the roll gap parameters of continuous casting.

2.1 Principal Component Analysis (PCA)

As a mainstream method of linear dimension reduction, PCA has the most linear reconstruction error because its theory is complete. So its application is very extensive [6,7,8,17,18].PCA is mainly looking for a series of orthogonal basis vectors that maximize the variance of all data.

Suppose the sample set $X_i = \{x_1, x_2, ... x_m\}$, in order to find a subspace to make the hyperplane projection of sample point xi is $W^T x_i$ in the subspace.

If the projection of all sample points is separated as much as possible, the variance of the projected sample point is the largest. The variance can be expressed as $\sum_i W^T x_i x_i^T W$; the optimization goal can be written as

$$\max_W tr(W^T XX^T W) s.t. W^T W = I$$

(1)

In equation (1) use the Lagrange multiplier method

$$XX^T W = \lambda W \quad (2)$$

Among them, W is the best projection matrix, λ is the eigenvalue.

Therefore, we only need eigenvalue decomposition of the covariance matrix XX^T, then the eigenvalues will be sorted: $\lambda_1 \geq \lambda_2 \geq ... \geq \lambda_n$ .Then, taking the first d eigenvalues corresponding to the eigenvectors $W = (w_1, w_2, ... w_d)$ is the solution of principal component analysis.

2.2 T-distribution Stochastic Neighbor Embedding (t-SNE)

2.1.1 Formatting the title

The title is set in bold 14-point Arial, flush left, unjustified. The first letter of the title should be capitalised with the rest in lower case. You should leave 35 mm of space above the title and 6 mm after the title.

2.1.2 Formatting author names and author affiliations

Author names should be typed in 9-point Arial. The style for the names is First Names then Last Name, with a comma after all but the last two names, which are separated by "and". Do not use academic titles.

Affiliations of authors should be typed in italic 8-point Arial. They should be preceded by a numerical superscri pt corresponding to the same superscript after the name o f the author concerned. Please ensure that affiliations are as full and complete as possible and include the country.

$$pj|i = \frac{\exp(-\|x_i - x_j\|^2 / 2\sigma_i^2)}{\sum_{k \neq i} \exp(-\|x_i - x_k\|^2 / 2\sigma_i^2)}$$

(3)

$$p_{ij} = \frac{p_j|ip_i|j}{2N}$$

(4)

Step2 Dimension reduction: learn probability distribution at low latitudes to define a d-dimensional map $y_i, ... y_N, y_i \in R^d$ .Assuming that the similarity between y_i and y_j is q_{ij}.

$$q_{ij} = \frac{\left(1 + \|y_i - y_j\|^2\right)^{-1}}{\sum_{k \neq i}\left(1 + \|y_k - y_l\|^2\right)^{-1}}$$

(5)

The relative entropy (Relative Entropy) is introduced to measure the similarity between Q and P.

$$C = KL(P\| \|) = \sum_{i \neq j} P_{ij} \log \frac{p_{ij}}{q_{ij}}$$

(6)

Using the gradient descent method makes the relative entropy minimum.

$$\frac{dC}{dy_i} = 4\sum_j (p_{ij} - q_{ij})(y_i - y_j)\left(1 + \|y_i - y_j\|^2\right)^{-1}$$

(7)

The advantage of t-SNE is that it takes a smaller distan ce to create a larger gradient that separates dissimilar poi nts, and this gradient does not go too far away from dissi milar points. Although t-SNE is almost impossible to sur pass the classic, there are some deficiencies. t-SNE is mai nly used for visualization and cannot be used for other pu rposes. t-SNE is mainly used to preserve the local feature s because it is difficult to map the high intrinsic dimensio n data set to 2-3 Dimensional space very well. t-SNE is a probabilistic distribution problem and is not possible to consider reconstructing a model after dimension reductio n.

2.3 Locally linear embedding (LLE)

Locally linear embedding (LLE) was proposed by S. T. Roweis in 2000. [22,23,24] It is a non-linear method of unsupervised dimension reduction. After dimension redu ction of low-dimensional, the data can maintain the topol ogy of the original data. This has been widely used in ima ge classification and clustering, face recognition, natural language processing, multidimensional data visualization and bioinformatics.

Step1 Establish the probability distribution of high di mensional objects: assume hypothetical high-dimensional observation dataset $X = \{x_1, x_2, \ldots x_n\}$, then search fo r k nearest neighbors $X_i = [x_{i1}, \ldots, x_{ik}], i = 1, \ldots N$.

Step2 Calculate the coefficients for each point which i s reconstructed by the domain point: The coefficient of e ach sample point x_i linearly reconstructed by its neighbo rhood X_i can be obtained by minimizing the reconstruct ion error.

$$e(f_i) = \left\| x_i - \sum_j f_{ij} x_{ij} \right\|^2 \quad (8)$$

$$s.t. \sum_j f_{ij} = 1 \quad (9)$$

Among them, $\sum_j f_{ij} x_{ij}$ represents the linear reconstruction of point x_i, $f_{ij} = f(x_i, x_{ij})$ indicates that x_i is a coefficient reconstructed by x_{ij}. The solution is $f_{ij} = \frac{\sum_h G_{jh}^{-1}}{\sum_{l,m} G_{lm}^{-1}}$.

Among them, G is the local Gram matrix: $G_{jh} = (x_i - x_{ij})^T (x_i - x_{ih})$. Reconstruction factor f_i describes the neighborhood geometry for rotation invariance and translation invariance for x_i and its neighbor point X_i.

Step3 Construct low dimensional embedding: Low-dimensional embedding is obtained by maintaining the weight of reconstructed data in each neighborhood in low-dimensional data; the objective function is

$$J(Y) = \sum_i \left\| yi - \sum_j f_{ij} y_j \right\|^2 \quad (10)$$

$$s.t. \frac{1}{N} \sum_i y_i y_i^T = I \,\&\, \sum_i y_i = 0 \quad (11)$$

So the low-dimensional data is still maintained at f_{ij}, the embedding result is completely characterized by the reconstruction weights that characterize the intrinsic geometry. The advantage of LLE algorithm is simpler calculation and less parameter selection. It can optimize the characteristic parameters. There is no local minimum problem. It can retain the essential characteristics of data and express the internal structure of the data. However, LLE is sensitive to signal noise, because it is limited by the local neighborhood parameter k and the target dimension of the low-dimensional data, affecting the dimension reduction effect of high-dimensional data.

2.4 Local tangent space alignment (LTSA)

LTSA was proposed by Zhang Zhenyue and others from Zhejiang University in 2004 [25,26]. The method is mainly to gradually align all the tangent planes when a manifold is unfolded. The local geometry of the low-dimensional data is constructed by approximating the tangent space of each sample point. The local low-dimensional coordinates are obtained by the projection of the observed data points in the local tangent space. The local low-dimensional coordinates of the overlap are locally transformed to obtain the global low-dimensional embedding coordinates.

Step1 Suppose the high-dimensional observation dataset $X = \{x_1, x_2, \ldots x_n\}$, then search for its k nearest neighbors $X_i = [x_{i1}, \ldots, x_{ik}], i = 1, \ldots N$.

Step2 Calculate the base vector of the d-dimensional local tangent space of the neighbor dataset X_i and the local tangent coordinate Θ_i of the observed data.

Step3 According to the N local projection coordinates $\Theta_i = [\Theta_{i1}, \ldots, \Theta_{ik}], i = 1, \ldots N$, the global coordinates $\{y\}_{i=1}^N$ are obtained by the local affine transformation $Li \in R^{d \times d}$.

LTSA can reflect the local geometric characteristics of the data well, but it cannot handle large sample data, and it does not work well for popular learning with high curvature.

2.5 Comparison of four methods for dimension reduction

The dimension reduction methods will be tested with the actual production of continuous casting data in this section.

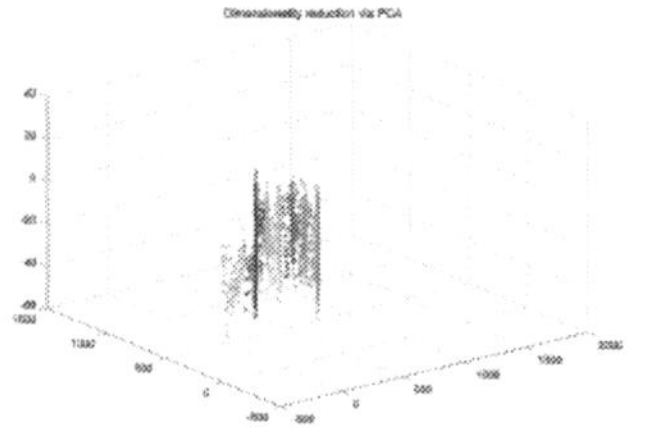

Figure. 1 Dimension reduction effect of PCA

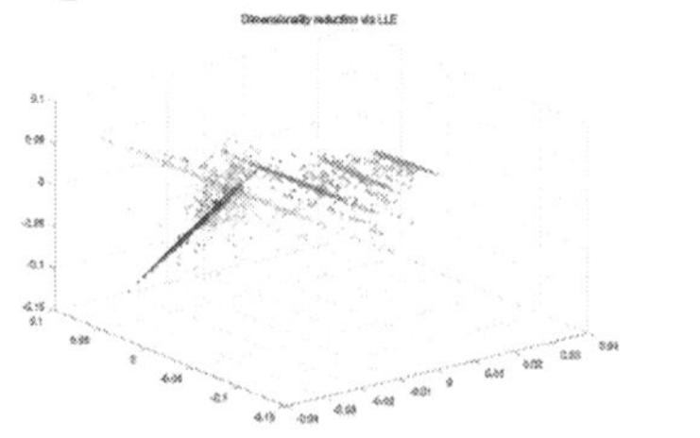

Corresponding author: Xianqi Liang: 1012363162@qq.com; Zhufeng Lei: 253574602@qq.com; Wenbin Su: 809355146@qq.com

Figure. 2 Dimension reduction effect of LLE

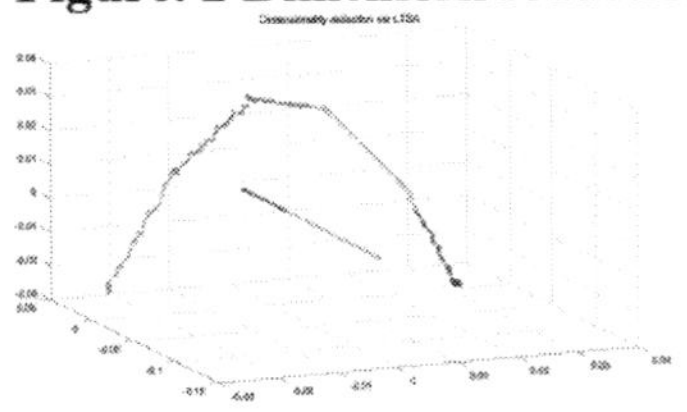

Figure. 3 Dimension reduction effect of LTSA

As can be seen from the figures, PCA dimension reduction effect is normal and can reduce the dimension of the data, but the data classification is not obvious after the dimension reduction. In contrast, LTSA dimension reduction effect is better. Effective dimension reduction and data classification is obvious; different types of data aggregation are obvious. The reason why the two types of methods produce different results is that, in principle, linear dimension reduction mainly analyzes the importance of a certain dimension data, reduces the dimension according to its importance and deletes the data of lesser importance. In order to achieve dimension reduction, linear dimension reduction method is mainly to carry out the data clustering, as well as high similarity data aggregation and projection in low-dimensional space.

In the roll gap prediction model of continuous casting, the characteristics of flow data from multi-information source determine that the feature reduction method should serve as the roll gap prediction model. When the nonlinear dimension reduction method is used, the original data is processed, resulting in better dimension reduction and improved operational efficiency. However, the data after dimension reduction cannot be evaluated; the prediction of the roll gap parameter cannot be realized. After repeated tests and comparison, the linear dimension reduction method is used to characterize the dimension of the continuous casting roll gap prediction model.

3 Experimental and Research on Improved PCA

In this section, PCA will be modified to make it suitable for feature reduction of continuous casting process parameters characterized by multi-information source data.

3.1 The theoretical basis of NRIPCA

The NRIPCA method updates the PCA model in real time based on sliding window thinking, replacing the oldest data in the window with the latest data.

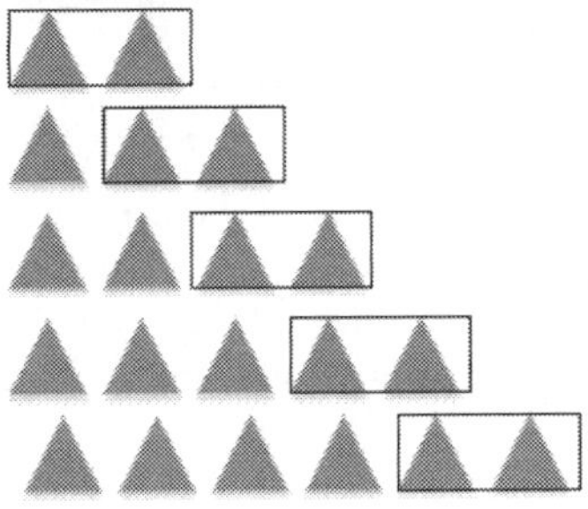

Figure. 4 Principle of sliding window

The main steps of this algorithm are:

(1) Data pre-process: use a certain amount of historical data for discrete point detection.

(2) The pre-process data are normalized to make the data mapped in a range, which makes the optimization process of the algorithm become more smooth and easier to correctly converge to the optimal solution.

(3) Principal component analysis: establish principal component analysis model and calculate the model parameters.

(4) Collect a new set of real-time data, and standardize the data with the current mean and variance.

(5) According to the current principal component analysis model, analyze whether the data is normal data, then go to the next step; otherwise, go back to the previous step.

(6) Add a new set of process parameters to the sliding window; meanwhile delete the oldest set of process parameters and update the principal component analysis model and related parameter values. Go to step 4.

3.2 data source

In order to verify the validity of the NRIPCA algorithm, a uniform parameter of a continuous casting production line of a steel mill was randomly collected in China for a week. These data include 32,154 dimensional data, of which 26,001 dimensional data are the switch data monitored by the control system, and 6153 dimensional data are valid process parameter monitoring data. The data in the data set cover important processes in the continuous casting process, including tundish, mould and sector segments. In order not to damage the sequence of data, Rada criterion is used to detect abnormal values of data, then the detected abnormal values are replaced by nearby values.

3.3 NRIPCA experiment on continuous casting flow data

Two important parameters of the NRIPCA algorithm are the size of both the sliding window and the sliding step. In order to study the size of both the sliding window and the step for the data set of this project, the total amount of data and the size of the step are changed respectively, and the number of stable principal components is obtained as the evaluation criterion. The result shows the ratio of sliding window size in the total amount of data.

Table 1 Sliding window size ratio of the total amount of data

The total amount of data	The ratio of sliding window size in each step to the total amount of data				
	30	60	120	240	480
One day (63700)	3/4	3/4	3/4	3/4	3/4
Two days (127400)	31/32	31/32	31/32	31/32	31/32
Three days (191100)	3/4	3/4	3/4	3/4	3/4

Four days (254800)	7/8	7/8	7/8	7/8	7/8
Five days (318500)	31/32	31/32	31/32	31/32	31/32
Six days (392200)	7/8	7/8	7/8	7/8	7/8
Seven days (445900)	3/4	3/4	3/4	3/4	3/4

From Table 1, it can be seen that the step size has little effect on the model within 480 (8 minutes), indicating that the change of continuous casting process parameters is a gradual process which can slow the model updating time in the subsequent online analysis. In other words, each group can reduce the computational cost without having to update the model parameters of the new process.

In this paper, a total of 446,000 samples are selected for sliding window PCA analysis in a week. The window size is 334500 and the step size is 60 (i.e. the principal component analysis model is updated once every minute). In order to calculate the reduction error, the latest model data is reconstructed with the current model parameters after each model update.

In order to quantitatively analyze the size of the reconstruction error, the average relative error between the reconstructed data of each parameter and the original data is calculated as shown in Table 2.

Table2 The average relative error of the process parameters

Process parameter name	Relative error (%)		Process parameter name	Relative error (%)	
	average value	Maximum		average value	Maximum
Tundish temperature of molten steel	0.11	0.65	Mold wide water flow	2.03	13.90
Overheating	4.89	22.32	Mold wide exit temperature	0.41	1.51
Pulling speed	1.51	6.71	Mold narrow surface pressure	2.94	21.28
Vibration frequency	1.39	9.03	Mold narrow surface water flow	0.23	2.52
Vibration amplitude	1.04	1.79	Narrow mold exit temperature	0.39	2.81
Drive roller pressure	4.24	22.22	Second cold water flow	4.89	42.69
Mold level	0.44	2.10	Second cold water pressure	7.73	54.24
Mold wide-face pressure	3.77	32.83			

As can be seen from Table 2, the average error between the original data obtained by PCA and sliding PCA is less than 5%, which is within the acceptable error range. Therefore, it is feasible to use the sliding window PCA to dynamically update the feature parameters of process flow data online.

4. Discussion

In this paper, fifteen kinds of process parameters that influence the roll gap value of continuous casting sector are reduced by one week's data. PCA and NRIPCA are used respectively. The former is to perform the feature reduction on the whole seven-day data; the latter uses part of the data as real-time data to simulate the process of online feature reduction. By observing the comparison between the reconstructed data and the original data, it can be found that both methods have a good fitting effect. In order to compare two methods more intuitively, the average reduction error of the reconstructed data and the original data of the two methods is shown in Table 3.

Table 3 The average relative error of reconstructed data of each parameter

Short for process parameters	PCA average relative error (%)	MWPCA average relative error (%)	Short for process parameters	PCA average relative error (%)	MWPCA average relative error (%)
GSWD	0.16	0.11	KMLL	1.26	2.03
GRD	4.25	4.89	KMWD	0.43	0.41
LS	1.53	1.51	ZMYL	4.35	2.94
ZDPL	1.12	1.39	ZMLL	0.68	0.23
ZDFZ	0.55	1.04	ZMWD	0.44	0.39
CDGYL	5.31	4.24	ELSLL	4.83	4.89
JJQYW	0.13	0.44	ELSYL	15.05	7.73
KMYL	2.58	3.77	Average	2.85	2.40

As can be seen from Table 3, NRIPCA reduction error is not small for each process parameter. And the reduction error of the two methods in the comparison of each parameter has its own advantages and disadvantages. But overall, the reduction error in NRIPCA is smaller than in PCA, meaning that NRIPCA has a small reduction error when feature reduction is performed on changing process parameters. From an algorithmic point of view, PCA calculates the global optimal reconstruction, while NRIPCA calculates the current local optimal reconstruction and has optimal reconstruction errors for each sliding window. Therefore, the overall reconstruction error in NRIPCA is smaller than in PCA.

5. Conclusion

In this paper, a standard robust incremental principal component analysis (NRIPCA) is proposed. The new method can reduce the robustness of real-time data by standardizing and pre-processing the data. Therefore, it is more suitable for the feature reduction of continuous casting production process parameters with multi-information source data characteristics. The experimental results prove the effectiveness of NRIPCA algorithm. The new method will provide a reliable source of data for the prediction of the roll gap parameters of the continuous casting and will be useful for establishing the prediction model for the parameters of the continuous casting. Further, we hope this feature reduction method can be applied to other industrial data area.

Acknowledgments

Corresponding author: Xianqi Liang: 1012363162@qq.com;Zhufeng Lei: 253574602@qq.com; Wenbin Su: 809355146@qq.com

This work was financially supported by the National Natural Science Foundation of China [NO. 51575429].

References

[1] ZHANG Guodong, SUN Jun. Application of Fuzzy Control on the Segment's Gap Control of Slab Continuous Casting Machine[J]. ELECTRIC DRIVE, 2009, 39, (05): 51-53.

[2]Fang Yiming, Hu Chunyang. Breakout Prediction Classifier for Continuous Casting Based on Active Learning GA-SVM[J]. China Mechanical.

[3]Carlos A Santos. A solidification heat transfer model and a neural network based algorithm applied to the continuous casting of steel billets and blooms[J]. Modelling and Simulation in Materials Science and Engineering. 2005, 10, (13): 1071-1087.

[4]Wang Nan, Dong Jianhong, Chen Min, YAO Yongkuan. Numerical Simulation on Thermal Stress Field in a Wide Slab Mould of Peritectic Steel Continuous Casting[J]. Journal of iron and steel research, 2012, (2): 909-912.

[5]Zhang Qunliang, Guo Zhaohui, Model control technologies of dynamic secondary cooling and soft reduction for slab continuous casting[J]. Baosteel technology research, 2012, 06, (3):61-64.

[6]WENG J Y, ZHANG Y L, HWANG W S. Candid Covariance — free Incremental Principal Component Analysis[J]. IEEE Transactions on Pattern Analysis and Machine Intelligence, 2003, 25(8): 1034-1040.

[7]JOSEPH A, TOKUMTO T, AZAWA S. Online Feature Extraction Based on Accelerated Kernel Principal Component Analysis for Data Stream[J]. Evolving Systems, 2015:1-13.

[8]Martins M, Santos C, Costa L, et al. Feature reduction with PCA/KPCA for gait classification with different assistive devices[J]. International Journal of Intelligent Computing and Cybernetics, 2015,8(4):363-382.

[9]Kaur M, Arora AS. Classification of ECG signals using LDA with factor analysis method as feature reduction technique[J]. Journal of Medical Engineering & Technology, 2012,3 6(8): 411-420.

[10]Wise B M, Gallagher N B. The Process Chemometrics Approach to Process Monitoring and Fault Detection[J]. Journal of Process Control (S0959—1524), 1996, 6(6): 329-348.

[11]Dung D, McAvoy T J. Nonlinear Principal Component Analysis Based on Principal Curves and Neural Networks[J]. Computers and Chemical Engineering(S0098—1354), 1996, 20(1): 65-78.

[12]You C K, Vanrolleghem P A, Lee I-B, Nonlinear Modeling and Adaptive Monitoring with Fuzzy and Multivariate Statistical Methods iu Biological Wastewater Trcatment Plants[J]. Journal of Biotechnology(S0168—1656), 2003, 105:135—163.

[13]Wang Xun, Kruger Uwe, Lennox Barry. Recursive Partial Least Squares Algorithms for Monitoring Complex Industrial Processes[J]. Control Engineering Practice(S0967-0661),2003, 11: 613—632 .

[14]Zhang Y, Dudzic M, Vaculik V. Integrated Monitoring Solution to Start — up and Run — time Operations for Continuous Casting[J]. Annual Review of Control(S1367—5788), 2003, 27: 141—149.

[15]Ivan Milefic, Shannon Quinn, Michael Dudzic, et al. An Industrial Perspective on Implementing On-line Applications of Multivariate Statistics[J]. Journal of Process Control(S0959-1524), 2004, 14: 821-836.

[16]L.J.P. van der Maaten, E.O. Postma, and H.J. van den Herik. Dimensionality Reduction: A Comparative Review. Tilburg University Technical Report, TiCC-TR 2009-005, 2009.

[17]Li W, Yue H, Valle — Cervantes S, Qin S. Recursive PCA for Adaptive Process Monitoring[J]. Journal of Process Control(S0959—1524), 2000, 10(5): 471—486.

[18]Wold S. Exponentially Weighted Moving Principal Components Analysis and Projections to Latent Structures[J]. Chemometrics and Intelligent Laboratory Systems(S0169—7439), 1994, 23: 149—161 .

[19]L.J.P. van der Maaten and G.E. Hinton. Visualizing High-Dimensional Data Using t-SNE. Journal of Machine Learning Research 9(Nov):2579-2605, 2008.

[20]N. Pezzotti, B.P.F. Lelieveldt, L.J.P. van der Maaten, T. Hollt, E. Eisemann, and A. Vilanova. Approximated and User Steerable t-SNE for Progressive Visual Analytics. IEEE Transactions on Visualization and Computer Graphics 23(7), 2017.

[21]L.J.P. van der Maaten. Accelerating t-SNE using Tree-Based Algorithms. Journal of Machine Learning Research 15(Oct):3221-3245, 2014.

[22]Roweis ST, Saul LK. Nonlinear dimensionality reduction by locally linear embedding. SCIENCE.5500(290): 2323-+,2000.

[23]Donoho DL, Grimes C. Hessian eigenmaps: Locally linear embedding techniques for high-dimensional data.PROCEEDINGS OF THE NATIONAL ACADEMY OF SCIENCES OF THE UNITED STATES OF AMERICA. 100(10): 5591-5596. 2003.

[24]Yao Chao, Liu Ya-Feng, Jiang Bo. LLE Score: A New Filter-Based Unsupervised Feature Selection Method Based on Nonlinear Manifold Embedding and Its Application to Image Recognition. IEEE TRANSACTIONS ON IMAGE PROCESSING.26(11):5257-5269.2017.

[25]Zhang ZY, Zha HY. Principal manifolds and nonlinear dimensionality reduction via tangent space alignment. SIAM JOURNAL ON SCIENTIFIC COMPUTING. 26(1): 313-338.2004.

[26]Zhang Tianhao, Yang Jie, Zhao Deli. Linear local tangent space alignment and application to face recognition.NEUROCOMPUTING. 70(7-9): 1547-1553. 2007.

Development and Application of Lightweight High Strength Organic Materials

Yang Zhang[1, a]

[1]*Jiangsu Province Xuzhou Technician Institute, Xuzhou, Jiangsu Province, China*

Abstract. This paper summarizes the commonly used organic materials with light weight and high strength characteristics, and introduces their types, characteristics, applications and current development status at home and abroad.

Introduction

According to statistics, 60% of automotive fuel is consumed in the vehicle's own weight [1], reducing the vehicle's own weight can effectively save energy and reduce emissions. Every 10% reduction in the vehicle's vehicle quality, the fuel consumption will decrease by 3%-7% [2]. The realization of lightweight not only is the automotive industry, but all energy-consuming equipment is an important way to save energy and reducing emissions and developing green and environmental protection. With the research and development of high-strength materials, materials that can meet the strength requirements are not limited to metal materials, and plastics gradually step into the stage of the times.

1 High-strength engineering plastics

Engineering plastics, especially high-performance engineering plastics, are widely used because of their good mechanical properties, comprehensive mechanical properties, heat resistance, acid resistance, long life, and good reliability. In order to achieve lightweight of equipment, Wang Tuyuan [3] and others put forward the concept of "plastic substitute steel". Polymers have excellent properties such as light weight, corrosion resistance, high strength, and easy processing. At present, high-strength engineering plastics are used in pipelines, automobiles, construction and other fields. Polyethylene chloride, rubber, polyoxymethylene, polyamide, polyurethane, polytetrafluoroethylene, nylon, and synthetic fibers are used for long time[4]. In order to ensure a certain degree of strength of the material, Zhang Chao[5] and others have proposed steel-plastic composites. The wire-wound mesh skeleton-reinforced plastic composite pipe is reinforced with a high-strength steel wire with a left-right inclination and is formed with a net-like skeleton reinforced with a thermoplastic plastic matrix. The high-performance adhesive resin is used to tightly connect the steel mesh skeleton with the inner and outer plastics using in the coal mine pipeline. This kind of pipe has the advantages of easy installation, easy maintenance, good corrosion resistance, long service life, no scaling, and good water conservancy performance. The wall climbing robot produced by Li Heng [6] chose POM as the body shell in light of the principle of miniaturization. Polyoxymethylene, as an engineering plastic with excellent performance, is known abroad as "steel-reducing" and "ultra-steel". It has similar hardness, strength and steel properties, and it has good performance in a wide range of temperature and humidity. Self-lubrication, good fatigue resistance, and flexibility, and it also has good chemical resistance and low cost. In addition, PET (polyethylene terephthalate) has excellent properties such as high strength, high rigidity, good heat resistance, excellent dimensional stability, outstanding chemical resistance, etc. Machinery, automotive parts and other fields have been widely used [7-9].

In the automotive field, the blending of polyethylene (PE) with PP (polypropylene) has high impact resistance. Peugeot bumpers are made from PP/HDPE blends [10, 11]. Polyamide (PA) has good mechanical properties and thermal stability, so it is an ideal material for producing high-strength, heat-resistant parts. Plastic accelerators are used on German Volkswagen's "Golf" and "Jetta" cars. The pedal, using DuPont's 180 G 33 PA, has low cost, high strength, and good reliability [12].

2 Lightweight high-strength fiber composites

Polymer materials have certain advantages, but their strengths are much different from those of metals. Steel-plastic composite materials do not fundamentally realize

[a]113503287@qq.com

© The Authors, published by EDP Sciences. This is an open access article distributed under the terms of the Creative Commons Attribution License 4.0 (http://creativecommons.org/licenses/by/4.0/).

the lightweight and high-strength characteristics of materials. The resin-based composite material is a kind of fiber-reinforced plastic. The low quality of plastic materials and its high-strength performance are currently the major trends.

2.1 Carbon Fiber Composites

Carbon fiber (CF) is a kind of high-performance fiber. It has excellent properties such as high specific strength, high specific modulus, fatigue and creep resistance, expansion coefficient, and small friction coefficient, becoming the most important enhancement in composite materials, being one of the most important materials [13]. High-strength carbon fibers such as T800, T1000, and T1100 carbon fibers have tensile strengths of 5.49 GPa, 6.37 GPa, and 6.6 GPa, respectively [14].

In the 40th and 50th centuries of the 20th century, carbon fiber composite materials (CFRP) were first manufactured in the United States. In 1950, CFRP was first produced by pulling rayon at a temperature of 2000°C. It has high specific strength, specific modulus, low density, good anti-fatigue properties, good anti-vibration performance, strong designability, and high temperature performance. With good moldability and other advantages, high-performance CFRP strength can reach more than ten times of steel [15, 16], and is widely used in aerospace and military fields. The specific strength of high modulus carbon fiber/epoxy resin composites is 5 times that of steel and 4 times that of aluminum alloys. Its specific modulus is 4 times that of aluminum and copper. The production of bearing members can exert unique and excellent mechanical properties. [17].

Bie Mingzhi [18] prepare carbon fiber/PA6 composites with high strength, high electrical conductivity and flame retardant properties, which can meet the safety and performance requirements of polymer products in coal mines.

Song Yanjiang [19] added carbon fiber to the thermoplastic polyimide resin for composite reinforcement. The effect of carbon fiber addition and molding method on the mechanical properties of the composite material was experimentally studied. The addition of carbon fiber can significantly improve the room temperature and high temperature mechanical properties of the material, and carbon fiber different types, the degree of improvement is also different; the tensile strength and bending strength of the composite material increase with the increase of the amount of carbon fiber; J. Karger-Kocsis [20], who used carbon fiber reinforced thermoplastic polymethyl methacrylate Ethyl acrylate composites and carbon fiber-reinforced thermoplastic polyetheretherketone composites, and their mechanical properties were studied. The results show that the mechanical properties of the composites are determined by the bonding of the resin matrix and the carbon fibers, the toughness of the resin, and the arrangement of the carbon fibers in the resin.

2.2 Fiberglass

FRP is a high-strength engineering plastics, its full name is - glass fiber reinforced plastic, which has the characteristics of corrosion resistance, good insulation, high specific strength, low cost, simple process, short production cycle, etc. [21], FRP the traditional products of composite materials are made by the glass fiber winding forming process, such as internal pressure vessels, pipes and tanks, which mainly utilize the properties of fibers with high tensile strength. The advantage of the filament winding method is that the fibers can be arranged according to the stress conditions, thereby improving the mechanical structural efficiency of the product [22-24].

FRP is often used in industrial pipes. The quality of FRP pipes is light. The same size is only 1/4 of the quality of steel pipes and 2/25 of concrete pipes. The pipes contain rich resin layers and the surface is extremely smooth, small resistance, no scaling. When the flow is the same, the pressure loss caused by the FRP pipe conveying medium is about 3/5 of the same inner diameter carbon steel pipe, the conveying capacity can be increased by more than 20%, and the power and energy of the pump can be saved by 30% to 40%. The resin-based composite material itself has good corrosion resistance. Unlike metal pipes, which require regular anti-corrosion and other maintenance, the service life is much higher than that of steel pipes. Especially in coal mine drainage systems with acidic water quality, corrosion resistance can be achieved, long-term safe and reliable operation, and comprehensive benefits are good. The use situation abroad and the research results of Henan Polytechnic University show that the high-strength double-resistant coal mine FRP pipe does not require maintenance within 30 years, even if it needs repairs, it is very convenient. If the pipe is damaged, it only needs to prepare the raw material with the prepared double-resistant glass steel. System can be repaired [25]. Zeng Xiantao [26] and other research and development of a new type of glass fiber reinforced plastic composite material tank, the tank path consists of two parts of rectangular steel core and glass fiber coating, this modified pipe has anti-static, flame retardant, wear resistance The performance partially replaces the steel, the quality is reduced, the surface corrosion resistance is far superior to steel, and the service life can reach 30 years.

2.3 Kevlar

"Kevlar" is an aramid composite material with dark yellow fibers, with a diameter of 0.01-0.02mm and a density of 1.43-1.44g/cm3. Its greatest feature is its high mechanical properties and good stability. Its strength is 3.6Gp, its modulus reaches 131GPa, and its elongation at break is only 2.8% and it is insensitive to temperature [27, 28]. It was made by DuPont in the 1960s. This new type of material has low density, high strength, good toughness, high temperature resistance, easy processing and molding and is widely used in military and other fields of defense. Compared with FRP, Kevlar bears the same pressure. The material weight of Kevlar can be reduced by half, and the toughness of Kevlar-laminated sheet is 3 times that of steel, which can withstand repeated impacts, including para-aramid. The specific strength is 6 times that of steel

Table 1 The main performance of Para aramid with the same brand number [29]

Product brand	Density/ (g/cm^3)	Tensile strengt/ (cN/dtex)	Tensile modulus/ (cNdtex)	Elongation/%	LOI/%	Decomposition temperature/°C	Moisture absorption rate/%
Kevlar29	1.44	20.3	490	3.6	29	500	7.0
Kevlar49	1.45	20.8	780	2.4	29	500	3.5
Kevlar119	1.44	21.2	380	4.4	29	500	7.0
Kevlar129	1.44	23.4	670	3.3	29	500	7.0
Kevlar149	1.47	16.8	1150	1.3	29	500	1.2
Twaron Reg	1.44	21	500	4.4	29	500	6.5
Twaron HM	1.45	21	750	2.5	29	500	3.5
Technora	1.39	24.7	520	4.6	25	500	2.0
Armos	1.43	35.0-39.0	1050	3.5-4.0	39-42	575	2.0-3.5
Rusar C	1.46	36.3	1074	2.6	35	575	2.25
Rusar HT	1.47	34.7	1200	2.6	45	575	1.35

wire, 3 times that of glass fiber, and 2 times that of high-strength nylon industrial yarn. After 100 hours at 200°C, it can still maintain 75% of its original strength. After 500 hours at 160°C, it can still To maintain the original strength of 95% [29], the main types and performance of the world para-aramid are shown in Table 1.

Summary

Lightweight is the eternal theme of material development, and it is an important way to realize green environmental protection. The quality of organic materials can be reduced to a certain extent compared to that of metal materials. What we need to do now is to increase the strength of organic materials. The effective method to increase material strength is using fiber to strengthen materials.

References

1. Tang Di, Mi Zhengli, Chen Yulai . Technical Requirements and Research and Development Status of New Type Automotive Steels in Foreign Countries [J]. Steel, 2005, 40(6):1.
2. Gao Jianhong, Wang Min, Wu Qingqing. Study on Resistance Spot Welding of TRIP590 High Strength Steel [M]. 2008 Light metal and high-strength materials welding International Forum Proceedings. Beijing. 2008: 133-9.
3. Wang Tongyuan, Wang Qiqiang. Application and Development of Polymer in Coal Mine Production [J] Technology and Innovation Management, 2011,1:1 (32)
4. Fang qian. Analysis of the Safety of Polymer Materials Used in Coal Mines [J]. Journal of North China Institute of Science and Technology. 2005, 3: 1(2)
5. Zhang Chao, Sun Yunguo, Han Qunliang, Peng Xinyun. Application of Steel Wire Mesh Framework Polyethylene Composite Pipe in Coal Mine [J]. Coal Technology. 2009, 3: 3(28)
6. Li Heng. Research on the Dynamic Characteristics of a Wall-climbing Robot and Its Simulation Analysis [D]. Qingdao University of Science and Technology . 2010, 4
7. Ulrich Thiele . Structural change in the polyester industry[J]. Man-made fiber yearbook ,2000:4
8. Papke N.,Karger-Kosis J . .Thermoplastic elastomers based on compatibilizedpoly (ethylene terephthalate) blends:effect of rubber type and dynamiccuring[J] . Polymer,2001,42:1109
9. Papke N.,Karger-Kosis J . .Thermoplastic elastomers based on compatibilizedpoly(ethylene terephthalate) blends:effect of rubber type and dynamiccuring[J] . Polymer,2001,42:1109
10. Plast, 1987,18(7):113-115

11. Gubitz Franz Kunststoffe, 1987,32(10):33-37
12. Kunststoffe,1987,21(9):30-32
13. Rong Huaiping. Study on the Fabrication of Nano-materials on Carbon Fiber Surfaces and Properties of Their Composites [D] Donghua University. 2013, 9
14. S.Chand. Carbon fibers for compos ites[J] . Journa l of Material science, 2000, 35 : 1303 ~ 1313 .
15. Li Wei, et al. Chinese Optics and Applied Optics Abstracts[J]. 2011, 4(3):201–212.
16. Zhang Hanying, Liu Ming. Development and Application of Carbon Fiber Composites [J]. Engineering plastics applications. 2015, 11:11(43)
17. Zhou Xiya . Composite materials [M] . Beijing: Chemical Industry Press, 2005
18. Bie Mingzhi, Li Youbing, Liang Quancai, Du Jing, Chen Jian. Development of Mine Carbon Fiber Composite Nylon 6 Impeller Material [J]. Plastic industry. 2013 , 12: 12 (41)
19. Song Yanjing, Wang Xiaodong, Wang Wei et al. Carbon fiber reinforced thermoplastic polyimide and composites thereof [J]. Journal of Materials Science and Engineering, 2007, 25(3): 363-366.
20. Karger-Kocsis J, Yuan Q, Mayer J, et al. Transverse impact behavior of knitted carbon-fiber fabric-reinforced thermoplastic composite sheets[J]. Journal of Thermoplastic Composite Materials, 1997, 10(2): 163-172.
21. John P.Busel .FRP Composites For Civil Infrastructure Applications. Plastics in Building Construction . VolumeXXII Number , 1998 , 11
22. Zheng Nongfeng , Howard G .Allen, Stuart S .J .Moy .Study of Stress Concentrations in Woven Composites.Journal of Reinforced Plastics and Composites, 1999 , 3 :18
23. B .Gommers , I .Verpoest, P.Houtte .Analysis of Knitted Fabric Reinforced Composites: PartII . Stiffness and Strength.Composites PartA 29A, 1998
24. M .Bannister , I .Herszbery .The Manufacture of Glass/ epoxy Composites With Multi-layer Woven Architectures.Volume29A, 1998.
25. Geng Yungui, Wang Yuqing, Li jiangzhong. Application of Resin Matrix Composites in Coal Mines [J]. Material Guide. 2014, 5 :(28)
26. Zeng Xiantao, Guo Jingpu. Coal Mine Shaft FRP Composite Material Tank Road [J]. China Engineering Science . 2002 , 12: 12(4)
27. Ian O'Connor, Hugh Hayden, Stephen O'Connor, Kevlar coated carbon nanotubes for reinforcement of polyvinylchloride[J], 2008, 18, 5585–5588.
28. Zhang Chengxiu, Guan Baoqiong. Application of Kevlar Fiber Composite Materials in Strengthening of Civil Engineering Structures [J]. Synthetic fiber, SFC, 2007(4).
29. Li Ye. Development Status, Technical Analysis and Prospect of Para-aramid [J]. Synthetic fiber , SFC, 2009(9)

Development and Application of Lightweight High-strength Metal Materials

He Pan[1, a]

[1] *Jiangsu Province Xuzhou Technician Institute, Xuzhou, Jiangsu Province, China*

Abstract. This article reviews the performance, alloy composition and the development of advanced lightweight high-strength materials such as high-strength steels, high-strength aluminum alloys, high-strength magnesium alloys, and titanium alloys.

Introduction

Lightweight and high-strength is the eternal theme of material development. It is an important foundation for promoting social progress, economic development, national defense construction, scientific and technological progress, and strengthening national competitiveness [1]. Today, with the rapid development of science and technology, the application of lightweight high-strength materials is not limited to cutting-edge technology. Its application is widespread to all aspects. In particular, self-contained energy equipment consumes more energy and uses more lightweight materials.

1 high-strength steel

According to the definition of high-strength steel in the world, steels with a tensile strength between 270-700 MPa are collectively classified as high-strength steel HSS (Conventional HSS) , steels with a tensile strength greater than700MPa classified as advanced high-strength steels AHSS (Advanced HSS) [2, 3], the use of high-strength is an important way to achieve lightweight equipment.

In the coal mining industry, due to the small space in the underground industry and complex geology, fires, gas explosions, floods, and other extremely hazardous accidents[4] are very likely to occur. Therefore, equipment requirements are more stringent, and light metal materials are restricted to use because of their poor frictional spark safety. Underground equipment often uses steel or copper materials. In order to ensure adequate strength and lightweight equipment, high-strength steels are generally used in explosion-proof devices. The material of the explosion-proof system for down hole operation robots uses steel of a certain grade, and its thickness is generally 4-6mm[5].

In the automotive field, according to research, 60% of automotive fuel is consumed in the vehicle's own weight[6], reducing the vehicle's own weight can effectively save energy and reduce emissions. Every 10% reduction in the vehicle's quality, fuel consumption will decrease by 3%-7%[7], so it has been to ensure that the reliability and function of the car are not affected, the realization of lightweight car body is an important direction for the development of the automotive industry. In Europe, high-strength steels and ultra-high-strength steels have been widely used in automotive parts[8] that have already been introduced. Currently, high-strength and ultra-high-strength dual-phase steels (DP steels) with low yield ratios are commonly used in automotive parts. Combining high-strength and good plasticity of phase-change induced plasticity steel (TRIP steel), twin-induced plasticity steel (TWIP steel), martensitic steel (M steel), Q & P steel (quenching and partitio-ning), hot stamping Forming steel (HF steel, boron steel) and nanoparticle precipitation strengthening steel (NANO-HITEN steel)[9], and Fe-Mn which combines the advantages of high strength, high toughness, low density, impact resistance, candle resistance, etc. -Al steel, tensile strength of 700-1100MPa, elongation after breaking reaches 60%, higher Mn, A1, C and other light alloy content can guarantee its excellent formability and impact resistance, its density can be reduced to 6.5-7.0g/cm3, which is 11%-18%[10] lower than that of pure iron, is one of the important materials for realizing vehicle weight reduction[11]. Among them, Fe-Mn-Al light-duty high-strength steels are becoming the development direction of automotive steels in the future due to their high tensile strength, good plasticity, and low density[12].

In the field of construction, with the development of economy, China's building construction is developing rapidly. High-rise and super-high-rise buildings have been scattered all over the country. High-strength steel is a necessary material for high-rise and super-high-rise buildings. In addition, the use of high-strength steel instead of ordinary steel can effectively save energy. The amount of steel used in the 20-story building, tensile strength from 490MPa to 590MPa can save 20% of steel, so the tensile strength of 590MPa ~ 780MPa has become a new trend of high-strength construction steel[13].

[a] 113503287@qq.com

© The Authors, published by EDP Sciences. This is an open access article distributed under the terms of the Creative Commons Attribution License 4.0 (http://creativecommons.org/licenses/by/4.0/).

Domestic and foreign commonly used high-strength steels are Japan-produced BL385, SA440, and SA630, whose tensile strengths can reach 550 MPa, 590 MPa, and 780 MPa. Structural steels used in high-rise buildings in some European countries include Q235GJ, Q345GJ, Q390GJ, and Q420GJ. In recent years, the development of high-rise steel structure buildings and large-span space structures in China has imposed higher requirements on steel strength and other indicators. For example, the National Stadium uses the Q460E-Z35, the National Swimming Pool (Water Cube) uses the Q420C. The CCTV's new site uses the Q420D-Z25, Q460E-Z35 high-strength steel. High-strength steel achieves lightness by increasing its strength, but its quality is reduced to a lesser extent, and the production of high-strength steels or over-strength steels is also expensive, and its application in certain specific fields has certain limitations. The use of lighter alloy materials for the manufacture of transportation equipment components is another way to achieve lightweight equipment.

2 high-strength aluminum alloy

In modern materials science, aluminum alloy materials have become second only used less to steel, and high-strength aluminum alloys are based on Al-Cu-Mg and Al-Zn-Mg-Cu alloys. With high strength, good processing performance and excellent welding performance, it has been widely used in aviation industry and civil industry, etc. [14]. Especially in the aviation industry, it occupies a very important position and is one of the main structural materials for the aviation industry. Al-Zn-Mg-Cu series ultra-high-strength aluminum alloys have been used to manufacture aircraft hubs, spars and wing-assisted key bearing components[15], its yield strength exceeds 500MPa[16], compared with the same yield strength of steel materials, its weight is greatly reduced. Cui Caiwei et al.[17] used 7A04 ultra-high strength aluminum alloy (7A04 aluminum alloy belongs to Al-Zn-Mg-Cu alloy) to replace steel. The tensile strength of the base is more than 540MPa, the yield strength is greater than 460MPa, and the elongation is greater than 8% and its quality is reduced by 30% compared with steel.

In addition, there are some promising high-strength aluminum alloys: Al-Li alloy—Lithium is the lightest metal element. For every 1% increase of lithium element in aluminum alloy, the density can be reduced by 3%, and the modulus can be increased by 5 %. In addition to the advantages of light weight, high modulus and high strength, aluminum-lithium alloys also have excellent fatigue resistance and good low-temperature toughness. In recent years, aluminum-lithium alloys have developed toward super-toughness, ultra-low density, high-strength weldability, low anisotropy, and strong thermal stability. However, the cost of aluminum-lithium alloys is high, the storage of metallic lithium is small, extraction is difficult, the production process is more complicated, the production process needs to be protected by argon gas and the cold hardening speed is fast, and when the temperature is high at a short time, the toughness declines seriously[18]; Aluminum matrix composites, with its small density, high specific strength and specific stiffness, specific elastic modulus, good electrical and thermal conductivity, high temperature resistance, oxidation resistance, corrosion resistance, flexible manufacturing process and many other advantages caused people The general concern is that the aluminum matrix composites currently being developed mainly include SiC/Al, B/Al, BC/AI, Al_2O_3/Al, etc.; Rare-earth aluminum alloys add some trace rare-earth raw materials to high-quality primary aluminum to improve the strength of primary aluminum, such as tensile strength, electrical conductivity, ductility, corrosion resistance and so on. Such as silicon, niobium and other minor elements added to the aluminum such as A356 aluminum alloy in a specific ratio, dedicated to the automotive aluminum wheel casting[19].

3 High-strength magnesium alloys

Magnesium alloy is the lightest metal structural material[20], its density is 1.75g/cm3, about 2/3 of aluminum alloy, 1/4 of steel. Magnesium alloy has low strength and elastic modulus, but it has high specific strength and specific rigidity. In the same weight of components, the use of magnesium alloy can make the component get higher rigidity. Moreover, magnesium alloys also have high damping capacity and good shock absorption performance, and they can withstand large shock and vibration loads. In addition, magnesium alloys are rich in resources, casting properties, excellent cutting performance, easy recycling, and no pollution. Widely used in aviation, aerospace, defense, automotive, communications electronics, computers, home appliances and other industrial fields, known as "21st century green environmental protection engineering materials" [21].

However, the application of magnesium alloys is far less widespread than that of aluminum alloys. The main reasons for this are that magnesium alloys have low absolute strength, poor room-temperature deformation processing capabilities, and are prone to oxidative combustion and susceptibility to corrosion, which limits their development as structural materials[22], so the development of high-strength magnesium alloys is the key to solving this series of problems. The addition of alloying elements to strengthen magnesium alloys is a simple, effective and economical method to increase the strength of magnesium alloys. It is more effective to study the addition of rare earth elements on the surface. Rare earth elements can increase the electron concentration in the alloy and enhance the bonding strength between the magnesium alloy atoms. It can also purify the alloy melt, refine the grain, improve the alloy structure and casting performance, thereby improving the overall performance of magnesium alloys, especially high temperature mechanical properties[23], the world's total of rare earth cast magnesium alloy brands account for the total number of magnesium alloys More than 50% [24], the earliest preparation of high-temperature tensile Mg-Ce alloy, after the researchers found that the heat resistance of rare earth magnesium alloys in accordance with the

Table 1 Titanium grades comparison of different countries

Alloy type	China(GB)	Soviet Union (TOCT)	United States (ASTM)	United Kingdom (IMI)	Gemany(BWB)	France(NF)	Japan(JIS)
Pure titanium	TA0	-	-	-	-	-	-
	TA1	BT1-0	Ti-35A	IMI115	LW3.7024	T-35	KS50
	TA2	BT1-1	Ti-50A	IMI125	LW3.7034	T-40	KS60
	TA3	BT1-2	Ti-65A	IMI135	-	-	KS85
αtitanium alloy	TA4	48-T2	-	-	-	-	-
	TA5	48-OT3	-	-	-	-	-
	TA6	BT5	-	-	-	-	-
	TA7	BT5-1	Ti-5Al-2.5Sn	-	-	-	KS115
	TA8	BT10	Ti-5Al-2.5Sn	IMI317	TA5E	-	AS
βtitanium alloy	TB1	BT15	-	-	-	-	-
	TB2	-	-	-	-	-	-
	TB3	-	-	-	-	-	-
α+βtitanium alloy	TC1	OT4-1	-	-	-	-	ST-A90
	TC2	OT4	-	-	-	-	-
	TC3	BT6C	-	-	-	-	-
	TC4	BT6	-	-	LW3.7164	-	-
	TC5	BT3	Ti-6Al-4V	-	-	T-A6V	-
	TC6	BT3-1	-	-	-	-	-
	TC7	AT6	-	IMI315	-	-	-
	TC8	BT8	-	-	-	-	-
	TC9	-	-	IMI318	-	T-A6V6S2	-
	TC10	-	Ti-6Al-6V-2Sn	-	-	-	-

sequence of La, Ce, Nd increased, can make the microstructure of magnesium alloy refinement, improve plastic shape. Magnesium alloys commonly used include: Mg-Re-Zn-Zr alloys, Mg-Nd-Zn-Zr alloys, Mg-Y-Zn-Zr alloys, and Mg-Re-Al alloys. At present, the Mg-Zn-Y alloy developed in Japan has the highest strength, 2.5 times that of ordinary magnesium alloys.

The rare earth-free magnesium alloys include: Mg-Al alloys, Mg-Zn alloys, Mg-Sn alloys, etc. Among them, AZ91 alloys (belonging to Mg-Al-Zn alloys) have excellent casting process performance and good resistance. Erosion, the addition of Zn will enhance the strengthening effect, but will increase the thermal cracking sensitivity; Si addition can effectively improve the high temperature creep properties of magnesium alloys; adding Bi, Sn, Sb alloy structure can be significantly refined[25].

4 Titanium

Titanium alloy is a new type of structural material. It has characteristics such as low density, light weight, high fracture toughness, high heat resistance, high strength ratio, good fatigue strength and crack propagation resistance, and excellent toughness and corrosion resistance, which is widely used in aviation, aerospace, military, weapons, shipbuilding, vehicle engineering, biomedicine and other fields [26]. Titanium alloy tensile strength of up to 1500MPa, comparable to ultra-high strength steel, while the density of pure titanium is only 4.5kg/cm3, only 50% of steel, its specific strength is the highest in commonly used engineering materials.

In the 1970s, a high-strength titanium alloy developed by a U.S. company was mainly used for Ti6-23-20d alloys. It has high strength, excellent toughness, tensile resistance, and good creep resistance[27]. The Beta-C alloy developed by the same company is widely used in the fields of petrochemicals and natural gas. The new type of casting alloy BT36 developed in Russia can be used in places where the tensile strength is 1200 MPa, and its plasticity is better than that of the common Ti-16-2 alloy. It is commonly used as the aircraft body. At 50°C, the ultimate tensile strength of this alloy was 1200 MPa, 700 MPa at 300°C, and 350 MPa at 500°C[28].

China has also made some achievements in the research and development of titanium alloys. The earliest TB5 and TB6 titanium alloys were developed. After the reform and opening up, TB9 was an ultrahigh-strength titanium alloy. Shanghai Metal Research Institute successfully produced Ti-563 alloy, and its mechanical properties and the fracture toughness has been greatly improved compared with the former kinds; since the 21st century, the TD22 titanium alloy developed by the Northwest Institute of Metal Research in China has laid a good foundation for the production of titanium alloy castings.

Summary

For metal materials, the method of achieving equipments' weight reducted and securing strength requirements and improving applicability is mainly metal alloying, and metal alloying can substantially improve metal properties. At present, light-weight high-strength materials with promising development are mainly titanium alloy materials, but due to their high cost, their widespread application is affected.

References

1. Yu Muhuo, Peng Yuqing, Tian yincai, Li Shuang, Han Keqing. High-performance fiber development status and reflection. 2011 National Conference on Polymer Academic Papers. 2011.9.24-28

2. Zhen Shu. Study on Microstructure and Mechanical Properties of Resistance Spot Welding Joint of High Strength Steel B340/590DP [D]. Jilin University, 2010
3. Tian Dacheng. Study on Dynamic Mechanical Behavior of DP780 High Strength Steel [D]. Shanghai Jiao tong University，2008
4. Chen Wenxue. Research on Key Risk Monitoring and Emergency Rescue Method System of Coal Mine [D]. Shandong University of Science and Technology. 2005
5. Wang Changlong, Li Xuewei, Xu Fang, Dong Shunji, Feng Yalei. Application of static positive pressure explosion-proof technology in mine search and rescue robots [J]. Coal Engineering. 2009, 11
6. Tang Di, Mi Zhenli, Chen Yulai . Technical requirements and research and development status of New foreign automobile steel [J]. Steel, 2005, 40(6):1.
7. Gao Jianhong, Wang Min, Wu Qingqing. Study on Resistance Spot Welding of High Strength Steel TRIP590 [M]. 2008 Proceedings of the International Forum on Welding of Light Metals and High Strength Materials. Bei jing. 2008: 133-9.
8. Ma Fengcang, Feng Weijun, Wang Li, et al. Research status of TWIP steel [J]. Bao steel Technology, 2009, 1(6): 62-66.
9. Bo Jianren. Steam car lightweight technology [J]. Microalloying technology, 2008, 8(3):1.
10. Yang Guofu. Preparation Process and Deformation Mechanism of Fe-Mn-Al Light-weight High-strength Steel for Automobile [D]. Beijing University of Science and Technology. 2015
11. Wang Li. Effect of Al Element on Microstructure and Mechanical Properties of Lightweight High Strength Fe-Mn-Al Steel and Its Action Mechanism [D]. Lanzhou University of Technology. 2016
12. Yang Fuqiang, Song Renbo, Zhang Leifeng, Zhao Chao. Research on Microstructure and Properties of 900MPa Grade Fe-Mn-Al Lightweight High Strength Steel. The 9th China Steel Conference Proceedings
13. Chang Yuefeng, Wang Zhubing, Zhao Wenzhong. Development Trend of Low Alloy High Strength Wide Steel Plates [J]. Steel, 2007, 42(8) : 1-6
14. Song Renguo. Research Status and Development Trend of High Strength Aluminum Alloys [J]. Material Guide. 2000, 1. 1 (14)
15. Zhang Jue. Application of Aluminum Alloy in Aerospace Aviation [J]. Aluminum processing. 2009, 3:50-53.
16. Fang Hongze. Development of over-strength aluminum alloys [D]. Shenyang University of Technology. 2014
17. Cui Caiwei. Research on Forming Process of Large High-bar Complex 7A04 Aluminum Alloy Base [D]. Taiyuan: North University of China, 2012.
18. Liu Bing, Peng Chaoqun, Wang Richu, Wang Xiaofeng, Li Tingting. Research Status and Prospect of Aluminum Alloys for Large Aircraft [J]. Chinese Journal of Nonferrous Metals. 2010 , 9. 9(20)
19. Feng Yunxiang, Liu Jiangan. The Development and Research Direction of Ultra High Strength Aluminum Alloys [J]. Material Guide, 2004, 18 (8A): 196−202.
20. Shen Guozhu. Research Status and Development Trend of High Strength Aluminum Alloys [J]. Technology Economy Market. 2016, 3
21. Li Hongzhan, Li Zhengxian, Du Jihong, Guo Guangrui, Wang Pei. Research status of high-strength magnesium alloys [J]. Material Development and Application. 2011, 12: 89-92
22. Zhang Dingfei, Zhang Hongju, Lan Wei, Ma Chunhua. Research Progress of High-strength Magnesium Alloys [J]. Journal of Materials Heat Treatment. 2012, 6.6(33)
23. Eliezer D，Aghion E，Froes F H. Magnesium science，technology and pplications[J]. Advanced Performance Materials，1998(5) : 201-212.
24. [24] Yu Wenbing, Liu Zhiyi, Chang Nanpu, et al . Research and Development of Rare Earth Metamorphic Magnesium Alloys [J] . Material Guide, 2006，20(11) : 65- 72.
25. Yang Bin, Huang Ruifen, Sun Quanxi, Wang Shoulong. Application and Development of Rare Earth in Magnesium Alloy [J]. Journal of Baotou Vocational and Technical College. 2006, 6: 7(2)
26. Xie Yongsheng, Xu Chunhui. Development Status and Prospects of High Strength Casting Titanium Alloys [J]. Casting technology. 2014, 2384-2386
27. Zhang Xuhu, Shan Qun, Chen Yonglai, et al. Application and Development of Titanium Alloy in Spacecraft [J]. Chinese material progress, 2011(6): 28-32, 63.
28. Yan Ping. Study on Microstructure and Properties of BT22 High Strength Casting Titanium Alloy [D]. Beijing: Machinery Research Institute, 2007.

Effect of impact energy on wear behavior of high chromium white iron produced by liquid die forging

B Qiu[1], S M Xing[1,a], Q Dong[1], H Liu[2]

[1] *School of Mechanical, Electronic and Control Engineering, Beijing Jiaotong University, China*

[2] *School of Economics and Management, Beijing Jiaotong University, China*

Abstract. Impact abrasive wear behavior of high chromium white iron (HCWI) produced by liquid die forging process were investigated. the wear tests were performed with the MLD-10 abrasive wear testing machine, using SiO_2 abrasive and with four impact energies of 1.5 J, 2.5 J, 3.5 J and 4.5 J for 120 min. The results indicated that the cumulative volume loss of HCWI sample increases with the growth of impact energy, and exhibits best wear resistance under low impact condition. For given impact energy, the volume loss increases with the increasing of wear time, which shown an approximately liner tendency. The macro-morphologies, SEM images of worn surface and cross-sectional images of wear samples were observed by optical microscope and SEM, and the wear mechanism and characteristics were analyzed. Results shown that the wear characteristics is mainly based on the shallow ploughing and accompanied by plastic deformation under lower impact energy, while the fatigue peeling and embedded abrasive become the most significant characteristics when the impact energy is higher.

1 Introduction

As one of the three failure forms of materials in engineering (the others are fatigue and corrosion), wear leads to both significant economic and environmental losses due to shortening the working life of equipment [1,2]. Typical wear can be divided into five categories as: abrasive, adhesive, erosive, fretting and chemical wear [3]. Among them, abrasive wear mainly exists in the transportation, mining and mineral processing industries, and accounts for roughly about 50% of industrial wear situations [4,5]. In general, abrasive wear is the process by which the surfaces of components are damaged by the gouging and scratching action of hard abrasives trapped between the contacting surfaces during the wear.

Impact abrasive wear, as one of the forms of abrasive wear, prominently increases the mass loss of materials as it combines friction and impact [6], has been gradually reported in recent years. Latest researches, however, start focus on advanced wear service materials such as metal matrix composites, coating, and cemented carbide [7-9]. But it is well-known that iron and steel material cannot be replaced in engineering, such as high chromium white iron (HCWI), has widely used in several structural applications as the third-generation wear-resistant materials, which attributed to the multiple advantages of HCWI such as high strength and wear resistance [10]. Little attention has been paid to the effect of impact energy on the wear mechanism and characteristics for the traditional HCWI material under impact wear condition. It would be of great importance to study the wear properties of HCWI as the composite components are often subject to wear loading conditions in service in various industries. In this paper, HCWI were prepared by liquid die forging process with conventional melting and casting route. The impact wear resistance of HCWI was investigated under casting conditions. The present work highlights the influence of impact energy on the wear characteristic and the wear mechanism of HCWI.

2 Materials and experimental procedures

The high chromium white iron (HCWI) used in the present study were melted in a 25 kg laboratory intermediate frequency induction furnace by using high-purity raw materials. Then the melt was rapidly superheated to 1560-1580 °C. After holding at the temperature for 5 min, the molten alloy was poured into the squeeze casting die placed on the bed of a 200 tones hydraulic press and was immediately squeezed at 115 MPa pressure. The die was made of heat treated H13 steel and was preheated to about 200 °C before casting. The chemical composition of HCWI is 26% Cr, 2.8% C, 1.8% Mn, 0.8% Si, 0.8% Mo, 0.4% Ni and balance Fe (all in wt.%).

Samples for metallography, impact and wear tests were cut from the cast bars, respectively. The metallographic samples (8mm×8mm×10mm) were cut with metallographic cutting machine and prepared according to standard mechanical polishing techniques, followed by etching with four percent Nital solution (4ml

[a] Corresponding author: smxing@bjtu.edu.cn

© The Authors, published by EDP Sciences. This is an open access article distributed under the terms of the Creative Commons Attribution License 4.0 (http://creativecommons.org/licenses/by/4.0/).

HNO_3+98ml ethanol) to reveal the microstructure. Optical microscope (OM, DM2000X, Leica) was used to observe the microstructure of samples. The hardness of HCWI is tested by a Rockwell hardness tester (TH300) with the load of 150 kgf and the holding time of 2 s. Ten readings are measured and take an average as the final result. The non-notched impact test samples (10mm×10mm×55mm) were prepared to measure the impact toughness of HCWI, and the test was performed on a JB-50 impact tester made in China. Three impact samples were tested for different composites, then take the average value as the result. The fracture surface of the impact samples was observed by scanning electron microscopy (SEM, EVO18, Carl Zeiss) equipped with an energy dispersive spectrometer (EDS).

The wear tests were performed on a dynamically loaded abrasive wear tester (MLD-10, produced by Zhangjiakou Cheng Xin Testing Equipment Manufacturing Co. Ltd., China) at room temperature, the schematic representation was exhibited in Figure 1 (a), the impact–abrasion wear of the samples is realized with the non-stop rotating of the lower sample and the vertical reciprocating motion of the upper sample. The experimental material (upper sample) were cut into 10 mm × 10 mm × 30 mm size, as shown in Figure 1 (a), and the surface of 10 mm × 10 mm was mechanically polished for impact-abrasive wear tests. The counterpart ring (lower sample) was made of 45# steel (58HRC) with the size of 50 mm in diameter and 20 mm in length, which was quenched at 840°C in water and tempered at 300°C for 1 h. During the test, the wear sample is fixed on a clamping chuck which connects with a hammer. The impact frequency of wear sample was 150 times/min, the rotation speed of contract sample was 200r/min, and the impact energy is set at 2 J, 3 J and 4 J, respectively. Refined quartz sands are used as the abrasive material (with particle size of 10-20 mesh, density of 2.6 g/cm^3 and hardness value of 800 HV–1200 HV). The process of the wear test was shown in Figure 1 (b). Each sample was subjected to pre-wear for 30min before the formal tests. The wear mass loss was measured every 10 minutes in the tests, before and after each test, the wear sample was cleaned ultrasonically, dried and then weighed by a precision electronic balance with an accuracy of 0.0001g, and take the average value of three measurements as the results. The wear tests were performed for 120 minutes, and the wear tests were repeated three times with new samples for each test condition to ensure the reliability. The wear surfaces of the tested samples were observed and examined by SEM.

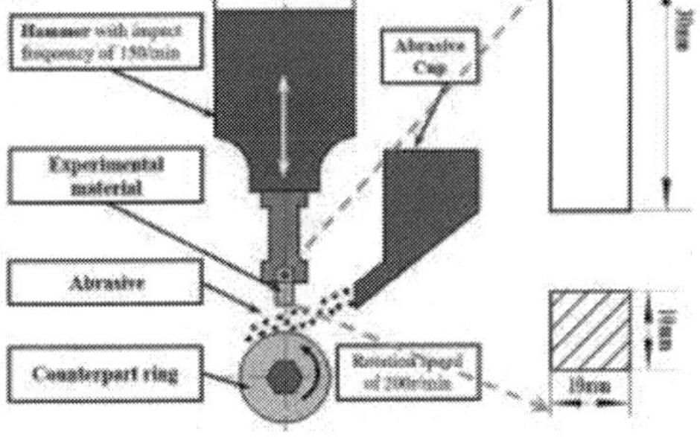

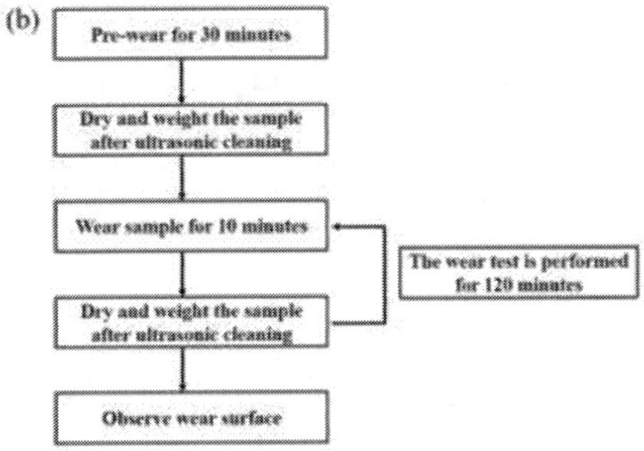

Figure 1. (a) The schematic of wear tester and the size of wear sample; (b) The flow chart of impact wear test.

3 Results and discussion

3.1 Microstructure and impact property of HCWI

The metallograph of HCWI under low and high times are show in Figure 2, the microstructure of HCWI exhibits a typical dendritic structure, which consists of the martensite and retained austenite in the form of off-white dendrites, as well as the eutectic structure between dendrites (Figure 2 (a)). The eutectic between dendrites are composed of alloy carbide, martensite and retained austenite, and the carbide, which shaped fish-bone, stripy or massive, distributed in the martensite and small amount of retained austenite matrix (Figure 2 (b)). The hardness test results show that the average macro-hardness of HCWI reaches to 48.86 HRC, and the impact test shows that the impact toughness of HCWI reaches to 9.43 J·cm^{-2}. The SEM micrograph of fractured surface of the HCWI is shown in Figure 3 (a). It can be seen that the fracture characteristics is river pattern, which formed by the confluence of cleavage steps. The fracture morphology indicates the fracture mechanism is dominated by cleavage fracture, which consistent with the fracture characteristics of HCWI. In addition, the energy spectrum analysis results of the regions 1 is shown in Figure 3 (b), It can be observed that the main elements of the region 1 are Fe and Cr.

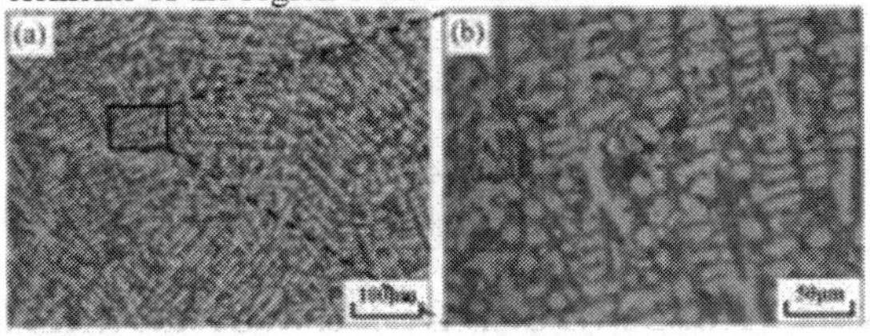

Figure 2. The microstructure of HCWI under (a) low and (b) high magnification

https://doi.org/10.1051/matecconf/201820703011

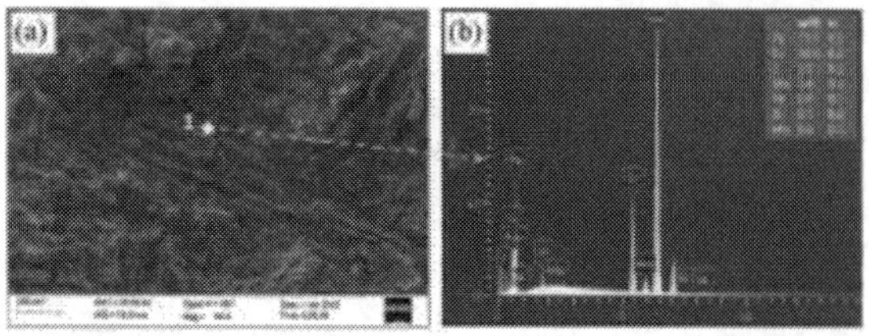

Figure 3. (a) The typical micrograph of fracture surface of HCWI (b) EDS analysis

3.2 Analysis of volume loss of wear samples

The volume loss of HCWI samples as a function of wear time under different impact energies is given in Figure 4 (a). For given impact energy, the volume loss increases with the increasing of wear time, which shown an approximately liner tendency. And the fitting curve equation between volume loss and wear time is $y = 0.0015x - 0.02175$, $y = 0.00154x - 0.01436$, $y = 0.00155x - 0.078$ and $y = 0.00158x + 0.00186$ for 1.5J, 2.5J, 3.5J and 4.5 J, respectively. Meanwhile, the confidence level of the linear fitting is all greater than 0.99. Fig. 4 (b) shows the influence of impact energy on the cumulative volume loss of the HCWI. It can be seen that the cumulative volume loss of tested material increases with the growth of impact energy, which illustrates that the minimum and maximum volume loss occur under the impact energy of 1.5 J and 4.5 J. Accordingly, the HCWI exhibits best wear resistance under low impact condition.

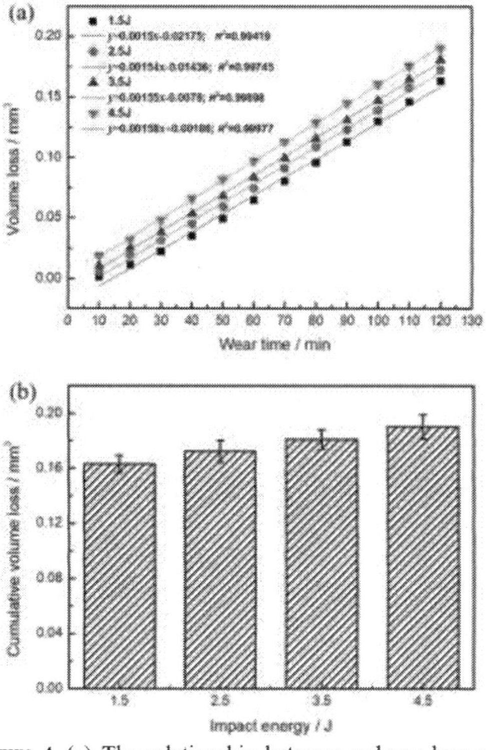

Figure 4. (a) The relationship between volume loss and wear time under different impact energy; (b) Variation of cumulative volume loss with impact energy for HCWI.

3.3 Analysis of worn surface of wear samples

In order to get a better understanding of impact-abrasive wear mechanism, as well as the effect of impact energy and wear time on wear behavior, investigation of the worn surface will be necessary. Figure 5 presents the macro-morphologies, SEM images of worn surface and cross-sectional images of the HCWI samples after a wear time of 120 min under the impact energies from 1.5 J to 4.5 J. It can be seen form the macro-morphologies of worn surface that the wear loss is gradually increasing with the impact energy. Correspondingly, the "wear pit" caused by impact and wear is bigger and deeper with the impact energies, which can be further illustrated from the cross-sectional images of worn surface. The depth of the "wear pit" in worn surface is 1.6 mm for 1.5 J, 1.9 mm for 2.5 J, 2.1 mm for 3.5 J and 2.2 mm for 4.5 J, respectively, as is seen in Figure 5. In addition, it can be observed form the SEM images of samples that the worn surfaces show three kinds of morphological characteristics, such as ploughing-cutting (as indicated by the arrow in Figure 5 (a) and (b)), fatigue peeling (as indicated by the arrow in Figure 5 (d)) and embedded abrasive (as indicated by the arrow in Figure 5 (c) and (d)).

During the impact abrasive wear process, the samples impacted the counterpart ring periodically, as well as the flowing abrasive between them. Under the lower impact energy (i.e. 1.5 J), the wear characteristics is mainly based on the shallow ploughing [11]. For the HCWI, the hard phase is M_7C_3 carbides, and the soft phase is martensite matrix, when the impact energy is low, the abrasive particles were squeezed into the softer martensite matrix at low contact pressure, the depth of indentation is shallow, and then the ploughing wear occurred on the matrix under the action of tangential stress [12,13], the ploughing is shallow and the wear loss is small. In addition, the plastic deformation is also observed on the worn surface, which caused by the existence of retained austenite in the matrix. When the impact energy increases to a medium value, like 2.5 J and 3.5 J, it can be seen that the ploughing is wider and deeper, and the cutting trace were clearly seen (Figure 5 (b)). That is because with the increase of impact energy, the acting force that abrasive to matrix is increase, and the depth of the pierce into the matrix increases. The cutting wear of the material is carried out under the action of tangential stress, and the cutting mechanism dominated the wear process. In addition, the embedded abrasive can be observed in worn surface when the impact energy is 3.5 J (Figure 5 (c)). When the impact energy increases to a high value (i.e. 4.5J), the fatigue peeling and embedded abrasive become the most significant characteristics. Under the high energy, massive carbides were subjected to larger impact force, resulting in local fatigue cracks, then the cracks expand rapidly, and following by the break of carbides and separate from the matrix. Besides that, the embedded abrasive can also be observed under the impact condition of 4.5 J. The embedded abrasive experienced the following steps: firstly, the abrasive with high brittleness was broken under the high impact effect and debris of abrasive occurred; then the sharp debris was squeezed into material under impact condition of the hammer and pierced into the material under the slide condition of the ring; next the crack produced by piercing expanded with exceeding the tensile strength limit; finally, the material that pierced into the debris separated from the surface and its debris mixed with abrasive caused further wear.

3.4 Analysis of morphologies of counterpart ring

In order to study the wear condition of the counterpart rings during the wear tests, the worn morphologies of the rings, which corresponding to the different impact energies were observed, and the obtained images are shown in Figure 6. It can be seen that the wear characteristic of the counterpart ring dominates by the continuous furrows that has occurred plastic deformation (F-PD) under the different impact energies. Since the condition of abrasives under the contact stress, the furrows firstly occurred in the surface of the ring, and further plastic deformation happened in the furrows due to the constant and rotary wear condition, thus formed the obvious and continuous F-PD, and the trace of F-PD can be clearly seen. In addition, the embedded abrasive particle under the high impact energy can also be observed (Figure 6 (d)).

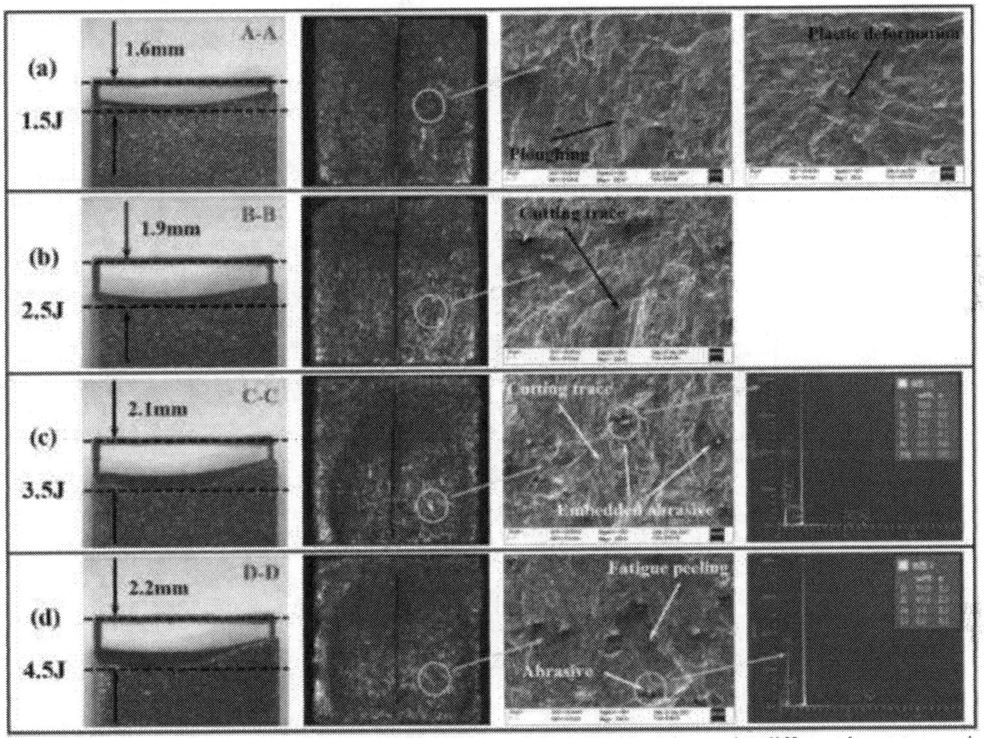

Figure 5. The worn surface and cross-sectional images of the wear samples under different impact energies: (a) 1.5 J; (b) 2.5 J; (c) 3.5 J; (d) 4.5 J.

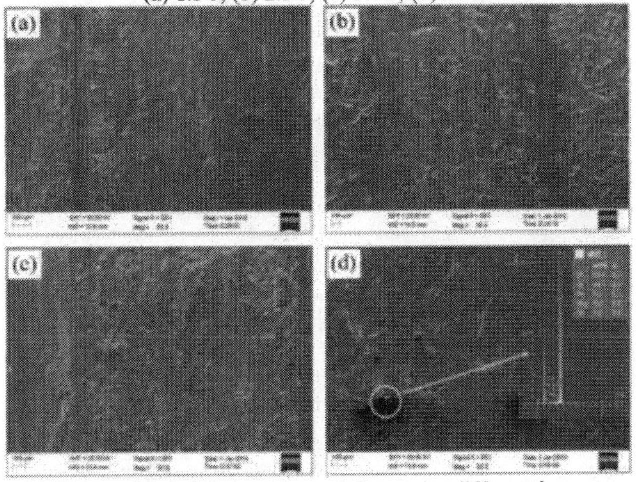

Figure 6. Worn surface of the counterpart ring under different impact energies: (a) 1.5 J; (b) 2.5 J; (c) 3.5 J; (d) 4.5 J.

4 Conclusions

In order to study the wear condition of the counterpart rings during the wear tests, the worn morphologies of the rings, which corresponding to the different impact energies were observed, and the obtained images are The impact abrasive wear behavior of HCWI produced by liquid die forging process were studied under the impact energies of 1.5 J, 2.5 J, 3.5 J and 4.5 J in this work. The main conclusions are drawn as follows:

(1) The microstructure of HCWI exhibits a typical dendritic structure, and the eutectic structure between dendrites are composed of carbide shaped fish-bone, stripy or massive, martensite and retained austenite. And the impact fracture characteristics of HCWI is river pattern, which formed by the confluence of cleavage steps.

(2) The cumulative volume loss of HCWI sample increases with the growth of impact energy, and for given impact energy, the volume loss increases with the increasing of wear time, which shown an approximately liner tendency.

(3) The wear characteristics of HCWI is mainly based on the shallow ploughing and accompanied by plastic deformation under low impact energy, when the impact energy increases to a medium value, the wear characteristics are mainly shown as wider and deeper ploughing and cutting, as well as some embedded abrasive. When the impact energy reach to a high value, however, the fatigue peeling and embedded abrasive become the most significant characteristics.

Acknowledgments

This work was supported by the Fundamental Research Funds for the Central Universities under Grant No. 2018YJS139

References

1. H.H. Ding, C.G. He, L. Ma, Wear **352**, 1 (2016)
2. V. Ratia, I. Miettunen, V.T. Kuokkala, Wear **301**, 94 (2013)
3. E.G. Moghaddam, N. Karimzadeh, N. Varahram, Mater. Sci. Eng. A **585**, 422 (2013)
4. B. Narayanaswamy, P. Hodgson, H. Beladi, Wear **350**, 155 (2016)
5. M. Lindroos, K. Valtonen, A. Kemppainen, Wear **322**, 32 (2015)
6. V.G. Efremenko, K. Shimizu, T. Noguchi, Wear **305**, 155 (2013)
7. T. Wang, M. Shozaki, M. Yamamoto, Mater. Des. **66**, 498 (2015)
8. W. Wang, C. Zhang, P. Xu, Mater. Des. **73**, 35 (2015)
9. M. Antonov, R. Veinthal, D.L. Yung, Wear **322**, 971 (2015)
10. S. Wang, L. Song, Y. Qiao, Tribol. Lett. **50**, 439 (2013)
11. Z. Zeng, L. Wang, L. Chen, Surf. Coat. Technol. **201**, 2282 (2006)
12. M.S. Bingley, S. Schnee, Wear **258**, 50 (2005)
13. P. Kulu, R. Veinthal, M. Saarna, Wear **263**, 463 (2007)

[a] Corresponding author: smxing@bjtu.edu.cn

Research on multidimensional loading device of material mechanical test

Guohua Zhao[1,a], Dan Wang[1,b], Liangbao Liu[1,c*], Rui Fan[1,d] and Rukun Mi[1,e]

[1]*School of Mechanical Engineering and Automation, Beihang University, 100191 Beijing, China*

Abstract. A multidimensional loading device for the material mechanical test based on *Stewart Platform* was proposed and manufactured in this paper. A determination method of mechanism parameters toward specific engineering requirements was analysed via inverse kinematics and contact interference boundary condition of mechanism components. A set of optimized mechanism parameters was achieved and then the pose space was obtained. Meanwhile, five repeated quasi-static standard tensile tests were performed on the multidimensional loading device and the standard tensile testing machine, respectively. The average yield strength and ultimate strength of the specimen achieved from the different testers were both merely 1.13%. Results indicated that material mechanical multidimensional loading could be conducted by the parallel mechanism.

1 Introduction

To guarantee engineering structure or machinery working in order, components should possess enough capacity to bear the loads on them [1]. Mechanics of material mainly focuses on deformation, force, and failure analysis. It provides related theory for optimizing component design, such as strength, stiffness, and stability [2]. Majority of formulas in mechanics of material were derived from mechanical tests and were established with many hypotheses and simplifications. Therefore, errors were inevitably generated and may deeply affect the validity and accuracy of the theoretical calculation. Yield strength derived by Von Miss were 15% greater than that obtained by Tresca in the pure torsional experiment [3]. Furthermore, many failure analyses, in practical application, could not be conducted by theoretical calculation [4]. Consequently, to pursue more accurate mechanical properties and the effective utilization of materials, a considerable amount of different material mechanical experiments were done. In addition to standard single axis tests, two-step uniaxial tests [5], plane biaxial tests [6], combined tension/torsion strain path change tests [7-8] were carried out. However, multidimensional loading experiments that are more in line with the practical engineering application were rarely performed. The main reason that these experiments were not conducted was the lack of multidimensional loading device (*MLD*) [9].

In this work, a *MLD* for material mechanical test based on *Stewart Platform* was proposed and manufactured. A brief architecture description and the inverse kinematics of the *MLD* were presented at the beginning and then the detailed method of designing and optimizing mechanism parameters were given. Finally, five repeated standard quasi-static tensile experiments were conducted on the *MLD* and Zwick Z100, respectively. Results indicated that it was feasible to perform material mechanical experiments on the *MLD*.

2 Architecture description of the *MLD*

The prototype of the *MLD*, shown in fig. 1, was composed of a fixed base, a mobile platform, and six identical motor-driven kinematic chains. Each limb, possessing a ball screw prismatic joint, connected the base with a *Hook* hinge and a spherical joint to the mobile platform. Fig. 2 was the schematic diagram of the *MLD*. A fixed *Cartesian* frame (Op-XYZ) and a moving *Cartesian* frame (Ob-xyz) were settled at the center of the base and the mobile platform, see fig. 2(a). According to the top view, see fig. 2(b), the coordinates of the *Hook* hinge center $\mathbf{P}_i$ (i=1~6) were denoted by eq.(2), while

Figure 1. Prototype of the *MLD*.

[c] Corresponding author: llb1985101@163.com

© The Authors, published by EDP Sciences. This is an open access article distributed under the terms of the Creative Commons Attribution License 4.0 (http://creativecommons.org/licenses/by/4.0/).

https://doi.org/10.1051/matecconf/201820703012

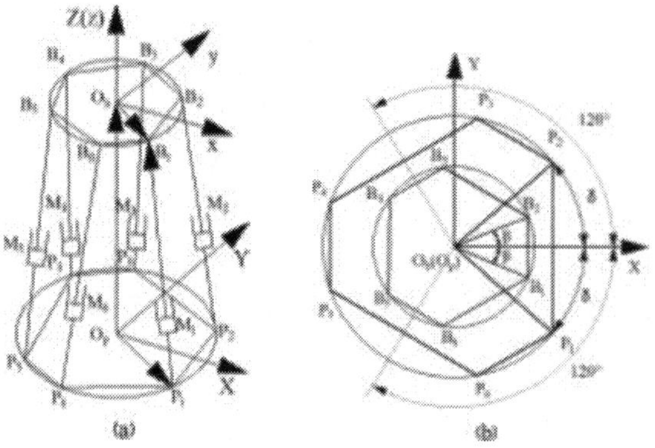

Figure 2. Schematic diagram of the *MLD*.

eq.(1) described the coordinates of the spherical joint center $\mathbf{B}_i$ (i=1~6).

$$\begin{aligned}
\mathbf{B}_{b1} &= [rcos(b), rsin(-b), 0] \\
\mathbf{B}_{b2} &= [rcos(b), rsin(b), 0] \\
\mathbf{B}_{b3} &= [rcos(2p/3-b), rsin(2p/3-b), 0] \\
\mathbf{B}_{b4} &= [rcos(2p/3+b), rsin(2p/3+b), 0] \\
\mathbf{B}_{b5} &= [rcos(4p/3-b), rsin(4p/3-b), 0] \\
\mathbf{B}_{b6} &= [rcos(4p/3+b), rsin(4p/3+b), 0]
\end{aligned} \quad (1)$$

$$\begin{aligned}
\mathbf{P}_1 &= [Rcos(d), Rsin(-d), 0] \\
\mathbf{P}_2 &= [Rcos(d), Rsin(d), 0] \\
\mathbf{P}_3 &= [Rcos(2p/3-d), Rsin(2p/3-d), 0] \\
\mathbf{P}_4 &= [Rcos(2p/3+d), Rsin(2p/3+d), 0] \\
\mathbf{P}_5 &= [Rcos(4p/3-d), Rsin(4p/3-d), 0] \\
\mathbf{P}_6 &= [Rcos(4p/3+d), Rsin(4p/3+d), 0]
\end{aligned} \quad (2)$$

3 Inverse kinematics analysis

To make the description of the moving *Cartesian* frame's pose, which was expressed in the fixed Cartesian frame, consistent with the rotational motion along axes of the fixed *Cartesian* frame, the *RPY* rotational matrix was selected, see eq.(3).

Where, c=cos, s=sin. φ1, φ2 and φ3 were rotational angles of the moving *Cartesian* frame corresponding to the X, Y and Z axes, respectively.

$$\mathbf{R}=\begin{bmatrix} c\varphi_1 c\varphi_2 & c\varphi_1 s\varphi_2 s\varphi_3 - s\varphi_1\, c\varphi_3 & c\varphi_1 s\varphi_2 c\varphi_3 + s\varphi_1 s\varphi_3 \\ s\varphi_1 c\varphi_2 & s\varphi_1 s\varphi_2 s\varphi_3 - c\varphi_1\, c\varphi_3 & s\varphi_1 s\varphi_2 c\varphi_3 + c\varphi_1 s\varphi_3 \\ -s\varphi_2 & c\varphi_2 s\varphi_3 & c\varphi_2 c\varphi_3 \end{bmatrix} \quad (3)$$

The inverse kinematics of this *MLD* was the solution of the limbs' length ($|\mathbf{P}_i\mathbf{B}_i|$) based on the mobile platform' pose. Take any limb, in fig. 2(a), as a research object. According to the vector algebra, eq.(4) could be expressed as below.

$$\mathbf{P}_i\mathbf{B}_i = \mathbf{O}_p\mathbf{O}_b + \mathbf{O}_b\mathbf{B}_i - \mathbf{O}_p\mathbf{P}_i \quad (i = 1 \sim 6) \quad (4)$$

$$\mathbf{B}_{pi} = \mathbf{R}\,\mathbf{B}_{bi} + \mathbf{O}_p\mathbf{O}_{pb} \quad (i = 1 \sim 6) \quad (5)$$

Substituting eq.(4) into eq.(5), the vector equation of the inverse kinematics was given bellow.

$$\mathbf{P}_i\mathbf{B}_{pi} = \mathbf{O}_p\mathbf{O}_{pb} + \mathbf{O}_{pb}\mathbf{B}_{pi} - \mathbf{O}_p\mathbf{P}_i \quad (i = 1 \sim 6) \quad (6)$$

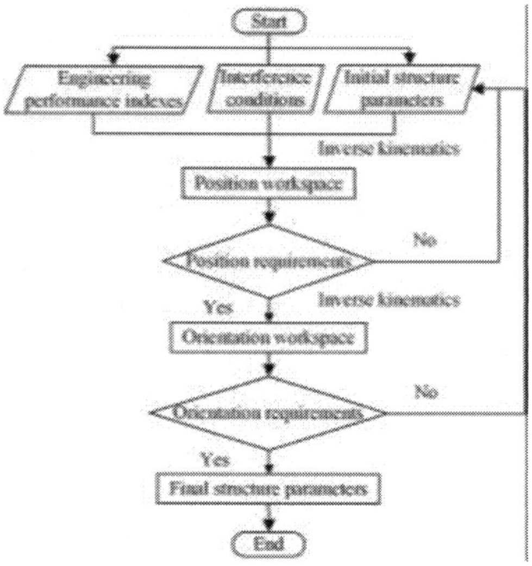

Figure 3. Two-step method of designing the *MLD* parameters.

3.1 Determination of the *MLD* structure parameters

The workspace of parallel mechanism including the position and orientation spaces was used to determine the structure parameters. Considering position and orientation simultaneously could result in a considerable calculation, and make it impossible to intuitively describe the pose space. A two-step method of the pose solution was introduced to design the *MLD* parameters. Firstly, selected a set of initial structure parameters based on the engineering performance indexes, and determined the interference conditions of the parallel mechanism. Secondly, using the inverse kinematics to obtain the position workspace based on the previous results; unless the initial parameters could meet the requirements of the engineering indexes, the program went down to the next step or went back. Thirdly, selected a subspace from the position workspace to verify whether or not the orientation workspace could meet the engineering indexes; the final structure parameters was obtained only if the orientation conditions was met. Detailed process of the two-step method was described in fig. 3.

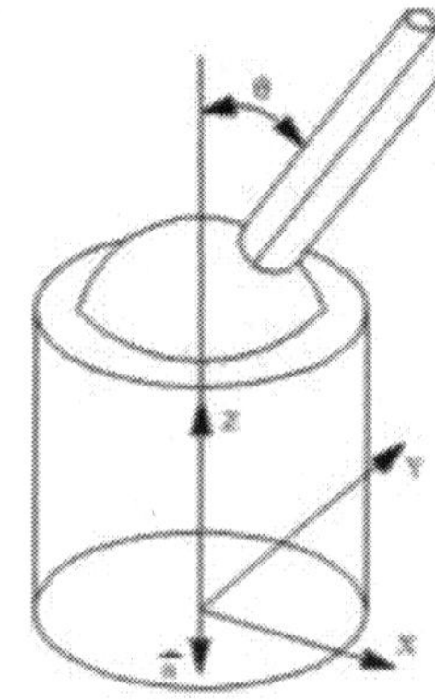

Figure 4. Spherical joint.

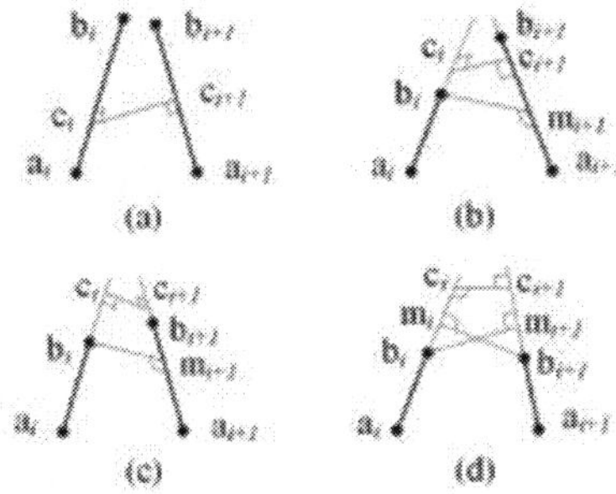

Figure 5. Four situations of two spatial lines.

3.1.1 Contact interference boundary conditions

Contact interference conditions of the mentioned *MLD* were composed of three kinds of interference conditions: the angle range of joints, limbs' length range, and contact interference conditions of components.

According to fig. 4, the spherical joint interference condition could be expressed by eq.(7). Note here that vector **n** denoted the unit vector along the z axis of the mobile *Cartesian* frame.

$$\theta_{pi} = \arccos\frac{\mathbf{B}_i\mathbf{P}_i g(\mathbf{T}\,\mathbf{n}_{pi})}{|\mathbf{B}_i\mathbf{P}_i|} \le \theta_{\max} \qquad (i=1\sim 6) \qquad (7)$$

Similar to the spherical joint, the interference conditions of the *Hook* hinge were given by eq.(8).

$$\gamma_{pi} = \arccos\frac{\mathbf{B}_{pi}\mathbf{P}_i g(\mathbf{T}\,\mathbf{N}_{pi})}{|\mathbf{B}_{pi}\mathbf{P}_i|} \le \gamma_{\max} \qquad (i=1\sim 6) \qquad (8)$$

Where $\mathbf{N}_{pi}$ represented the unit vector along the connecting line of two adjacent *Hook* hinges' rotary center, see fig. 1.

Each length range of limbs could be calculated from eq.(9), as bellow.

$$l_{\min} \le |\boldsymbol{l}_i| \le l_{\max} \qquad (i=1\sim 6) \qquad (9)$$

According to the structure of the *MLD*, the contact interferences only could occur at the end of the motors or at positions where limbs were close to the upper grip. Since the principle of these two kinds of interferences was the same, only the interferences at the end of the motors were analysed in this paper. To simplify the interference analysis, assuming that the motor was a cylinder whose radius was *r*. Therefore, the contact interference analysis between two motors was transferred to the problem of the contact of two cylinders. Let *dmin* be the minimum distance between two cylinders, *d* be the length of the common normal line of these two cylinders' centreline, shown in fig. 5. The radius of the circumscribed circle of the motor, denoted by r_1, could be obtained by eq. (10). Let *s* be the side length of the motor and $r'=r_1$. If the thick line segment a_ib_i and $a_{i+1}b_{i+1}$, shown in fig. 5, did not contact with each other, adjacent motors would not occur contact interference.

$$r_1 = \sqrt{2}s/2 \qquad (10)$$

It could be divided into two main situations according to the relative position of the two thick lines: (1) If $d > 2r'$, motors would not contact each other; (2) if $d \leqslant 2r'$, contact interference could be subdivided into four cases, shown in fig. 5. Detailed analysis of the cases, in fig. 6, were summarized:

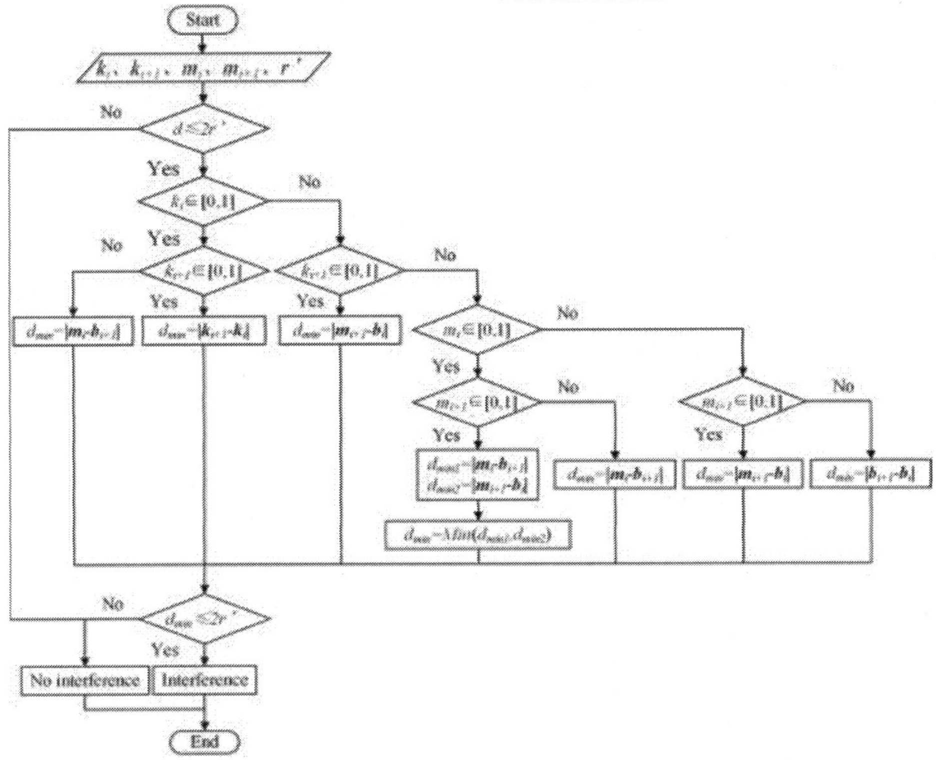

Figure 6. Flow chart of the analysis of contact interference.

1) $dmin = d$, for case fig. 5(a);
2) $dmin = b_i m_{i+1}$, for case fig. 5(b);
3) $dmin = b_i m_{i+1}$, for case fig. 5(c);
4) $dmin = b_i b_{i+1}$, for case fig. 5(d).

The common normal line ($c_i c_{i+1}$) of line segment $a_i b_i$ and $a_{i+1} b_{i+1}$ could be expressed by eq.(11).

$$d = \left| (\mathbf{b}_i - \mathbf{b}_{i+1}) \cdot \frac{(\mathbf{b}_i - \mathbf{a}_i) \times (\mathbf{b}_{i+1} - \mathbf{a}_{i+1})}{\left\| (\mathbf{b}_i - \mathbf{a}_i) \times (\mathbf{b}_{i+1} - \mathbf{a}_{i+1}) \right\|} \right| \qquad (i = 1 \sim 6) \tag{11}$$

Where $\boldsymbol{a}_i$, $\boldsymbol{b}_i$, $\boldsymbol{a}_{i+1}$ and $\boldsymbol{b}_{i+1}$ were the vectors described in the fix *Cartesian* frame.

Let **m** be the vector along $\boldsymbol{a}_i \boldsymbol{b}_i$. It could be expressed by the parametric equation, see eq.(12). Meanwhile, the feet of perpendicular c_i and c_{i+1} could be obtained by solving eq.(13).

$$\mathbf{m}_i = \mathbf{a}_i + k_i (\mathbf{b}_i - \mathbf{a}_i) \qquad (i = 1 \sim 6) \tag{12}$$

$$\begin{cases} [\mathbf{a}_i + k_i(\mathbf{b}_i - \mathbf{a}_i) - \mathbf{a}_{i+1} - k_{i+1}(\mathbf{b}_{i+1} - \mathbf{a}_{i+1})] \cdot (\mathbf{a}_i - \mathbf{b}_i) = 0 \\ [\mathbf{a}_i + k_i(\mathbf{b}_i - \mathbf{a}_i) - \mathbf{a}_{i+1} - k_{i+1}(\mathbf{b}_{i+1} - \mathbf{a}_{i+1})] \cdot (\mathbf{a}_{i+1} - \mathbf{b}_{i+1}) = 0 \end{cases} \qquad (i = 1 \sim 6) \tag{13}$$

Unless $k_i \in [0, 1]$, point m_i lied between the points a_i and b_i; unless $k_{i+1} \in [0,1]$, point m_{i+1} lied between the points a_{i+1} and b_{i+1}. Let $b_i M_{i+1}$ be the perpendicular line of $a_{i+1} b_{i+1}$. m_{i+1} could be derived from eq. (14). Detailed flow chart of the contact interference was shown in fig. 6.

$$[\mathbf{b}_i - \mathbf{a}_{i+1} - m_{i+1}(\mathbf{b}_{i+1} - \mathbf{a}_{i+1})] \cdot (\mathbf{b}_{i+1} - \mathbf{a}_{i+1}) = 0 \qquad (i = 1 \sim 6) \tag{14}$$

3.1.2 Structure parameters and workplace of MLD

Based on the inverse kinematics and the interference boundary condition, structure parameters and the position workspace were obtained. Optimized parameters were summarized in table 1; different views of the position workspace were plotted in fig. 7, respectively. Different from the serial mechanism, the position and orientation workspaces of the parallel mechanism possessed strong coupled effects with each other. Consequently, the pose workspace usually be expressed by specific engineering performance indexes. As for this *MLD*, the pose workspace could be described that the orientation space should be as large as possible under the premise of meeting the requirements of the position space, see table 2.

Table 1. Structure parameters of the *MLD*.

Structure parameters	Value
r [mm]	120.00
R [mm]	175.86
β [°]	22.50
δ [°]	34.46
Initial length of L[mm]	370.00

Table 2. Pose workspace of the *MLD*.

Pose workspace parameters	Value
X [mm]	±40
Y [mm]	±40
Z [mm]	410~440
X [°]	-8~8
Y [°]	-8~8
Z [°]	-8~8

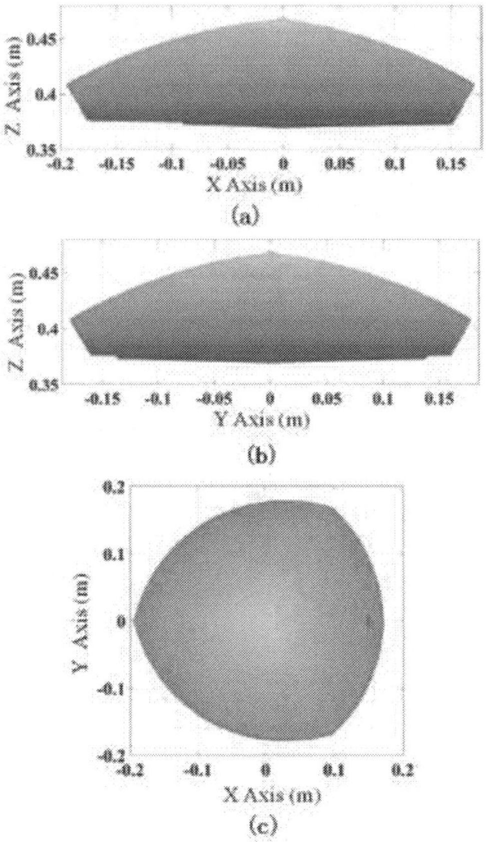

Figure 7. Different views of the position space.

3. Material mechanical experiments

Since the mobile platform of this *MLD* had six degrees of freedom, it could act more general loads on the material specimen, such as F_x, F_y, F_z, M_x, M_y, and M_z. As for this mechanism, as long as the *MLD* possessed enough stiffness, strength, motion space, and measurement accuracy, it could be used as a material testing machine. Translations and rotations along the three axes of the fixed *Cartesian* frame were the same in mechanism theory. Therefore, only tensile tests were performed to verify the loading capacity in this paper. In this work, specimen was manufactured using *AISI* 1045 steel widely used in practical application. All samples were referenced

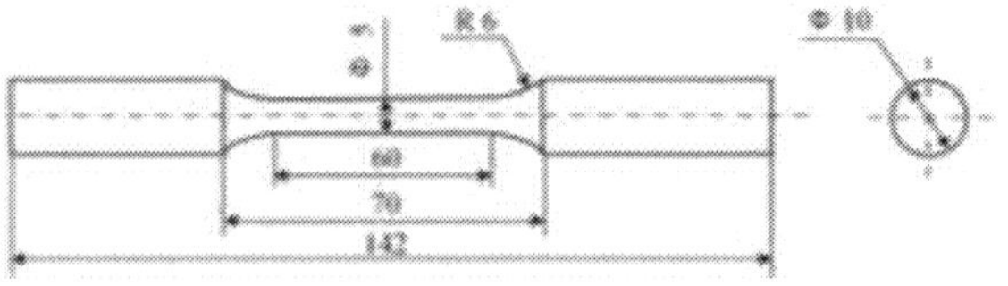

Figure 8. Specimen geometry.

to the *Chinese* national standard *GB/T* 10128-2007, see fig. 8. Chemical composition of the specimen was 97.2%Fe, 0.46% C, 0.24% Si, 0.45% Mn, 0.015% Cr, 0.08% P, 0.003% S, ≥0.01% Co, and 0.01% Ni. The measured average hardness of these specimen was 209.17HB.

Five repeated tensile tests were conducted on the *MLD* and the standard tensile testing machine Zwick Z100 according to the *Chinese* national standard *GB/T* 10128-2007, respectively. Experimental photos of these two devices were shown in fig. 9, and fig. 10. Stress-train curves obtained from these tests were plotted in fig. 11 and fig. 12, respectively.

Figure 9. Tensile tests on the *MLD*.

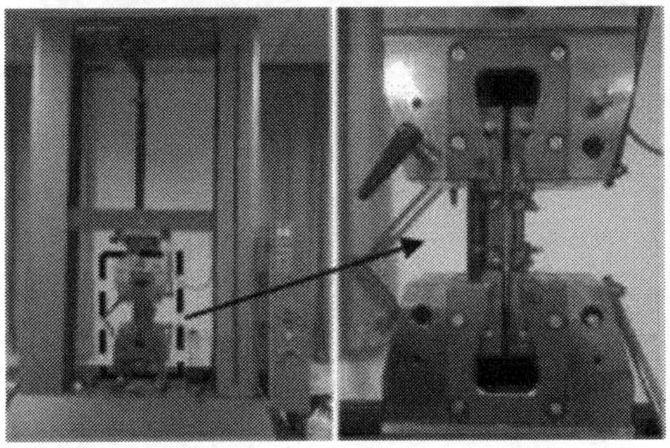

Figure 10. Tensile tests on the Zwick Z100.

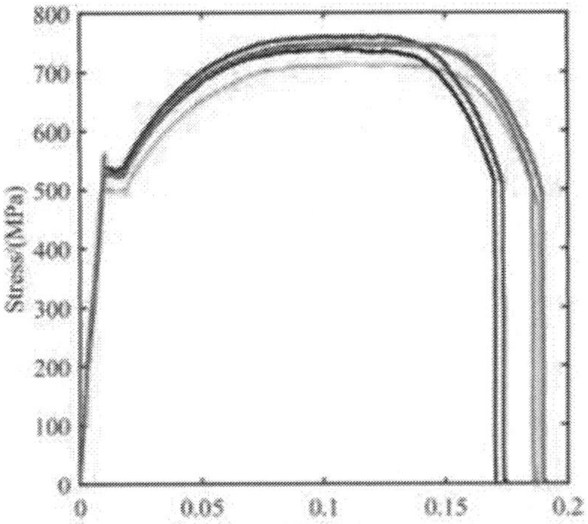

Figure 11. Stress-strain curves based on *MLD*.

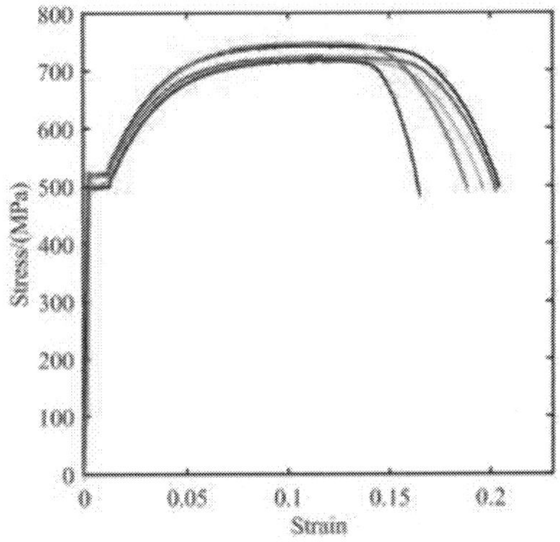

Figure 12. Stress-strain curves based on Zwick Z100.

Table 3. Comparison of mechanical properties derived from *MLD* and Zwick Z100.

Mechanical properties	*MLD* tests	Zwick Z100 tests	Error [%]
Average yield strength σ_s [Mpa]	521.36	515.52	1.13
Average ultimate strength σ_b [Mpa]	737.64	729.40	1.13

Mechanical properties derived from the two testers were compared and summarized in table 3.

Errors of the elastic segment in fig. 11 and fig. 12 were caused by that extensometer was not installed on the *MLD* while Zwick Z100 possessed an extensometer. Meanwhile, table 3 manifested that the maximum errors of σ_s and σ_b were merely both 1.13%. Therefore, results indicated that it was feasible to conduct material mechanical experiments on the *MLD*.

4. Summary

Aiming at the defects of the traditional material testers, on which the more general loads could not be applied, a new material mechanical *MLD* based on the parallel mechanism was proposed. Meanwhile, pure tensile experiments of *AISI* 1045 steel were performed on the *MLD* and on the standard tensile tester Zwick Z100, respectively. Main conclusions could be drawn as follows:

(1) Detailed contact interference boundary conditions were analysed which could be applied to the design of the parallel mechanism;

(2) The method of designing and optimizing structure parameters of the parallel mechanism was proposed based on the inverse kinematics and the contact interference boundary condition;

(3) Errors of the average yield and ultimate strengths obtained from the two devices were merely both 1.13%.

Results indicated that the *MLD* could meet the demands of material mechanical testing machine. It was feasible to perform material mechanical experiments via parallel mechanism.

5. Acknowledgements

All of the experiments described herein were supported by the AVIC industry-academy-research cooperation project (CXY2013BH05).

References

[1] F.Y. Zhou, Z.Y. Mao, *Engineering materials and application*, second ed., Huazhong University of science and technology, Wuhan, (2002).
[2] Z.H. Shan, *Mechanics of material*, third ed., Higher education press, Beijing, (2009).
[3] W.M. Lu, *Advanced strength and applied stress analysis*., Tsinghua university, Beijing, (2001).
[4] G.P.Z, *Experimental foundation of material mechanics*, first ed., Harbin institute technology, Harbin, (2010).
[5] J.J. H, M.G. Lee, F. Barlat, Strain hardening response and modelling of EDDQ and DP780 steel sheet under non-linear strain path, Mech Mater. **64**(2013) 11-26.
[6] S. Bruschi, T. Altan, Testing and modelling of material behaviour and formability in sheet metal forming, CIRP Annals. **63**(2014) 727-749.
[7] M.M. Rodríguez, R.A. Vaz, A. Arias, Failure behaviour of 2024-T3 aluminium under tension-torsion conditions, J. Mech Sci Technol. **29**(2015) 4657-4663.
[8] G.H. Zhao, L.B. Liu, Mechanical properties of AISI 1045 steel subjected to combined loads of tension and torsion, Exp Tech. **42** (2018) 393-406.
[9] J.Z. Guo, D. Wang, Development of a material testing machine with multi-dimensional loading capability, JSME. **10**(2016) JAMDSM0017-JAMDSM0017.

Research on Capability Evaluation System for Elevator Inspector

Kexi Fei[1], Qinqiao Zhou[1] and Kai Huang[1,a]

[1]*Jiangsu Province Special Equipment Safety Supervision Inspection Institute, Nanjing, China*

Abstract. This paper analyzes the current situation of elevator inspectors' training based on the QFD method. Then, an elevator inspector's capability evaluation system for elevator inspector is established. The evaluation index and the weight determination are presented in detail. The research can favor to assess the comprehensive capabilities of elevator inspectors, and provide a theoretical reference of appropriate training activities for elevator inspectors.

1 Introduction

With the rapid development of society and economy, the number of special equipment has grown at a rate of nearly 10% each year. Especially, the elevators closely related to daily work and life of people, have received much attention from society and the government [1]. Therefore, the establishment of a comprehensive evaluation system to identify and improve the elevator inspector's business level, so as to ensure the safe operation of elevator equipment, is of great significance and has become the development trend in this field [2].

2 QFD-based demand analysis of capability evaluation system for elevator inspector

A QFD -based demand analysis is conducted to explore the relevance and sequencing of the evaluation needs. Through the progressive analysis at each level, the corresponding training objectives and training methods for elevator inspectors are established from the strategic and organizational levels, as well as the corresponding policy mechanisms[3]. The QFD-based requirements analysis flow chart is shown in Figure 1.

According to the demand analysis, it can be found that a complete and standard capability evaluation system needs to be established as an evaluation benchmark for elevator inspector performance[4]. At the same time, elevator inspectors need to be regularly assessed, evaluated, educated and trained to promote their comprehensive capabilities.

3 Establishment of elevator inspector capability evaluation system

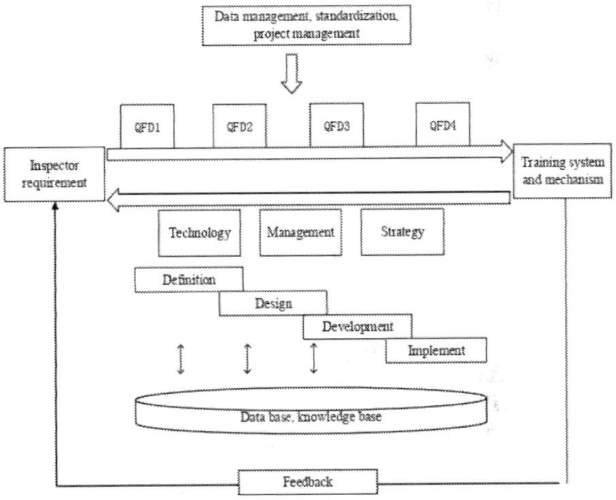

Figure 1. Flow chart of requirements analysis based on QFD.

(1) The establishment of elevator inspector skills evaluation criteria based on TSG Z8002-2013, GB/T 7024-2008

According to the results of the needs analysis, a set of evaluation criteria for the qualification of elevator inspectors from the aspects of qualification, inspection ability, learning and training, technical achievements, practical operation, and annual assessment shall be established [5].

This standard is based on <elevators, escalators, moving sidewalk terminology> (TSG Z8002-2013), <special equipment inspection personnel assessment rules> GB/T 7024-2008.The terminology definitions, evaluation indicators, calculation methods, and skill grading of skills are applicable to the skill evaluation of elevator in-service inspection personnel [6]. Details are as in Figure 2.

[a] Corresponding author: js_nj_hk@163.com

© The Authors, published by EDP Sciences. This is an open access article distributed under the terms of the Creative Commons Attribution License 4.0 (http://creativecommons.org/licenses/by/4.0/).

 https://doi.org/10.1051/matecconf/201820703013

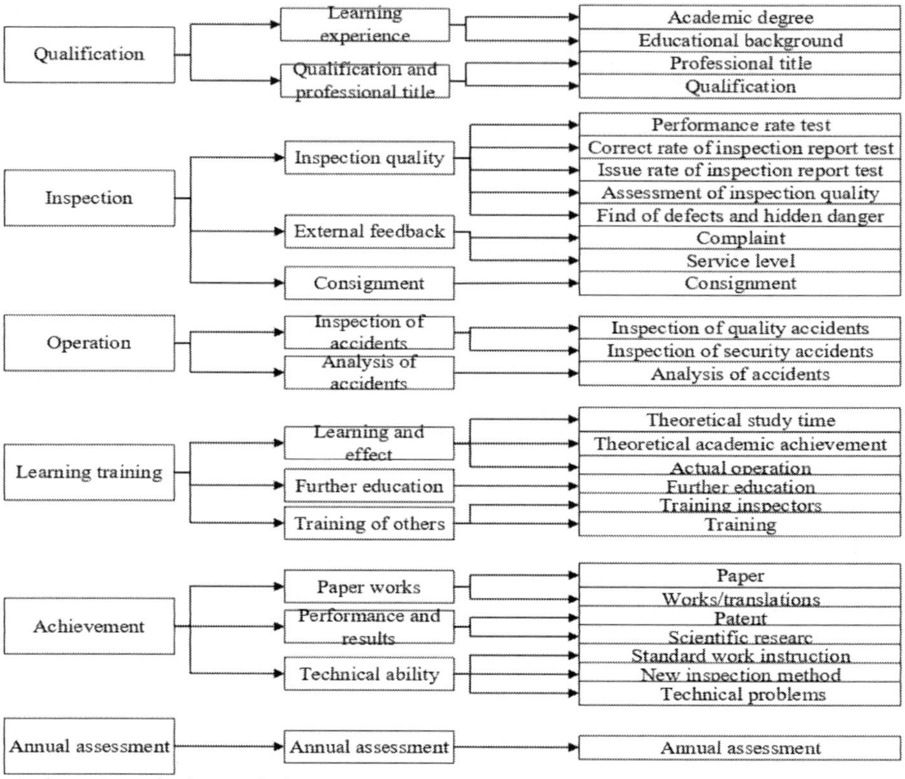

Figure 2. Elevator inspector evaluation standard system.

(2) Evaluation of elevator inspector ability based on fuzzy comprehensive evaluation

For the elevator inspector evaluation problem, most of the indicators are qualitative. By introducing the concept of membership degree, the boundary of membership function classification index is constructed to solve the problem. Finally, a fuzzy comprehensive evaluation can be performed [7]. The specific flow chart is shown in Figure 3.

Through the above process, a complete elevator inspector capability evaluation system has been established, which can comprehensively evaluate the elevator inspector's theoretical knowledge, methods, skills and other overall qualities [8]. The evaluation system established is shown in Table 1.

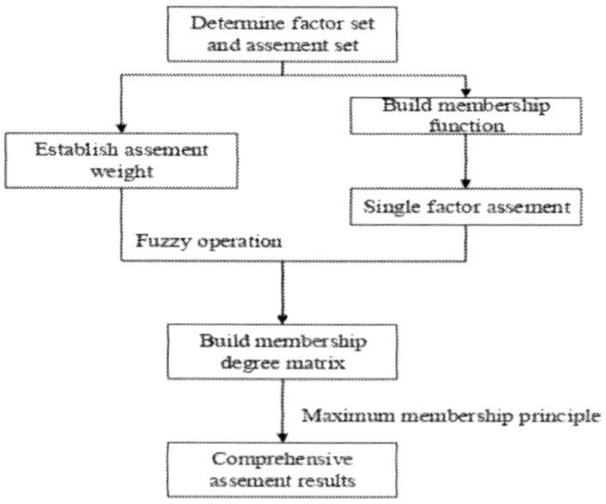

Figure 3. Fuzzy evaluation flowchart.

Table 1. Elevator Inspector Skills Evaluation System

One class index	Weight	Two class index	Weight	Three class index	Weight
1 Qualification	0.10			Major	
		1.1 Learning experience	0.50	Academic degree	0.50
				Educational background	0.50
		1.2 Qualification and professional title	0.50	Professional title	0.50
				Qualification	0.50

[a] Corresponding author: js_nj_hk@163.com

2 Inspection ability	0.50	2.1 Inspection quality	0.60	Performance rate test	0.05
				Correct rate of inspection report test	0.20
				Issue rate of inspection report test	0.15
				Assessment of inspection quality	0.30
				Defects and hidden danger find	0.30
		2.2 External feedback	0.20	Complaint	0.50
				Service level	0.50
		2.3 Consignment	0.20	Consignment	1.00
3 Accident	0.05	3.1 Accidents inspection	0.50	Quality accidents inspection	0.50
				Security accidents inspection	0.50
		3.2 Accidents analysis	0.50	Accidents analysis	1.00
4 Learning and Training	0.20	4.1 Learning and effect	0.50	Theoretical study time	0.30
				Theoretical academic achievement	0.30
				Practical operation	0.40
		4.2 Further education	0.10	Further education	1.00
		4.3 Others training	0.40	Training inspectors	0.50
				Training	0.50
5 Technical achievements	0.10	5.1 Paper works	0.30	Paper	0.50
				Works/translations	0.50
		5.2 Performance and results	0.40	Patent	0.50
				Scientific research	0.50
		5.3 Technical ability	0.30	Standard work instruction	0.40
				New inspection method	0.30
				Major defects problem	0.30
6 Annual assessment	0.05	6.1 Annual assessment	1.00	Annual assessment	1.00

4 Summary

This study focuses on the construction of an elevator inspector capability assessment system. From system and actual needs analysis to standardized evaluation process management, an elevator inspector capacity evaluation system is established. The platform has the ability to assess the comprehensive capabilities of elevator inspectors and can favor the appropriate training activities for elevator inspectors.

Acknowledgment

This work was financially supported by the Program of Jiangsu Province Special Equipment Safety Supervision Inspection Institute (KJ(Y)2014032). The supports are gratefully acknowledged.

References

1. C. Hepler, Q.B. Belt, G. Mazur, Q.R. Belt, *18th Symposium on QFD*, (2006)

2. J.L. Jin, Y.M. Wei, J. Ding, Journal of Hydraulic Engineering, 144-147 (2004)
3. T. Zhang, *Proceedings of the 2012 International Conference on Cybernetics and Informatics*. (Springer New York, 1203-1210 2014). DOI 10.1007/978-1-4614-3872-4_154
4. D.U. Baojiang, Y. Wang, N.I. Jiawei, D.U. Qiuyu, Electronic Science & Technology, (2017)
5. M.L., Siikonen, *Planning and Control Models for Elevators in High-Rise Buildings*, (1997)
6. M.D. Yang, C F. Chao, K.S. Huang, L.Y. Lu, Y.P. Chen, Automation in Construction **33**, 48-60 (2013). DOI: 10.1016/j.autcon.2012.09.017
7. A. Kaneko, *Collision test apparatus and collision test method* (US, US7886574, 2011)
8. Y.M. Xiong, Y.M. Chen, Advanced Materials Research **204-210**, 1041-1044 (2011). DOI: 10.4028/www.scientific.net/AMR.204-210.1041

Safety evaluation for bridge crane based on FTA and AHP

Shuai Fei[1], Li Xiangdong[1,a] and Fan Yali[2]

[1]Jiangsu Province Special Equipment Safety Supervision Inspection Institute, Nanjing, China
[2] Nanjing University of Science and Technology, Nanjing, China

Abstract. Starting from the development status of the lifting machinery industry, this paper analyzes the data about safety accidents of bridge cranes, and identifies the assessment indicators affecting the safety performance of bridge cranes based on fault tree analysis (FTA), including fatigue factors, human factors, environmental factors, and management factors. Then, the weights of safety evaluation indicators are determined based on analytic hierarchy process (AHP). Finally, the safety evaluation system is established and demonstrated with use case diagram and role flow chart.

1 Introduction

There are millions of bridge cranes in China, a large number of which have exceeded the service life and the metal structures of many cranes have various crack defects 错误！未找到引用源。. If an accident occurs, it will cause personal injury and property loss错误！未找到引用源。.With the development of modern science and technology, bridge cranes need not only high load capacity under complex working environment, but also to reduce the cost of raw materials for manufacturing. In a word, it is necessary to evaluate the safety performance in order to prevent crane accidents. Therefore, this paper aims to establish a safety evaluation system for bridge crane.

2 Establishment of safety evaluation system for bridge crane

Fatigue crack damage is one of the main failure forms of crane metal structures 错误！未找到引用源。. In daily use, parts of hoisting machinery will cause problems such as abrasions and failures of safety protection devices 错误！未找到引用源。.

During the service period, the bridge cranes are subjected to environmental impacts. Long-term overload and other unexpected situations, and the fatigue of metal materials will weaken the function of bridge cranes. When the damage is accumulated to a certain extent, the girders will fail 错误！未找到引用源。.Therefore, this paper combines fault tree analysis (FTA) to determine the fatigue factors that affect the safety of bridge cranes, as shown in Figure 1.

Besides the fatigue factors, human factors, environmental factors, and management factors also affect the safety performance of bridge cranes. The safety assessment indicators of bridge cranes are proposed in Table 1.

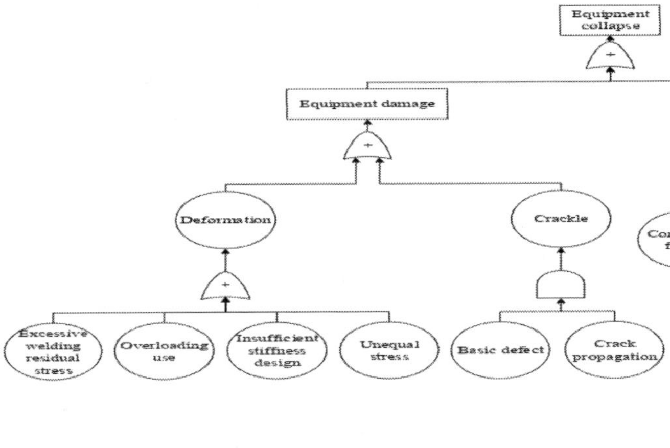

[a] Corresponding author: 252363775@qq.com
© The Authors, published by EDP Sciences. This is an open access article distributed under the terms of the Creative Commons Attribution License 4.0 (http://creativecommons.org/licenses/by/4.0/).

 https://doi.org/10.1051/matecconf/201820703014

Figure 1. Equipment faultanalysis diagram.

Table 1. Bridge crane safety assessment indicators.

Fatigue factors A1(0.522)	Metal mechanisms B1(0.528)	Surface wear degree C1(0.31)
		Crack deformation degree C2(0.20)
		Weld defect degree C3(0.16)
		The fatigue life of crack C4(0.13)
		Vibration conditions C5(0.09)
		Flaw detection conditions C6(0.11)
	Operating mechanisms B2(0.332)	Retarder performances C7(0.5)
		Brake performances C8(0.5)
	Safety devices B3(0.140)	Service life C9(0.5)
		Reliability C10(0.5)
Human factors A2(0.239)	Operators B4(0.5)	Mechanical workers C11(0.637)
		Commanders C12(0.258)
		Managers C13(0.105)
	Quality of personnel B5(0.5)	Psychological quality C14(0.5)
		Physiological quality C15(0.5)
Environmental factors A3(0.153)	Weather B6(0.341)	Visibility C16(0.5)
		Wind load C17(0.5)
	Foundation B7(0.659)	Ground deformation C18(0.582)
		Flatness of the ground C19(0.109)
		Safe distance C20(0.309)
Management factors A4(0.086)	Maintenance situation B8(0.5)	Maintenance cycle C21(0.5)
		Maintenance records C22(0.5)
	Inspection B9(0.5)	Inspection cycle C23(0.5)
		Inspection records C24(0.5)

3 Determination of the weights for safety assessment indicators

The weight of each factor is obtained by using AHP. The procedure is as follows: (1) Model a hierarchy decision model; (2) Establish a judgment matrix; (3) Synthesize these judgments to yield a set of overall priorities for the hierarchy; (4) Check the consistency; (5) Draw a final decision[错误！未找到引用源。], as shown in Figure 2.

Set up the hierarchy structure of safety evaluation index system. The hierarchy architecture is divided into three levels: the target layer, the rule layer, and the scheme layer as shown in Figure 3.

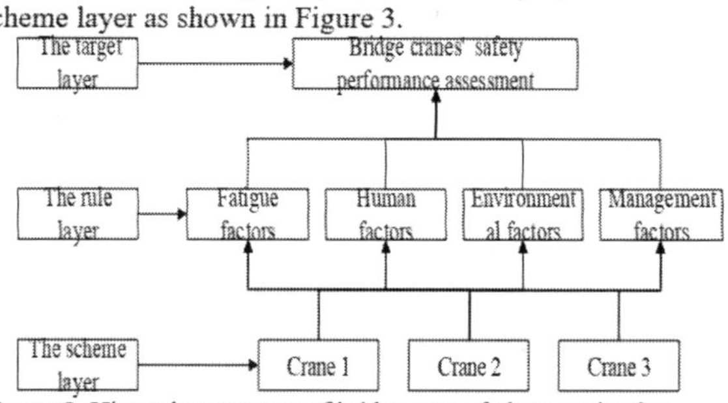

Figure 3. Hierarchy structure of bridge crane fatigue evaluation index system.

Establish a judgment matrix. First, this paper uses AHP to build a judgment matrix for peer indicators. Suppose there are n indicators, {A1, A2, ...,Ai,...,Aj, ...,An}, a_{ij} indicates the ratio of the importance of the index Ai to the index Aj, that is, the element in judgement matrix A. a_{ij} takes 1, 3, 5, 7, 9 five scales: 1 means that the index Ai is as important as the index Aj, the importance of index Ai is increasing than that of index Aj with the increase of the numbers.

Synthesize these judgments to yield a set of overall priorities for the hierarchy. After the completion of the judgement matrix, the second-level indexes are sorted and the root weight method is applied to normalize the index to get the initial weights of the indexes. This step includes calculating single-level weights and multi-level weights. This paper chooses the root weight method to calculate the maximum eigenvector of the judgment matrix. The process is as follows.

(1) Calculate the product of each row element of the judgment matrix A, the formula is as follows.

$$m_i = \prod_{j=1}^{n} a_i \qquad (1)$$

(2) Calculate the n roots of each row element of the judgment matrix separately, and the initial weights of the indexes at all levels are obtained after normalization, the formulas are as follow[错误！未找到引用源。].

$$\overline{w_i} = \sqrt[n]{m_i} \qquad (2)$$

$$W_i = \overline{W_i} \Big/ \sum_{j=1}^{n} \overline{W_j} \quad (3)$$

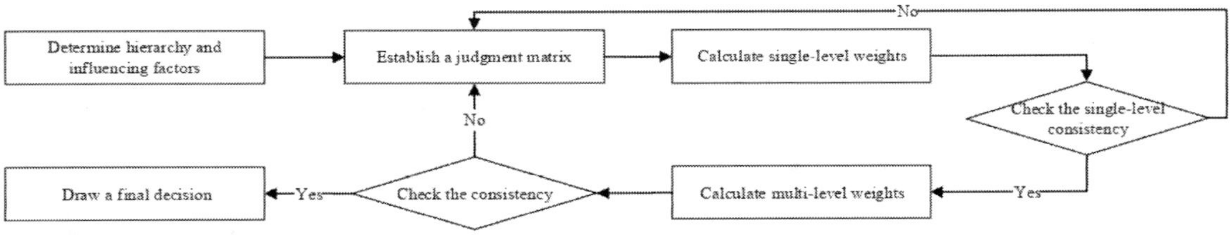

Figure 2. AHP's flow chart.

Check the consistency. It is necessary to test the random consistency of the index to ensure the logic accuracy when the initial weights are obtained. The process is as follows.

(1) Calculate the maximum eigenvalue based on the constructed judgement matrix A, the formula is as follows.

$$\lambda_{max} = \sum_{i=1}^{n} \frac{(AW)}{nW} \quad (4)$$

(2) Calculate the consistency ratio C_R according to the maximum characteristic root, and test the random consistency of the indexes at all levels, the formulas are as follow.

$$C = \frac{\lambda_{max} - n}{n-1} \quad (5)$$

$$R = \frac{\overline{\lambda}_{max} - n}{n-1} \quad (6)$$

(3) Calculate the proportion of random consistency, the formula is as follows.

$$C_R = \frac{C}{R} \quad (7)$$

When the consistency ratio of the random consistency test satisfies the index, the original judgment matrix is maintained, and the initial weights are the final weights, otherwise, the original judgment matrix is adjusted and the initial weights are recalculated until the index is satisfied.

Results. Here is a demonstration of the pairwise comparison and priority calculation, as shown in Table 2 (Take metal mechanism, operation mechanism and safety device as an example).

The influencing factors and weights of the safety assessment indicators are shown in Table 2.

4 Design of safety evaluation system for bridge crane

Software function requirements and performance requirements are determined based on the above analyses. The functional modules of the bridge crane safety evaluation system are divided into five modules: information acquisition module, information storage module, information query module, parameter change module and system management module. Each module has a specific functional division and different functional requirements for different user identities.

According to the established model, the users and administrators' permissions are different, and the corresponding application processes are also different. Administrators can log in directly without registration, and can modify indicator parameters. The users enter the main interface by logging in or registration, and can view, modify, add, delete the stored bridge crane information, and so on. The specific process is shown in Figure 4.

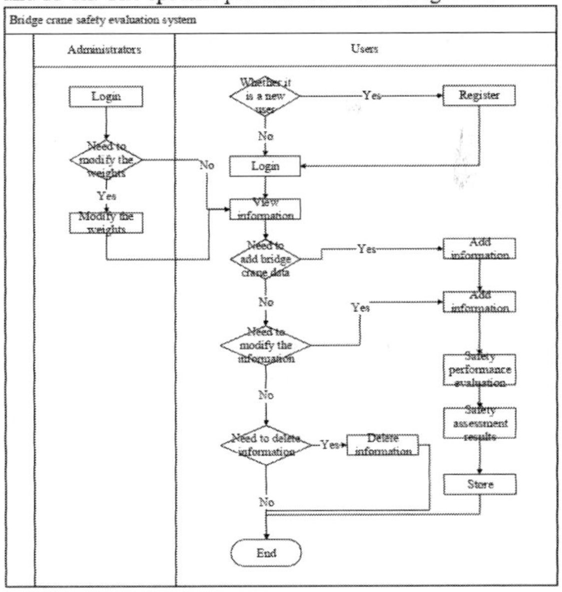

Figure 4. System flow chart.

Table 2. A case of the pairwise comparison and priority calculation (CR=0.046).

	Metal mechanisms	Operation mechanism	safety device	Geometric means	Normalized priorities
Metal mechanisms	1	2	3	$\sqrt[3]{1 \times 2 \times 3}$=1. 817	0.528

https://doi.org/10.1051/matecconf/201820703014

operation mechanism	1/2	1	3	1.145	0.332
safety device	1/3	1/3	1	0.481	0.140

5 Summary

This paper presents a safety performance assessment model for bridge cranes and uses the analytic hierarchy process (AHP) to determine the weights of the indicators in the model. At the same time, this article uses information software to achieve the safety assessment of bridge cranes. The software can make the assessment of bridge cranes more convenient and faster, and provides a basis for the management and maintenance of bridge cranes.

6 Acknowledgment

This work was financially supported by the Program of Science Foundation of General Administration of Quality Supervision and Inspection of Jiangsu Province (KJ175940). The supports are gratefully acknowledged.

References

1. Zhan Weigang, Fatigue life prediction based on crack propagation and its Application in crane metal structures, Harbin Institute of Technology (2014).
2. Xia Lu, Analysis and Preventive Measures of Common Mechanical Accidents of Cranes, Brand and standardization. 10 (2015) 83-84.
3. Li Yidong, Safety Evaluation of Metal Structure of Shipbuilding Crane Based on Fatigue Failure, Zhejiang University (2012).
4. Chen Aiqing, Statistics on safety risks and defects of cranes, China Chemical Trade. 7 (2015) 110.
5. He Yali, Study on fatigue life analysis and safety evaluation method of bridge crane, Beijing University of Chemical Technology (2014).
6. Wei C C, Chien C F, Wang M J J. An AHP-based approach to ERP system selection, J. International Journal of Production Economics, 2005, 96(1):47-62.
7. Ramanathan R, Ganesh L S. Group preference aggregation methods employed in AHP: An evaluation and an intrinsic process for deriving members' weightages, J. European Journal of Operational Research, 1994, 79(2):249-265.

Investigation of tribo-mechanical properties and structure of specimens obtained by selective laser melting of stainless steel powders

Leonid Belyaev [1], Aleksey Zhdanov [1,a] and Valentin Morozov [1]

1 Department of the mechanical engineering, Vladimir State University, Gorky str. 87, Vladimir, Russia

Abstract. The article presents experimental data about the quality parameters of the stainless steel three-dimensional parts obtained by the selective laser sintering procedure. The results of the porosity, tribological properties, tensile test, hardness measurement and density for different values of the focal point position during the engineering process of the three-dimensional parts sintering were given. For parts made from steel powder CL20ES at a laser radiation power of 200 W in a different focal length range, it is demonstrate that the specimens with the highest density and microhardness values are obtained during the synthesis by using a converging laser beam.

1 Introduction

Selective laser sintering is one of the most popular process in the production of finished three-dimensional parts in a tight schedule [1-4]. However, in order to ensure the required parameters of strength and durability, the most of finished parts should have a defectless microstructure, the required roughness and have a minimal porosity [5, 6]. This work presents the experimental studies results of the tribo-mechanical properties of the specimens obtained by selective laser sintering.

2 Materials and methods

Specimens synthesis with a 30×30×6 mm size in a quantity of 10 pieces were performed on the installation Concept Laser M2 (Germany) in a protective atmosphere of nitrogen (Fig. 1). The system is equipped with a Yb:YAG fiber optic diode-pumped laser, with a capacity of 400 W, manufactured by IPG (Russia). The wavelength of the laser radiation is 1070 nm. The specimens were obtained on the modes: power 200 W, scanning speed 800 mm/s. The position of the focal laser radiation point was changed from -4 to +5 mm. As a baseline, the upper level of the spread powder was taken. Besides, the spot diameter in the focus was 50 μm.
CL20ES powder was used, its chemical composition is close to Chrome Nickel Molybdenum Steel grade "03X17H14M3" (GOST 5632-2014) and steel AISI 316L. This powder is widely used for additive manufacturing, acid and corrosion-resistant specimens, single or small-scale production in engineering industry, automobile industry and others [7].

The study of the structure was carried out by using the scanning probe microscope NTEGRA Aura and the scanning electron microscope Quanta 3D 200. The x-ray micro-chemical analysis was carried out using the sequential analysis spectrometer ARL ADVANT-X. The density of the specimens was determined by the hydrostatic weighing method in accordance with GOST 25281-82 (ASTM B962-08). Microhardness was measured using the microhardness tester PMT-3 by the indentation of a diamond shaped like a pyramid according to GOST R 8.748-2011 (ISO 14577-1:2002).

For the tensile tests and the evaluation of tribological properties, the specimens obtained at the focal spot position at a distance of -4 mm; 0 mm; +4 mm have been selected. The tensile tests were carried out on a tensile-testing machine WDW-100. For this purpose, plane specimens with a total length head of 30 mm were manufactured (Fig. 2). The study of tribological properties (dry friction coefficients) of the specimens was performed using a CSM tribometer. The study was carried out under normal conditions with a back and forth motion of the counterbody at a load in the range of 10-100 N. The maximal linear velocity is 5 cm/s, the amplitude of the motion is 22 mm and a total displacement length of 100 m using «Pin-On-Disk» methodology.

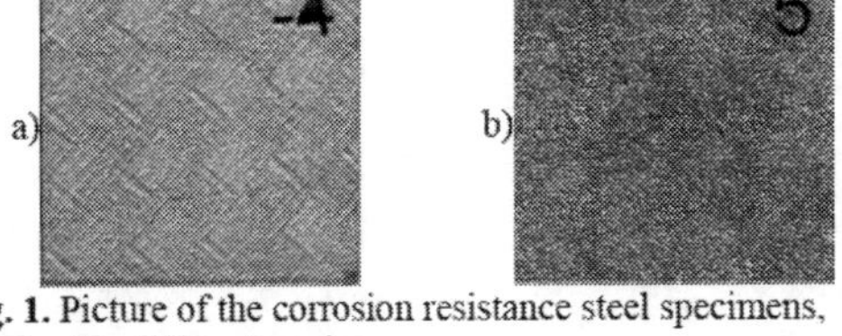

Fig. 1. Picture of the corrosion resistance steel specimens, obtained in different modes:
a – position of the focal point (-4), b – (+5)

[a] Corresponding author: zhdanov@vlsu.ru

© The Authors, published by EDP Sciences. This is an open access article distributed under the terms of the Creative Commons Attribution License 4.0 (http://creativecommons.org/licenses/by/4.0/).

 https://doi.org/10.1051/mateconf/201820703015

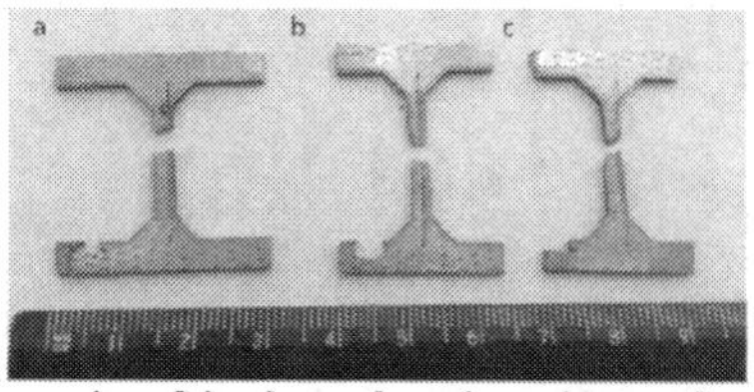

Fig. 2. Examples of dog bone shaped specimens after tensile tests

3 Results and discussion

Fig. 3 shows the image of the CL20ES steel powder in initial state obtained with the scanning electron microscope. On the surface of the particles there are satellites, which can reduce the fluidity and stability of bulk density. The investigated steel powders are polycrystals with an average diameter of about 30 µm and a microhardness 239HV.

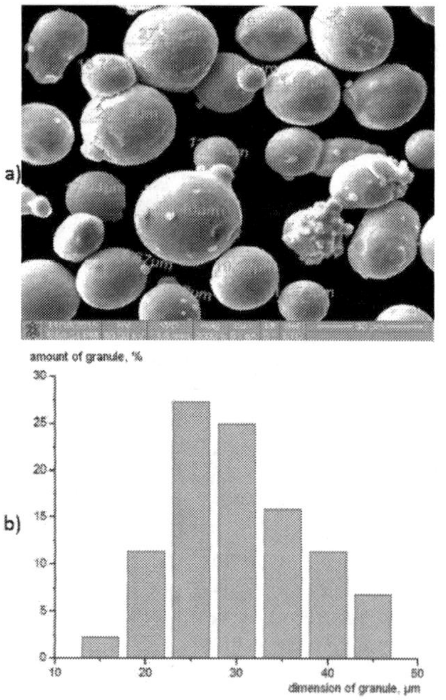

Fig. 3. Steel powder particles of corrosion resistant steel CL20ES (a) and their distribution by size (b)

The results of the comparative x-ray micro-chemical analysis of individual steel powder particles and the chemical composition of the powder according to the manufacturer's data are presented in Table 1. As studies have shown, the content of alloying components in individual steel powder particles differs slightly and corresponds to the composition declared by the manufacturer.

Figure 4 shows the optical images of the specimen at different magnifications, from the CL 20ES powder in its various regions. As the analysis of the microstructure of the specimen shows, the specimen is uniform, with small pores.

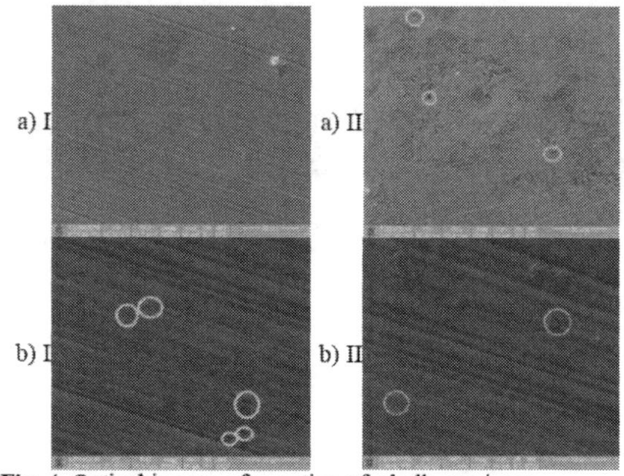

Fig. 4. Optical images of a section of a bulk specimen synthesized from a CL20ES powder:
I - the central region and II - the edge region, with different magnification: a) - 1000x, b) - 5000x

During layered laser synthesis, a porous structure with a low density of 7.15 g/cm^3 was formed (Fig. 4). Pores in the form of irregularly shaped cavities with a length of up to 600 µm. When measuring the microhardness, the pore size significantly exceeds the diameter of the indentation. Therefore, the microhardness was measured at the metal sites between the pores.

Reducing the position of the focal point leads to a decrease in the diameter of the spot, an increase in the power density of the laser radiation, and the formation of an alloy with a higher density (Fig. 5).

The points are the measured values of the microhardness of the specimens, the line is the

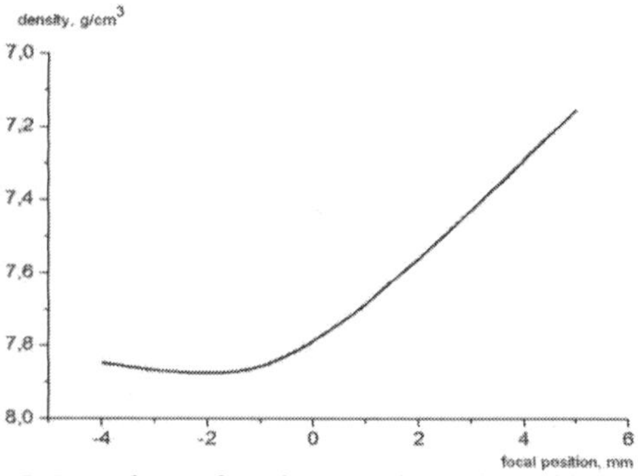

Fig. 5. Dependence of specimen density on the position of the focal point

Table 1. Chemical composition of corrosion resistant steel powder CL20ES

	Percentage composition of elements by mass				
	Si	Cr	Mn	Ni	Mo
Microchemical analysis	0.68 ± 0.12	18.53 ±0.19	1.65 ± 0.12	12.17 ±0.30	2.30 ± 0.88
Manufacturer's data [8]	0…1.0	16.5… 18.5	0… 2.0	10.0…13.0	2.0…2.5

approximation of the experimental data on the Boltzmann function using the computer program Origin. An abrupt change in the microhardness is observed when the laser treatment scheme is changed (Fig. 6). When melting steel powder particles with a converging laser beam, a high density and microhardness structure is formed at 400...500 MPa higher than when melting the steel powder particles with a divergent beam.

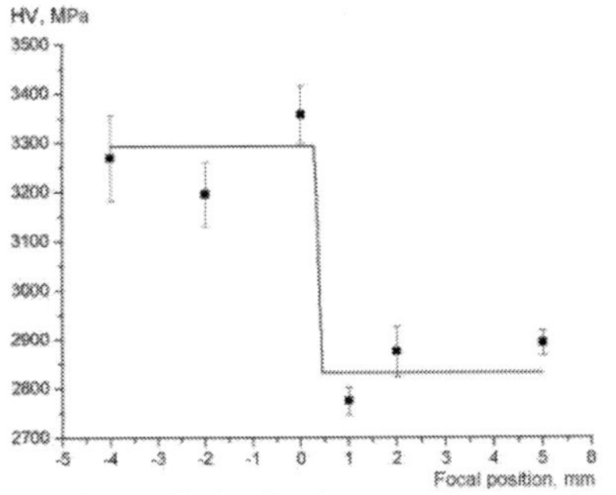

Fig. 6. Dependence of microhardness specimen on the position of the focal point

When focusing laser radiation above the surface to be treated or under it, different mechanisms occur for the formation of a recrystallized layer. In the case of surface treatment with a converging beam of radiation (Fig. 7, a), the power density in the upper parts of the steel powder particles is lower than in the lower ones. As a result, the melt crystallizes sequentially. Conditions are created for the formation of a dense without porous structure with high hardness. An inverse situation is observed in the case of surface treatment by a divergent beam (Fig. 7, b).

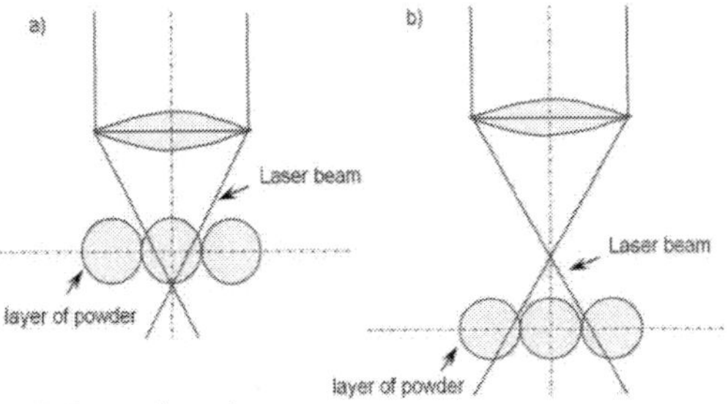

Fig. 7. Processing schemes:
a - melting of steel powder particles with a converging beam,
b - melting of steel powder particles by a divergent beam

Images of the structure of the specimens at different positions of the focal point are used in Figure 8.

The pore sizes are shown in Table 2. The power density on the protruding parts of the steel powder particles is greater than in the depressions. The upper part of the powder layer begins to melt and crystallize earlier than the lower part. As a result, a porous material with a low hardness is formed.

Table 2. Pore sizes in specimens obtained at different positions of the focal point

Focal point position	Pore size, μm	
	min	max
+2	7.23	11.97
+3	0.72	1.6
+5	0.73	2.72
+6	1.46	7.36

During the tensile tests, "stress-strain" diagrams have been obtained for the studied specimens (Fig. 9). And also a comparison of the above characteristics with the characteristics of a similar curve for the casted and hot rolled reference steel of Chrome Nickel Molybdenum Steel grade "03X17H14M3".

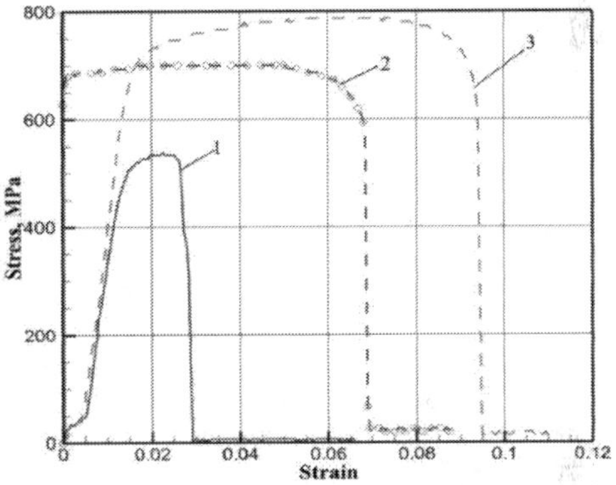

Fig. 9. Stress-strain diagram for dog bone shaped specimens obtained at different positions of the focal point: 1 - +4 mm, 2 - 0 mm and 3 - +4 mm.

The analysis of these diagrams shows that the obtained "stress-strain" curves are qualitatively similar to the same curve for the casted and hot rolled reference steel of Chrome Nickel Molybdenum Steel grade "03X17H14M3". Rupture occurs at the following load values: 534.6 MPa; 778.8 MPa and 699.6 MPa for specimens placed at the focal point +4 mm; 0 mm and -4 mm respectively.

Below the graphs of coefficients of friction obtained for back and forth motion of the counterbody and the

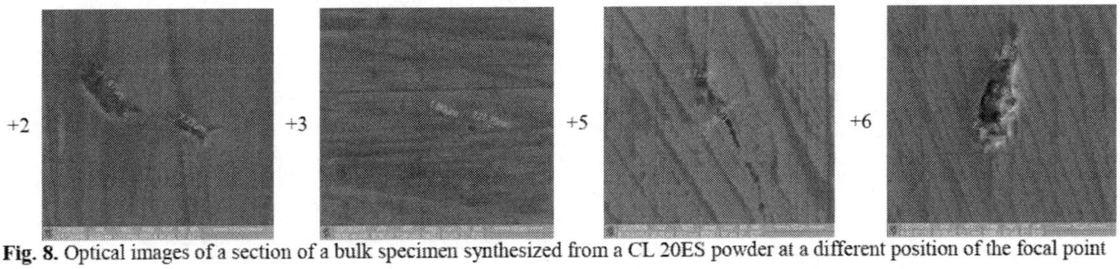

Fig. 8. Optical images of a section of a bulk specimen synthesized from a CL 20ES powder at a different position of the focal point

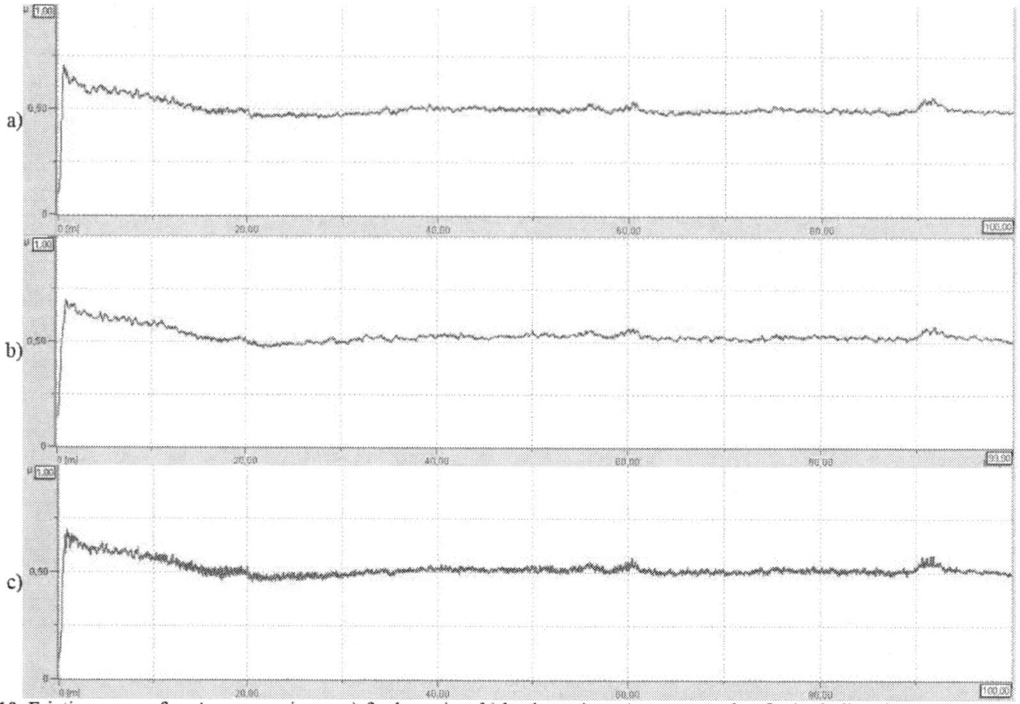

Fig. 10. Friction curves for -4 mm. specimen: a) forth motion; b) back motion; c) average value for both directions

graph of the average coefficient of friction for a specimen obtained at a focal spot position at a distance of -4 mm are presented (Fig. 10). For other specimens, similar graphs were obtained. The values of the coefficients of friction for all the specimens obtained are summarized in a table (Table 3), which also gives comparative values for the Chrome Nickel Molybdenum Steel grade "03X17H14M3", obtained with the same values of the experimental parameters.

The analysis of the presented data shows that the average values of the coefficients of friction for the indicated focal point positions and the "sample-metal" friction pair are 0.519; 0.539 and 0.531, respectively. The obtained values for the CL20ES steel are similar to the average coefficient of friction of the Chrome Nickel Molybdenum Steel grade "03X17H14M3" which is 0.505.

4 Conclusion

The technological parameters of selective laser melting are important for obtaining high-quality products with high density, low porosity and hardness. When manufacturing specimens from corrosion resistant steel CL20ES, the best values of density, friction coefficient, tensile test, porosity and microhardness were obtained by recrystallization of the powder layer with a converging beam of laser radiation.

Table 3. Friction coefficients of the CL20ES steel specimens obtained at different positions of the focal point compared with Chrome Nickel Molybdenum Steel grade "03X17H14M3"

Focal point positions, mm / Friction coefficient, μ	-4	0	+4	Chrome Nickel Molybdenum Steel grade "03X17H14M3"
forth motion	0.506	0.525	0.510	0.480
back motion	0.532	0.552	0.552	0.531
average value for both directions	0.519	0.539	0.531	0.505

References

1. E.C. Hussam, C. Bruno, S. Branchu, H. Xiaowei, J.Y. Hasco, R. Guille, OPT LASER ENG **50** (2012), pp. 1779–1784.
2. K. Guan, Z. Wang, M. Gao, X. Li, X. Zeng, Effects of processing parameters on tensile properties of selective laser melted 304 stainless steel, Materials & Design **50** (2013), 581-586.
3. C. Hauser Selective laser sintering of a stainless steel powder, *diss.*, University of Leeds (2003). 279 p.
4. K. Kempen, L Thijs, B. Vrancken, S. Buls, J. Van Humbeeck, J.P. Kruth, Producing crack-free, high density M2 HSS parts by selective laser melting: pre-heating the baseplate, *Proceedings of the 24th international SFF symposium. Laboratory for freeform fabrication* (2013) pp. 131-139.
5. E. Bashkirov, K. Khorkov, D. Kochuev, A. Zhdanov, S.Arakelyan, *Fundamental Research*, **3** (2017).
6. F. Xie, X. He, S. Cao, X. Qu, Structural and mechanical characteristics of porous 316L stainless steel fabricated by indirect selective laser sintering. JMPTE **213** (2013). 838–843.
7. T. Dvoryaninova, A. Zhdanov, L. Belyaev, Scientific and technical messenger Volga region 3 (2015).
8. Datasheet CL20ES, ww.conceptlaserinc.com, pdf

MATEC Web of Conferences **207**, 03016 (2018) https://doi.org/10.1051/matecconf/201820703016
ICMMPM 2018

The hot flow behavior of Al-2wt.%Cu-1wt.%Mn alloy based on Gleebe-3800 hot compression simulation

JinLong Chen, Hengcheng Liao* and Heting Xu

School of Materials Science and Engineering, Southeast University, Jiangsu Key Laboratory for Advanced Metallic Materials, Nanjing 211189, China

: Corresponding author, professor of Southeast University, hengchengliao@seu.edu.cn

Abstract: Flow behavior of Al-2.0wt.%Cu-1.0wt.%Mn aluminum alloy was investigated by hot compressive test using a Gleebe-3800 thermal simulator. The test parameters about temperature and strain rate range from 573K to 773 K and from $0.01s^{-1}$ to $5s^{-1}$, respectively, and the true strain is up to 0.6. It can be seen that the peak flow stress increases with increasing the strain rate and decreases by deformation temperature. And the deformation activation energy of Al-2.0wt.%Cu-1.0wt.%Mn alloy is 183.02KJ/mol. A constitutive equation has been constructed to predict the hot deformation behavior, and correlate the peak flow stress predicted with strain rate and deformation temperature.

© The Authors, published by EDP Sciences. This is an open access article distributed under the terms of the Creative Commons Attribution License 4.0 (http://creativecommons.org/licenses/by/4.0/).

1. Introduction

Al-Cu-Mn series alloys have high strength at elevated temperature and high strength to weight ratio, so it has become a kind of prospective armor material to be used in chariot, aero-craft and other commercial machines. With the development of technology, the aircraft flight velocity is further increased, thus, further improvement of strength at elevated temperatures of aluminum alloy is needed.

Many plastic forming methods are used to further improve the strength and formability of aluminum alloys such as accumulative roll bonding(ARB), equal-channel angular pressing (ECAP)[1], cross-channel extrusion(CCE)[2], severe cold rolling, hot rolling[3], high pressure torsion(HPT) and so on in addition to the alloying. For examples, Vaseghi and Kim [4] obtained the high hardness of Al-Mg-Si alloy from 86 HV (as solution treatment) to 138 HV by ECAP. Chou et al.[5] reported that after solutionizing the tensile strength of Al-Mg-Si alloy was improved to 364 MPa by 8 passes of EACP at 448 K and then artificial aged, 13% higher than that in T6 state. Krishnaiah et al.[5] studied that the grain size of pure Al could be reduced from 38 μm to almost 1 μm by groove pressing. Jayaganthan et al. [3] investigated that the Al 6061 alloy by cryorolling followed by warm rolling can get a significant improvement in tensile strength (376MPa) and partial improvement in ductility (5%) as measured from tensile testing.

The peak flow stresses and the activation energy (Q) are measure indexes of difficulty for deformation. Deng et.al [6] studied the flow behaviors of tension and compression considering the influence of deformation temperature and strain of 2024 aluminum alloy. Bardi et.al [7] investigates the correlation between the flow stresses and dynamic precipitation of 2618 aluminum alloy by means of torsion tests at temperatures between 423 and 573 K. Li et.al [8] used a Gleeble-1500 machine to investigate the flow stress features of 2519A aluminum alloy and to determine the optimum hot-working condition. Li et.al [9] studied the hot deformation behaviors of Ag-containing 2519 aluminum alloy by isothermal compression at 300 °C-500 °C with strain rates from $0.01s^{-1}$ to $10s^{-1}$. Liu et.al [10] studied the flow behavior of Al-Cu-Mg-Ag alloy and obtained the value of hot deformation activation energy of the alloy. Few researchers study the hot flow behaviors of Al-Cu-Mn alloys with low Cu content, and relation of the precipitation phase T_{Mn} to the flow behavior.

In this article, the hot deformation behavior of Al-2.0wt.%Cu-1.0wt.%Mn alloy is investigated at various temperatures and strain rates by Gleeble-3800 thermal simulator. The constitutive equation for the alloy is constructed relating flow stress, deformation temperature, strain rate, Zener–Hollomon parameter, and activation energy to deformation.

2. Experimental procedures

Al-2.0wt.%Cu-1.0wt.%Mn alloy was prepared in an electric resistance furnace using pure aluminum ingot and Al-20wt.%Cu and Al-10wt.%Mn master alloys. The chemical composition of the prepared alloy (Marked as A_1 alloy), measured by an MAX ×LMF15 spectrum, is listed in Table 1. After processing, the melts were poured into a metal mold with a size cavity size of 170mm×20mm×100mm (preheated at 250°C for at least 5 hrs). Then small cylindrical specimens with 12 mm in height and 8 mm in diameter were cut from the castings, then solution-treated at 525 °C for 6 hrs +535 °C for 6hrs and terminated by a water quench.

A Gleebe-3800 thermal simulator is used to accomplish the isothermal compression test. In order to minimize friction and to make the specimen deform uniformly during the test, graphite foils were used as lubricant. Five temperatures (573, 623,673, 723 and 773 K) and four strain rates (10^{-2}, 0.1, 1 and 5 s^{-1}) were chosen as the experimental parameters. Specimens were heated to the test temperature at a heating rate of 5 K/s and held for 1.5 minutes. The specimens were compressed to a final strain of 0.6. The true strain–stress curves were recorded automatically. After deformation, the specimens were quenched immediately by water to preserve the as-deformed microstructure.

Table1 Chemical composition of the studied alloy (wt.%)

Alloy	Cu	Mn	Si	Fe	Al
A_1	2.04	0.998	0.064	0.131	Bal.

3. Results and discussion

3.1Flow stress characteristics

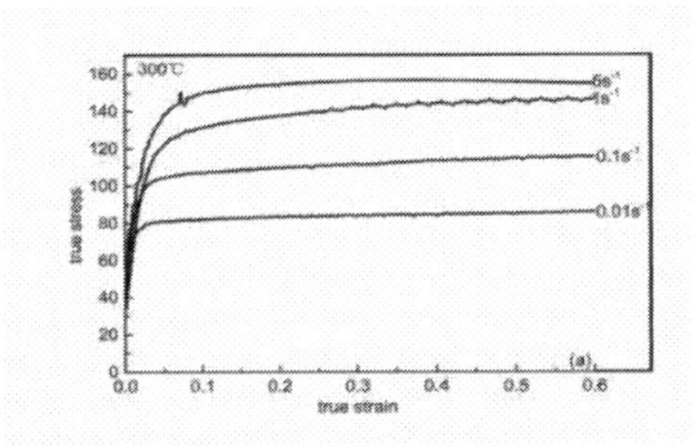

(a)

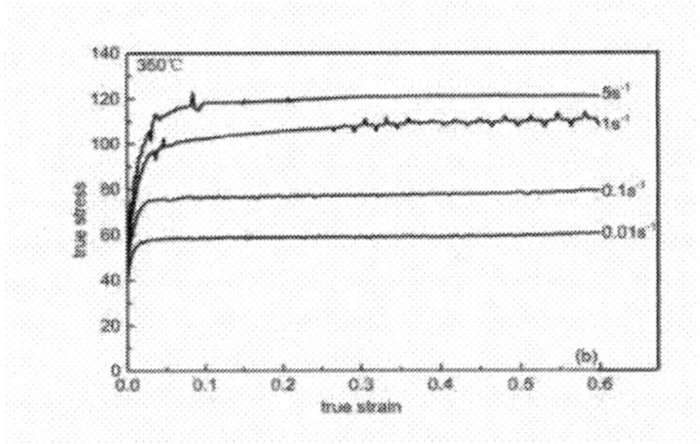

(b)

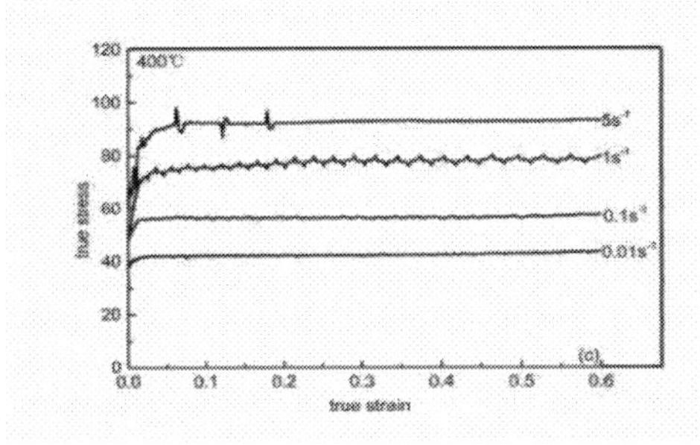

(c)

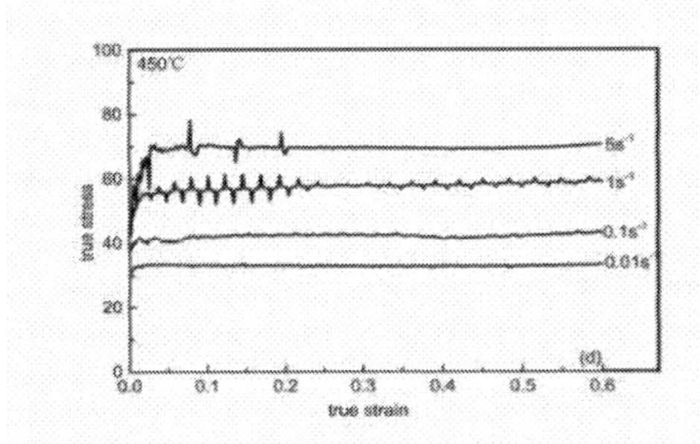

(d)

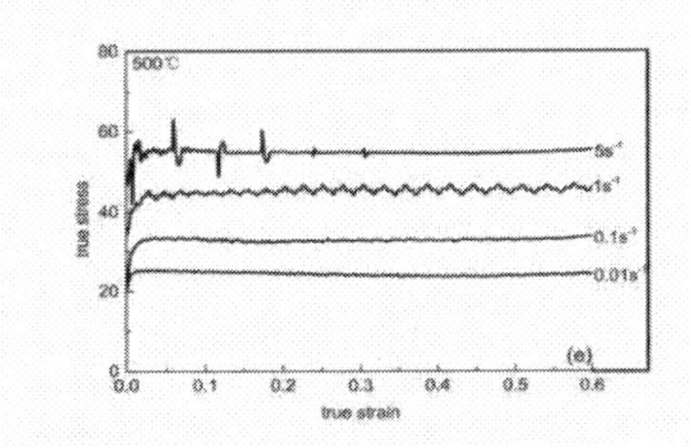

(e)

Fig.1 True stress–true strain curves of A_1 alloy obtained from hot compression tests at deformation temperatures of (a) 300 °C (b) 350 °C (c) 400 °C (d) 450 °C (e) 500 °C

The flow curves (σ vs ε) for Al-2.0wt.%Cu-1.0wt.%Mn alloy obtained at various strain rates ($\dot{\varepsilon}$) and deformation temperatures (T) are shown in Fig.1 through hot compressive tests. The stress values corresponding to the deformation temperature at 573K(300°C) and strain rate at $5s^{-1}$ are the highest, and the peak stress increases with increasing the strain rate for a given strain because there is not enough time to fully recrystallize in high strain rate, and decreases by deformation temperature due to the recovery or recrystallization. It is consistent with the results reported by Li et al.[11]and Ou et al.[12].

Fig.1 also shows that at 300 °C and 350°C with a high strain rate ($5s^{-1}$), the flow stress is increased considerably with strain when the material has just yields. This is due to the strain hardening caused by multiplication and interaction between dislocations and particles. With further increase in strain, the flow stress remains almost constant during the whole compression process, owing to the balance of strain hardening and softening from dynamic recovery or recrystallization. At the high deformation temperature above 350 °C and low strain rate of $0.1s^{-1}$ and $1s^{-1}$, the flow stress keeps constant after yielding. It is worth noting that at all the deformation temperatures with a strain rate of $1s^{-1}$, serrated wave lines exhibit, which may be the typical character of dynamic recrystallization. At the deformation temperature of 300 °C and 350 °C, the waves present only at the high strain, but at the higher deformation temperatures, waves pass through the plastic deformation stage. Especially at 450 °C, the waves fluctuate violently first, and then become weakened.

3.2 Activation energy for deformation

It is usually accepted the flow behavior of material is controlled by thermal activation at elevated temperature, which is always described using Arrhenius type equation, in which a variable of activation energy is contained that stands for the difficulty for deformation.

The flow stress of hot deformation material depends on the deformation temperature, strain rate and true strain. When the true strain is fixed, the flow stress just depends on the deformation temperature and strain rate and several equations are used to relate the flow stress to strain rate,

$$\dot{\varepsilon} = A_1\sigma^{n_1}\exp(-\frac{Q}{RT}) \text{ when } \alpha\sigma<1.2 \quad (1)$$

$$\dot{\varepsilon} = A_2\exp(\beta\sigma)\exp(-\frac{Q}{RT}) \text{ when } \alpha\sigma>1.2 \quad (2)$$

$$\dot{\varepsilon} = B[\sinh(\alpha\sigma)]^n\exp(-\frac{Q}{RT}) \text{ for all } \sigma \quad (3)$$

in which R is the universal gas constant (8.314 J mol^{-1} K^{-1}), T is the absolute temperature (K), Q is the activation energy of hot deformation(kJ.mol^{-1}). Moreover, A_1, $A2$, A, α, β, n_1, n are material constants, and $\alpha = \beta/n_1$.

The material constants in constitutive equations afore-mentioned are calculated through extracting the true stress-true strain data from the compression tests. Steady state true stress data at true strain of 0.3 are usually taken to evaluate the material constants above. For low and high stress levels, natural logarithm is taken at both sides of Eqs. (1) and (2), and then the following equations can be obtained:

$$\ln\dot{\varepsilon} = \ln A_1 + n_1\ln\sigma - \frac{Q}{RT} \quad (4)$$

$$\ln\dot{\varepsilon} = \ln A_2 + \beta\sigma - \frac{Q}{RT} \quad (5)$$

Then, the values of the flow stresses and the corresponding strain rates under the strain of 0.3 are substituted into Eqs. (4) and (5) , the data of σ, lnσ and ln$(\dot{\varepsilon})$ are computed, then plots ln$(\dot{\varepsilon})$-lnσ and ln$(\dot{\varepsilon})$-σ linearly according to the figures as shown in Fig 2. The average value of slopes are used to derive the values of n_1 and β , 8.267 and 0.132MPa^{-1}, respectively, so α=0.016MPa^{-1}.

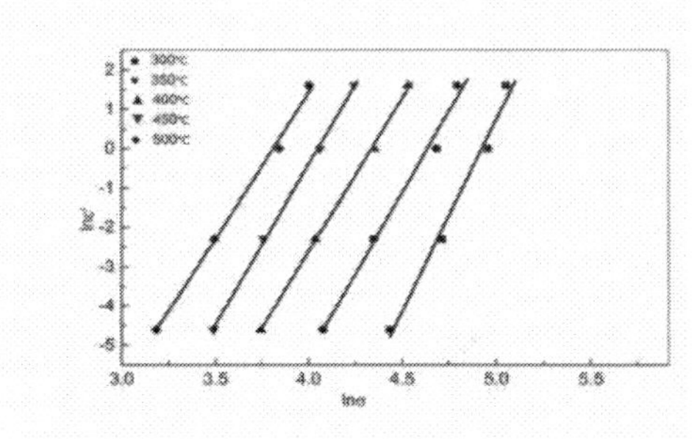

(a)

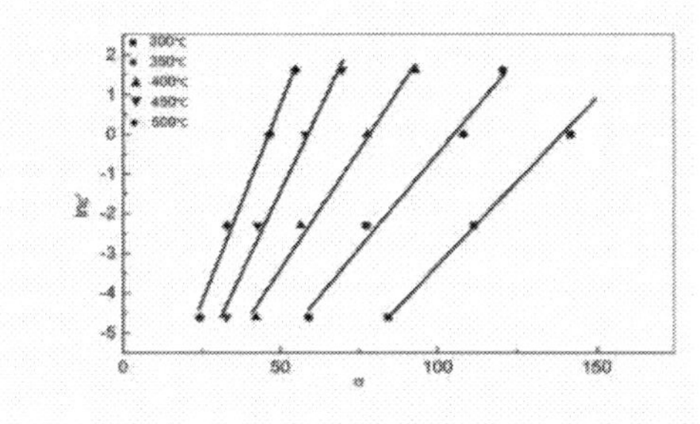

(b)

Fig.2 Relationship of (a) ln$(\dot{\varepsilon})$*vs* lnσ for alloy A_1, (b) ln$(\dot{\varepsilon})$ *vs* σ for the studied alloy during hot compression

Taking partial differential equation of Eq. (4) into consideration yields:

$$Q = R\left[\frac{\partial\ln\dot{\varepsilon}}{\partial\ln[\sinh(\alpha\sigma)]}\right]\left[\frac{\partial\ln[\sinh(\alpha\sigma)]}{\partial(1/T)}\right]_T = Rn_2s \quad (6)$$

The relationships of ln $\dot{\varepsilon}$ -ln[sinh$(\alpha\sigma)$] and ln[sinh$(\alpha\sigma)$]-$1/T$ are plotted in Fig.3. As a result, the value of n_2 and s are determined by calculating the average slope of two groups of lines. So n_2=5.88, s=3.744 and Q=183.02kJ/mol.

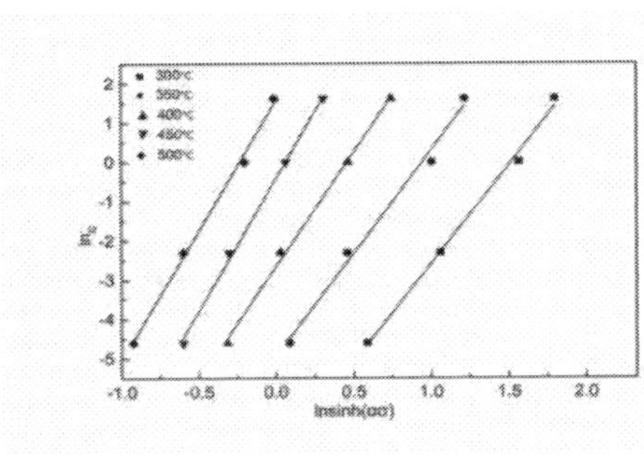

(a)

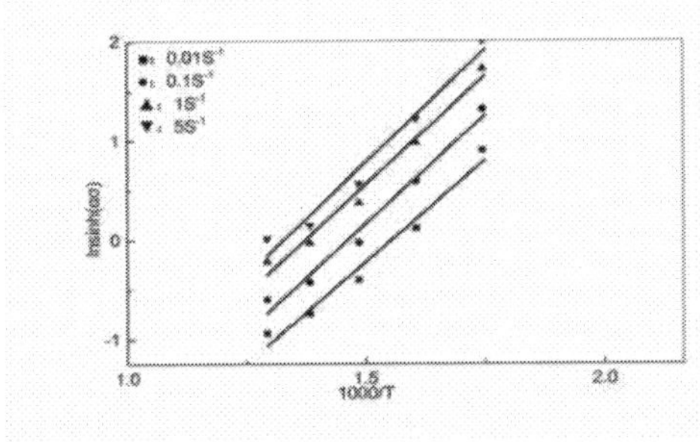

(b)

Fig.3 Relationship of (a) lnε-ln[sinh(ασ)] for the studied alloy (b) $\ln[\sinh(\alpha\sigma)] - 1/T$ for A_1 alloy

3.3 Constitutive equations

The Zener-Hollomon parameter representing the temperature modified strain rate is used to illustrate the variation of flow stress with deformation conditions. The correlation between the Zener-Hollomon parameter with the strain rates, deformation temperature, and activation energy is expressed by the equation as follows:

$$Z = \dot{\varepsilon} \exp(\frac{Q}{RT}) \tag{7}$$

Combining Eqs.1-3 with (7), then, taking the logarithms, yields:

$$\ln Z = \ln B + n \ln[\sinh(\alpha\sigma)] \tag{8}$$

Where B and n are derived from the ln*Z* *vs*. ln*[sinh(ασ)]* plots. By substituting the values of *Q* and different hot deformation conditions into Eq. (1), *Z* values are calculated subsequently. The values of ln*(Z)* for various strain rates and deformation temperatures can be evaluated for all the alloys studied. It is evident that *Z* increases with the increased strain rate and decreases by deformation temperature, which is similar to the variation of the peak flow stress *(σ)* shown in Figs.1. Liu et al. [13] reported that dynamic recrystallization is generally favored at low *Z* values, which corresponds to low strain rates and high deformation temperatures.

Then the relationship between ln*Z* and ln[sinh*(ασ)*] is shown in Fig.4. So we can derive the values of ln*B* and *n* are 30.03 and 6.65 for A_1 alloy, respectively.

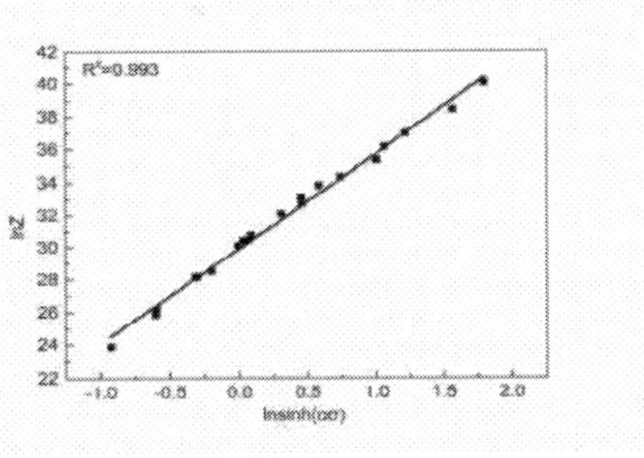

Fig.4 Plot of ln*Z* vs ln[sinh*(ασ)*] for Alloy-A_1

We can get the values of materials constants n_1, *β*, *α*, n_2, *Q*, ln*B* and *n* in constitutive equations of A_1 alloy at *ε*= 0.3 are 8.857, 0.107, 0.0122, 6.156, 222.84, 36.95 and 6.03, respectively. So the constitutive equation of the studied alloy (A_1) for *ε*=0.3 is given as following:

$$\dot{\varepsilon} = 9.48 \times 10^{12} [\sinh(0.016\sigma)]^{5.88} \exp(-\frac{183020}{RT}) \tag{9}$$

4. Conclusions

Hot deformation behaviors of Al-2Cu-1Mn alloy were researched by isothermal compression using a Gleebe-3800 thermal simulator. The main results are summarized as follows:

(1)The deformation activation energy for Al-2Cu-1Mn alloy is obtained as 183.02KJ/mol through hot compressive test using a Gleebe-3800 thermal simulator.

(2)The constitutive equation for A_1 alloy has been established relating flow stress to deformation temperature and strain rate.

(3)It is easier to start dynamic recrystallization at higher temperature with strain rate $1s^{-1}$ than other conditions.

Acknowledgements

This work is supported by Jiangsu Key Laboratory Metallic Materials (Grant no. BM2007204) and the Fundamental Research Funds for the Central

Universities (Grant no.2242016k40011). The authors are also thankful to Sha-steel Iron and Steel Research Institute of Jiangsu Province for their useful assistance.

References

[1]S. Nicolae, C. Vasile-Danut, B. Mihai, JOM. **64**, 607-614 (2012).

[2]C.Y. Chou, C.W. Hsu, S.L.Lee, K.W.Wang, J.C.Lin, J.Mater. Process.Technol. **202**, 1-6 (2008).

[3]P.Nageswara Rao, R.Jayaganthan, Mater. Des.**39**, 226-233 (2012).

[4]M.Vaseghi, H.S. Kim, Mater. Des. **36**, 735-740 (2012).

[5]A.Krishnaiah, U.Chakkingal, P. Venugopal, Scripta Mater. **52**, 1229-1233 (2005).

[6] L. Deng, T. Zhao, J.S. Jin, X.Y. Wang, 11th International Conference on Technology of Plasticity, ICTP 2014, 19-24 October 2014,Nagoya Congress Center, Nagoya, Japan.

[7]F. Bardi, M. Cabibbo, S. Spigarelli, Mater.Sci.Eng.A **334**, 87-95 (2002).

[8]H.Z. Li, H.J. Wang, X.P. Liang, H.T. Liu, Y. Liu, X.M. Zhang, Mater.Sci.Eng.A **528**, 1548-1552 (2011).

[9]H.Z. Li, Z. Li, M. Song, X.P. Liang, F.F. Guo, Mater.Des.**31**, 2171-2176 (2010).

[10]X.Y. Liu, Q.L. Pan, Y.B. He, W.B. Li, W.J. Liang, Z.M. Yin, Mater.Sci.Eng.A **500**, 150-154(2009).

[11]B.L, Q.L.Pan, Z.Y.Zhang, J Mater Eng Perform. **22**, 536-540 (2013).

[12]L.Ou, Y.F.Nie, Z.Q.Zheng, J Mater Eng Perform. **23**, 25-30 (2014).

[13]X.Y. Liu, Q.L. Pan, Y.B. He, W.B. Li, W.J. Liang, Z.M. Yin, Mater.Sci.Eng.A **500**, 150-154 (2009).

 https://doi.org/10.1051/matecconf/201820703017

Magnetic properties of silicon steel sheet core with different processing stress

Toshiki Matsui[1], and *Kyyoul* Yun[1,*]

[1]Gifu University, Faculty of Engineering, 1-1 Yanagido Gifu, Japan

Abstract. The motor core including a rotor core and a stator core, which is made from silicon steel sheets. Iron loss increases during fixation of the stator core, e.g., by interlocking, welding, and shrink fitting installation. In this paper, the magnetic properties changes by each processing such as wire cutting, punching, interlocking and shrink fitting are investigated. Iron loss of the toroidal cores using punching, interlocking and shrink fitting are increased from 1.16 W/kg to 1.56 W/kg (34.4 % increased) at 50Hz, and from 21.1 W/kg to 27.5 W/kg (30.3 % increased) at 400 Hz compared with iron loss of wire cut toroidal core.

1 Introduction

Modern society requires a huge amount of energy, especially clean and highly adaptable electrical energy. The highest electricity consumers in modern society are rotating machines, which have an electricity usage rate of over 57.3%. Rotating machines have industrial, commercial, and household applications, because of their high efficiency and handiness. Therefore, an increase in rotating machine efficiency leads to huge amounts of electrical energy savings realized in megawatts [1]. Magnetic hysteresis curves are affected by the residual stress (mainly compressive stress) of the material [2, 3]. The motor core includes a rotor core and a stator core, which are made from electrical steel sheets. Iron loss increases during fixation of the stator core, e.g., by interlocking (mechanical bonding), welding, and thermal insert installations [4-9]. These methods produce a residual stress in the motor core, leading to an increase in the building factor and iron loss of the motor. To realize a high-efficiency motor, the residual stress of the motor core must be figured the process out of the interlocking, welding or the shrink fitting of the core for the motor. In this paper, the magnetic properties affected by residual stress such as wire cutting, punching, interlocking and shrink fitting.

2 Experimental procedure

2.1. Specimen

Ring-shaped silicon steel sheets were prepared by four types of process such as (i) wire cutting, (ii) punching, (iii) interlocking and (iv) shrink fitting after (iii). The material used is 35A270 (Nippon steel & Sumitomo metal), and each silicon steel sheets has 50 mm in outer diameter and 40 mm in inner diameter. And then toroidal cores were prepared by 10 sheets lamination. Magnetic properties are affected by various stress. To prevent it, toroidal cores were set on in a frame, which made from Bakelite 0.2 mm diameter larger than outer diameter of ring shaped silicon steel sheet as shown in Fig 1 (a). Thickness of Bakelite frame was 1 mm. Excitation coil was 200 turns and B coil was 100 turns. Excitation wave form was sinusoidal curve. Maximum magnetic flux density was 1.0 T. In order to separate iron loss DC hysteresis loss and eddy current loss from iron loss. DC magnetization equipment and AC magnetization equipment were used. Frequency of AC measurement was 50 Hz and 400 Hz. To realize the effect of shrink fit, SUS 304 frame was used as shown in Fig.1 (b). Diameter of outer SUS frame was 0.1 mm radially shorter than a dimeter of toroidal core. Both frames have 6 mm in height and 52.2 mm in outer diameter. To investigate an insulation between SUS frame and toroidal core, insulating tape (12 μm, thickness) was used in boundary surface.

Magnetic properties of silicon steel sheets are easy to deteriorate by residual stress. Therefore, it is important to suppress the effect of the stress from coil winding. To prevent of coil winding effect, Bakelite cases were prepared as shown in Fig. 1. Thickness of the Bakelite cases is 1 mm, and diameter of the case are 50.2 mm in inner diameter of upper case and 39.9 mm in outer dimeter of lower case.

Figure 2 shows the diagram to reproduce the processing of thermal insert. Nonmagnetic material of case is SUS304 to prevent the additional permeability from case. To add the compressive stress, diameters of the case are 49.9 mm in inner diameter of upper case and 39.8 mm in outer dimeter of lower case.

* Corresponding author: yun7110@gifu-u.ac.jp

© The Authors, published by EDP Sciences. This is an open access article distributed under the terms of the Creative Commons Attribution License 4.0 (http://creativecommons.org/licenses/by/4.0/).

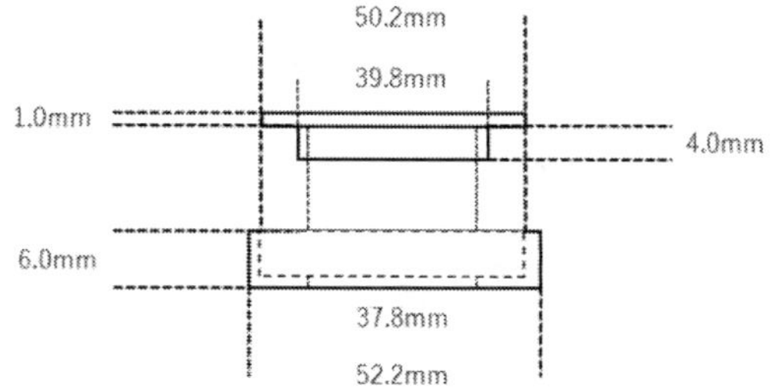

(a) Bakelite.

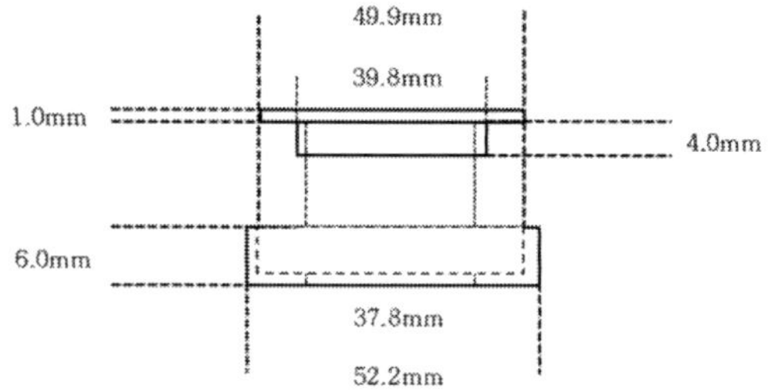

(b) SUS304.

Fig. 1. Frame to prevent stress and shrink fitting.

2.2 Experimental equipment

In this time, to separate the total iron loss with DC hysteresis loss and eddy current loss, two different equipment are used. One of them is DC magnetization equipment to measure the DC magnetic properties. The other one is AC magnetization equipment. The diagram of AC experimental equipment is represented in Fig. 2. The exciting coil with a 1 mm wire diameter is wound in two layers with 200 turns. The iron loss (W) was calculated as follows:

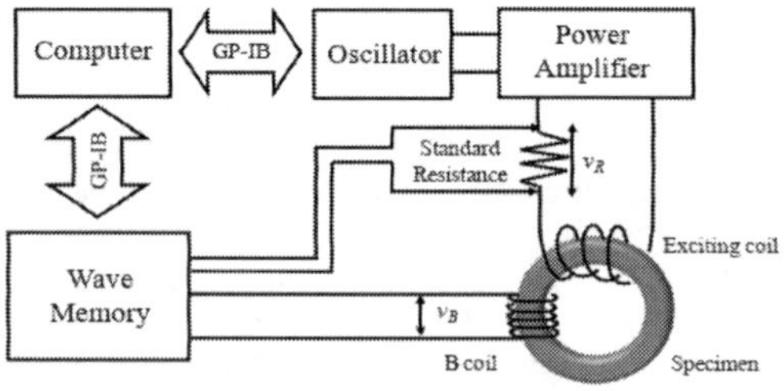

(a) AC measurement system.

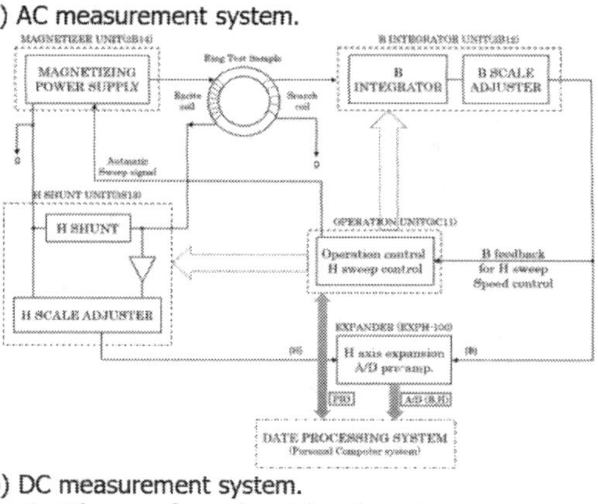

(b) DC measurement system.

Fig. 2. Diagram of experimental equipment.

$$W[W/kg] = \frac{1}{\rho T}\int_0^T H\frac{dB}{dt}dt \tag{1}$$

where ρ was the density of the specimen and T was the cycle. Form factor of excitation wave form was 1.11 ± 0.01.

3 Results and discussions

From Fig. 3, shrink fitting has the largest hysteresis loop and the largest iron loss value. Then the iron loss value is decreases in the order of shrink fitting, interlocking, punching and wire cutting. The coercively force (H_c), the maximum magnetic field (H_m) and relative permeability (μ_r) were compared at 50 Hz. The wire cutting was H_c = 48 [A/m], H_m = 131.84 [A/m] and μ_r = 6041. The punching was H_c = 58 [A/m], H_m = 325.21 [A/m] and μ_r = 2453. The interlocking was H_c = 63 [A/m], H_m =

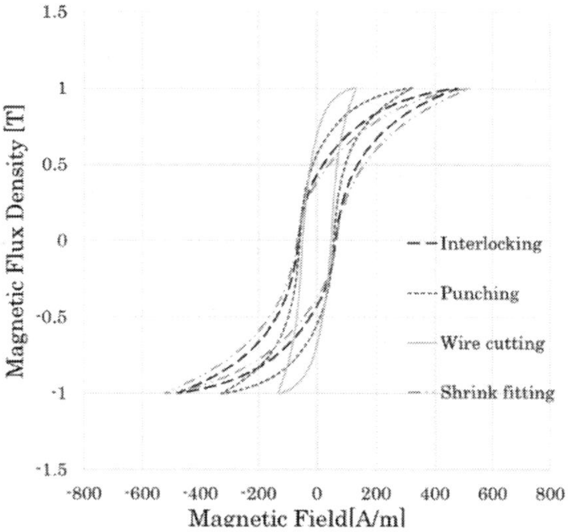

(a) Magnetic hysteresis curves at 50 Hz.

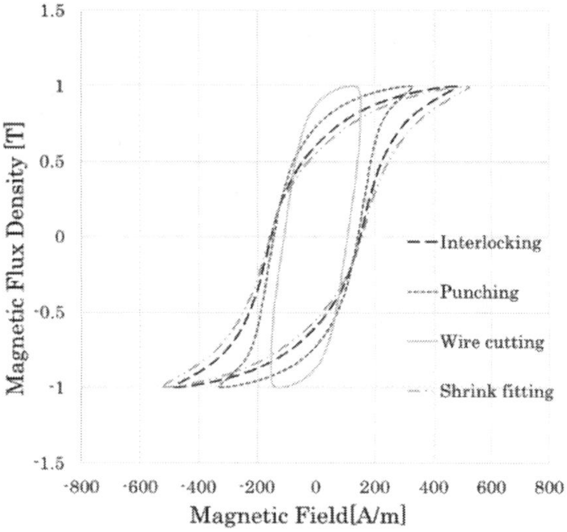

(b) Magnetic hysteresis curves at 400 Hz.

Fig. 3. Magnetic hysteresis curves with 50 Hz magnetization.

https://doi.org/10.1051/matecconf/201820703017

477.66 [A/m] and μ_r = 1667. The shrink fitting was H_c = 65.25 [A/m], H_m = 521.44 [A/m] and μ_r = 1525. From the results, magnetic properties using wire cutting show the smallest coercive force and maximum magnetic field, maximum permeability is the highest. It can be seen that the coercive force is increased with each processing and the characteristics of the soft magnetic material deteriorate.

Iron loss of wire cutting shows the lowest value. The eddy current loss of the wire cutting at 50 Hz is larger than that of the punching specimen, but at 400 Hz the wire cutting specimen has the smallest value. As a result, deterioration of magnetic characteristics with wire cutting is smaller than that with punching. Also, it can be seen that hysteresis loss due to shrink fitting is smaller than interlocking. Due to stress caused by interlocking deterioration can be reduced by shrink fitting the punching specimen. What can be seen from the above graphs and tables is that the processing stress due to the cutting method when forming a ring sample has the most influence on the magnetic characteristics, and the iron loss is increased by 24% in the punching process as compared with the wire cutting. Toroidal core of interlocking has an iron loss of 5% higher than that of punched core. Although the crimped sample was used as the shrinkage fitting sample, the iron loss value

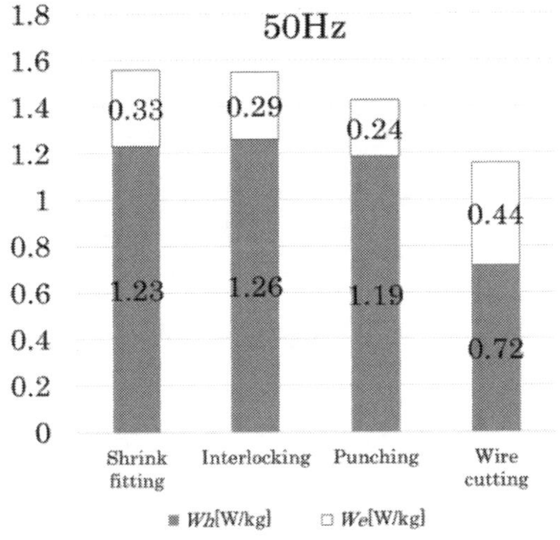

(a) Iron loss saperation at 50 Hz.

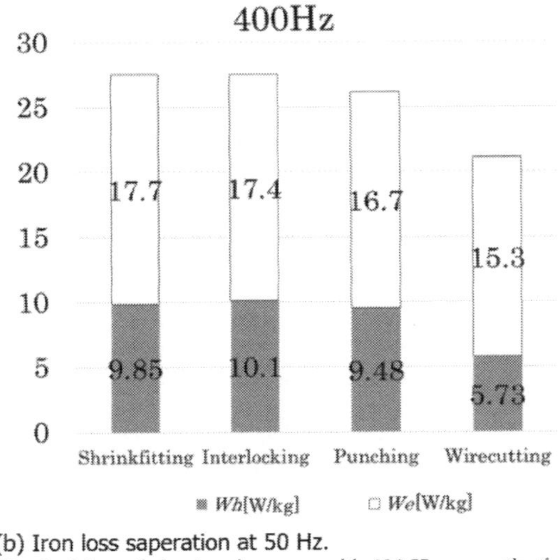

(b) Iron loss saperation at 50 Hz.

Fig. 5. Magnetic hysteresis curves with 400 Hz magnetization.

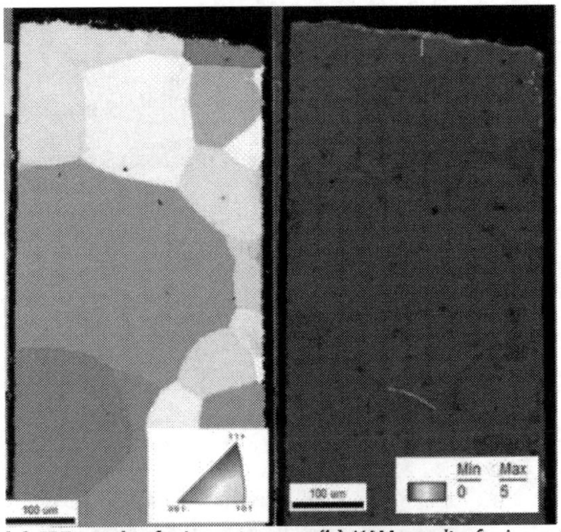

(a) IPF result of wire cut. (b) KAM result of wire cut.

(c) IPF result of punching. (d) KAM result of punching.

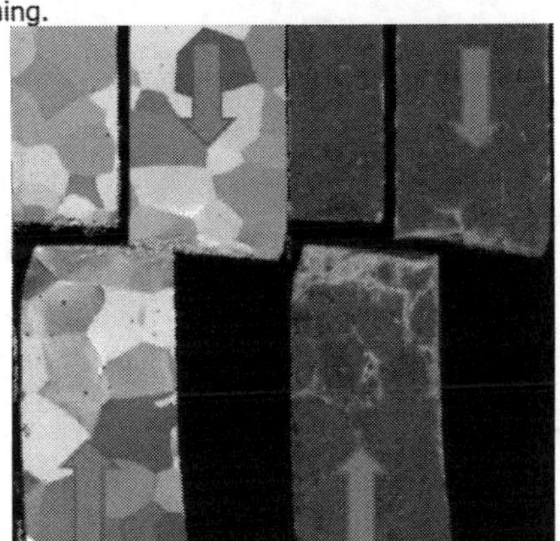

(e) IPF result of interlocking. (f) KAM result of interlocking.

Fig. 5. Iron loss of each process.

increased by 7%. There is an increase in core loss value of 12% by carrying out all the steps from punching to shrink fitting. Considering hysteresis loss and eddy current loss 50Hz due to iron loss separation with reference to wire cut. Hysteresis loss of wire cutting toroidal core is 65.5% and 76.1% smaller than that of punched toroidal core and interlocked toroidal core, respectively.

IPF and KAM results of each processing stress are shown in Fig.5. In the IPF diagram of the wire cutting (fig.5 (a)), the crystal orientation of the crystal grain at the cutting portion are not changed by wire cutting. Therefore, the iron loss of the wire cutting toroidal core shows the minimum value. Also in the KAM diagram (fig.5 (b)), the distortion at the incision is several μm at the sample edge. It means that the residual stress is not occured by wire cutting. In the IPF diagram of punching (fig.5 (c)), and residual stress was confirmed in crystal grains of about 200 μm from the edge of the sample. Also in the KAM diagram (fig.5 (d)), it can be seen that a strain of about 200 μm is generated from the sample end. The crystal grains around the cutting portion by the punching shows a fine-grained crystals compare with that by wire cut. Therefore, magnetic path is possible to change and get longer by the fine crystal grain. It means that the magnetic properties are deteriorated by punching. It is clear from the IPF diagram (fig.5 (e)) that the crystal grains of the interlocking portion are also getting smaller and finer. There are the convex and the concave portion of the interlocked area. It is similar with the result of (c), however it is more complecated compare with that of punching. From the KAM diagram (fig.5 (f)), many complecated distortions were observed in the end of seperated area by interlocking. Moreover, interlocked area is possible to contact with upper silicon steel sheet. It makes the increment of the eddy current loss. From the results, the hysteresis loss is increased by punching or interlocking after punching.

4 Conclusions

The magnetic characteristics affected by the residual stress of the silicon steel sheet are investigated. It was found from the IPF diagram that the crystal grains are refined by punching and interlocking. It can be seen from the KAM drawing that distortion of about 200 μm is generated from the edge of the sample by punching. This is the same as the conventional knowledge that the hardness increases in the region about 1/2 to 1 times the thickness at the time of punching and shearing[9]. The magnetic properties due to the residual stress of the silicon steel sheet were investigated. In particular, iron loss increased from 21.06 W / kg to 29.28 W / kg (39.0% increase) at 400 Hz from wire cutting to shrink fitting.

References

1. M. Croucher, Energy Policy, **45**, 304 (2012)
2. M. Oka, T. Ogasawara, N. Kawano, M. Enokizono, IEEE Trans. Magn., **50**, DOI: 10.1109/TMAG.2014.2319237 (2014)
3. R. Siebert, J. Schneider, E. Beyer, IEEE Trans. Magn., **50**, DOI: 10.1109/TMAG.2013.2285256 (2014)
4. D. Miyagi, N. Maeda, Y. Ozeki, K. Miki, N. Takahashi, IEEE Trans. Magn., **45**, 1704 (2009)
5. K. Fujisaki, R. Hirayama, T. Kawachi, S. Satou, C. Kaidou, M. Yabumoto, T. Kubota, IEEE Trans. Magn., **43**, 1950 (2007)
6. K. Senda, H. Toda, M. Kawano, IEEJ Trans. Ind. Appl., **4**, 496 (2015)
7. M. Yabumoto, C. Kaido, T. Wakisaka, T. Kubota, N. Suzuki, Nippon Steel Tech. Rep., **87**, 57 (2003)
8. T. Nakayama, H. Kojima, JMEPEG, **16**, 7 (2006)
9. K. Yun, S. Yanase, T. Ban, Y. Ohya, IEEE Trans. Mag., **53**, DOI: 10.1109/TMAG.2017.2720139 (2017)

A New Method of D-TDOA Time Measurement Based on RTT

Junjie Zhou[1], LiangJie Shen[1],Zhenlong Sun*

[1]*Department of Info Sci.& Tech., Wenhua College, Wuhan, China*

Abstract. In this paper, a novel difference-time difference of arrival (D-TDOA) time measurement method based on round-trip transmission (RTT) is proposed. The accuracy of time measurement is critical for high accuracy positioning, yet the traditional time difference of arrival (TDOA) time measurement method cannot deal well with the problem of time synchronization error, such as the initial time offset and transmission cumulate time error caused by clock frequency offset and drift. We formulate a D-TDOA time measurement mechanism to overcome the shortcoming of the traditional TDOA method in eliminating the influence of clock error. It is proved that the D-TDOA method can effectively eliminate the influence of initial time offset by analyzing the simulation data, which can also reduce the cumulate time error remarkably. The simulation results demonstrate that the D-TDOA method can significantly enhance the accuracy of time measurement in the circumstance that the AP clock synchronization is not required seriously.

Key Words: D-TDOA;RTT;time measurement; clock errors

1 Introduction

In recent years, with the rapid development of communication and navigation technology, location services as a new strategic industry has become an indispensable part. A variety of indoor positioning technologies have been proposed, such as RFID,Bluetooth, UWB, WIFI, A-GPS, ZigBee, etc..

At present, the common indoor positioning methods are generally defined as the following four categories: Time of arrival (TOA), Time Difference of Arrival (TDOA),Arrival of Angle (AOA) and Received Signal Strength Indication (RSSI). Among them, TOA and TDOA are required to get the accuracy time parameters. The TOA method determines the location of the unknown node by measuring the signal transmission and reception time of the location node and the unknown node, and the TDOA method measures the distance by measuring the time difference of the same signal from the unknown node to the difference location nodes. Compared with other positioning methods, the positioning method based on time measurement has the advantages of simple and universal, so it has a wide application prospect. However,due to the indoor environment is complex and changeable that the measurement precision of time is influenced by many factors, such as the multipath[1], the interference of the same frequency and the error of the node clock. Therefore, how to get the accurate time information is very significant for the realization of high precision indoor positioning. Authors in[2] achieved the high accuracy positioning method whose accuracy is 1m by using timing resolution of 1μs. However, their final measurement result is the average of 500 times repeated measurements. Obviously, it is not desirable in practical applications. Reference[3] has been proposed through the integrated application of multi-hop latency compensation and asymmetric compensation algorithm to ensure nanosecond clock synchronization between the various WLAN access points (AP) accuracy. The traditional TDOA [4-5] positioning method cannot deal well with the initial time offset and the cumulate time error caused by clock oscillator frequency drift, so it' unable to meet the demand of indoor high accuracy positioning. In order to overcome the time measurement error,Difference-Time Difference of Arrival (D-TDOA) time measuring method is proposed in this paper. Theoretical analysis and experimental results show that the method can greatly improve the accuracy of time measurement.

This paper is organized as follows. In Section II, this part illustrates the main sources of time measurement errors and the necessity of eliminating clock errors. The description of the D-TDOA time measurement method based on RTT as described in Section III. In Section IV, the simulation results show that the D-TDOA method can effectively mitigate the influence of the main clock error on time measurement accuracy. Finally, In Section V conclusions are provided.

2 TIME MEASUREMENT ERRORS

*Corresponding author e-mail: zjj0718@126.com
Acknowledgments: This work was financially supported by the following fund projects:: 2017 CERNET Innovation Project（NGII201707085）; Wenhua College 2018 School Research Fund Project（No.2018Y04）.

© The Authors, published by EDP Sciences. This is an open access article distributed under the terms of the Creative Commons Attribution License 4.0 (http://creativecommons.org/licenses/by/4.0/).

 https://doi.org/10.1051/matecconf/201820703018

2.1 Formatting the title, authors and affiliations

The main source of time measurement errors can be summarized as the following three aspects.

Firstly, the additional time delay caused by the non-line of sight (NLOS) propagation and the multipath effect is the main factor to affect the accuracy of time measurement. Existing research shows that the error can be mitigated to a certain extent by using biased Kalman filtering[6].Authors in [7] proposed a new positioning algorithm (N-CHAN) to overcome an obvious shortcoming that the accuracy of traditional CHAN algorithm effected by no-line-of-sight (NLOS).

Secondly, timing resolution plays an important role in the accuracy of time measurement. Usually if the resolution of the timer is low, its measurement time will be smaller than the real time. For example, assuming the real time is 333.33ns, then 333ns will be getting when the timing resolution is 1ns. Therefore, with the enhancement of timing resolution, the more precise time information and accurate location information can be obtained.

Thirdly, the reason for the stability of the clock will cause time error, while the difference of clocks between the different nodes also presents a great challenge to the accuracy of the timing, which is not negligible for high precision indoor positioning.Obviously, how to effectively restrain the clock errors is worth in-depth study.

2.2 Clock errors

It's very difficult for the normal TDOA positioning mechanism to realize clock synchronization [8-9]among AP nodes. GPS, the most accurate time synchronization system, can achieve the accuracy of 15-30ns. Usually clock synchronization mainly depending on the standard clock to periodically sends clock synchronization signal to other slave clocks. Then, the slave clocks would calculate the path delay, time offset and frequency offset by using the information from the signal, and adjust their time and frequency offset close to the standard clock as much as possible. However, the clock error still has a negative impact on time measurement accuracy.

The main sources of clock errors are concluded as follows. On the one hand, the clocks of the AP still have small time offset compared to the standard clock after the time compensation. It is called the initial time offset. On the other hand, the crystal frequency offset and drift of clocks will cause transmission cumulate time offset that named as cumulate time error. Therefore, the clock errors discussed in this paper are consisted of initial time offset and cumulate time error.

In the actual situation, although most of the crystal frequency offset is usually as small as dozens of ppm and the cumulate time error caused by crystal frequency drift can be ignored in minimal period, it is difficult to overlook the impact of the errors on timing measurement accuracy, during a relatively long period. Meanwhile, the initial time offset also seriously affects the time accuracy. The above problems can be dealt well with the D-TDOA time measurement method.

3 D-TDOA METHOD

3.1 Measuring mechanism

D-TDOA measurement mechanism is based on the traditional TDOA signal transmission mechanism combined with RTT to transmit signal. Regard AP_0 as the central reference point, AP_1, AP_2 and AP_3 as auxiliary reference points, respectively.Taking the signal transmission process that the signal transmits from the MT to AP_0 and AP_1 for example.

The MT sends the WIFI signal at t_0, AP_0 receives the signal at t_1 and AP_1 receives the signal at t_2. After that AP_1 immediately has a RTT with AP_0. The frame leaves AP_1 at t_3 and reaches AP_0 at t_4, then leaves AP_0 at t_5 and reaches AP_1 at t_6. AP1 sends data frame composed of t_2, t_3 and t_6 to AP_0. The transmission process above shows as in Figure 1. In the case that clock synchronization is not required, in order to eliminate the initial time offset and reduce the cumulate time error caused by clock crystal oscillator frequency drift, the following method is used to calculate the time difference of the MT to the center reference point AP_0 and the auxiliary reference point AP_1.

In the case depicted in Figure 1, the transmission time from MT to AP_0:

$$\Delta_1 = t_1 - t_0 \tag{1}$$

The transmission time from AP_1 to AP_0:

$$\Delta_3 = \frac{(t_6 - t_3) - (t_5 - t_4)}{2} \tag{2}$$

The transmission time from MT to AP_1:

$$\Delta_2 = (t_4 - t_0) - (t_3 - t_2) - \Delta_3 \tag{3}$$

The TDOA of MT to AP_0 and MT to AP_1:

$$T_{TDOA} = \Delta_2 - \Delta_1 = (t_4 - t_1) - (t_3 - t_2) - \frac{(t_6 - t_3) - (t_5 - t_4)}{2} \tag{4}$$

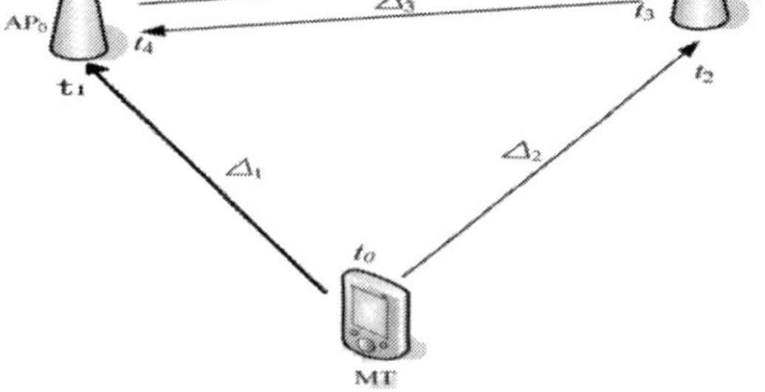

Figure 1. Schematic diagram of D-TDOA time measurement method.

3.2 Analysis of the D-TDOA time measurement method

In this paper, a simplified time model is used. Suppose that the moment of the standard clock is M when the last clock synchronization is finished. And after a period of time $T_i(i=0,1,2...)$, the standard time is $t_i=M+T_i$.The corresponding time of AP_i clock is:

$$t_i = t_i^{'} + u_i + \eta_i(t_i^{'} - M) \quad (5)$$

u_i is the initial time offset between standard clock and AP_i clock after the last clock synchronization. η_i is crystal frequency drift of AP_i. Thus $\eta_i(t_i'-M)$ is the cumulate time error of crystal oscillator frequency offset of AP_i from standard time M to t_i'.Then the following expression can be obtained:

$$t_i = t_i^{'}(1+\eta_i) + u_i - \eta_i M = (M + T_i)(1+\eta_i) + u_i - \eta_i M = M + u + T_i(1+\eta_i) \quad (6)$$

Suppose the TDOA is TTDOA in the traditional TDOA time measurement method, so the following expression can be obtained:

$$T_{TDOA} = (t_2 - t_0) - (t_1 - t_0) \quad (7)$$

And according to expression (6), the following expression can be obtained

$(t_0^{'} = M + T_0; T_1 = T_0 + \Delta_1; T_2 = T_0 + \Delta_2)$:

$$T_{TDOA} = u_1 - u_0 + T_2 - T_1 + T_2\eta_1 - T_1\eta_0 \quad (8)$$
$$= u_1 - u_0 + \Delta_2 - \Delta_1 + \Delta_2\eta_1 - \Delta_1\eta_0 + T_0(\eta_1 - \eta_0)$$

While according to the expression (4) of D-TDOA method, the following expression can be obtained:

$$T_{TDOA} = (t_4 - t_1) - (t_3 - t_2) - \frac{(t_6 - t_3) - (t_5 - t_4)}{2} = a - b - \frac{c - d}{2} \quad (9)$$

Where $a=(1+\eta_0)(T_4-T_1)$, $b=(1+\eta_1)(T_3-T_2)$, $c=(1+\eta_1)(T_6-T_3)$ and $d=(1+\eta_0)(T_5-T_4)$, we can see that the parameters of each equation are from the same clock, and it contains the difference of time parameters, so this method can eliminate the partial errors of clock.

As can be seen from Figure 1,

$T_3=T_2+p_1; T_4=T_3+\Delta_3; T_5=T_4+p_0; T_6=T_5+\Delta_3; p_0$ represents t_5-t_4; p_1 represents t_3-t_2, expression (9) can be represents as:

$$T_{D-TOA} = \Delta_2 - \Delta_1 + \eta_0(\Delta_2 - \Delta_1) + (\eta_0 - \eta_1)(\frac{1}{2}P_0 + P_1 + \Delta_3) \quad (10)$$

In the expression (10), assuming the crystal frequency drift coefficient η_0 and η_1 are ±25ppm . The covering radius of WIFI signals is smaller than 100m, thus the maximum time difference $(\Delta_2 - \Delta_1)$ is approximately 333ns. So the error of $\eta_0(\Delta_2 - \Delta_1)$ is at 10^{-11} order magnitude, and it is negligible for the time measurement resolution of 1ns. In the term of $(\eta_0 - \eta_1)(\frac{1}{2}p_0 + p_1 + \Delta_3)$, the order of the first part is at 10^{-5} order magnitude. In the second part, assuming p0 and p1 are consist of the time of processing frames and waiting period (the node needs to monitor whether the channel is idle before starting to send data) in the case of neglecting other influencing factors, Meanwhile, the frame processing time can be also ignored relative to the frame waiting time. The IEEE802.11 MAC provides three kinds of access priority, including DIFS, PIFS and SIFS, where DIFS is the longest frame interval that be set at 50 microseconds in IEEE802.11b. So the part $(\frac{1}{2}p_0 + p_1 + \Delta_3)$ is at 10^{-4} order magnitude. Thus, the multiply of the two parts might bring several nanoseconds to the time measurement error.

Comparing the expression (8) and (10), it can be found that u0 and u1 are counteracted by D-TDOA measurement method of the same clock, so the initial time offset between the AP clock and the standard clock can be eliminated. Meanwhile, the terms containing T_0 is also counteracted so that the main time measurement errors in the process of clock synchronization have been avoided. That is to say, the time measurement cumulate error of D-TDOA method will not increase with the increase of the time after the previous synchronization. Consequently, the accuracy of time difference which is measured by D-TDOA measurement method can be improved substantially, thus the positioning accuracy will be greatly enhanced.

4 SIMULATION RESULTS

This section consists of two groups of simulations and a summary of results. The simulation A compared the influence of initial time offset to the time measurement accuracy on traditional TDOA method and the D-TDOA method. Simulation B illustrates the influence of cumulate time error to the time measurement accuracy on traditional TDOA method and the D-TDOA method.

Assume that the channel environment is ideal in all simulations. The experimental environment is set in a square region $(x\in[1,100], y\in[1,100])$, the default clock timing resolution of each AP is 1ns. Suppose η_0 and η_1 are taken from a uniform distribution with range of [-25×10^{-6}, 25×10^{-6}], where a negative value indicates the crystal frequency is slower than the standard clock frequency, and positive value indicates that the crystal frequency is faster than the standard clock frequency.

4.1 Simulation about initial time offset

In this simulation, the AP_0 is set at (36, 28) and AP1 is set at (71, 84). Assume the $T_0 = 0$; u_1- u_0=u; u_0 and u_1 are taken from a discrete uniform distribution whose range is ±[15,30] ns, and can only be integers.

Figure 2 shows the distribution of the average absolute error when u=8ns and the computer simulation is performed 1,000 times that MT at each point. The mean absolute error of TDOA method is fluctuating near 8ns. However, the mean absolute error of D-TDOA method is just between 1.17ns-1.54ns.

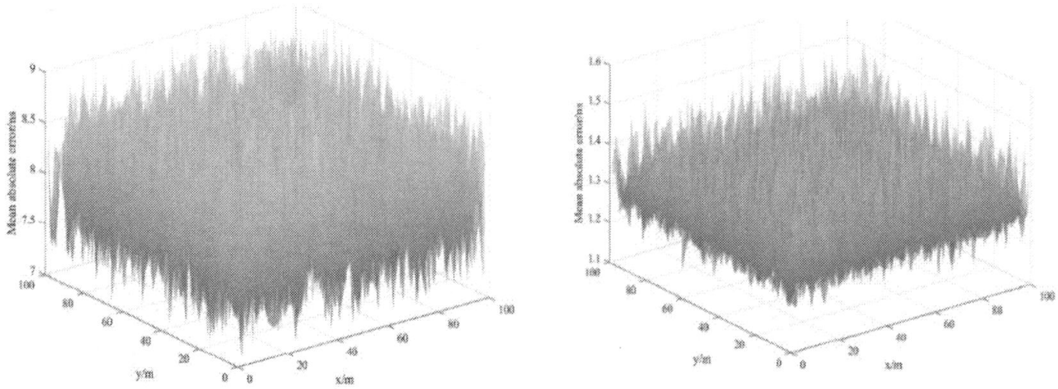

(a) TDOA method (b) D-TDOA method

Figure 2.Distribution of the mean absolute error when u =8ns

The mean absolute error curve, we show in Figure 3, the mean absolute error of TDOA method is proportional to u and it is closely related to the value of u, while the mean absolute error of D-TDOA method fluctuates between 1.32ns-1.43ns.

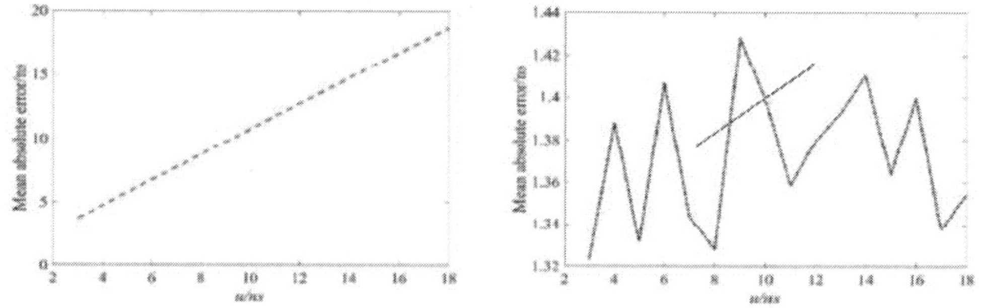

(a)TDOAmethod (b) D-TDOA method

Figure 3. Mean absolute error curve with the change of u.

Compared to the results of the traditional TDOA method with D-TDOA method in the case of T_0 =0, the mean absolute errors of D-TDOA method are less than the traditional TDOA method. Theoretical analysis and simulation results show, the mean absolute error of the traditional TDOA method is mainly caused by the initial clock error, and it is close to the initial time offset difference between AP_1's clock and AP_0's clock. For the D-TDOA method, the error is primarily due to the processing time of the AP, and it basically remains invariant. It can be possible to conclude that the D-TDOA method can significantly eliminate initial synchronization error and extremely improve the accuracy of time measurement.

4.2 Simulation about cumulate time error

In this simulation, mainly compare the influence of two timing measurement methods on cumulate time error. Assume the u_0=u_1=0ns, the AP_0 is set at (25, 25), the AP_1 is set at (75, 75) and the MT is set at (60,45). Because of crystal oscillator frequency drift coefficient remaining unchanged in a short time, the value of η_0 and η_1 in each group remains unchanged. Usually the value range of T_0 is from 0 to synchronization period, and T_0 is defined as 0-2.0ms in this simulation. We define P=MSETDOA/MSED-TDOA as gain reference value and repeated 10000 times to ensure the accuracy of the experiments. Then we have the simulation results in Table 1.

Table 1 shows the variation of the mean absolute error, mean square error and the gain with the change of T_0. The simulation results indicate that, under the traditional TDOA method, with the increase of T_0, the mean and mean square value of cumulative time error of the crystal frequency drift increases sharply, and the mean absolute error is proportional to the time T_0. However, the mean absolute error of D-TDOA measurement method remains the same value, which is due to the time measurement results under this method is independent of T_0. Therefore, the crystal frequency shift for traditional TDOA method will have a bad impact on the precision of time measurement, but the D-TDOA method can overcome the negative effects effectively. Considering the results, the proposed method has better performance than the traditional TDOA method.

Table 1 The influence of cumulate time error on two methods

$T_0(ms)$	$E[TDOA]$ (ns)	$E[D\text{-}TDOA]$ (ns)	MSE_{TDOA} (ns^2)	$MSE_{D\text{-}TDOA}$ (ns^2)	P
0	0.4324	1.3049	0.1870	2.5629	0.0730
0.5	8.3407		104.4024		40.7360
1.0	16.6701		417.1569		162.7675
1.5	25.0020		938.4505		366.1674
2.0	33.3347		1668.2832		650.9357

4.3 Summary of results

From the results of the two simulations, we can get the following:

1) Initial time offset after previous clock synchronization has no effect on D-TDOA measurement method.

2) D-TDOA method can effectively overcome cumulate time error caused by clock crystal oscillator frequency drift.

5 CONCLUSION

The D-TDOA time measurement method presented in this paper is simple and does not require additional cost, while it can also overcome major clock errors in the traditional TDOA method. The D-TDOA time measurement mechanism can enhances the accuracy of timing and avoids the problem of clock synchronization. Obviously,the initial time offset and cumulate time error has been eliminated effectively. The theoretical analysis and simulations results prove that the D-TDOA method significantly improves the precision of the time measurement.

With extensive coverage of wireless signal, D-TDOA method can be widely employed in indoor and outdoor positioning system. Simultaneously, D-TDOA method can also be used in the traditional 3-dementional TDOA positioning system to improve the positioning accuracy. In addition, the system which fuses the method with error correction and adaptive positioning algorithm can achieve the positioning accuracy of less than 1m. Apparently, D-TDOA method has a good application prospects.

References

1. Zhang J, Dong F, Feng G, et al. Analysis of the NLOS channel environment of TDOA multiple algorithms[C]// Sensors. IEEE, 2015.
2. Mccrady D D, Doyle L, Forstrom H, et al. Mobile ranging using low-accuracy clocks[J]. IEEE Transactions on Microwave Theory & Techniques, 2000,48(6):951-958.
3. Zhou J, Ke Y, Yu K, et al. A solution of High-precision WLAN positioning based on TDOA and PTP[J]. 2016, 61.
4. Uebayashi S, Shimizu M, Fujiwara T. A study of TDOA positioning using UWB reflected waves[J]. 2013:1-5.
5. Bocquet M, Loyez C, Benlarbi-Delai A. Using enhanced-TDOA measurement for indoor positioning[J]. IEEE Microwave & Wireless Components Letters, 2005,15(10):612-614.
6. Wann C D, Hsueh C S. Non-line of Sight Error Mitigation in Ultra-wideband Ranging Systems Using Biased Kalman Filtering[J]. Journal of Signal Processing Systems, 2011, 64(3):389-400.
7. Dong J Z, Wang Y W, Wei F, et al. A Localization Algorithm Research to Eliminate the Indoor NLOS Error[J]. Advanced Materials Research, 2014,989-994:2232-2236.
8. F. Sivrikaya and B. Yener. Time synchronization in sensor networks: A survey.IEEE Networks[J], 2004, 18(4): pp. 45-50.
9. B. Sundararaman, U. Buy, A.D. Kshemkalyani. Clock Synchronization in Wireless Sensor Networks: A Survey. Ad-Hoc Networks[J], 2005, 3(3) pp.:281-323.

Investigation on Damping Capacities of Spherical Mg_2Si Magnesium Matrix Composites Prepared by Super-Gravity Method

Yun-si Wang[1,2,3], Rui-long Niu[3,4], Fang-jia Yan[3,4], Dong-ping Duan[*3,4] and Xue-min Yang[*4]

[1]*Key Laboratory of Comprehensive and Highly Efficient Utilization of Salt Lake Resources, Qinghai Institute of Salt Lakes, Chinese Academy of Sciences, 810008 Xining, China*
[2]*Qinghai Engineering and Technology Research Center of Comprehensive Utilization of Salt Lake Resources*
[3]*University of Chinese Academy of Sciences, 100049 Beijing, China*
[4]*Key Laboratory of Green Process and Engineering, Institute of Process Engineering, Chinese Academy of Sciences, 100190 Beijing, China*

Abstract. Due to inherent high damping capacities of Mg, Mg-based alloys or composites can absorb enormous elastic energy, thus, can be used as damping materials playing important role in reducing vibration and noise. Spherical Mg_2Si magnesium matrix composites were prepared by super-gravity method with fly ash floating beads with particle sizes of 80, 125 and 500 μm. The microstructures and damping capacities of prepared samples were discussed. Spherical and fragment-shaped Mg_2Si phases were observed in the matrix by SEM micrograph. Increasing particle size of fly ash floating beads can lead to an obviously decreasing tendency of internal friction $\tan\varphi$. The critical strain $\varepsilon_{cr}=1.9\times10^{-2}$ of $\varepsilon-\tan\varphi$ curve and critical temperature $T_{cr}=150$ °C of $T-\tan\varphi$ curve for $\tan\varphi$ changing from slowly to greatly increasing. Damping peaks occur in the vicinity of strain $\varepsilon=5.2\times10^{-2}$ in $\varepsilon-\tan\varphi$ curve whereas no damping peaks occur in $T-\tan\varphi$ curve.

1 Introduction

Due to inherent high damping capacities of Mg[1], Mg-based alloys or composites can absorb enormous elastic energy, thus, can be used as damping materials playing important role in reducing vibration and noise. Relatively speaking, Mg-based composite material have better damping capacities due to either an addition of the second phase resulting in much more crystal defects, or inherent high damping capacities of the second phase itself, or absorption of vibration energy by phase interface[2,3]. The common preparation methods of Mg-based composite material are by usage of particles such as SiC, Al_2O_3, TiC_p particles and so on [4,5]. Liao *et al.*[6] successfully synthesized in-situ Mg_2Si/Mg-Al composites with Si particles, and obtained increasing Mg_2Si content can effectively increase the damping capacities of the composites. However, little hard research has been done on how the second phase with hollow spatial structure can influence the damping capacities of Mg-based composite material.

In the present work, in-situ spherical Mg_2Si magnesium matrix composites were successfully prepared in centrifugal apparatus by super-gravity method with hollow fly ash floating beads in sizes of 80, 125 and 500 μm, whose damping capacities has been investigated under conditions of changing temperature from ambient temperature T_{am} to 300 °C, strain ε from 1.0×10^{-3} to 1.0×10^{-1}.

2 Experimental

2.1 Preparation of Spherical Mg_2Si Magnesium Matrix Composites

Spherical Mg_2Si magnesium matrix composites were prepared in centrifugal apparatus in a type of DL-8M equipped with a heating furnace[7] using pure Mg ingot with 99.95 mass % purity and fly ash floating beads with particle sizes of 80, 125 and 500 μm shown in Figure 1..

The preparation procedures of spherical Mg_2Si magnesium matrix composites shown in Figure 2. can be summarized as follows: (1) Several pieces of magnesium about 30 g were placed into the upper graphite crucible and fly ash floating beads were placed into the lower one, where the upper and lower graphite crucible were tied together tightly as whole special crucible with substantial sieve pores between them[7]. Afterwards, the whole crucible were put in the heating furnace of the centrifugal apparatus; (2) The heating furnace was heated to 660 °C at a heating rate of 10 °C/min, and maintained for 10 min to ensure magnesium completely melted; (3) The whole crucible was rotated at a speed of 1338 r/min for

* Corresponding author: douglass@ipe.ac.cn (D. P. Duan), yangxm71@ipe.ac.cn (X. M. Yang)

© The Authors, published by EDP Sciences. This is an open access article distributed under the terms of the Creative Commons Attribution License 4.0 (http://creativecommons.org/licenses/by/4.0/).

promoting molten magnesium to percolate down through the sieve pores into the space among fly ash floating beads and their hollow space. During rotation process, the heating furnace was held at 660 °C for 1 min and then cooled to 600 °C; (4) The whole crucible including magnesium matrix composites sample was taken out from the heating furnace and then cooled to ambient temperature T_{am} in air atmosphere.

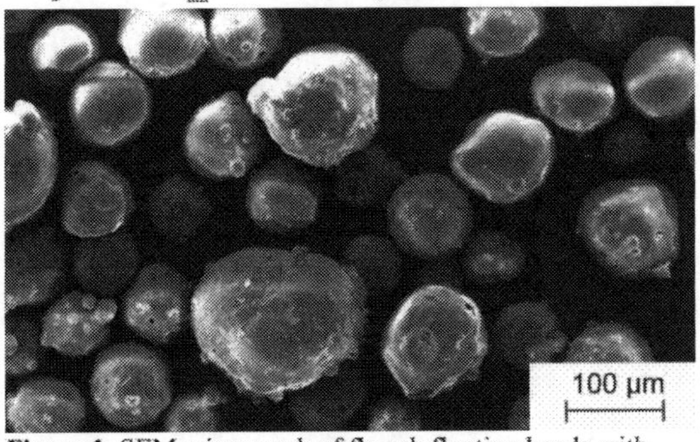

Figure 1. SEM micrograph of fly ash floating beads with particle sizes of 125 µm

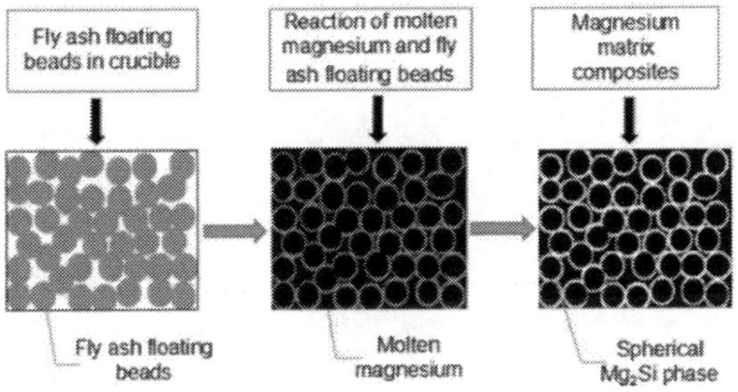

Figure 2. Preparation procedures of spherical Mg_2Si magnesium matrix composites by super-gravity method

2.2 Characterization of Microstructures and Damping Capacities

The microstructures of Mg_2Si magnesium matrix composites were determined by scanning electron microscopy (SEM, JSM−5610LV) equipped with energy dispersive X−ray spectrometer (EDS) operating at 20 kV. The dynamic mechanical analyzer (DMA, Q800, TA Instruments, USA) was used in the characterization of damping properties of spherical Mg_2Si magnesium matrix composites specimens under forced vibration by the test mode of single cantilever beam in this study. The damping capacities of spherical Mg_2Si magnesium matrix composites was measured by the method[8] of tangent phase angle φ expressed by $\tan\varphi$.

The measured specimens of prepared spherical Mg_2Si magnesium matrix composites were cut into the size of 10 mm×10 mm×10 mm for characterization of microstructures using SEM and 30 mm×10 mm×2 mm for determination of damping capacities using DMA by wire cutting electrical discharge machining (WEDM), and then ground with SiC abrasive paper in size of 800, 1600, and 2000 grits for successively polishing the surface of specimens, polished by silica suspension, and cleansed by distilled water and ethanol.

Two relationships have been studied between strain ε and internal friction $\tan\varphi$ of spherical Mg_2Si magnesium matrix composites specimens at ambient temperature, between temperature T and internal friction $\tan\varphi$ at constant strain ε. Then the damping properties of spherical Mg_2Si magnesium matrix composites specimen have been discussed.

The effect of strain ε from 1.0×10^{-3} to 1.0×10^{-1} on internal friction $\tan\varphi$ is determined at vibration frequency f =1.0 Hz. The influence of temperature T from ambient temperature T_{am} to 300 °C on internal friction $\tan\varphi$ is under conditions of the heating rate as 5 °C /min, and strain $\varepsilon=7.0\times10^{-2}$.

3 Experimental Results and Discussion

3.1 Microstructures of Mg_2Si magnesium matrix composite

The microstructure of the spherical Mg_2Si magnesium matrix composite prepared with 125 µm fly ash floating beads is shown in Figure 3.. It can be clearly observed that there are spherical phase and fragment-shaped phase in the matrix shown in blue circles. Due to effect of centrifugal force, the molten magnesium can be percolated into hollow space of fly ash floating beads with open pore structure in this study and formed spherical phase. However, some fly ash floating beads with lower strength and hardness were fragmented because of influence of centrifugal force and formed fragment-shaped phase.

Further EDS analysis was performed on the matrix region labeled A and the spherical region labeled B. The point EDS analysis result illustrates chemical compositions of region A are almost pure Mg phase containing traces of Al and O, and that of region B contains considerable Mg and Si with an atomic ratio of Mg:Si of approximately 2:1, which is confirmed as Mg_2Si phase.

Due to the fact that main composition of the fly ash floating beads is SiO_2, the reaction between Mg and SiO_2 at high temperatures can occur:

$$2Mg + SiO_2 \rightarrow 2MgO + Si \tag{1}$$

Excess magnesium can continue to react with the generated Si and form Mg_2Si:

$$2Mg + Si \rightarrow Mg_2Si \tag{2}$$

The formation of the Mg_2Si phase can be well explained by Eq. (1) and Eq. (2).

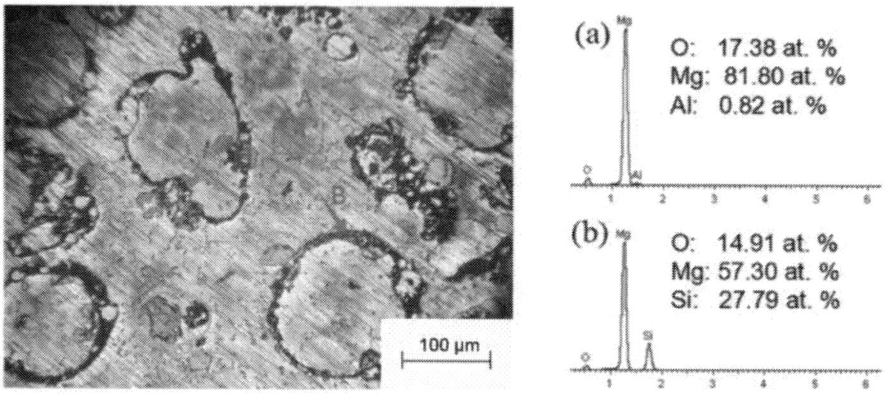

Figure 3. SEM micrograph of spherical Mg_2Si magnesium matrix composites and the corresponding point EDS spectra (a) area A, (b) area B

3.2 Damping Capacities of Mg_2Si magnesium matrix composites

The effect of strain ε from 1.0×10^{-3} to 1.0×10^{-1} on internal friction $\tan\varphi$ of Mg_2Si magnesium matrix composites prepared by 80, 125, and 500 μm fly ash floating beads under the condition of vibration frequency f =1.0 Hz shown in Figure 4. It can be evidently observed that damping capacities of three specimens have similar tendency with an increase of strain ε . This implies that the internal friction $\tan\varphi$ of three samples increase slowly with strain ε increasing in a range of 1.0×10^{-3} to 1.9×10^{-2}; however, the internal frictions $\tan\varphi$ exhibit a greatly increasing trend with an increase of strain ε from 1.9×10^{-2} to 1.0×10^{-1} and reach the maximum value of $\tan\varphi$ followed by a slight decline. Moreover, three specimens have higher damping capacities in the full strain range and reach the damping peaks in the vicinity of strain ε =5.2×10^{-2}, which can be attributed to the spherical structure of Mg_2Si in the matrix. It also can be observed that increasing particle size of fly ash floating beads can lead to an obviously decreasing tendency of internal friction $\tan\varphi$. It can be ascribed to the fact that the larger the particle size of fly ash floating beads with hollow structures, the less Mg_2Si phase formed resulting in lower dislocation density, less interface area and smaller damping value.

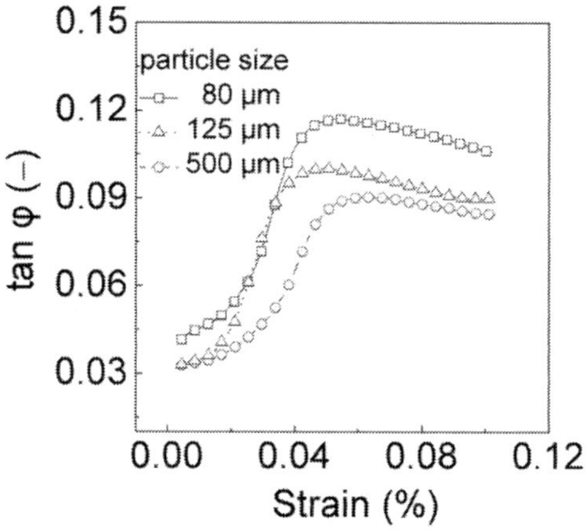

Figure 4. Effect of strain ε on damping capacities $\tan\varphi$ of spherical Mg_2Si magnesium matrix composites specimens at vibration frequency f =1.0 Hz

The influence of temperature T from ambient temperature to 300 °C on internal friction $\tan\varphi$ of Mg_2Si magnesium matrix composites prepared by 80, 125, and 500 μm fly ash floating beads under conditions of vibration frequency f =1.0 Hz and strain ε =7.0×10^{-2} is shown in Figure 5. It can be obviously observed that three specimens of magnesium matrix composites have higher damping capacities in whole temperature range especially $\tan\varphi$ =0.09 of the specimen with a particle size of 80 μm at T_{am} . In addition, the internal friction $\tan\varphi$ of three samples increase slowly with temperature T increasing from T_{am} to 150 °C; however, the internal frictions $\tan\varphi$ exhibit a greatly increasing trend with an increase of temperature T from 150 °C to 300 °C but no damping peaks occur. It also can be observed that increasing particle size of fly ash floating beads can lead to an obviously decreasing tendency of internal friction $\tan\varphi$ due to aforementioned reasons.

[a] Corresponding author: douglass@ipe.ac.cn (D. P. Duan), yangxm71@ipe.ac.cn (X. M. Yang)

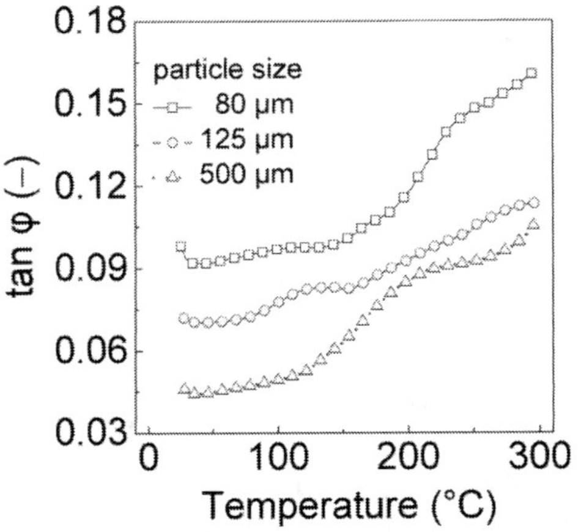

Figure 5. Effect of temperature T on damping capacities $\tan\varphi$ of spherical Mg_2Si magnesium matrix composites specimens at vibration frequency f =1.0 Hz and strain ε =7.0×10^{-2}

4 Conclusions

According to the experimental results of microstructure and damping capacities expressed by internal friction $\tan\varphi$ of spherical Mg_2Si magnesium matrix composite and related discussion, the main summary remarks can be obtained as follows:

(a) Spherical and fragment-shaped Mg_2Si phases were observed in the matrix, where fragment-shaped phases were formed owing to the reason that some fly ash floating beads with lower strength and hardness were fragmented because of influence of centrifugal force.

(b) Increasing particle size of fly ash floating beads can lead to an obviously decreasing tendency of internal friction $\tan\varphi$ ascribed to the fact that the larger the particle size of fly ash floating beads with hollow structures, the less Mg_2Si phase formed resulting in lower dislocation density,less interface area and smaller damping value.

(c) The critical strain ε_{cr} =1.9×10^{-2} of $\varepsilon-\tan\varphi$ curve and critical temperature T_{cr} =150 °C of $T-\tan\varphi$ curve for $\tan\varphi$ changing from slowly to greatly increasing. Damping peaks occur in the vicinity of strain ε =5.2×10^{-2} in $\varepsilon-\tan\varphi$ curve whereas no damping peaks occur in $T-\tan\varphi$ curve. Spherical Mg_2Si magnesium matrix composite have high damping capacities in the full strain and temperature range possibly ascribed to spherical structure of Mg_2Si in the matrix.

References

[1] B. L. Mordike and T. Ebert: Magnesium: Properties−application−potential. Mat. Sci. Eng. A, **302** (2001), pp. 37−45.

[2] J. Zhang, R. J. Perez and E. J. Lavernia: Effect of SiC and graphite particulates on the damping behavior of metal matrix composites. Acta. Metall. Mater., **42** (1994), pp. 395−409.

[3] J. Zhang, R. J. Perez: Dislocation-induced damping in metal matrix composites. J. Mater. Sci., **28** (1993), pp. 835−846.

[4] S. Narasimalu, and G. Manoj: Damping characterization of Mg-SiC composites using an integrated suspended beam method and new circle-fit approach. Mater. Res. Bull., **37** (2002), pp. 1149−1162.

[5] W. Cao, C. F. Zhang, T. X. Fan, and D. Zhang: In situ synthesis and damping capacities of TiC reinforced magnesium matrix composites. Mater. Sci. Eng. A, **496** (2008), pp.242−246.

[6] L. H. Liao, H. W. Wang, X. F. Li, and N. H. Ma: Research on the dislocation damping of Mg_2Si/Mg−9Al composite materials. Mater. Lett., **61** (2007), pp. 2518−2522.

[7] Z. Wang, J. T. Gao, K. Chang, L. Meng, N. Zhang and Z. C. Guo. He: Manufacturing of open-cell aluminum foams via infiltration casting in super-gravity fields and mechanical properties. RSC. Adv., **8** (2018), pp. 15933−15939.

[8] R. L. Niu, F. J. Yan, Y. S. Wang, D. P. Duan, and X. M. Yang: Effect of Zr content on damping property of Mg−Zr binary alloys. Mater. Sci. Eng. A, **718** (2018), pp. 418−426.

Deep Drawing of a Ni-Ti Shape Memory Alloy Sheet

Kuang-Jau Fann[1,a] and Chun-Hao Chang[1]

[1]*National Chung Hsing University, Department of Mechanical Engineering, 402 Taichung, Taiwan, R.O.C*

Abstract. This study is aimed to investigate the properties of Ni-rich Ni-Ti shape memory alloy sheet parts during and after deep drawing process at room temperature, in which the metal is under plane tension-compression stress state. After annealing at different elevated temperature, shape memory alloy sheet is formed at room temperature by a circular cylindrical punch in different forming depths. Thereafter the deep drawn sheet part is subjected to aging for shape memory treatment. A free shape recovery test is further conducted onto the deep drawn shape memory alloy sheet part. As a result, the sheet metal has austenitic phase at room temperature, which shows superelasticity. Too small forming amount causes severe springback and the form of the deep drawn sheet part is very hard to detect with the naked eye. Only if a forming amount greater than or equal to 4 mm is applied, a form can be achieved on to the sheet part. The subsequent aging shape memory treatment can induce further springback of the deep drawn sheet part. The higher the temperature of the previous annealing, the less the springback by aging. The free shape recovery test shows that the deep drawn sheet part only undergoing annealing without shape memory treatment still possesses shape memory effect. Nevertheless, the deep drawn sheet part aged for shape memory can show a better shape recovery.

1 Introduction

It was found in 1963 that equiatomic Ni-Ti alloys carry shape memory effect in association with superelasticity [1]. The shape memory effect shows that after severe deformation the alloy can recover its original shape prior to the deformation by raising its temperature [2]. It is caused by phase transformation between austenite at high temperature and martensite at low temperature [3], while superelasticity is a similar behaviour when the material undergoes those phase transformations during the elastic deformation between loading and unloading [4]. The Ni-Ti alloys are the most widely used shape memory alloy because of their excellent mechanical properties, distinguished corrosion inhibition, brilliant wear resistance, and perfect biocompatibility [5, 6]. The application of Ni-Ti shape memory alloy is primarily made of tube or wire, particularly because the alloy does not have good processing and machining characteristics. It is often made by plastic forming [7]. However the Ni-Ti shape memory alloy has the superelastic property at room temperature, making it easy to springback to its original shape, so that the thermomechanical processing technology is usually applied [8], in that the alloys are formed with a pair of dies in hot state. To develop the shape memory effect, the alloys and the dies are then kept in a constraint state inside a furnace and followed by aging treatment. However, this process demands large volume of dies in an extraordinary quality showing good resistance to high temperature, which results in huge cost of dies and equipment. Furthermore, sheet metal forming process, which is broadly used in mass production, is distinctly operated in cold because of geometry and heat transfer characteristic of metal sheets. These two hurdles result in the application of Ni-Ti shape memory alloy being limited to the tube or wire as raw material. Fann and Huang [9] have developed a cold forming process for Ni-Ti shape memory alloy wires to eliminate the use of expensive die sets, not only in the forming, but also in the constraint aging process at elevated temperature, which can help to extend the application of Ni-Ti shape memory alloys to the sheet forming process. However, the stress state appearing in sheet metal forming is a plane stress state, which is different to the wire bending process. This study is thus aimed to investigate the feasibility of the processes for shape memory alloy sheet part, cold forming with only one die set and aging without any further die set, and to investigate the shape memory effect of Ni-Ti shape memory alloy sheets during and after deep drawing process, which is under plane tension-compression stress state.

2 Setups

Ni-rich Ni-Ti shape memory alloy sheets having a thickness of 0.8mm were commercially acquired from a local manufacturer (Zhongrui Material Technology Corp., Taiwan). After waterjet cutting the alloy sheets into circular shapes with a diameter of 36 mm, the blank sheets are annealed at different temperatures, 650°C, 750°C, and 850°C, respectively, for one hour and

[a] Corresponding author: author@e-mail.org

© The Authors, published by EDP Sciences. This is an open access article distributed under the terms of the Creative Commons Attribution License 4.0 (http://creativecommons.org/licenses/by/4.0/).

thereafter cooled in the furnace until it reaches 400°C and taken out for cooling in the air to the room temperature.

The metal sheet blank was put onto the bottom die without further surface treatment and any lubrication. The metal sheet blank was then clamped by the blankholder with four bolts. Each was preloaded by four stacks of disc springs between the blankholder and the flange, as shown in Figure 1. The bottom die had a corner radius of 2.5 mm and the four bolts could provide a 680 N blankholder force on to the sheet blank, which was determined according the handbook [10]. The whole die set was then mounted on the test bench, which was a universal material test machine driven by a servo motor. The cylindrical punch in a diameter of 20 mm having a corner radius of 4 mm was then installed on to the cross head of the test machine. The approaching speed of the punch to the blank was set as 6 mm/min.

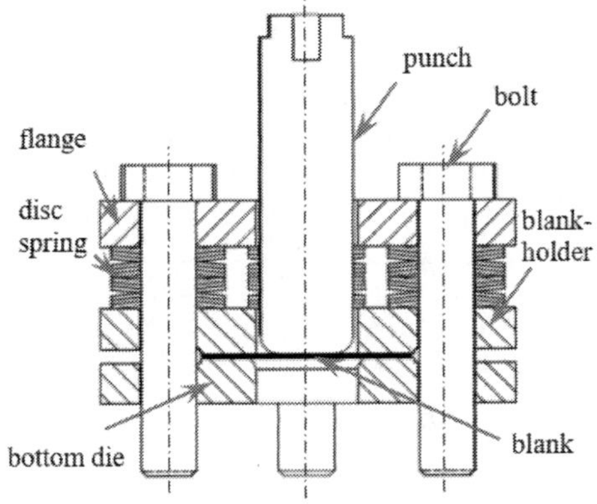

Figure 1. Sectional view of experiment setup (die set).

Because the experiment was conducted at room temperature, at which the material used in this study was at a state in the stable austenite phase, the sheet blank possessed superelasticity. It could be expected that there would be considerable springback after deep drawing process. For this purpose, the springback rate p was defined as a ratio of the variation of the height Δh of the axis center of the sheet part between the down stroke and the up stroke of the punch during the deep drawing process to the forming depth s. Namely,

$$p = \Delta h / s \times 100\% \quad (1)$$

3 Results and Discussion

A simulation of deep drawing Ni-Ti shape memory alloy sheets had been conducted with the commercial finite element software DEFORM 2D [11] under elastic-plastic deformation, because the formulation for an elastic-superelastic constitution is similar to elastic-plastic one [12]. The stress-strain relation used in the previous study [13] on stretch forming, in which the material had a Young's modulus of 41 GPa with a plateau strength of about 530 MPa and a tangent modulus of 1,286 MPa during loading and 21 GPa during unloading after a 12% strain, was applied to this study as well.

Figure 2 shows the load over the stroke of the punch. It can be observed that all the deformation made by a punch stroke or a forming depth of 2 mm (solid blue line) sprung completely back as the punch load fully vanished and the punch just returned to its starting position (stroke 0 mm, lower curve). As the deep drawing process was performed by a forming depth of 4 mm (solid green line), the punch was at the stroke of 2.69 mm from its starting position at the time that the punch load disappeared. The height of the formed part had a springback of 1.31 mm. In case of a forming depth in 6 mm (solid red line), the springback of the formed part was only 1.09 mm. It seems that the more the stroke applied, the deeper the forming depth, the lower the springback. It can be attributed to that the shape of formed cup might lead to the material not easily to springback.

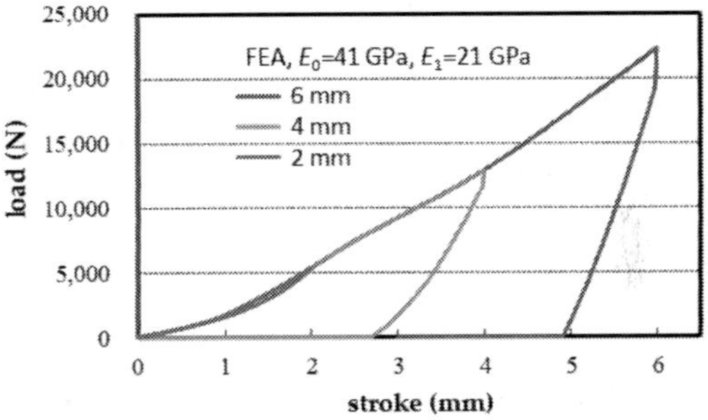

Figure 2. Punch load over stroke of deep drawing Ni-Ti shape memory alloy sheet determined by finite element analysis.

Figure 3 shows the above mentioned formed parts before unloading (dashed green line) at the bottom dead center and after unloading (solid red line) in comparison with the original blank sheet shape (dashed blue line).

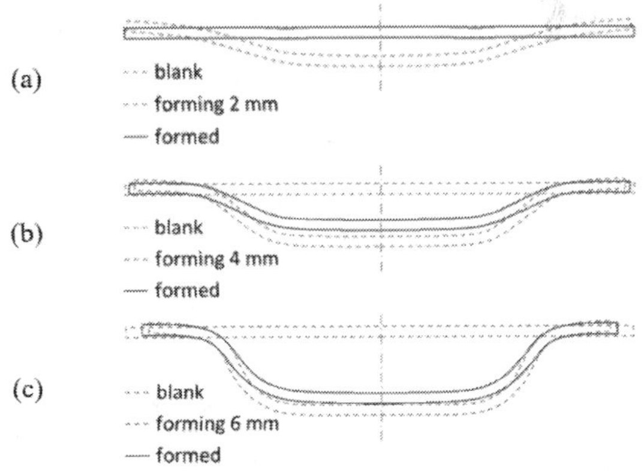

Figure 3. Sectional view of the blank (dashed blue line) and the part at the bottom dead center of the punch (dashed green line) and after unloading (solid red line) obtained from a finite element analysis of deep drawing Ni-Ti shape memory alloy sheet with a forming depth of (a) 2 mm, (b) 4 mm, and (c) 6 mm.

It can be seen that there is almost no difference between the formed part after deep drawing with a forming depth of 2 mm (Figure 3 (a)) and the original blank sheet. The only one variation is located around the radius corner of the punch, where an indistinguishable about 0.05 mm deformation can be exposed. Furthermore, the distance between the bottom die and the blankholder is not equal to the thickness of the blank sheet, while the punch reaches its bottom dead center, so that the deformation of the blank sheet is not thoroughly as expected. According to the equation (1), the springback rate for this case is 97.4%.

If the forming depth is extended to 4 mm as shown in Figure 3 (b), the height difference at the axis center of the part obtained at the punch bottom dead center and after unloading is 1.26 mm, which deviates from 1.31 mm obtained with load stroke curve (Figure 2). It might be associated to that the latter only considers the contact condition between the tools and the blank sheet, while the former counts all the deformation on the blank sheet. If the deformation on the outer ring of the flange is considered as well, the springback is then 1.37 mm and the springback rate is thus 34.2%, which is distinctly reduced in comparison to the forming depth of 2 mm.

When the forming depth is increased to 6 mm, the height at the axis center of the sheet part has its maximum either at the bottom dead center of the punch stroke or after unloading the punch. The difference of these depths is then 0.89 mm. If the deformation on the outer ring of the flange is regarded as well, the springback is as large as 1.09 mm, which is the same as the result obtained from the load stroke curve (Figure 2). The springback rate is further turned down to 27.2%. If more attention is paid on the outer ring of the flange, it can be noticed that the diameter of the outer ring is extended from 32.72 mm at the bottom dead center of the punch stroke to 33.75 mm after unloading, totally 1.03 mm, which is larger than that observed from the smaller punch strokes.

If the observation of the aforementioned distance between the blankholder and the lower die is moved onto the stroke of the press plate, the result of Figure 4 can be obtained, wherein the stroke shown in the ordinate refers to the upward movement of the blankholder forced by the deformation of the blank flange during deep drawing the blank sheet. It can be seen that the first 2 mm of the punch stroke the blankholder keeps rising until its stroke reaches 0.54 mm. It means that it is obvious that the preload of the blankholder, 680 N, is not enough to keep the blankholder at its bottom dead center. Thereafter, during the subsequent deep drawing stroke, the stroke of the blankholder begins to decrease. This might be due to the continuous reduction in the area of the flange part, which results in an increase in the average pressure on the blankholder, which in turn causes the flange part to press down and the blankholder stroke to decrease. When the punch stroke is at 4.8 mm (at time 48 s), the blankholder stroke begins to rise again. This might be due to the fact that the blankholder on the flanges needs more load because of strain hardening. However, the blankholder force was set as a constant, so that the blankholder rises again.

Figure 5 shows the deep drawn parts from the sheet blank annealed at different temperatures and formed at different depths. It can be seen that there is still an oxide layer remaining visible on the surface of the sheet after the annealing process and that the larger the forming depth, the more obvious the shape of the deep drawing. That means that the higher the side wall formed, the smaller the springback, and the higher the dimensional accuracy. At the forming depth of 2 mm, the formed shape is almost completely sprung back. Although it is hard to directly convert the area change on the formed part with the depth of 2 mm to the strain, it can be explained by the strain distribution shown in Figure 6 along with the stress-strain curve mentioned above that all deformation will completely springback after the force is released if the strain is below 1%. If the average strain is about 5%, which is in the early stage of the plateau region, such deformation is likely to springback with superelasticity, so that the residual plastic deformation is not easy to see. For the part formed with a depth of 4 mm, its strain around the punch corner has exceeded 10%, which is in the late stage in the plateau region, so that there should be more plastic deformation on the part and the superelastic area is then contracted. Therefore, the springback is reduced and a noticeable formed shape can be seen. For the part formed with the depth of 6 mm, the average strain on the sidewall is more than 10%, so that the whole formed region is full plastically deformed. A springback caused by superelastic deformation does almost not exist and the formed shape is extremely clear.

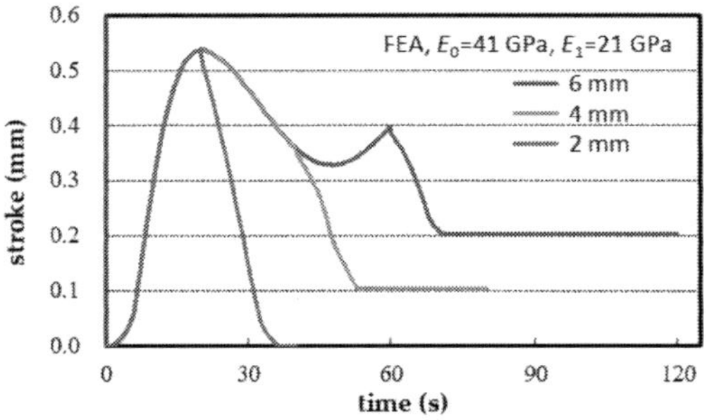

Figure 4. Stroke evolution of the blankholder obtained from a finite element analysis of deep drawing a Ni-Ti shape memory alloy sheet.

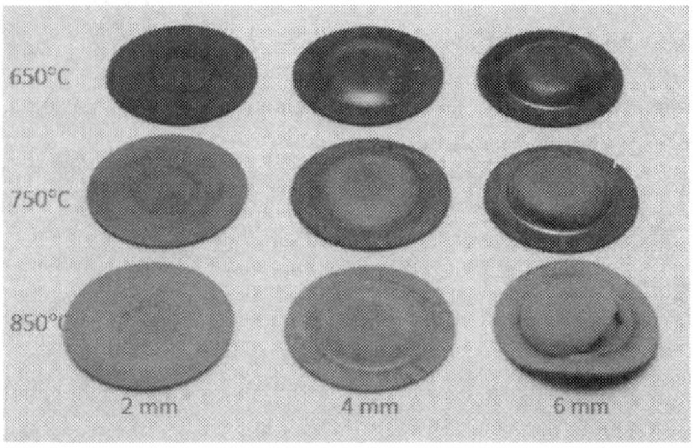

Figure 5. Parts of Ni-Ti shape memory alloy sheet annealed at different temperatures and formed in diverse depths.

The above-mentioned formed parts were placed in a furnace treated at 300°C for one hour as recommended in [9], and water quenched thereafter. This precipitation or aging served as a shape memory treatment. However, after the shape memory treatment, the residual stress generated by the cold forming on the sheet parts was then eliminated, and the parts sprung back further, which caused the height at the axis center to fall again. Figure 7 shows the height at the axis center of the parts after deep drawing with a depth of 4 mm and after shape memory treatment.

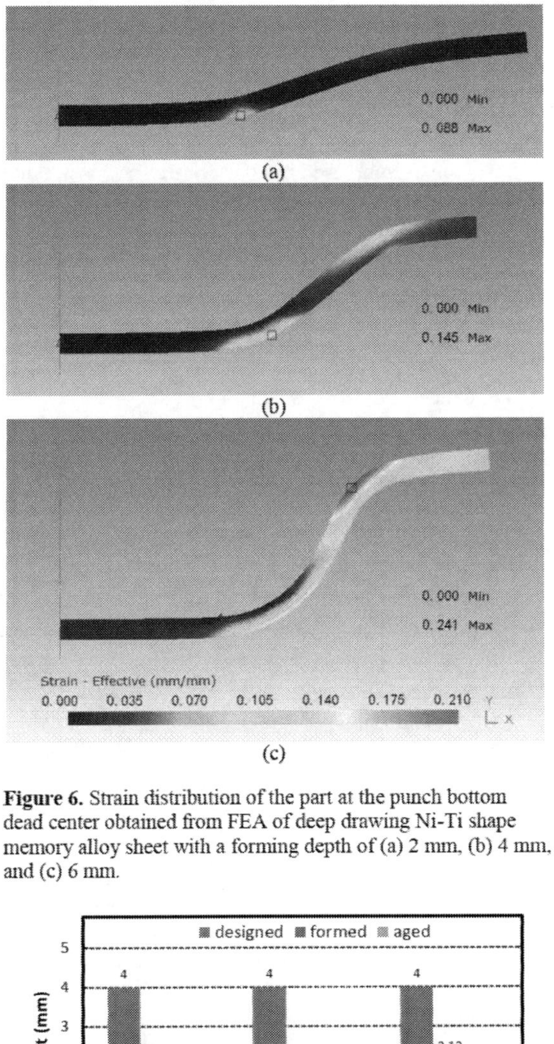

Figure 6. Strain distribution of the part at the punch bottom dead center obtained from FEA of deep drawing Ni-Ti shape memory alloy sheet with a forming depth of (a) 2 mm, (b) 4 mm, and (c) 6 mm.

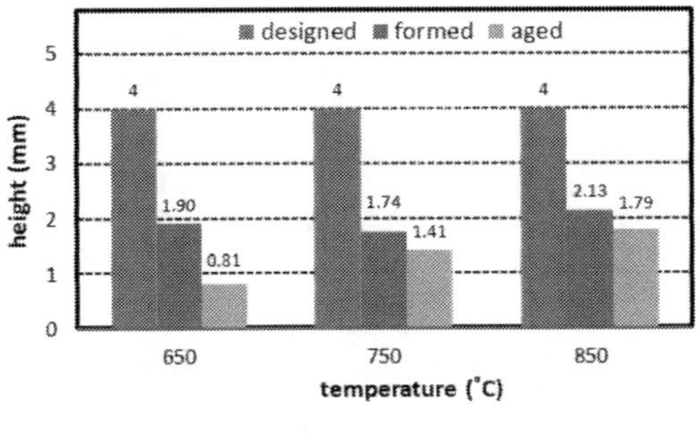

Figure 7. Height at the axis center in parts of Ni-Ti shape memory alloy sheet annealed at different temperatures after forming with depth of 4 mm and after shape memory treatment.

The height shown in blue bar is the forming depth, and the red bar shows the average height from the four deep drawing experiments. The green bar shows the average height of the two parts measured after shape memory treatment. It can be seen that the aging treatment would create a considerable springback, particularly for the sheet annealed at 650°C, in which the shape memory treatment caused a springback rate is as high as 57.4% (calculated from 1 − 0.81/1.90). It might be attributed to that the annealing temperature 650°C did not exceed the solvus. There was no solid solution effect activated, which caused in turn that the subsequent treatment at 300°C had no aging effect. The subsequent treatment could only be regarded as an annealing treatment after a cold deep drawing process, so that the cold work characteristics disappear and the formed height was lost further. For the sheets annealed at 750°C and at 850°C, the springback rate after aging (19.2% and 16.1%, respectively) is much lower than that after forming (56.5% and 46.8%, respectively). It is obvious that although the annealing followed by furnace cooling and air cooling is different from the solid solution treatment, which is followed by water quench, there is still an aging effect, so that the original cold formed shape is not easily lost.

The formed part was placed between the two plates and afterwards 95% ethanol soaked with dry ice, showing a temperature of -70±2°C, was injected there between, so that the Ni-Ti shape memory alloy is indeed in martensitic phase. Applying force to the two plates to let the part back into a flat shape and to maintain the state for more than 30 seconds, the flat part was taken out and gradually returned to room temperature from low temperature and its shape came back to the formed cup shape. Because there was no any external force involved, it is called free recovery.

The principle of the shape memory effect and its free recovery test is that the shape memory alloy part is first subjected to aging at a high temperature (room temperature for this study) in austenitic phase for shape memory and then its temperature drops to a low temperature (for this study, a 95% ethanol environment impregnated with dry ice) having martensitic phase in a crystal lattice of twin, but the shape obtained at the high temperature remains. At this low temperature, applying a load on the part, the part will be deformed by changing the crystal lattice structure in the way of de-twinning. When the load is removed, the de-twinned crystal lattice does not return to the twin form and keeps the deformed shape. This is also the reason why the part in the aforementioned test maintains its flat state after removing the clamping plates at a low temperature. Once the temperature rises to a high temperature, the internal stress of the de-twinned crystal lattice is released and the twin crystal lattice is restored for the martensitic phase. At the same time, as temperature rises, the austenitic phase is converted from the martensitic phase and the previous part shape obtained at the high temperature is then restored.

If the height at the axis center of the part before the shape recovery test h1 is compared with that after the test h0, the shape recovery rate r is then defined as follows.

$$r = h_1 / h_0 \times 100\% \qquad (2)$$

For different forming depth, the shape recovery rate can be calculated from the experiment results and plotted in a bar graph, as shown in Figure 8.

The shape recovery rate shown in red bar is taken from the recovery test, which was conducted directly with the part just formed with a depth of 4 mm without further aging shape memory treatment, while the green bar shows the result of the recovery test from the part after shape memory treatment. It is obvious that the shape recovery rate of the formed part without shape memory treatment is not as good as that with the shape memory treatment. However, the shape recovery rates are similar and show that the formed part without shape memory treatment has an excellent shape recovery ability. The higher the annealing temperature, the higher the shape recovery rate, regardless whether the part was aged for shape memory or not.

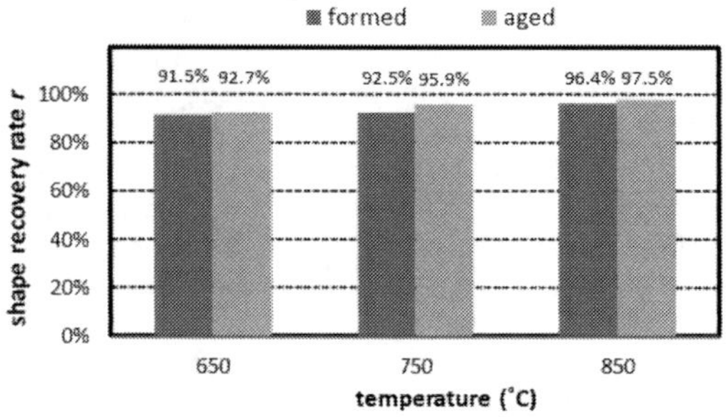

Figure 8. Free recovery rate determined from parts of Ni-Ti shape memory alloy sheet annealed at different temperatures after forming with depth of 4 mm and after shape memory treatment.

4 Conclusions

This study attempted to investigate how Ni-rich Ni-Ti shape memory alloy sheet deforms during a deep drawing process and how is the shape memory effect of the deep drawn part. Ni-rich Ni-Ti shape memory alloy sheets having a circular shape in a diameter of 36 mm, annealed at 650°C, 750°C and 850°C, respectively, have been drawn with depths of 2 mm, 4 mm, and 6 mm, respectively. It is followed with an aging process at 300°C as a shape memory treatment, and a shape recovery test is thereafter conducted. The material shows its austenitic phase at room temperature, while it shows its martensitic phase in an ethanol environment impregnated with dry ice.

The experiment results show that the higher the forming depth, the higher the deformation and the smaller the springback rate. The formed part with a drawing depth of 2 mm has a small amount of deformation, and the springback rate is nearly 100%, so that it is hard to see the formed shape. But the formed parts with a drawing depth of 4 mm and 6 mm have a higher deformation, so that the springback rate is also less than 50% and the parts have a distinct formed shape.

When the formed part is subjected to aging shape memory treatment, it will springback further. The higher the annealing temperature before forming, the smaller the springback during the aging heat treatment, but this springback rate is at least 16%.

The shape recovery test shows that the deep drawn sheet part without the aging shape memory treatment still has the shape recovery ability, and its shape recovery rate is a little smaller than that with the aging shape memory treatment but at least more than 90%. The higher the annealing temperature, the higher the shape recovery rate, regardless whether the part was aged for shape memory or not.

Acknowledgements

The authors would like to acknowledge that this study was financially supported by the Ministry of Science and Technology of the Republic of China with the grant MOST 104-2221-E-005-022.

References

1. W.J. Buehler, J.V. Gilfrich, R.C. Wiley, J. Appl. Phys., **34**, 1475 (1963)
2. A. Ölander, Z. Krist., **83A**, 145 (1932)
3. L.C. Chang, T.A. Read, T. Metall. Soc. AIME J. Metals, **189**, 47 (1951)
4. J.W. Christian, *The Theory of Transformations in Metals and Alloys* (J.W. Christian, Elsevier, Oxford, UK, 1965) 1102
5. C.M. Wayman, T.W. Duerig, *Engineering Aspects of Shape Memory Alloys* (T.W. Duerig, K.N. Melton, D. Stöckel, C.M. Wayman, Butterworth-Heinemann, London, UK, 1990) 3
6. T. Duerig, A. Pelton, D. Stöckel, Mater. Sci. Eng., **A273-275**, 149 (1999)
7. M.H. Wu, Mater. Sci. Forum, **394-395**, 285 (2002)
8. K.-J. Fann, H.-C. Hsu, Adv. Mater. Res., **939**, 430 (2014)
9. K.-J. Fann, P.-M. Huang, Key Eng. Mater., **661**, 98 (2015)
10. H. Tschätsch, *Metal Forming Practise Processes - Machines - Tools* (Springer, Berlin, Germany, 2006)
11. X. Gao, W. Huang, J. Zhu, *IUTAM Symposium on Smart Structures and Structronic Systems* (U. Gabbert, H.S. Tzou, Kluwer Academic Publishers, Dordrecht, The Netherlands, 2001) 137
12. J. Fluhrer, *Deform 2D Version 8.1 User's Manual* (Scientific Forming Technologies Corporation, Columbus, Ohio, USA. 2004)
13. K.-J. Fann, J.-Y. Su, IOP Conf. Ser.: Mater. Sci. Eng., **317**, 012079 (2018)

Synthesis and characterization of the novel color-tunable Eu/Tb(BPA)$_3$phen composites

Guangbo Xie[1], Zijun Zhang[1], and Jingjing Zhang[1,2,a]

[1]*School of Textile and Material Engineering, Dalian Polytechnic University, Dalian, 116034, P. R. China*
[2]*School of Materials Science and Engineering, Dalian University of Technology, Dalian, 116024, P. R. China*

Abstract. The new color-tunable luminescent Eu/Tb(BPA)$_3$phen composites have been successfully fabricated by the solvothermal reactions. The characterization of the final products have been investigated by field emission scanning electron microscopy (FESEM), Fourier transform infrared spectroscopy (FTIR), Thermogravimetric analysis (TG-DSC) and Ultraviolet and visible (UV-Vis) spectrophotometer. It is found that the geometry of the composite particles is regular by scanning electron microscopy image. Thermogravimetric analysis indicates that the initial decomposition temperature of the final products is 334 ℃, approximately. The results demonstrate that Eu/Tb(BPA)$_3$phen composites have excellent thermal stability. And the products after decomposition are stable oxides (Eu_2O_3 and Tb_2O_3). Furthermore, Eu and Tb ions complexes exhibit strong red and green luminescence, respectively. Emitting color of physically blended Eu(BPA)$_3$phen and Tb(BPA)$_3$phen composites can be tuned in a wide range from red to yellow to green under the excitation of 350 nm single-wavelength ultraviolet light. The CIE coordinates of C_{Tb}, C1, C2, C3, C4, C5, C6 and C_{Eu} are calculated as (0.24, 0.60), (0.31, 0.56), (0.33, 0.55), (0.38, 0.52), (0.40, 0.51), (0.43, 0.48), (0.54, 0.41) and (0.65, 0.34), respectively. The color-tunable luminescent materials can be widely applications in many fields, such as the optical and electronic devices, fluorescent probe and labels.
Keywords: Luminescence; Color-tunable; Thermal stability; 1, 10-Phenanthroline; Bisphenol A

1 Introduction

Recently, the color-tunable luminescent materials for their potential applications in flat panel display systems, full-color displays [1], next-generation lighting sources [2], biomakers, imaging [3, 4] and so on, has been paid much attention. Many kinds of color-tunable materials have been investigated, including inorganic phosphors [5, 6], metal-organic frameworks [7, 8], functional organic materials [9, 10], rare earth (RE) organic complexes [11], etc. Among which, rare earth ions possess luminescence characteristics like narrow spectral widths, long luminescence lifetimes, large Stokes shifts and large strong binding with biological molecules. Luminescent rare-earth complexes have potential technological applications in fluorescence devices from the intense luminous properties of rare-earth complexes [12, 13]. Their application has also been reported in electroluminescent devices [14] and as fluorescence probes and labels in a variety of biological systems [15].

In order to enhance the fluorescence properties of the rare-earth complexes, co-luminescence system with two kinds of rare-earth elements has been successfully fabricated. Several authors have also indicated that the addition of the certain second rare-earth ions, such as Gd (III) and La (III), could change the fluorescence properties of the rare-earth complexes [16], for example, SHAH [17] and Zhao [18] found Gd^{3+} ions in the mixed complex could remarkably increased the luminescent intensities of the terbium complex; Zhai [19] found the intrinsic fluorescence of Eu^{3+} could be enhanced by add Tb^{3+} ions and this is primarily due to an energy transfer from the $^5D_4$ level of Tb(III) to the $^5D_0$ level of Eu(III) ions. The luminescence mechanism is an intra-molecular energy transfer between Gd^{3+} and the central Tb^{3+} ions and from Tb^{3+} to Eu^{3+}, that is called "antenna effect". However, in the above mentioned co-luminescence system, the color of the emission light was single. Therefore, physically blended of the complexes with different RE (III) was conducted to realize multicolor [20, 21]. In the previous literature [22, 23], benzoic acid was widely used as the first ligand. In this paper, Bisphenol A (BPA) was used as the first ligand. As we know, BPA has not been mentioned in previous articles. The luminescence efficiency of BPA to Eu^{3+} and Tb^{3+} was studied, results show that it has excellent coordination ability to Eu^{3+} and Tb^{3+}.

In the present paper, Eu(BPA)$_3$phen and Tb(BPA)$_3$phen were firstly prepared, in which Bisphenol A (BPA) was used as the first ligand and 1, 10-phenanthroline (phen) as the cooperative ligand. Besides, the novel color-tunable [Eu(BPA)$_3$phen + Tb(BPA)$_3$phen] composite were also fabricated by physical mixture action. The emission color of the as-prepared complexes

[a] Corresponding author: zhangjj@dlpu.edu.cn

© The Authors, published by EDP Sciences. This is an open access article distributed under the terms of the Creative Commons Attribution License 4.0 (http://creativecommons.org/licenses/by/4.0/).

are tuned by adjusting the mass ratio of Eu(BPA)$_3$phen to Tb(BPA)$_3$phen. The morphology and fluorescence properties of the composites were systematically investigated. The resulting compounds are promising materials to fabricate various optoelectronic devices such as OLED.

2 Experimental sections

2.1 Materials

All experiment manipulations were performed in air using commercial grade solvents. Europium(III) nitrate hexahydrate ($Eu(NO_3)_3 \cdot 6H_2O$), Terbium(III) nitrate hexahydrate ($Tb(NO_3)_3 \cdot 6H_2O$), Bisphenol A (BPA), phenanthroline (phen), Ethanol absolute were used. All the reagents were of analytical grade and directly used as received without further purification. The purity of $Eu(NO_3)_3 \cdot 6H_2O$ and $Tb(NO_3)_3 \cdot 6H_2O$ was 99.99%.

2.2 Synthesis of rare earth complexes

Eu(BPA)$_3$phen powders were synthesized according to the solvothemal reactions. 0.0025 mol of $Eu(NO_3)_3 \cdot 6H_2O$ was dissolved in 20 ml of anhydrous ethanol at room-temperature. Then 0.03 mol of BPA and 0.01 mol of phen were dissolved in 30 ml of ethanol. $Eu(NO_3)_3$ ethanol solution was then added to the mixture solution of BPA and phen under magnetic stirring for 3 hour at 60 °C. The precipitates were collected by filtration and dried for 12 hour at 60 °C. The synthesis method of Tb(BPA)$_3$phen complex was similar to the above method. Eu(BPA)$_3$phen and Tb(BPA)$_3$phen complexes were marked as C_{Eu} and C_{Tb}, respectively.

2.3 Preparation of [Eu(BPA)$_3$phen + Tb(BPA)$_3$phen] composite

By adjusting different mass ratios of Eu(BPA)$_3$phen to Tb(BPA)$_3$phen, a series of [Eu(BPA)$_3$phen + Tb(BPA)$_3$phen] composite were prepared, and the mass ratios of Eu(BPA)$_3$phen to Tb(BPA)$_3$phen were 1: 30, 1: 15, 1: 8, 1: 6, 1: 3 and 1: 1, in sequence. And these samples were respectively marked as C1, C2, C3, C4, C5 and C6. All samples were listed in Table 1.

Table 1. Composition of the composites.

Composite	$m(C_{Eu})$:$m(C_{Tb})$	$m(C_{Eu})$/g	$m(C_{Tb})$/g
C_1	1:30	0.0333	1.0000
C_2	1:15	0.0667	1.0000
C_3	1:8	0.1250	1.0000
C_4	1:6	0.1667	1.0000
C_5	1:3	0.3333	1.0000
C_6	1:1	1.0000	1.0000

2.4 Characterization methods

The morphologies of samples were observed by field emission scanning electron microscope (FESEM, JSM-6460LV). The infrared spectra were recorded by a Fourier infrared spectrometer (FTIR, Waltham) in the range of 4000-400 cm^{-1}. Thermogravimetry and differnetial scanning calorimetry (TG-DSC) were carried out by a differential analyzer at the rate of 10 °C min^{-1} from room temperature to 700 °C. The fluorescent properties of the samples were investigated by a Hitachi fluorescence spectrophotometer (F-7000).

3 Results and Discussion

3.1 Characterization

The morphologies of the sample C_{Tb}, C2 and C_{Eu} were presented in figure 1(a), figure (b) and figure 1(c), respectively. It was shown that the three kinds of the samples are similar in morphology and they are similar to the quadrangular prism unit structure, the structure is relatively uniform. The above characteristics indicate that the preparation of the complex is crystal.

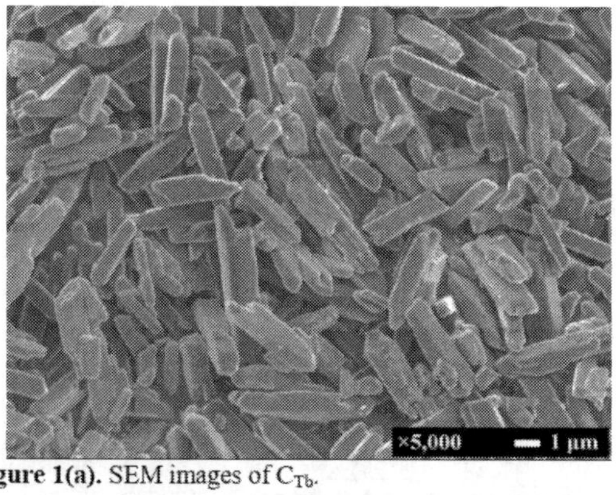

Figure 1(a). SEM images of C_{Tb}.

Figure 1(b). SEM images of C2.

Figure 1(c). SEM images of C_{Eu}.

FTIR spectra of the ligands BPA and phen, the complexes C_{Tb}, C2 and C_{Eu} were recorded in the range of 4000-400 cm^{-1}, as shown in figure 2. It can be easily seen from figure 2 that, the disappearance of absorption peak at 551 cm^{-1} (ν_{O-H}) for the pure BPA, and the appearance of new peak at 638 cm^{-1} of the title complex assign to symmetric absorption peaks indicate that the oxygen atoms of BPA coordinate to Eu(III) and Tb(III) ions. Besides, the appearance of absorption band for ν(Eu-O) and ν(Tb-O) at 420 cm^{-1} accounts for above-mentioned conclusion. In addition, the absorption bands of ν_{CN}(1585 cm^{-1}), δ_{CC}(1085 cm^{-1}) and δ_{CH}(840 cm^{-1}) of phen ligand shift to higher wave number around at 1625 cm^{-1}, 1106 cm^{-1} and 855 cm^{-1} in the spectra of the complex, respectively, suggesting that the nitrogen atoms of a phen ligand coordinate to Eu^{3+} or Tb^{3+} ions [12, 24]. It was also proved by the appearance of the absorption band at 472 cm^{-1}. On the basis of the above analyses and referencing other articles [25], the coordination structure of the as-prepared complexes C_{Eu} and C_{Tb} are shown in figure 3 and figure 4, respectively.

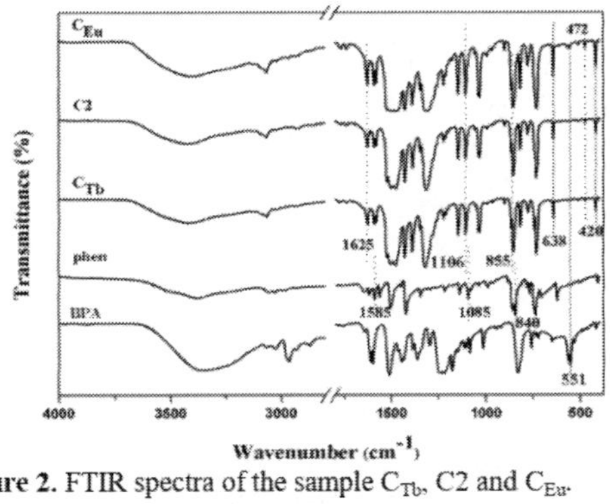

Figure 2. FTIR spectra of the sample C_{Tb}, C2 and C_{Eu}.

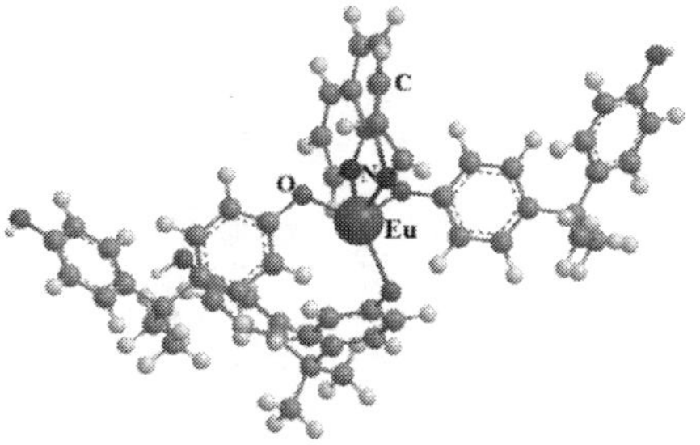

Figure 3. Coordination structure of C_{Eu}.

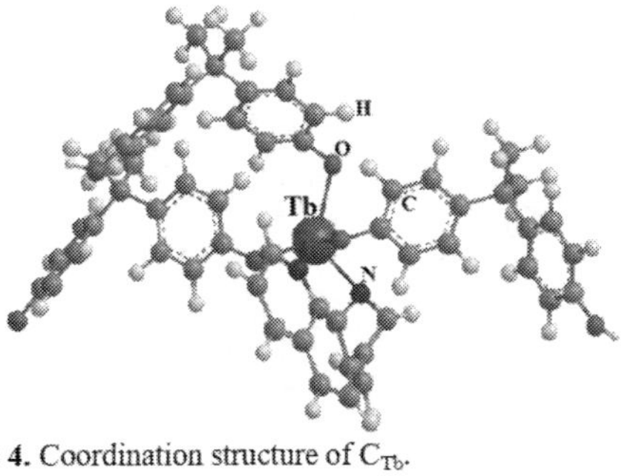

Figure 4. Coordination structure of C_{Tb}.

3.2 Thermal stabilities

The thermal stability of the sample C_{Tb}, C2 and C_{Eu} were investigated by analyzing the TG-DSC traces. The TG-DSC curves of $Eu(BPA)_3phen$ complexes are shown in figure 5. Thermogravimetric analysis indicates that $Eu(BPA)_3phen$ decomposed into two stages. The first weight loss is 51.09 % between the temperature of 334 and 400 °C, which may be due to the release of BPA molecules. As shown in the IR spectra of the residues, the bands of the asymmetric vibrations ν_s (O-Eu) at 638 cm^{-1} disappeared. In addition, the process is an endothermic reaction. The second weight loss is 22.83 % from 400 to 500 °C, which was due to the release of phen molecules. The IR spectra of the residue shows that the absorption band of C=N at 1625 cm^{-1} disappears [26]. Furthermore, it can also be concluded that the melting point of $Eu(BPA)_3phen$ is about 480 °C. Due to the formation of thermally stable europium oxide (Eu_2O_3), the phenomenon of weight reduction does not occur in the next process of temperature rise [27].

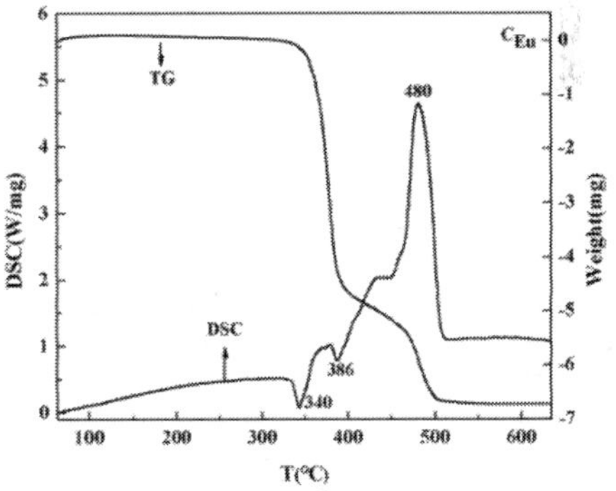

Figure 5. The TG-DSC curves of C_{Eu}.

According to the TG-DSC curves (as shown in figure 6), $Tb(BPA)_3phen$ complexes is stable up to 334 °C, afterwards it begin to decompose. When the temperature reaches 456 °C, the decomposition is terminated. The losing weight is 67.37 % with a high speed. The total process was decomposed in one-step, which may be due to the specially coordination between Tb^{3+} ions and the ligands. There is an exothermic peak on the DSC curves, and the melting point is about 412 °C.

The TG-DSC curves of C2 are shown in figure 7, TG analysis indicates that sample C2 decomposes into two stages. The initial decomposition temperature was 334 °C, it was attributed to the release of Tb(BPA)$_3$phen and the release of BPA molecules in Eu(BPA)$_3$phen. With the temperature rising to 420 °C, the decomposition was terminated. The second weight loss was observed from the temperatures of 420 to 500 °C, which may be due to the release of phen molecules. Moreover, one exothermic peak appears on the DSC curves and the melting point is also 480 °C. Therefore, figure 7 indicated that Eu(BPA)$_3$phen and Tb(BPA)$_3$phen were physically blended and without chemical reaction with each other.

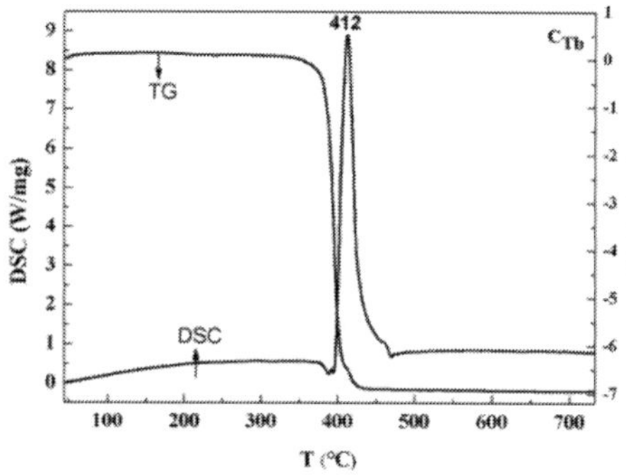

Figure 6. The TG-DSC curves of C_{Tb}.

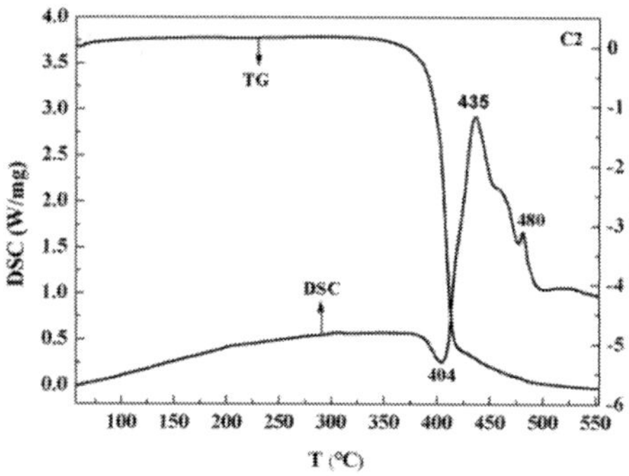

Figure 7. The TG-DSC curves of C2.

3.3 Fluorescence properties

The luminescent characteristics of terbium (III) complexes and europium (III) complexes are investigated, firstly. The excitation and emission spectra of Tb(BPA)$_3$phen (C_{Tb}) are demonstrated in figure 8. When the monitoring wavelength of sample C_{Tb} is set as 544nm, a broad excitation band extending from 200 to 380 nm with the maximum at 350 nm can be observed, which is assigned to the $\pi \rightarrow \pi^*$ electron transition of the ligands. Under the excitation of 350-nm ultraviolet light, characteristic emission peaks of Tb^{3+} ions are observed, which are ascribed to the energy levels transitions of $^5D_4 \rightarrow {}^7F_6$ (490 nm), $^5D_4 \rightarrow {}^7F_5$ (544 nm), $^5D_4 \rightarrow {}^7F_4$ (583 nm), and the $^5D_4 \rightarrow {}^7F_5$ and $^5D_4 \rightarrow {}^7F_6$ hypersensitive transitions at 544 and 490 nm are the predominant emission peaks.

The excitation and emission spectra of Eu(BPA)$_3$phen (C_{Eu}) are indicated in figure 9. When the monitoring wavelength of sample C_{Eu} is set as 615 nm, a broad excitation band extending from 200 to 380 nm with the maximum at 290 nm is observed, which is assigned to the $\pi \rightarrow \pi^*$ electron transition of the ligands. Under the excitation of 350-nm ultraviolet light, it is found that characteristic emission peaks at 593 nm and 615 nm which attributed to the $^5D_0 \rightarrow {}^7F_1$ and $^5D_0 \rightarrow {}^7F_2$ energy levels transitions of Eu^{3+} ions, respectively.

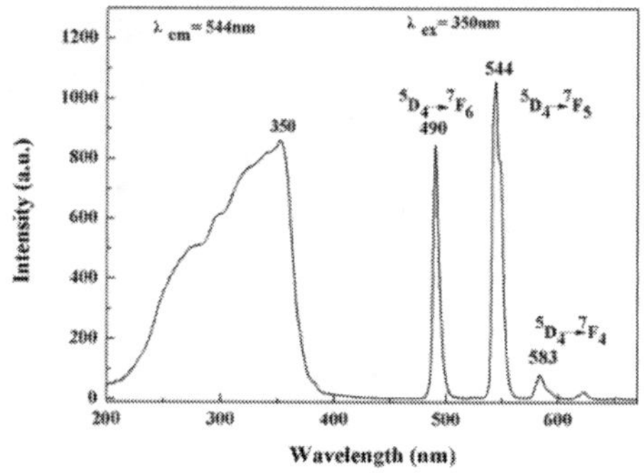

Figure 8. Excitation (left) and emission (right) spectra of C_{Tb}.

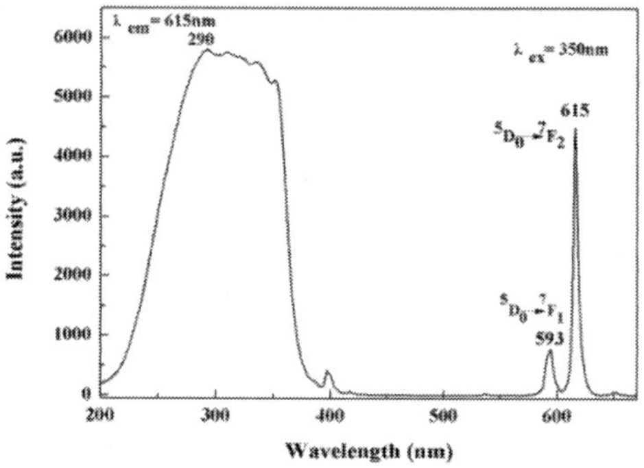

Figure 9. Excitation (left) and emission (right) spectra of C_{Eu}.

Figure 10 is the emission spectrum of all the composites. Four predominant peaks are observed for the [Eu(BPA)$_3$phen + Tb(BPA)$_3$phen] composite. The green, orange and red fluorescence emissions of predominant peaks at 490 nm, 544 nm, 593 nm and 615 nm can be observed, simultaneously. And with the decrease of the mass of Eu(BPA)$_3$phen, the orange and red fluorescence emissions at 593 nm and 615 nm gradually decrease. The variation of the PL intensity of Eu^{3+} and Tb^{3+} can be attributed to the energy distribution. Since the energy that the matrix absorbs and the content of RE(BPA)$_3$phen (RE= Eu and Tb) are constant, more energy is assigned to Eu^{3+} with the increase of the Eu(BPA)$_3$phen content, thus leading to stronger fluorescence peaks at 615 nm. Therefore, it is evidently found that the emission spectra of the samples could be tuned by adjusting the mass ratio of Eu(BPA)$_3$phen.

Generally, color can be represented by the Commission Internationale de L'Eclairage (CIE) 1931 chromaticity coordinates. Figure 10 reveals the CIE coordinate diagram of all the samples under the excitation of 350-nm ultraviolet light. It is clearly to see that the emitting color of samples could be tuned by adjusting the mass ratios of Eu(BPA)$_3$phen to Tb(BPA)$_3$phen complexes, the color is from red to yellow, yellow green, and green.

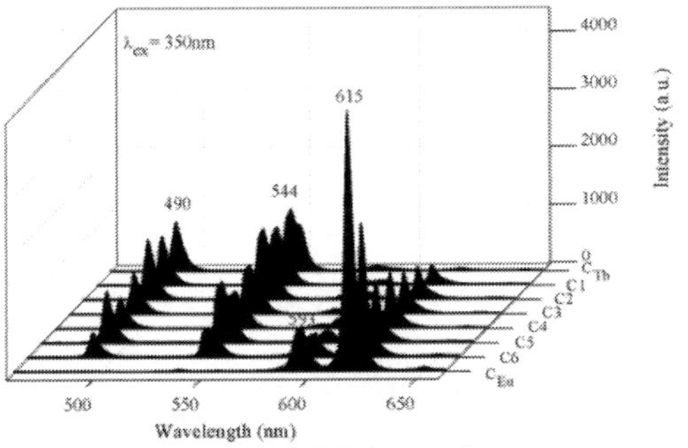

Figure 10. Emission spectra of all the samples.

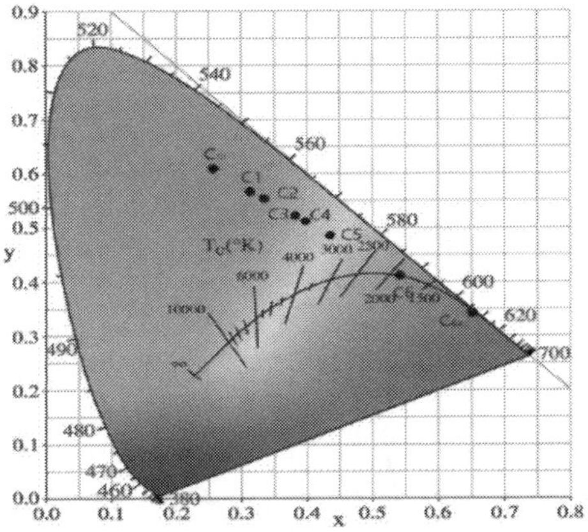

Figure 11. CIE chromaticity coordinates diagram of all the samples

4 Conclusions

In summary, in this work, the red emission of Eu(BPA)$_3$phen and the green emission of Tb(BPA)$_3$phen have been successfully synthesized, using BPA as the first ligand and phen as the cooperative ligand. Then novel color-tunable system of [Eu(BPA)$_3$phen + Tb(BPA)$_3$phen] composite were synthesized by adjusting the mass ratio of Eu(BPA)$_3$phen to Tb(BPA)$_3$phen complexes. The morphology and properties of final products were characterized in detail by SEM, FTIR, TG-DSC and fluorescence spectral. Through a serious of representations that [Eu(BPA)$_3$phen + Tb(BPA)$_3$phen] composite, a physical blend of Eu(BPA)$_3$phen and Tb(BPA)$_3$phen, can emission fluorescent from red to yellow to green under the excitation of 350-nm single-wavelength ultraviolet light. The initial decomposition temperature was 334 ℃, approximately. The products after decomposition are stable oxides (Eu_2O_3 and Tb_2O_3). The color-tunable luminescent materials have potential applications in many fields, such as the lighting, fluorescent probe and biological labeling.

Acknowledgments

This work was supported by the Doctor Startup Foundation of Liaoning Province, China (No. 201601277), the Scientific Research Foundation for Universities from Education Bureau of Liaoning Province (2017J039). It was also supported by China Postdoctoral Science Foundation funded project (No. 2017M621128).

References

1. M.M. Shang, J. Fan, Y. Zhang, H.Z. Lian, J. Lin, Inorg. Chem. Commun. **52**, 73 (2015)
2. W. Huang, L. Sun, Z.W. Zheng, J.H. Su, H. Tian, Chem. Commun. **51**, 4462 (2015)
3. L.D. Sun, Y.F. Wang, C.H. Yan, Acc. Chem. Res. **47**, 1001 (2014)
4. H.Y. Ko, J. Lee, Y.S. Lee, H.N. Gu, B.A. Ali, A. Al-Khedhairy, H. Heo, S. Cho, S. Kim, Chem. Commun. **51**, 2159 (2015).
5. K. Li, Y. Zhang, X.J. Li, M.M. Shang, H.Z. Lian, J. Lin, Phys. Chem. Chem. Phys. **17**, 4283 (2015)
6. X.G. Zhang, J.L. Zhang, M.L. Gong, Mater. Lett. **143**, 71 (2015)
7. X.Y. Xu, B. Yan, Dalton Trans. **44**, 1178 (2015)
8. Y. Wang, J. Yang, Y.Y. Liu, J.F. Ma, Chem. Eur. J. **19**, 14591 (2013)
9. M.Y. Liu, X.Q. Zhang, B. Yang, F.J. Deng, J.Z. Ji, Y. Yang, Z.F. Huang, X.Y. Zhang, Y. Wei, RSC Adv. **4**, 22294 (2014)
10. A.R. Ramya, S. Varughese, M.L.P. Reddy, Dalton Trans. **43**, 10940 (2014)
11. X. Shen, B. Yan, RSC Adv. **5**, 6752 (2015)
12. N.Q. Sun, L.P. Li, Y.M. Yang, A.Q. Zhang, H.S. Jia, X.G. Liu, B.S. Xu, Opt. mater. **49**, 39 (2015)
13. A.C. Sabrina, H.B.A. Pedro, J.L.C. Constantino, M.P.J. Ana, Lumin. **153**, 272 (2016)
14. A. Kukhta, E. Kolesnik, I. Grabchev, S. Sali, J. Fluoresc. **16**, 375 (2006)
15. T. Nishioka, J. Yuan, Y. Yamamoto, K. Sumitomo, Z. Wang, K. Hashino, C. Hosoya, K. Ikawa, Inorg. Chem. **45**, 4088 (2006)
16. F.F. Wang, B. Yan, J. Photoch. Photobio. A. **194**, 238 (2008)
17. M.D. Shah, B. Want, Mater. Sci-Poland. **33**, 685 (2015)
18. Z.S. Zhao, J.H. Yang, X. Wu, B.Y. Su, C.X. Sun, S.F. Liu, Anal. Bioanal. Chem. **380**, 104 (2004)
19. Y.F. Zhai, H.D. Xie, H.J. Cai, P.Q. Cai, H.J. Seo, Opt. Mater. **3**, 27 (2015)
20. Q.L. Ma, J.X. Wang, X.T. Dong, W.S. Yu, G.X. Liu, ChemPlusChem **79**, 290 (2014)
21. H.J. Xue, X.P. Sun, J.W. Bi, T.Y. Wang, J. Han, Q.L. Ma, L. Han, X.T. Dong, J. Mater. Sci. **26**, 8774 (2015)
22. B.J. Gao, N. Shi, Z.W. Qiao, Spectrochim. Acta. A Mol. Biomol. Spectrosc. **150**, 565 (2015)
23. J. Zhang, L.X. Wang, L. Zhang, Y. Chen, Q.T. Zhang, Rare Metals. **32**, 599 (2013)
24. W.J. Zhang, H.Y. Wang, Opt. Mater. **50**, 208 (2015)
25. M.T. Metlin, S.A. Ambrozevich, D.A. Metlina, A.G. Vitukhnovsky, I.V. Taydakov, J. Lumin. **188**, 365 (2017)
26. H.Y. Zhang, K.Z. Wu, J.J. Zhang, S.L. Xu, N. Ren, J.H. Bai, L. Tian, Synthetic Met. **158**, 157 (2008)
27. N. Goel, J. Mol. Struct. **1080**, 1 (2015)

Design and Fabrication of Friction Stir Welding Machine

Jithin Ambarayil Joy[1], Muhammad Sajjad[1], Dong-Won Jung[1,a]

[1]*Department of Mechanical Engineering, Jeju National University, 1 Ara 1-dong, Jeju-si, Jeju 690-756, Republic of Korea*

Abstract: The aim of the project is to manufacture a welding machine which simplifies the work and improve the accuracy. The working principle of this machine is different other that of other welding machine. The aluminum work piece is held in the vice. The two pieces of aluminum work piece is welded with the help high speed drilling head with friction tool. The working principle is very easy and at the same time production time is very much reduced. This machine is best suitable for mass production

Keywords: Friction, Welding, Joints, Motor, Bearings

1. INTRODUCTION

Welding is a process of joining similar metals by the application of heat. Welding can be done with or without the application of pressure. While welding, the edges of metal pieces are either melted or brought to plastic condition. Welding can be done with the addition off filler materials or without it welding is used of making permanent joints. It is used in the manufacture of automobile bodies, aircraft frames, railways wagons, machine frames, structural work, tanks, furniture, boilers, general repair work and ship building; At most in all metal working industries welding is used.

In Friction Stir Welding, the part is to weld and joined by forcing a rotating tool to penetrate into the joint and then moving across the entire joint. The solid-state joining process is been promoted by the movement of a non-consumable tool through the welding joint. It consists mainly of three phases, in which each one has been described as a period where the welding tool and workpiece to move about each other. In the first step, rotating tool is vertically displacing into the joint line (plunge period). Then period is followed by the dwell period in which the tool is held steady above workpiece rotating. The velocity between a rotating tool and the stable workpiece produce heat on mechanical interaction which results in friction work and plastic material deformation. The heat is dissipated to the neighboring material, promoting an increase in temperature and consequent material softening. After these two initial phases, welding operation can be carried out by moving either the tool or the workpiece about each other along the joint line. Figure.1 gives a schematic representation of

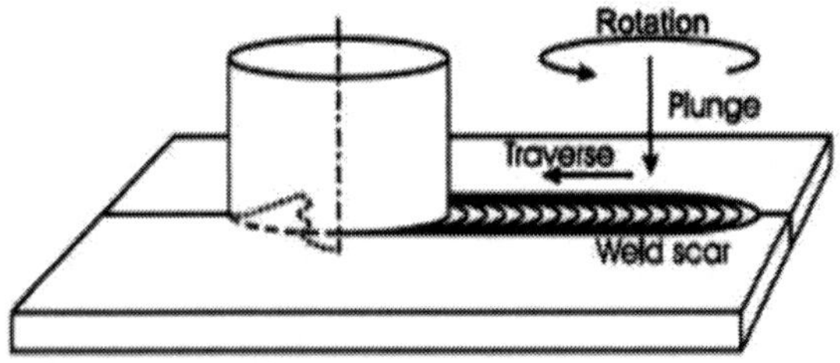

the FSW setup and operation.

Figure 1: Friction stir welding setup

2. COMPONENTS AND DESCRIPTION

The major components are used in this project:

- SINGLE PHASE AC MOTOR
- BEARING WITH BEARING CAP
- PULLEY
- BELT

[a] Professor Dong-Won Jung, Department of Mechanical Engineering, Jeju National University, Jeju, Republic of Korea : jdwcheju@jejunu.ac.kr

© The Authors, published by EDP Sciences. This is an open access article distributed under the terms of the Creative Commons Attribution License 4.0 (http://creativecommons.org/licenses/by/4.0/).

2.1 Single Phase Ac Motor

2.1.1 Single Phase Induction Motor With Pulley

This is used to drive the wheel by using two pulleys with belt drive mechanism. Because it has but a single alternating current source, a single-phase motor can only produce an alternating field: one that pulls first in one direction, then in the opposite as the polarity of the field switches. A squirrel-cage rotor placed in this field would merely twitch, since there would be no moment upon it. If pushed in one direction, however, it would spin.
The major distinction between the different types of single-phase AC motors is how they go about starting the rotor in a particular direction such that the alternating field will produce rotary motion in the desired direction. This is usually done by some device that introduces a phase-shifted magnetic field on one side of the rotor.

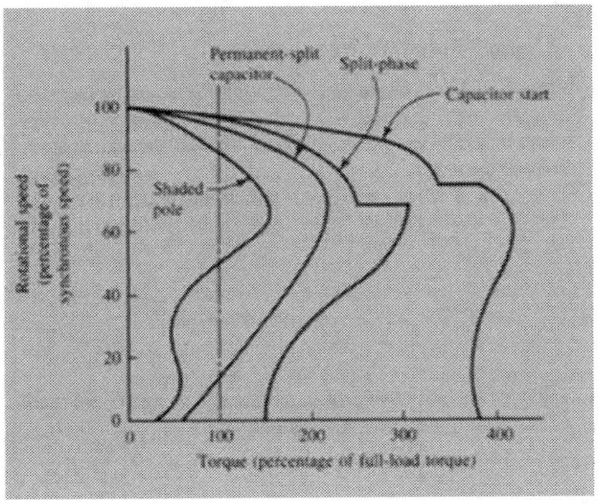

Figure 2: The performance curves of the four major types of single-phase AC motors

2.1.2 Split-Phase Motors

The split phase motor achieves its starting capability by having two separate windings wound in the stator. The two windings are separated from each other. One winding is used only for starting and it is wound with a smaller wire size having higher electrical resistance than the main windings. From the rotor's point of view, this time delay coupled with the physical location of the starting winding produces a field that appears to rotate.

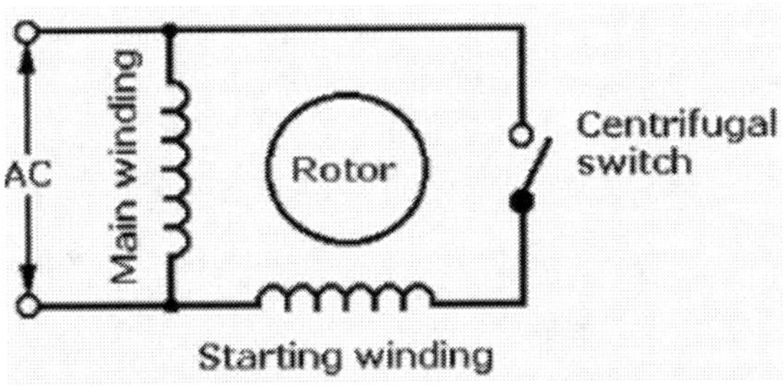

Figure 3: Circuit diagram of Split – Phase Motor

The apparent rotation causes the motor to start. A centrifugal switch is used to disconnect the starting winding when the motor reaches approximately 75% of rated speed. The motor then continues to run on the basis of normal induction motor principles.

Figure 4: Split-phase motors

2.1.3 Capacitor-Start Motors

Capacitor start motors form the largest single grouping of general purpose single phase motors. These motors are available in a range of sizes from fractional through 3HP. The winding and centrifugal switch arrangement is very similar to that used in a split phase motor. The main difference being that the starting winding does not have to have high resistance. The addition of this capacitor produces a slight time delay between the magnetization of starting poles and the running poles. This moderately priced motor produces relatively high starting torque, 225 to 400% of full load torque. The capacitor start motor is ideally suited for hard to start loads such as conveyors, air compressors and refrigeration compressors. Due to its general overall desirable characteristics, it also used for many applications where high starting torque may not be required.

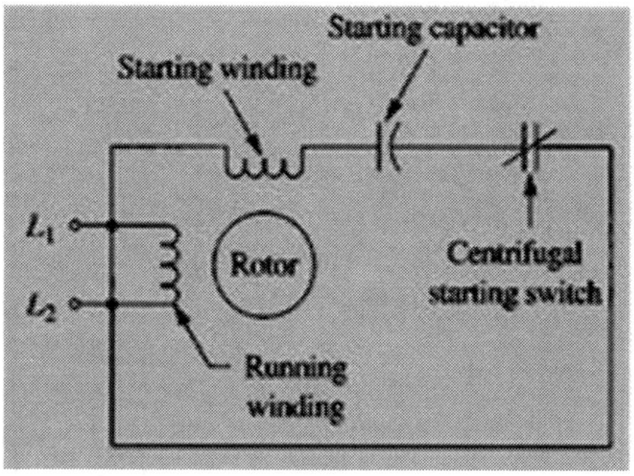

Figure 5: Circuit diagram of Capacitor-start motor

2.1.4 Permanent-Split Capacitor Motors

The starting torque is quite low, roughly 40% of full-load, so low-inertia loads such as fans and blowers make common applications. Running performance and speed regulation can be tailored by selecting an appropriate capacitor value. No centrifugal switch is required.

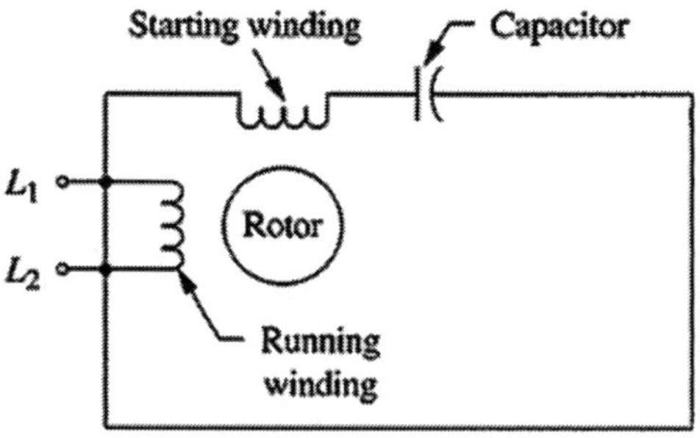

Figure 6: Circuit diagram of Permanent-Split Capacitor Motors

2.1.5 Shaded-Pole Motors

The shaded pole motor is the simplest of all single phase starting methods. In the shaded pole motor, the stator poles are notched and a copper short circuiting ring is installed around a small section of the poles. The shaded pole motor is simple and inexpensive, but has low efficiency and a very low starting torque. Speed regulation is poor, and it must be fan-cooled during normal operation. Shaded-pole motors are thus used in shaft-mounted fans and blowers, and also small pumps, toys, and intermittently used household items.

3. CONSTRUCTION AND TYPES OF BALL BEARINGS

A ball bearing usually consists of four parts: an inner ring, an outer ring, the balls and the cage or separator. To increase the contact area and permit larger loads to be carried, the balls run in curvilinear grooves in the rings. The radius of the groove is slightly larger than the radius of the ball, and a very slight amount of radial play must be provided. The bearing is thus permitted to adjust itself to small amounts of angular misalignment between the assembled shaft and mounting. The separator keeps the balls evenly spaced and prevents them from touching each other on the sides where their relative velocities are the greatest. Ball bearings are made in a wide variety of types and sizes. Single-row radial bearings are made in four series, extra light, light, medium, and heavy, for each bore, as illustrated in Fig. 1-3(a), (b), and (c)

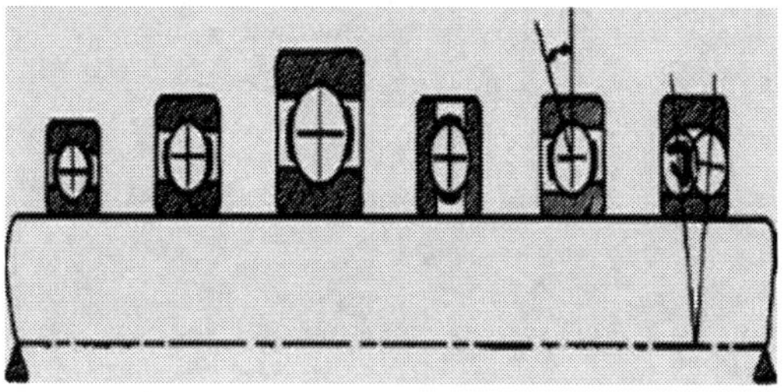

Figure 7: Types of Ball Bearings

The heavy series of bearings is designated by 400. Most, but not all, manufacturers use a numbering system so devised that if the last two digits are multiplied by 5, the result will be the bore in millimeters. The digit in the third place from the right indicates the series number. Thus, bearing 307 signifies a medium-series bearing of 35-mm bore. For additional digits, which may be present in the catalog number of a bearing, refer to manufacturer's details.

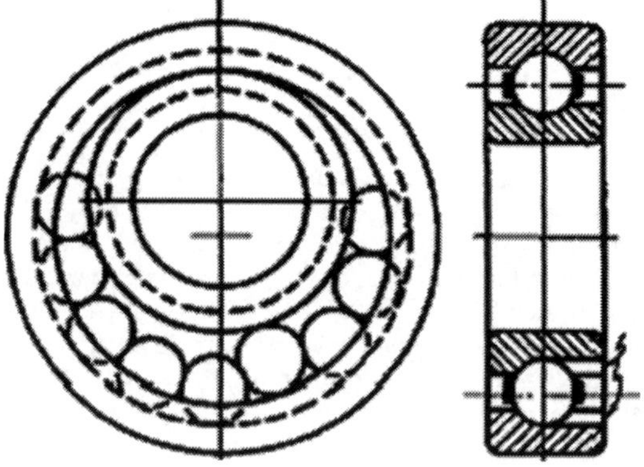

Figure 8: Conrad bearing

Some makers list deep groove bearings and bearings with two rows of balls. For bearing designations of Quality Bearings & Components (QBC), see special pages devoted to this purpose. The radial bearing is able to carry a considerable amount of axial thrust.
However, when the load is directed entirely along the axis, the thrust type of bearing should be used. The angular contact bearing will take care of both radial and axial loads. The self-aligning ball bearing will take care of large amounts of angular misalignment. An increase in radial capacity may be secured by using rings with deep grooves, or by employing a double-row radial bearing. Radial bearings are divided into two general classes, depending on the method of assembly. These are the Conrad, or non-filling notch type, and the maximum or filling-notch type.

3.1 Bearing With Bearing Cap

The bearings are pressed smoothly to fit into the shafts because if hammered the bearing may develop cracks. In our project, the 6202 bearing with bearing cap is used. The bearings are pressed smoothly to fit into the shafts because if hammered the bearing may develop cracks. Bearing is made up of steel material and bearing cap is mild steel.
Ball and roller bearings are used widely in instruments and machines in order to minimize friction and power loss. While the concept of the ball bearing dates back at least to Leonardo da Vinci, their design and manufacture has become remarkably sophisticated. This technology was brought to its present state of p erfection only after a long period of research and development. The benefits of such specialized research can be obtained when it is possible to use a standardized bearing of the proper size and type.

3.1 Pulley

There are two pulleys are used in our project. One is coupled with motor shaft and another one is coupled with spur gear shaft. These two pulleys are connected by belt drive.

3.3 Belt

This is used to transmit the power from one place to another place. In our project this is transmit power from motor to the spur gear. The belt drive mechanism is used for this project. This is used to transmit the power from the motor shaft to the brush shaft. The greater the shaft center distance, the more practical the use of chain and belt, rather than gears.

Table 1: Comparison of Roller chain, Tooth belt, V belt, Spur gear

Type	Roller Chain	Tooth Belt	V Belt	Spur Gear
Synchronization	◎	◎	×	◎
Transmission Efficiency	◎	◎	×	◎
Anti-Shock	△	○	◎	×
Noise/Vibration	△	○	◎	×
Surrounding Condition	Avoid Water, Dust	Avoid Heat, Oil, Water, Dust	Avoid Heat, Oil, Water, Dust	Avoid Water, Dust
Space Saving (High Speed/ Low Load)	×	◎	○	○
Space Saving (Low Speed/ High Load)	◎ Compact	△ Heavy Pulley	× Wider Pulley	○ Less Durability Due to Less Engagement
Lubrication	× Required	◎ No Lube	◎ No Lube	× Required
Layout Flexibility	◎	○	△	×
Excess Load onto Bearing	◎	△	×	◎
	◎ Excellent	○ Good	△ Fair	× Poor

Conveyor systems use chains, belts, or rollers, depending on the application. The general guidelines for suitability are shown in below Table.
Belt conveyors are most suitable for large-volume movement of bulk materials. Except for this situation, chains, belts, and rollers are generally difficult to compare in terms of capacity, speed, or distance of conveyance of unit materials.

NOTE: In this discussion, bulk materials refer to items like grain or cement that may shift during conveyance. Unit materials, such as automobiles or cardboard, are stable when conveyed.

Table 2: Conveyor Type

Conveyor Type	**Chain**	**Belt**	**Roller**
Bulk Handling	◎	◎	×
Unit Handling	◎	○ Only for light conveyor	◎
Dust in Conveying Bulky Goods	◎	× / ○ (○for closed conveyor)	
Space Required	Small	Large	Large
	◎Excellent	○Good	×Poor

4. DESIGN CONCEPT

4.1 CAD/CAE

Computer aided design or CAD has very broad meaning and can be defined as the use of computers in creation, modification, analysis and optimization of a design. CAD/CAE is employed in numerous industries like manufacturing, automotive, aerospace, casting, mold making, plastic, electronics and other general-purpose industries. It is used for the complete modelling, analysis and manufacturing of products. The design and development of products, which took years in the past to complete, is now made in days with the help of high-end CAD/CAE systems. The main advantage of using Cad is we create the model before the manufacturing of the product, which is the easy and economical way to get the perfect product.

4.2 Modeling

Model is a Representation of an object, a system, or an idea in some form other than that of the entity itself. Modeling is the process of producing a model; a model is a representation of the construction and working of some system of interest. A model is similar to but simpler than the system it represents. One purpose of a model is to enable the analyst to predict the effect of changes to the system. On the one hand, a model should be a close approximation to the real system and incorporate most of its salient features. On the other hand, it should not be so complex that it is impossible to understand and experiment with it. A good model is a judicious tradeoff between realism and simplicity. Simulation practitioners recommend increasing the complexity of a model iteratively. An important issue in modeling is model validity. Model validation techniques include simulating the model under known input conditions and comparing model output with system output. Generally this model which is intended for a simulation study, is a mathematical model developed with the help of solid works.

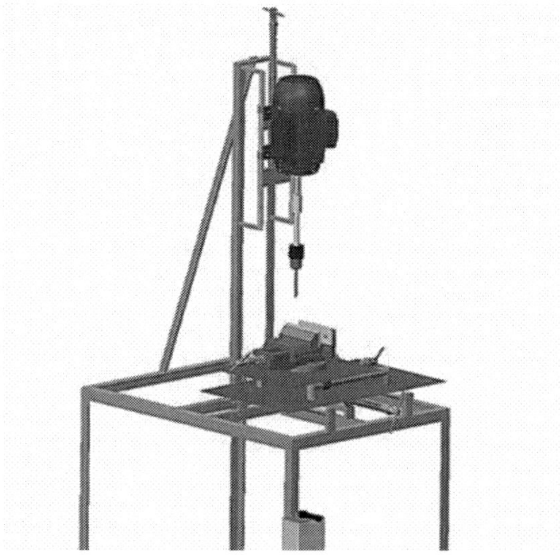

Figure 9: Model view of friction stir welding machine

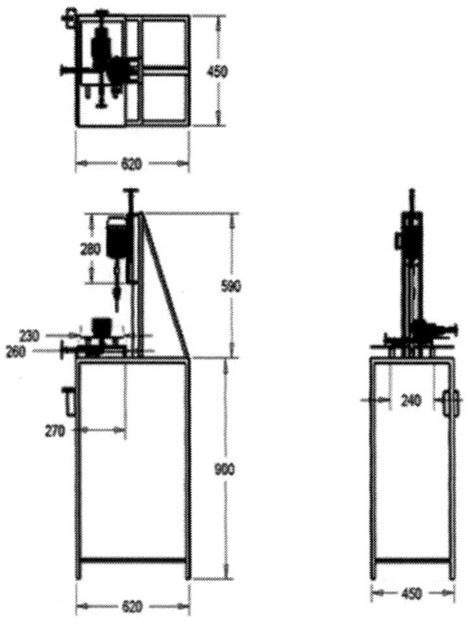

Figure 10: sketch of friction stir welding machine

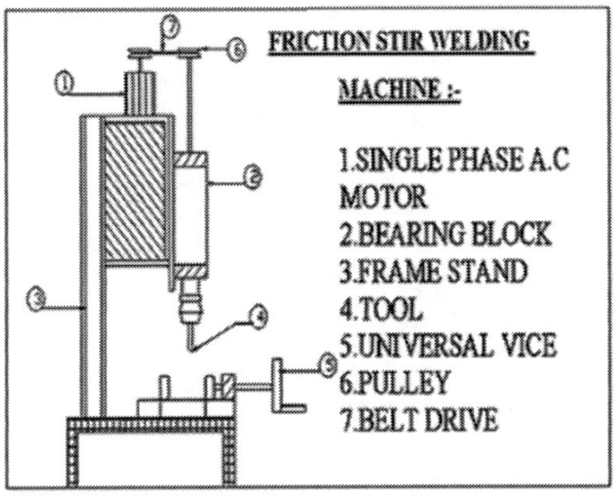

Figure 11: Components of friction stir welding machine

5. FABRICATION OF THE UNIT

The fabrication of unit consist of almost all the standard welding processes such as welding, fitting, assembling etc..
The unit necessitates the manufacturing of following parts.

- Vertical Movable bed
- Horizontal Moving Bed
- Vice
- Friction Tool
- Motor
- Frame stand

The components manufactured in process involved in manufacturing in detail in the report else. The manufacturing and assembly of this arrangement is made as rigid as possible.

5.1 Vertical Movable Bed (Upper Arm)

Upper arm is also called as movable Bed. As the arm can move up and down, it is called as movable arm. The upper arm is connected to the frame stand. The motor is fixed on this moving bed with suitable bolt and nut arrangement.

5.2 Horizontal Moving Bed (Lower Arm)

Horizontal moving bed also called as Lower arm. As the arm can move linear it is called as movable arm. The lower arm is connected to the frame stand. The vice is fixed on this moving bed with suitable bolt and nut arrangement.

5.3 Vice

The vice is found over the base of the machine and from the bottom of the moving bed. This type of vice is used to hold the work piece in a straight manner as well as in a tilted manner. Such a way the jaws are found in the vice. The various part of the vice is supporting jaw, screw rod with self-tilting jaw.

5.5 Supporting Jaw

It is found on the either end of the vice which is fixed on the base plate by the help of bolt and nut. Its supports the work piece can be moved of max angle fixing the jaw the work piece to be cut is placed to an angle whose edge. Handle is rotated the self-tilting jaw, when approaches the other side of the work piece automatically moves to angle and grips the work piece.

5.1 Friction Tool

Friction tool is made up of mild steel. It is fixed to the three phase induction motor which is held on the vertical moving bed.

6. MOTOR

6.1 Three-phase AC induction motors

Disassembled 250W motor from a washing machine, the 12 stator windings are in the housing on the left. Next to it is the "squirrel cage" rotor on its shaft.
Where a poly-phase electrical supply is available, the three-phase (or poly-phase) AC induction motor is commonly used, especially for higher-powered motors. The phase differences between the three phases of the poly-phase electrical supply create a rotating electromagnetic field in the motor.
Through electromagnetic induction, the rotating magnetic field induces a current in the conductors in the rotor, which in turn sets up a counterbalancing magnetic field that causes the rotor to turn in the direction the field is rotating. The rotor must always rotate slower than the rotating magnetic field produced by the poly-phase electrical supply; otherwise, no counterbalancing field will be produced in the rotor.
Induction motors are the workhorses of industry and motors up to about 500 kW (670 horsepower) in output are produced in highly standardized frame sizes, making them nearly completely interchangeable between manufacturers (although European and North American standard dimensions are different).
Very large synchronous motors are capable of tens of thousands of kW in output, for pipeline compressors, wind-tunnel drives and overland convertor systems.
There are two types of rotors used in induction motors.

6.2 Squirrel Cage rotors

Most common AC motors use the squirrel cage rotor, which will be found in virtually all domestic and light industrial alternating current motors. The squirrel cage takes its name from its shape - a ring at either end of the rotor, with bars connecting the rings running the length of the rotor. It is typically cast aluminum or copper poured between the iron laminates of the rotor, and usually only the end rings will be visible. The vast majority of the rotor currents will flow through the bars rather than the higher-resistance and usually varnished laminates. Very low voltages at very high currents are typical in the bars and end rings; high efficiency motors will often use cast copper in order to reduce the resistance in the rotor.

In operation, the squirrel cage motor may be viewed as a transformer with a rotating secondary - when the rotor is not rotating in sync with the magnetic field, large rotor currents are induced; the large rotor currents magnetize the rotor and interact with the stator's magnetic fields to bring the rotor into synchronization with the stator's field. An unloaded squirrel cage motor at synchronous speed will consume electrical power only to maintain rotor speed against friction and resistance losses; as the mechanical load increases, so will the electrical load - the electrical load is inherently related to the mechanical load. Furthermore, a stalled squirrel cage motor (overloaded or with a jammed shaft) will consume current limited only by circuit resistance as it attempts to start. Unless something else limits the current (or cuts it off completely) overheating and destruction of the winding insulation is the likely outcome. Virtually every washing machine, dishwasher, standalone fan, record player, etc. uses some variant of a squirrel cage motor.

6.3 Wound Rotor

An alternate design, called the wound rotor, is used when variable speed is required. In this case, the rotor has the same number of poles as the stator and the windings are made of wire, connected to slip rings on the shaft.

Compared to squirrel cage rotors, wound rotor motors are expensive and require maintenance of the slip rings and brushes, but they were the standard form for variable speed control before the advent of compact power electronic devices. Transistorized inverters with variable-frequency drive can now be used for speed control, and wound rotor motors are becoming less common. (Transistorized inverter drives also allow the more-efficient three-phase motors to be used when only single-phase mains current is available, but this is never used in household appliances, because it can cause electrical interference and because of high power requirements.)

Several methods of starting a poly-phase motor are used. Where the large inrush current and high starting torque can be permitted, the motor can be started across the line, by applying full line voltage to the terminals (Direct-on-line, DOL). Where it is necessary to limit the starting inrush current (where the motor is large compared with the short-circuit capacity of the supply), reduced voltage starting using series inductors, an autotransformer, thermistor, or other devices are used.

The speed of the AC motor is determined primarily by the frequency of the AC supply and the number of poles in the stator winding, according to the relation

$$N_s = 120F / p$$

Where,
N_s = Synchronous speed, in revolutions per minute
F = AC power frequency
p = Number of poles per phase winding

Actual RPM for an induction motor will be less than this calculated synchronous speed by an amount known as *slip*, that increases with the torque produced. With no load, the speed will be very close to synchronous.

When loaded, standard motors have between 2-3% slip, special motors may have up to 7% slip, and a class of motors known as *torque motors* are rated to operate at 100% slip (0 RPM/full stall).

The slip of the AC motor is calculated by
$S = (N_s - Nr) / N_s$
$Percentages\ lip = (N_s - Nr) / N_s * 100$
Where
N_r = Rotational speed, in revolutions per minute.
S = Normalized Slip, 0 to 1.

As an example, a typical four-pole motor running on 60 Hz might have a nameplate rating of 1725 RPM at full load, while its calculated speed is 1800.

The speed in this type of motor has traditionally been altered by having additional sets of coils or poles in the motor that can be switched on and off to change the speed of magnetic field rotation. However, developments in power electronics mean that the frequency of the power supply can also now be varied to provide a smoother control of the motor speed.

6.4 Three-phase AC synchronous motors

If connections to the rotor coils of a three-phase motor are taken out on Slip-rings and fed a separate field current to create a continuous magnetic field (or if the rotor consists of a permanent magnet), the result is called a synchronous motor because the rotor will rotate in synchronism with the rotating magnetic field produced by the poly-phase electrical supply. The synchronous motor can also be used as an alternator.

Nowadays, synchronous motors are frequently driven by transistorized variable-frequency drives. This greatly eases the problem of starting the massive rotor of a large synchronous motor. Once the motor reaches synchronous speed, no current is induced in the squirrel-cage winding so it has little effect on the synchronous operation of the motor, aside from stabilizing the motor speed on load changes.

Synchronous motors are occasionally used as traction motors; the TGV may be the best-known example of such use. One apparently unusual use for this type of motor was its use in a power factor correction scheme. This exploited a feature of the machine where it consumed power at a leading power factor when its rotor was over excited.

It thus appeared to the supply to be a capacitor, and could thus be used to correct the lagging power factor that was usually presented to the electric supply. The excitation was adjusted until a near unity power factor was obtained

 https://doi.org/10.1051/matecconf/201820703022

(often automatically). Machines used for this purpose were easily identified as they had no shaft extensions. They were referred to as *synchronous capacitors*.

6.5 Two-phase AC servo motors

A typical two-phase AC servo motor has a squirrel-cage rotor and a field consisting of two windings:1) a constant-voltage (AC) main winding, and 2) a control-voltage (AC) winding in quadrature with the main winding so as to produce a rotating magnetic field. The electrical resistance of the rotor is made high intentionally so that the speed-torque curve is fairly linear. Two-phase servo motors are inherently high-speed, low-torque devices, heavily geared down to drive the load.

7. CONSTRUCTION AND WORKING PRINCIPLE

Two pieces are welded together due to the pressure exerted by the two materials, which are connected to the lower and upper arms, where the upper arm is movable and the lower arm is also movable.

7.1 Working

Traditionally, friction welding is carried out by moving one component relative to the other along a common interface, while applying a compressive force across the joint. The friction heating generated at the interface softens both components, and when they become plasticized the interface material is extruded out of the edges of the joint so that clean material from each component is left along the original interface. The relative motion is then stopped, and a higher final compressive force may be applied before the joint is allowed to cool. The key to friction welding is that no molten material is generated, the weld being formed in the solid state.

Figure 12: Welded plates sample

8. ADVANTAGES

Friction welding has been used by the automotive industry for decades in the manufacture of a range of components. The process is attractive for several reasons.

- The friction heating is generated locally, so there is no widespread softening of the assembly
- The weld is formed across the entire cross-sectional area of the interface in a single shot process
- The technique is capable of joining dissimilar materials
- The process is completed in a few seconds with very high reproducibility - an essential requirement for a mass production industry.

Now, a new variant from the same stable of friction processes, known as 'friction stir welding', is finding increasing use in the fabrication of automotive components, even though its processing speed is not yet as rapid.

9. DISADVANTAGES

- High speed motor is required
- Applicable for particular Products only

10. APPLICATIONS

The original application for friction welding was the welding of long lengths of material in the aerospace, shipbuilding and railway industries. Examples include large fuel tanks and other containers for space launch vehicles, cargo decks for high-speed ferries, and roofs for railway carriages. In the automotive sector the drive to build more fuel efficient vehicles has led to the increased use of aluminum in an effort save weight, which also improves recyclability when the vehicles a scrapped. Such has been the interest in friction welding, which was patented no so long ago, that considerable effort is being made in transferring the technological benefits from aluminum and magnesium to higher temperature materials such as copper, titanium and steels.

11. CONCLUSION

We feel very happy and proud in fabricating this project "FRICTION WELDING MACHINE". Fixture is adjustable and easy to operate which leads to no shifting of welding workpiece during FSW. It's have good vibration absorbing capacity with good axial load carrying capacity which leads to significant defect free friction stir welding. On variable condition sound FSW found without any distortion of weld material due to no excess heat effect. A performance feasibility test should be carried out to ensure that the proposed FSW joints have strength greater than or equal to that of the corresponding MIG welded joints. Functional feasibility tests and performance feasibility tests should be completed to determine the extent to which FSW may be used in automobile chassis fabrication.

REFERENCES

1. Mishra, R. S. and Ma, Z. Y.: Friction stir welding and processing, Materials Science and Engineering, **50**, 1–78, 2005.
2. Baghel, P. K. and Siddiquee, A. N.: Design And Development of Fixture For Friction Stir Welding, IISTE, **3**, 40–47, 2012.
3. Singh, R., Rizvi, S. A. and Tewari, S. P.: Design and Fabrication of FSW Fixture With A New Approach, Elixir Mech. Engg, **86**, 34844–34846, 2015.
4. Raman, S., Wadke, A., Trichy, M., Niver, N. and Way, C.: Fixture Design Criteria: **Phase I, CASI**, 1–14, 2003.
5. Piccini, J. M. and Svobod, H. G.: Effect of the tool penetration depth in Friction Stir Spot Welding (FSSW) of dissimilar aluminium alloys, Procedia Materials Science, **8**, 868–877, 2015.
6. Shah, P. H. and Badhek, V.: An experimental investigation of temperature distribution and joint properties of Al 7075 T651 friction stir welded aluminum alloys, Procedia Technology, **23**, 543–550, 2016.
7. Sivakumar, Bose, V., Raguraman, D. and Muruganandam, D.: Review Paper on Friction Stir Welding of various Aluminium Alloys, IOSR-JMCE, **e ISSN**: 2278- 1684, **p-ISSN**: 2320–334X, 46–52, 2014.

Surface modification of Cr–Mo steel by new technology using high speed jet in water under ultrasonic irradiation

Masataka Ijiri[1,*] and *Toshihiko* Yoshimura[1]

[1]Tokyo University of Science, Yamaguchi, 1–1–1 Daigaku–Dori, Sanyo–Onoda, Yamaguchi 756–0884, Japan

Abstract. In this study, to further improve current multifunction cavitation (MFC) techniques, the surface modification of Cr–Mo steel was further investigated using 1200 W ultrasonic power. In MFC using 1200 W ultrasonic power, the corrosion resistance, and compressive residual stress of the specimens were improved when the processing time was 10 min; however, decarburization occurred at longer processing times, causing these characteristics to worsen. The decarburization that occurs at high ultrasonic outputs may be caused by an increase in the water temperature, and of the heating of the specimen surface.

1 Introduction

Alloy steel for machine structural has been used for a wide range of machine parts mainly automobile parts. The improvement of fatigue characteristics isn't always satisfactory with the conventional surface modification at present, because of the use environment of this material is harsh year by year. To solve this problem, development of surface treatment and its practical application are expected.

Among them, there is water jet peening (WJP) technology [1–5]. This technology is a peening technology utilizing the cavitation phenomenon, and it is applied as a preventive maintenance technology [2] of a nuclear power plant. It is possible to reduce tensile residual stress generated in the structure by welding or machining. When this technique is applied to a structure, cracking due to stress corrosion cracking [1] and metal fatigue [1][6] can be prevented. When in WJP injects high pressure water underwater, cavitation jets occur. At that time, a high–speed water jet containing fine microscopic cavitation is injected. This cavitation intermittently repeats occurrence and disappearance in turbulent flow containing surrounding water and generates extremely large impact force simultaneously with sound. Also, with the stress improving effect of this technique, the grown cavitation shrinks at high speed due to the surrounding pressure and generates large shock waves and micro jets and disappears. WJP is a processing method of slightly plastic deforming the material surface by this impact pressure. When the minute portion on the material surface is stretched and plastically deformed, the deformed portion is elastically restrained from the surroundings, so that compressive residual stress is imparted to the pressing portion. Therefore, when WJP is applied to the material surface where tensile residual stress remains, it is possible to improve from tensile residual stress to compressive residual stress. However, when the residual stress improves, it is reported that voids and cracks tend to form inside the specimen due to the pressure increase applied to the surface in the sample processed with WJP [3–5].

Recently, Yoshimura et al have developed multifunction cavitation (MFC) [7]. MFC processing is a cavitation technique that applies ultrasonic waves to WJP. It is theoretically proposed [7–9]. It is possible reforming of material surface in the same way as WJP. However, the temperature in the cavitation is different. In MFC processing, the sound pressure due to ultrasonic irradiation exceeds the break threshold before the cavitation collides with the material surface from the water jet nozzle, so high–temperature and high–pressure cavitation including hot spots is generated by repetition of isothermal expansion and adiabatic compression. When cavitation begins to collapse, it approaches the sample surface at the same time, the volume in the bubble decreases, forming a microjet that pierces the surface of the object. Impact force at the time of collapse is characterized by high temperature and high pressure because the bubble of high pressure (about 1000 MPa) generated by WJP and high temperature (several 1000 °C) generated by ultrasonic wave is applied to the material surface. Various functional properties [4][5][7–10] are generated in MFC–processed materials so far.

In this study, to further improve current MFC techniques, the surface modification of Cr–Mo steel was investigated using an ultrasonic power of 1200 W, which is higher than the value of 225 W used in a similar previous report [4].

* Corresponding author: ijiri@rs.tusy.ac.jp

© The Authors, published by EDP Sciences. This is an open access article distributed under the terms of the Creative Commons Attribution License 4.0 (http://creativecommons.org/licenses/by/4.0/).

2 Experimental methods

The material used for these tests was Cr–Mo steel, the chemical composition of which is shown in Table 1. Round bar (rod) specimens were heated at 860 °C as a solution treatment, followed by quenching. Tempering was performed at 600 °C. The specimens were subsequently cut into rectangular specimens with dimensions of 100 mm×100 mm×3 mm.

Fig. 1 shows the MFC processing equipment, in which an ultrasonic transducer irradiates the water jet as it emerges from the nozzle. A swirl flow nozzle [11] was used at the tip of the WJ nozzle to increase the number and size of cavitation bubbles. The discharge pressure of the pump was about 35 MPa, the nozzle diameter was 0.8 mm, and the distance between the nozzle and the specimen was assumed to be 65 mm. The considered processing times were 2, 10, 20, and 30 min.

As shown in Fig. 2, water jet cavitation isothermally expands when the sonic pressure due to ultrasonic irradiation exceeds the Break threshold, and Rayleigh shrinkage then occurs rapidly (adiabatic compression) after the cavitation expands to a certain size. The flow cavitation including the hot spot becomes MFC by the repetition of isothermal expansion and adiabatic compression. At the same time that the MFC begins to collapse, it approaches the specimen surface.

Fig. 3 shows the shape change of a photograph taken by a high-speed camera during bubble collapse into an aspheric shape [12]. The volume of the bubble decreases, and a MJ with a columnar shape that impinges the specimen surface is formed. This phenomenon is referred to as micro forging because the MJ provides high-temperature and high-pressure processing in a microscale area. Conventional WJP produces large high-pressure (several 100 μm and approximately 1000 MPa, respectively) bubbles, whereas those produced with conventional ultrasonic waves are small (several micrometers) with high temperature (several 1000 °C). In contrast, the MJ that occurs in a bubble of MFC is a deforming liquid phase body similar to a liquid jet (column) at the terminal stage of bubble collapse, and the interior of the bubble becomes high temperature (several 1000 °C) and high pressure (approximately 1000 MPa).

In the present work, the microstructure observations were conducted using optical microscopy (OM). In preparation for observations, specimens were corroded in a 5 vol% Nital etchant solution. Prior to all post–processing characterizations, specimens were cut into 1×1 cm test samples. For the measurement of the surface potential, Kelvin probe force microscopy (KFM) was performed using a multi–compatible miniature probe microscope (AFM5200S, Hitachi High–Tech Sciences, Co., Ltd.). The measurements were carried out in the atmosphere. Residual stress was measured using an X–ray stress analyzer (MSF–3M, Rigaku Corporation) with the peak top method after measurement of the (211)

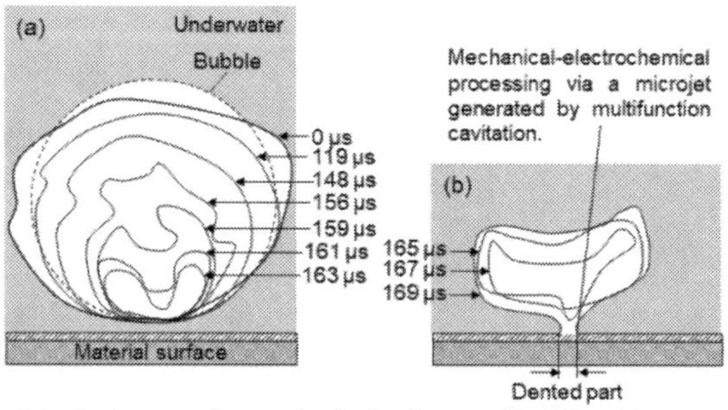

Fig. 3. A state of non-spherical collapse of bubbles. (a) 0-163 μs, (b) 165 μs~.

Table 1. Chemical composition of the Cr-Mo steel.

(mass%)

C	Si	Mn	P	Ni	Cr	Mo	Cu	Fe
0.37	0.32	0.81	0.014	0.012	0.95	0.15	0.14	Bal.

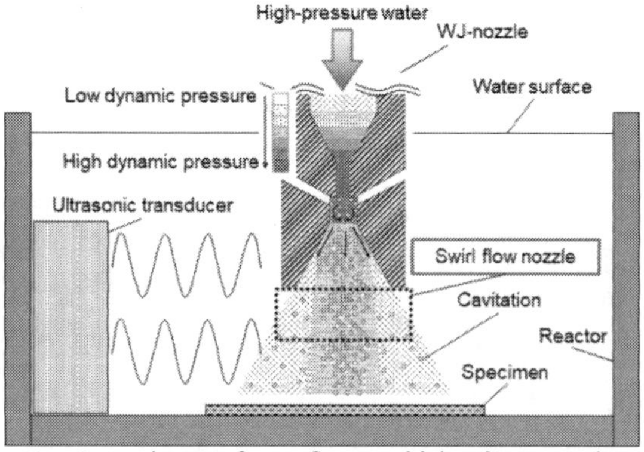

Fig. 1. Equipment for surface machining by water jet peening under ultrasonication.

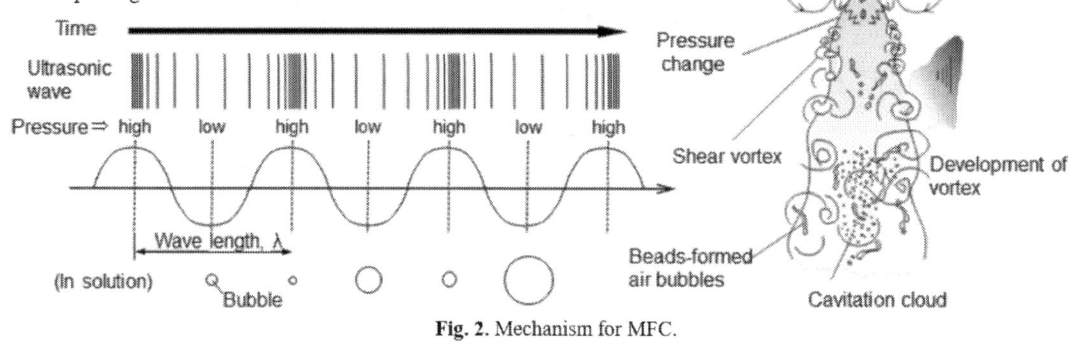

Fig. 2. Mechanism for MFC.

strain between lattice planes with the Cr Kα line generated at 30 kV–10 mA. The dissolved oxygen (DO) values of water in which cavitation had occurred in an SUS310S reactor for 30 min were obtained using a portable analyzer (OM – 71, Horiba, Ltd.).

3 Results discussion

To evaluate the effectiveness of the MFC processing, the residual stress was measured, and the results are shown in Fig. 4. To ascertain the effects of MFC processing, high tensile residual stress was imparted to the specimen surface by applying grinding before processing. Tensile residual stress (+179 MPa) was applied to the abraded material in the direction parallel to grinding, and it was confirmed that a nonzero compressive residual stress (−158 MPa) was present in the direction perpendicular to grinding. This phenomenon indicates that rust abnormally formed on the specimen surface when processing was applied for longer than 20 min.

Photographs of the specimen surface after MFC processing are shown in Fig. 5. To specify the position being peened, oil-based ink was applied to the specimen surface prior to processing.

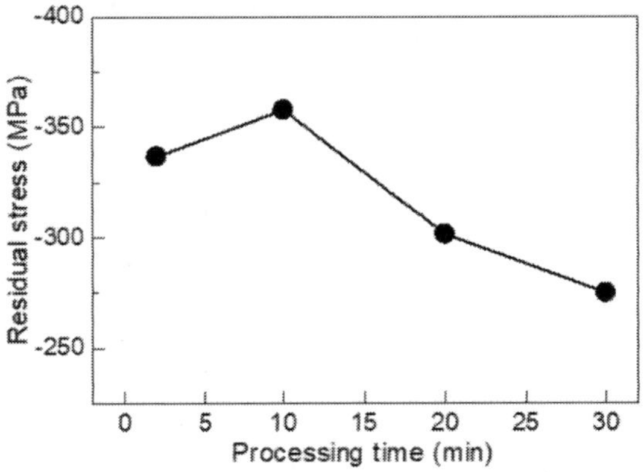

Fig. 4. Relationship between processing time and residual stress.

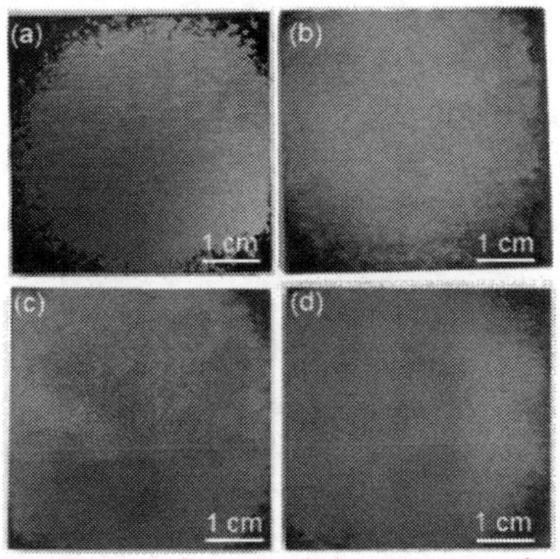

Fig. 5. Change in surface morphology as a result of MFC treatment for (a) 2 min, (b) 10 min, (c) 20 min, and (d) 30 min.

Grinding scratches were observed on the surface of specimens with a processing time of 2 min, but such grinding flaws were nearly absent for specimens processed for longer than 10 min. Furthermore, the specimen surface became bright yellow after processing for longer than 20 min, and 30 min of processing generated a large amount of the bright yellow color products on the specimen surface.

The KFM results, which give an indication of the surface condition of the specimens, are shown in Fig. 6. Because KFM involves the detection of the displacement of a cantilever due to the electrostatic attractive force on the specimen surface, it is possible to obtain the work function difference (contact potential difference) between the needle and the specimen surface. The electrostatic force received by the probe from the fixed charge distributed on the specimen surface can be detected using KFM to obtain the surface potential. The surface potential of the abraded specimen prior to MFC treatment was 234 mV. The surface potential of the specimens increased with increasing processing time until reaching a maximum value of 648 mV at 10 min and subsequently decreased with further increases to the processing time. At the longest considered processing time of 30 min, the surface potential of the treated specimen was lower than that of the abraded untreated specimen.

Fig. 7 shows optical microscopy (OM) images of the cross section of specimens after abrasion and after MFC treatment for each considered processing time.

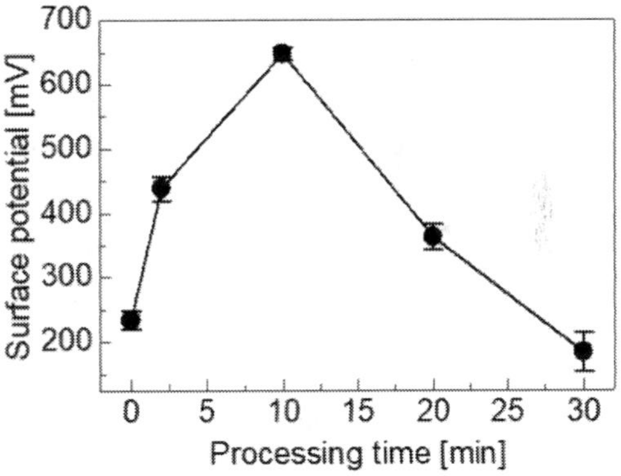

Fig. 6. Relationship between surface potential and processing time.

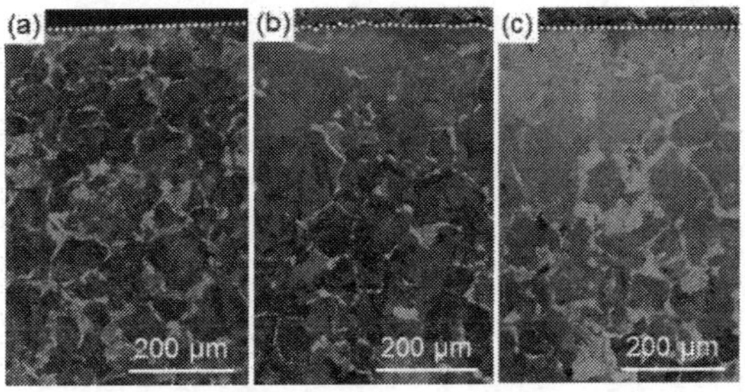

Fig. 7. Optical micrographs of the cross sections of (a) an abraded specimen and specimens treated by MFC for (b) 2 min and (c)30 min. The dashed line indicates the specimen surface.

The grain size in the depth direction from the surface after processing at each considered processing duration did not change greatly from case to case. However, in the vicinity of the surface, when the processing time was prolonged, it was observed that the color of the surface became white, with the color change progressing from the surface inward. Regarding the color of the crystal grains in the OM photograph, black grains are perlite, and white grains are ferrite. Thus, the whitening of the specimens shows that MFC processing causes pearlite existing near the processed surface to disappear. The likely cause of this is considered to be diffusion decarburization. In general decarburization, oxygen in the atmosphere and carbon in the steel combine on the steel surface during heating. This is caused by carbon scattering from the steel in the form of carbon monoxide and carbon dioxide gas. The occurrence of decarburization affects the hardness of the specimen surface. In decarburization, carbon and oxygen in steel easily bond with each other.

Fig. 8 shows the measured water temperature and DO concentrations in the water tank during MFC processing at 1200 W along with the corresponding results at 225 W obtained in a previous study [5]. When the ultrasonic output was high, the water temperature increased, and the DO concentration decreased, as shown in Fig. 8(a) and (b), respectively.

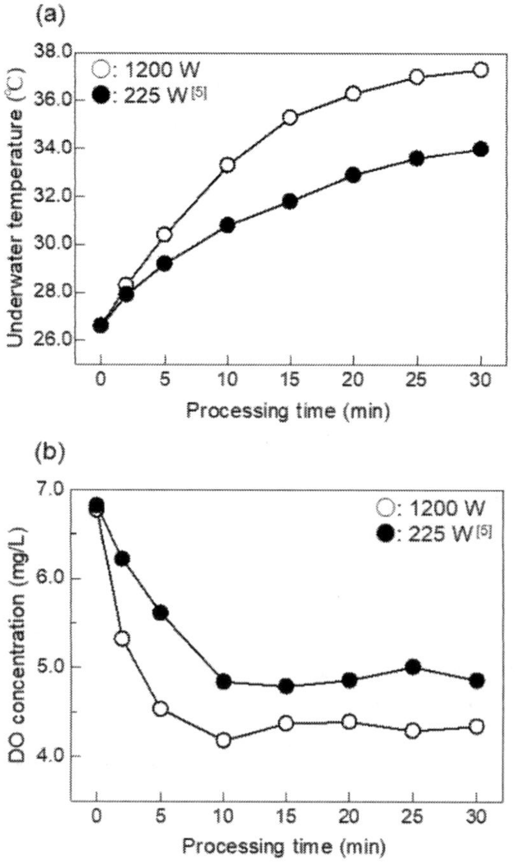

Fig. 8. Effects of processing time on (a) underwater temperature and (b) DO concentration.

The increase in the water temperature is attributable to the pressurization of the water by a high-pressure pump, which produces energy in the water flow that is partially converted into thermal energy. Regarding the reduction of the DO concentration in the water, selective oxidation occurs because of the increase in the surface temperature that occurred after 2 or 10 min of processing, causing the Cr concentration on the specimen surface to increase. It is considered that the DO concentration decreased because the Cr combined with the DO and formed an oxide film. After MFC processing for at least 20 min, the oxidized film was destroyed by cavitation bubbles. It is thus considered that the concentration of iron oxide (rust) on the surface increased as a result of the binding of DO and iron due to the Cr deficiency on the surface. The reason for the decrease in the carbon concentration on the surface after processing is considered to be as follows. As shown in Fig. 7, the color of the surface of the specimen processed for 2 min was slightly whiter than that of the unprocessed abraded specimen. It is considered that the DO and carbon on the specimen surface combined and escaped in the form of carbon monoxide or carbon dioxide gas, causing a reduction in the amount of cementite and an increase in the amount of ferrite on the surface. When diffusion decarburization occurs, the surface hardness decreases, and the fatigue strength also decreases. When applying MFC processing to various materials, it is necessary to evaluate the surface condition of the specimens after processing.

Conclusions

When applying MFC at an ultrasonic output of 1200 W, it was found that the corrosion resistance and compressive residual stress of Cr–Mo specimens were optimized by using a processing time of 10 min. However, all three of these factors decreased with increased processing time beyond this point. The reason for the deterioration of these characteristics is the occurrence of decarburization on the specimen surface. When the output power of the ultrasonic waves is high, the DO concentration decreases, the water temperature increases, and the temperature of the specimen surface increases, and these factors are thought to contribute to the occurrence of decarburization. When MFC technology is used, the amount of DO in the water, the water temperature, and the temperature of the specimen surface are affected by MFC processing, necessitating the evaluation of the specimen surface after MFC processing.

Acknowledgement

This work was supported by the Innovative Science and Technology Initiative for Security program of the Acquisition, Technology & Logistics Agency (ATLA) of Japan.

References

1. K. Hirano, K. Enomoto, E. Hayashi and K. Kurosawa, J. Soc. Mat. Sci **45** (1996) 740–745.
2. N. Saitou, K. Enomoto, K. Kurosawa, R. Morinaka, E. Hayashi, T. Ishikawa, T. Yoshimura, J. Jet. Flow Eng, **20** (2003) 4–12.
3. M. Ijiri, D. Shimonishi, D. Nakagawa and T. Yoshimura, Mater. Sc, Appl., **8** (2017) 708–715.
4. M. Ijiri, T. Yoshimura, J. Mater. Process. Tecg., **251** (2018) 160–167.
5. M. Ijiri, D. Shimonishi, D. Nakagawa, K. Tanaka and T. Yoshimura, J. Mater. Sci. Eng A., **7** **(11-12)** (2017) 290–296.
6. D. Odhiambo, H. Soyama, Inter. J. Fatifue, **25** (2003) 1217–1222.
7. T. Yoshimura, K. Tanaka, N. Yoshinaga, J. Jet. Flow Eng **32** (2016) 10–17.
8. T. Yoshimura, K. Tanaka and N. Yoshinaga, Nanosci. Nanotechnol.–Asia **8** (2018) 41–54.
9. T. Yoshimura, K. Tanaka and N. Yoshinaga, BHR Group 2016 Water Jetting (2016) 223–235.
10. M. Ijiri, T. Yoshimura, 2018 IOP Conf. Ser.: Mater. Sci. Eng. **307** (2018) 012040.
11. M. Ijiri, D. Shimonishi, D. Nakagawa and T. Yoshimura, Int. J. Lightweight Mater. Manufac. **1** (2018) 12–20.
12. C. L. Kling and F. G. Hammitt, J. Basic Eng. **94** (1972) 825-833.

https://doi.org/10.1051/matecconf/201820703024

Effective Recycling of Spent Auto Catalytic Converters by Using Electrochlorination Method

Nikolay Samotaev[1,a], Andrey Antonov[1], Grigory Tsarev[1] and Andreas Tietz [2]

[1]*National Research Nuclear University MEPhI (Moscow Engineering Physics Institute), 115409, Kasirskoe highway 31, Moscow, Russia*
[2]*ReMetall Deutschland AG, 01994, Hauptstrasse 2a, Drochow, Germany*

Abstract. Modern methods of recycling spent automotive catalysts and their main disadvantages in industrial practice are considered. The electrochlorination method is proposed as the basis of the platinum-group metals (PGM) recycling technology. As the test of proposed technology a few tons of spent automotive catalysts were processed. The results of the work on the extraction of platinum, palladium, rhodium are analyzed. The extraction rates during experiments were reached for Pt - 97%, Pd - 97% and Rh - 80%.

1 Introduction

The main raw material source of PGM scrap is spent automobile catalysts. The annual consumption of PGM used in the production of automotive catalysts is: for Platinum (Pt) around 50 tons, for Palladium (Pd) 120 tons and for Rhodium (Rh) 15 tons. The percentage of using platinum group metals (PGM) in automotive catalysts from the total annual volume produced is: for Pt - 30%, Pd - 60% and Rh - 85%.

At present, spent automobile catalysts is accumulating in the companies that are processing black and colour scrap and participating in the car recycling program. Collectors of this type of raw materials are interested in creating an independent production of the processing of spent automobile catalysts, which significantly increases the profitability of their business.

Hydrometallurgical methods with various author's chemical-technological variations, including protected by patents, have a standard set of shortcomings. First, it is a multi-stage process - activation of PGM by thermal stabilization (Fig.1) of the ceramic carrier (ɤ- Al_2O_3 in α-Al_2O_3), acid leaching or hydrochlorination of PGM, filtration and washing of pulp, precipitation of the concentrate from the solutions obtained, again filtration of the finished concentrate and the like.

Each operation takes place in a separate installation, which requires multiple overloads of raw materials. In addition, the acceptable recovery of PGM is achieved by circulating the acid or mixing the ceramic carrier. This, under conditions of the fineness of the raw state of raw materials, considerably complicates the hardware design process. Environmental problems are significant. In the process, large quantities of waste solutions are generated that require disposal. As a result, such a technological "luxury" can be afforded by large nobel metal refineries, but not by small enterprises that are raw material collectors.

Figure 1. Photo of the furnace for activation of PGM by thermal stabilization of spent catalyst carrier.

Increasingly, in the choice of technological methods for processing catalysts in world practice gives preference to a two-stage scheme. The first stage is metallurgical (induction, arc, plasma) melting of a ceramic material with the addition of fluxes to a metal-collector. In this case, the PGM dissolve in the collector and in the form of metal of the ingot comes out of the process with an extraction of no higher than 97%, but such a final extraction is achieved through a separate, and sometimes repeated, melting of the associated slag to isolate the residual PGM in collector. In addition, sometimes melting part of the PGM "hangs" in the dust filter and directly into the lining of the furnace, which requires their separate accounting and extractiontion. Thus, the direct output of PGM to the collector reaches, at best, 90-92%. The main material of the collector is iron (or copper), which has a high solubility of PGM. The

[a] Corresponding author: nnsamotaev@mephi.ru

© The Authors, published by EDP Sciences. This is an open access article distributed under the terms of the Creative Commons Attribution License 4.0 (http://creativecommons.org/licenses/by/4.0/).

https://doi.org/10.1051/matecconf/201820703024

content of PGM in iron concentrate does not exceed 10%. The low enrichment of this method neutralizes its main advantage, namely productivity, and certainly implies the presence of a stage of subsequent enrichment, which is a limiting scheme for this scheme. Besides, moreover, the energy costs of this process are significant (2.5-3 kWh / kg for raw materials, without taking into account the melting of slag), the hardware design is expensive.

At the second stage, the iron collector is refined (concentrate), where traditional hydrometallurgical schemes are used, where accompanied by the formation of a large number of liquid waste solutions. The extraction in this case is 98%. Thus, the final the extraction of PGM from the spent automobile catalysts does not exceed 95% (Rh ≈ 85%), which is a low indicator. Each stage is performed, as a rule, at different enterprises and requires from the supplier of is spent automobile catalysts temporary and material costs, as well as the procedure for sampling and matching input analysis.

Figure 2. Photo of first generation of the horizontal electrochemical modules for recycling PGM.

2 Experiments

Our team solved the problem of developing in principle new technology, which has no analogues in the world industrial practice recycling of spent automobile catalysts containing PGM and combining the following advantages:

- Conducting a full cycle of processing in one hardware module;
- Achievement of direct output of PGM Pt – 98%, Pd - 98% and Rh - 95% in concentrate not lower than 40% by amount of PGM;
- High productivity of one device (400 kg / per day) at low energy costs (1-1.5 kWh / kg);
- Ecological compatibility due to recycling of circulating fluids and effective utilization of gas emissions;
- Use of solid ceramic waste in the production of new products;
- The possibility of increasing productivity through the installation of additional processing modules.

Achieving such parameters makes this technology breakthrough in the secondary PGM industry. The method is based on the method of electrochlorination [1], which in industrial processes of recovery of secondary PGM (0.05-0.2% by weight) has not been applied and has limited use in certain processes of gold recovery from pulps of mineral raw materials [2], affinity of rich concentrates PGM [3].

Figure 3. Photo of first generation of the vertical electrochemical module for recycling PGM.

During the development of the technology and the search for an optimal solution, the installations of the first generation [4, 5] have been created, 10 tons of catalysts (mainly chemical and petrochemical catalysts). Extraction rates for Pt and Pd were achieved for spent catalysts 97%, however for Rh, not more than 80%. Control of the residual content of PGM on a carrier was carried out at the ReMetall AG (Germany) using X-ray photoelectron spectroscopy and inductively coupled plasma atomic emission spectrometry. The analytical laboratory of the company conducted about 600 studies.

For depassivation of PGM in spent and grind catalysts carrier, (bulk density of raw materials of the order of 0, 8 kg / l.) carried out their thermal activation in a specially designed furnace with an argon-hydrogen gas mixture temperature 700 ^{0}C for 2-4 hours. Fig.1 shows the general view of the furnace. After activation, the catalyst was pour forth into the interelectrode space of the electrolyser.

The hardware design at this stage of development had primitive constructive solution, the purpose of which was to confirm the effectives of the classical method of electrochlorination for raw material processing containing PGM. Figure 2 and 3 shows the horizontal and vertical installation of the first generation. The electrolyte installation of the first generation installations is water a solution of hydrochloric acid, which circulated in a closed loop (anode-cathode). Decomposition on the anode of hydrochloric acid due to the transmission of a constant current leads to the formation of reactive Chlorine, which enters in reaction with PGM, transferring them into a water-soluble form. As a result extraction of PGM into the electrolyte proceeds with their subsequent electrodepositing on the cathode. The resulting cathode deposit (concentrate based on the sum of PGM 40-70%) was readily affinity in the laboratory. However, in addition to low extraction of Rh, the technology and installations of the first generation had significant shortcomings, which limited the industrial introduction of the process in mass production.

3 Discussions

The limiting stage of the electrochemical process, which determines its productivity and acceptable values of extraction of Pt, Pd and Rh, is a long-term the electrodeposition PGM from a depleted solution, the duration of which was 24-50 hours. Circulation of electrolyte through finely dispersed raw materials constantly led to blockages of the separation membrane, which acts as a filter. The spent electrolyte after each process must be disposed of or subjected to regeneration. In addition, without the stage of preliminary thermal depassivation of PGM, the classical electrochlorination method allowed to achieve extraction no higher than 80%.

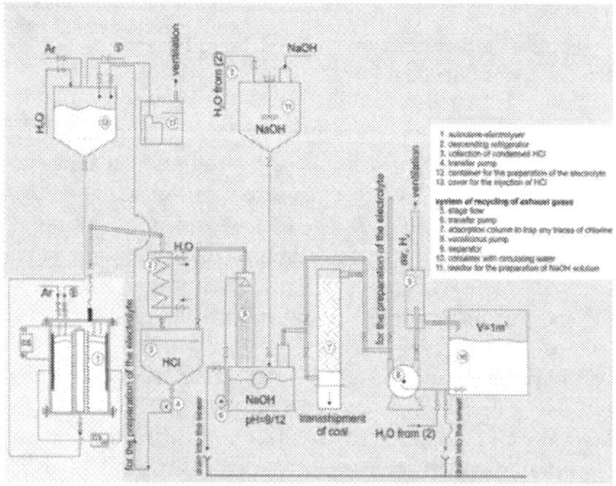

Figure 4. The hardware scheme for third generation installation. Yellow is indicates the part of autoclave solution.

For remedy founded shortages, modernization of electrochemical module for recycling PGM from the spent automobile catalysts based on the described electrochemical method, which allows to achieve high extraction of PGM without stage of thermal activation and provide high-performance and environmental a safe process scheme for its industrial implementation was developed [6]. The scientific approach of the proposed method is the superposition of two independent processes of electrolysis in DC and AC currents under conditions of forced, continuous and complete condensation of electrolyte. Method is protected by two patents [7,8]. In the experimental setup of the second generation technical solutions for achievements of high extraction, which are formulated on the basis of a complex of physicochemical investigations, including phase co-standing PGM residuals on the ceramic carrier in the process of electrochlorination. As shown by the research data, the residual metal of PGM on the carrier is present in three main forms. The first is metal on wet ceramics, whose concentration is determined by the content of PGM in the electrolyte, i.e., an open chloride form, is not which has passed into the electrolyte solution. The second is a metal in active form, but not entered into interaction. And the third is a passive unopened form. The typical percentage ratio of the listed forms is the following - 10%, 20%, 70%. The residual content of the first two forms is determined by the kinetics electrolytic process and with its optimal is reduced to a minimum. Thus, the residual content of PGM on the carrier is mainly determined by the efficiency of the chlorination and opening of the metal in the active and passive forms.

It is important that the installation involves the imposition of electrochemical processes in a DC and AC currents, and not the use of the result of the classical superposition of an alternating current to a constant one in a single source. This method allowed avoiding a long preliminary stage of thermal activation of PGM. The mechanism of this process is next - the presence of even a small amount of a metal phase (up to 0.2 % mass) in a ceramic raw material under conditions of an external AC current leads to a variable polarization of the active metal centres, turning them into a dipole. At the same time, the entire mass of the bulk material begins to represent a volumetric multipolar electrode, which provides anodic dissolution of the metal in the entire volume of the raw material and subsequent isolation at the cathode upon application of a DC current. Moreover, at a low frequency of the current, the polarity of the dipole alternates with time in time. During the existence of the particle as an anode on it, the electrolyte decomposes and atomic Chlorine is released. Active Chlorine opens the passive form of PGM and dissolves it. However, during the existence of a PGM particle as a cathode, the separation of dissolved PGMs does not have time to occur. In the case of high-frequency alternating current, this effect is absent, and the energy is mainly spent on heating the installation.

Figure 5. Image of electronic project of third generation installation with closed recycling PGM cycle. Installation has productivity of 50 kg per working cycle and adapted to location in technical area of 47 sq. meters.

The DC current from a separate power source (alternative current) provides separation of the dissolved metal on the main cathode and generation on the anode of the atomic Chlorine, which immediately recombines into the molecular one. The latter is the main chlorinating agent already pre-opened with active atomic Chlorine of the bulk of PGM. To effectively remove carrier water-soluble forms of PGM, as well as from electrolyte, platinum electroplating is carried out under conditions constant increase in the concentration of PGM in the electrolyte due to the volume ratio of the solid and liquid phases. This is achieved through forced, continuous evaporation of the electrolyte and its condensation into pure form without impurities in a separate container. This made it possible to increase the precipitation of PGM and to obtain regenerated solutions for their subsequent re-use. Thus, during the process there are absolutely no solutions containing residual PGM and recycling. In addition to the above, this circumstance made it possible to avoid the circulation of the electrolyte, which is difficult to ensure in the conditions of the finely dispersed phase, and to precede the static regime, with an increase in the kinetic parameters of the process. It also made the installation simpler, which is important for industrial implementation.

Under these conditions, the limiting stage of the whole process is to become the evaporation of the electrolyte, the speed of which is easily increased by forced evacuation of its vapours by chemical pumps, and the process of electroprecipitation of PGM in continuously enriched electrolyte occurs rapidly. The complete cycle of processing raw materials under these conditions is 6-8 hours, with direct extraction of PGM Pt-98 %, Pd – 98% and Rh 90%. Simultaneously with condensation of electrolyte vapours (H_2O-HCl) the capture and neutralization of gaseous products takes place during decomposition (Cl_2 and H_2).

The hardware scheme for installation of third generation is shown in Fig. 4. The The process is carried out with a starting volume ratio of solid and liquid phases which is 1 / 1-1.5. That makes possible to reduce the amount of technical solutions and is an additional advantage of this method. Now days, work is under way to establish the installation of the third generation, which makes it possible to achieve extraction of rhodium up to 95% and higher performance (full cycle is 4-5 hours). The main principle of this work is the use of the above factors in the design, and also the ability to conduct the process of electrolysis at elevated temperatures up to 150 ^{0}C. The increase in temperature makes it possible additionally and effectively carry out depassivation and chlorination of PGM (mainly Rh) during minimum time. The hardware design (installation design) for the implementation of the technology being developed presupposes an autoclave solution that allows to conduct electrolysis processes at temperatures up to 150 ^{0}C and a pressure up to 5 atm. One of the main technological requirements for design of the system being developed is the reduction of raw material processing in the device, rational (fast and convenient) raw material transfer, explosion safety, remote process control. The installation will have a modular solution for replicating one.

4 Conclusions

The proposed technology for recycling catalysts based on the method electrochlorination, has the main advantages:
• High PGM recovery rate in the condition of the direct output with obtaining high grade and easy affined concentrate,
• Minimum number of technological operations and quantity of instrumental devices,
• Stability of obtaining cycling solutions, null waste solutions,
• Ability of using ceramic wastes in the production of heat-insulating materials,

Now the development of a pilot autoclave plant of the third generation with a productivity of 50 kg per working cycle is completed with its further scaling into an industrial version with a capacity of 400 kg per working cycle.

Also to increase the efficiency of the recycling process were developed a technology for obtaining moulding high-temperature heat-insulating products by pressing with using ceramic waste processing. Substitution of basic cordierite powder (ceramic carrier free from PGM) has allowed to reduce the per unit shrinkage of the product by 50-60%, the increase in thermal stability in 1,5-2 times and cost reduction by 15-20%. In addition, the price of materials for heat-insulating products is several times higher than the price of residual PGM in the cordierite.

Acknowledgement

This work was supported by the Ministry Education and Science Russian Federation (Grant No. 14.575.21.0153 from 26.09.2017, unique identifier RFMEFI57517X0153).

References

1. V. Maksimov, *Electrochlorination as a method of complex extraction of methal,* Moscow: Metallurgy, (1955)
2. M. Kiselev *Research laws and technology development extraction of gold by electrochemical chlorination from gold-containing sulfid products,* PhD thesis, Ekaterinburg, (2012)
3. K. Konik, E. Kuzas, S. Lobko, S. Naboychenko, Tsvetnyye metally, **12**. (2016)
4. A. Antonov and etc., Proc. 7th Int.conf. Russian precious metals and gemstones market. (2004)
5. A. Antonov, Proc. Conf. Committee for Ecology Statetive Duma of the Russian Federation, (2004).
6. A. Antonov and etc., Proc. 30-th ISTC Korea Workshop (2013).
7. A. Antonov, A. Morozov, K. Kryschenko, Russian Patent №198947, (2003)
8. A. Antonov, A. Morozov, A. Novikov, V. Sapylkin, Russian Patent №2540251, (2015)

Integrated Product and Process Development of Clutch Friction Material Using Quality Function Deployment Processes

Zhengfeng Yan[1, a], Guangping Tu[2, b,*] and Jiajun He[3, c]

[1] *School of Mechanical Engineering, Hefei University of Technology, China, 230009*

[2] *Getrag group, Nanchang, Jiangxi, China, 330013*

[3] *School of Mechanical Engineering, University of Jilin, China, 130012*

* *Corresponding author, [a]zf.yan@hfut.edu.cn, [b]tuguangping@126.com, [c]jiajun.he@jlu.edu.cn*

Abstract. Using an improved quality function deployment (QFD) process, an integrated product and process development method was proposed for the clutch friction material, one key part of driveline. The importance ratings of customer requirements for the clutch friction material was calculated and synthesized through the reasonable selection of respondents and the method to determine customer requirements based on importance of rough analytic hierarchy process (AHP) in QFD. Using improved QFD decomposition model, the quality house between customer requirements and part features, process matrix and the key process control sample table were established. The method can be widely used in the development of other key parts.

1 Introduction

As it's expected for the design team to achieve product function design, part feature design, selection of processing method, manufacturability design, choice of technical scheme and many other objectives in the whole process of product development, system engineering method is required to help materialize integrated product and process development (IPPD). IPPD is a system engineering method in the context from product concept across production and on-site support activities, and an important characteristic of IPPD is close integration of product in manufacturing process [1,2]. Quality function deployment (QFD), however, is a rounded scientific methodology for systematic quality development based on business practical experiences from product design and manufacturing activities. In the development course of QFD theory, integrated QFD mode (Yoji Akao Mode), ASI (American Supplier Institute) four-stage mode and GOAL/QPC mode (Growth Opportunity Alliance of Lawrence, Inc.) have been evolved. More specific four-stage process is invented by J.R.Hauser, D.Clausing et al[3], which starts with customer requirement and then goes through four stages, i.e. product planning stage, part configuration stage, process design stage and production control stage.

In the face of proliferating product specialization, many small and medium-sized enterprises (SMEs) focus on production of a special category of key part for an assembly, such as clutch fiction material (CFM). CFM, as a key part of driveline, is a part material for activation of transmission function on the principle of friction effect. Friction of clutch facing is a complex problem, which, in view of its highly sensitive property, is affected by such factors as test force, temperature, rotating speed, medium and the like[4], together with notable impact of process. Due to technical research of assembly manufacturers in integrated product process of key parts is far from in-depth, traditional ASI four-stage mode is no longer effective in changing customer requirement to part feature. Here CFM is chosen as the subject for related research, with which analysis is conducted with the impact of product property and reliability life-cycle by many procedures, including material preparation, mixing, coating, winding, hot pressing, thermal stabilization and mechanical machining, and an improved quality function deployment process is suggested to establish integrated product process design method for CFM, which pinpoints customer requirement through reasonable selection of research objects and research method, in the aim to integrate abundant expertise of part manufacturer and customer requirement in a more effective manner. Effectual integration of QFD and IPPD technology will bring innovative change in product development, and it is advisable to apply this method in design and manufacturing processes of the manufacturers of other key parts, so as to realize effective and systematic upgrading of product development quality.

[a] Corresponding author: *tuguangping@126.com*

© The Authors, published by EDP Sciences. This is an open access article distributed under the terms of the Creative Commons Attribution License 4.0 (http://creativecommons.org/licenses/by/4.0/).

2 Integrated Product and Process Development Methodology

2.1 Clutch friction material product and process analysis

Automotive clutch is a part for disconnecting and transmitting power between the engine and the driveline, Fig. 1 is a schematic diagram of vehicle driveline.

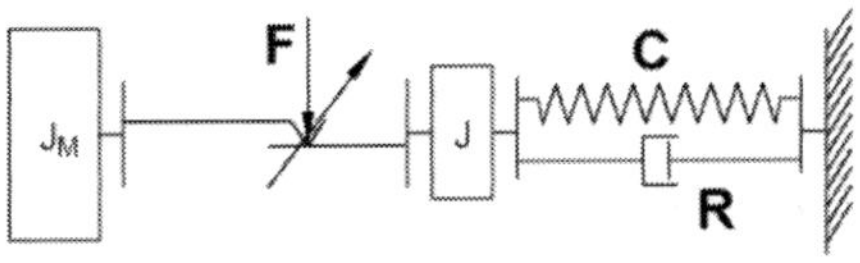

Fig.1 Simple model of vehicle driveline

Clutch torque calculation is shown as equation 1.

$$M_{(\varphi)} = 2 * F * \mu_{(\varphi)} * r \quad (1)$$

F-clamping force;
$\mu_{(\varphi)}$-friction coefficient;
r-effective radius.

Kinetic formula is shown as equation 2.

$$J * \varphi + R * \varphi + C * \varphi = M_{(\varphi)} = 2 * F * \mu_{(\varphi)} * r \quad (2)$$

J, R, C-the moments of inertia;
φ-angular displacement. [5]

CFM is a key part in this type of driveline, and its main functions are to transmit kinetic force through friction to enable safe and reliable operation of the vehicle. CFM is a widely used key material.

Essential requirements for clutch facing: stable coefficient, perfect comfort, good wear performance, withstanding high temperature and minimal thermal deformation; high strength, withstand impact, high speed stability; no adhesion; long lifetime and cost effective, and so on. Moreover, friction material should be environmental friendly and harmless to human health.

Production process of friction facing is illustrated in Fig. 2. [6]

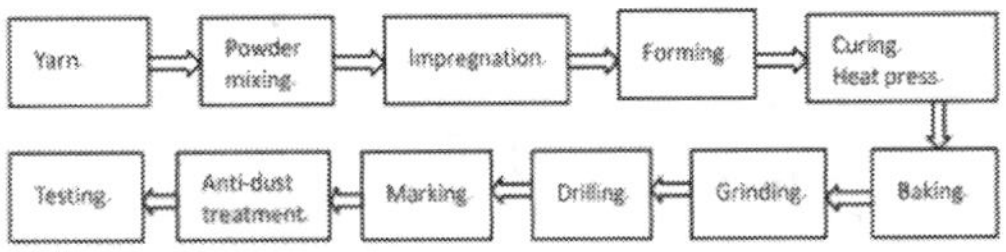

Fig. 2 Production process flow diagram of CFM

2.2 Integrated product and process development of CFM using improved quality function deployment process

Base on the characteristics of key parts manufacturers, an improved quality function deployment process, with following particular steps:

Step 1: Select reasonable research respondents, identify customer requirement and importance rating;
Step 2: Carry out IPPD;
Step 3: Build the house of quality for customer requirement and part feature;
Step 4: Construct process matrix, establish process control parameters;
Step 5: Establish key process production control table.

3 Determining the importance rating of customer requirement

In this article, Delphi Method is employed in designing questionnaire and identifying customer requirement. With this method, hierarchy of customer requirement is derived from affinity diagram, tree diagram and cluster analysis on the basis of original customer requirement on clutch friction material obtained. In classifying customer requirements, the study took eight basic quality factors, namely, property, feature, reliability, durability, serviceability, consistency, aesthetics and perceived quality, as the criteria for the completeness of customer requirement data by which the design team will be guided in the product development process (PDP)[7], and 14 customer requirements are identified as the key in the design and development of clutch friction material. Moreover, hierarchical chart of customer requirements is derived from cluster analysis and affinity diagram based upon actual situation of CFM, as shown in Fig. 3.

 https://doi.org/10.1051/matecconf/201820703025

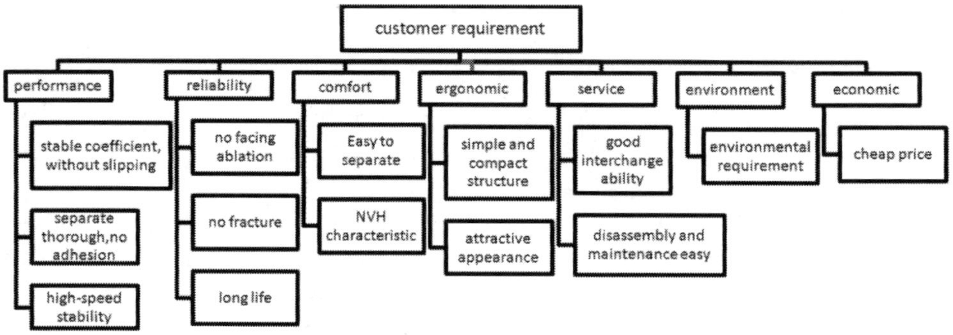

Fig.3 the hierarchy of customer requirements

Base on expert assessment method, five types of "experts" ("car manufacturer's driveline professionals", "professionals in vehicle clutch sector", "CFM product designers & testers", "CFM process designer", "vehicle clutch servicemen") are selected to participate in the judgment of importance rating of customer requirement, and one AHP matrix is constructed in terms of judgment result for each group, five AHP judgment results are obtained, with five AHP pairwise comparison matrices derived.

To determine the importance rating of customer requirements, many scholars have adopted different processing methods[8,9]. Base on determination of importance rating of customer requirements with rough analytic hierarchy process of QFD[10], importance rating of customer requirements in each hierarchy is calculated, and importance rating of CFM customer requirement is derived, as shown in Fig. 4.

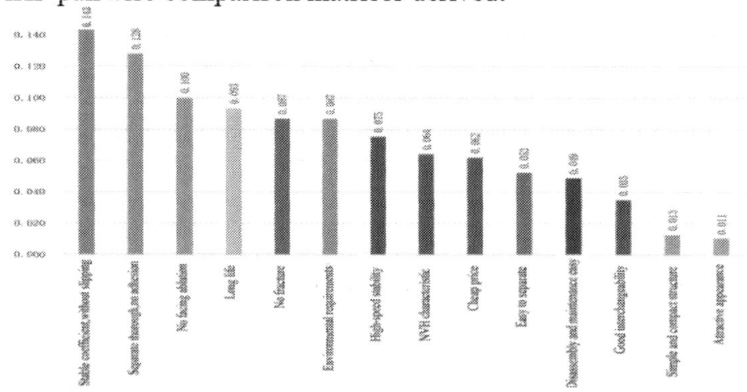

Fig.4 the importance ranking of customer requirements for clutch friction material

4 Integrated Product and Process Development of Clutch Friction Material

As nearly each use/each loading condition or each customer requires different CFM, and due to the property and life-cycle of CFM necessitate selection of different product material and process, we need to weigh and conduct integrated material and process design based on material performance. For development of new CFM product, we should start with describing from the aspects of friction coefficient, wear, judder, burst speed, rupture, interchangeability and cost, then carry out integrated product and process design according to the requirements, determine material and processing technology, and obtain better performance advantage by screening method[11], as shown in Fig. 5.

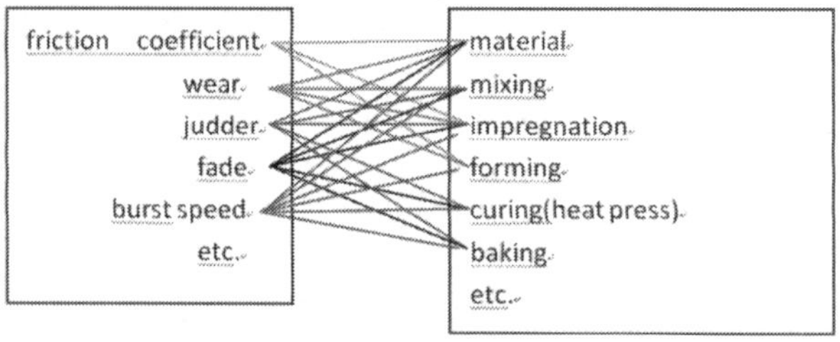

Fig. 5 Relational graph of performance of CFM versus composition material and key Process

5 Build House of Quality for Customer Requirements & Part Features

[a] Corresponding author: *tuguangping@126.com*

With the house of quality, customer requirements are converted to part features, and the same type of products in the market is assessed from the perspective of the customer, from which target value of CFM characteristics is determined.

A matrix of relation between customer requirements and part features is constructed, where ◎ , ○ and Δ represent the intensity of correlation between customer requirements and each part feature, respectively, standing for Strong, Intermediate and Weak relations.

Two leading transnational enterprises are selected for the assessment of competitiveness by means of expert scoring, digits 1～5 are used to represent the degree of customer satisfaction to a given customer requirement, the higher the score, the better the customer satisfaction.With autocorrelation matrix and matrix of relation of customer requirements versus part features determined, assessment of competitiveness is conducted, and the house of quality is derived by adding the contents in various parts in corresponding areas of the diagram, as shown in Fig. 6.

Correlation matrix notation:
* Strong positive correlation
× Postive correlation
— Negative

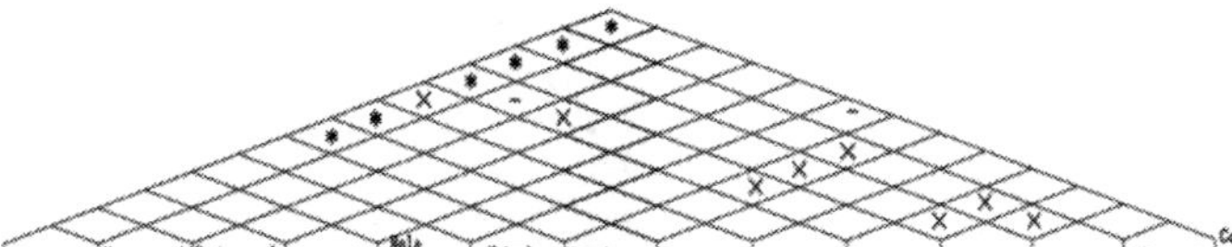

Customer requirments		Material	External diameter	Internal diameter	Thickness/parallelism	Hole diameter /position	Sink hole depth	Friction coefficient	Density	Release torque	Wear	Judder	Fade	Burst speed	Y company	R company	N company
Stable coefficient	14.3	◎						◎				○	◎	◎	5	4	4
Separate thorough	12.8	◎			○					◎			○		5	5	5
No facing ablation	10	◎									○		○		5	5	5
Long life	9.3	◎			○		○	○			◎		◎	○	5	5	4
No fracture	8.7	◎					○						○	◎	5	5	5
Environmental requiements	8.7	◎													5	5	5
High-speed stability	7.8	◎					○							◎	5	4	4
NVH characteristic	6.4	◎			△							◎	○		4	5	4
Cheap price	6.2	◎	○	○	○				○		○	○	○	○	1	2	3
Easy to separate	5.3	◎							△				○		4	5	4
Disassembly, maintenance	4.9		△	△	○	○	○								5	5	5
Good interchangebility	3.5		○	○	○	○	○		△						4	4	5
Simple, compact structure	1.3	○	○	○	○	○	○		△			◎			5	5	5
Attractive appearance	1.1	○	○	○	○									◎	5	5	4
Important degree of RCs		453.2	41.2	41.2	123.7	29.1	105.6	99.4	28.7	64	95.1	100	266.2	204.5			
Goal		Design requirments					Thickness ≥1mm	L:≥0.27 H:≥0.20	≤2g/cm3	≤12Nm	L:≤15 H:≤ 250mm3/MJ	≤ 0.1Nms	u=0.2, T≥ 350° C	≥ 10000rpm			

Correlation matrix notation:; ◎Strong correlation=5, ; ○Medium correlation=3; △Weak correlation=1

Fig. 6 Quality house of customer requirements versus part features

	process	yarn	powder mixing		impregnation	forming		curing(heat press)			baking			grinding			drilling		marking	anti-dust treatment		testing
key part feature	key process parameter	material compositon	mixing time	powder compositon	movement speed	weight	Forming petal number	temperature	pressure	holding time	raised temperature speed	temperature	holding time	speed	depth	feed rate	speed	feed rate	marking liquid	cleaning fluid	cleaning time	testing specification
material	design requirment	◎	◎	◎	◎			◎	◎	◎	◎	◎	◎							○	○	◎
External diameter						○		○	○	○			○									
Internal diameter						○		○	○	○												
thickness /parallelism						○		○	○	○	○	◎	◎	○	◎	○						○
Hole diameter																	○					
Sink hole	≥1mm											◎					○	◎				
Friction coefficient	L:≥0.27;H:≥0.20	◎	◎	◎	◎																	
density	≤2g/cm3	◎	◎	◎	◎			○	○	○												
Release	≤12Nm	◎	◎	◎	◎			◎	◎	◎	◎	◎	◎							◎	○	◎
wear	L:≤ 15mm3/MJ; H:≤	◎	◎	◎	◎		○	◎	◎	◎	◎	◎	◎									◎
judder	≤0.1Nms	◎	◎	◎	◎			◎	◎	◎	◎	◎	◎									◎
fade	w=0.2, T≥350 ℃	◎	◎	◎	◎		○	◎	◎	◎	◎	◎	◎									◎
Burst speed	≥10000rpm	◎	◎	◎	◎		◎	◎	◎	◎	◎	◎	◎									◎
	Process specification	design requirment	20-30min	design requirment	1-3m/min	difference 1-3%		160	15-18Mpa	4-6min	1-2℃/min	150-160℃	≥6Hour	25m/s	≤ 0.05mm	6m/min			ink	sodium nitrite		testing requirement

Fig. 7 Process matrix

7 Establishing Key Process Production Control Table

The output of the house of process design quality is manufacturing process characteristic requirements; in order to meet these requirements, corresponding production requirements should be established and production method be designed to establish the house of QFD production control quality. The input of the house of production control quality is manufacturing process characteristic, while the output is production requirements and control method.

Because each enterprise is unique in terms of equipment condition, production size, technical strength etc, certain difference exists between the structures of the house of production control .Table 1 shows control method determined by a given enterprise in terms of its own production situation.

Table 1 Sample CFM key process production control table (partial)

Process	Process parameter	Production requirements	Control method	Inspection approaches	Evaluation
Powder mixing	Mixing time	20-30min	PLC	Time recorder	Control chart
	powder composition	Design requirement	Automatic machine	composition analyzer	Sampling check
Impregnation	Movement speed	1-3m/min	PLC	Speed sensor	Calibration
Forming	weight	Difference 1-3%	NC machine	Electronic scale	Static process analysis
	Forming petal number	3—5	Winded mould	Naked eye	Sampling check
Curing(heat press)	Temperature	160℃	PLC	Temperature controller	Calibration
	Pressure	15-18Mpa	PLC	Pressure sensor	Calibration
	Holding time	4-5min	PLC	Time recorder	Control chart
Baking	Temperature raised speed	1-2℃/min	PLC	Temperature controller/time recorder	Calibration
	Temperature	150-160℃	PLC	Temperature controller	Calibration
	Holding time	≥6Hour	PLC	Time recorder	Control chart

8 Conclusion

From what has discussed above all, we can draw the conclusions:

（1）The study, by incorporating IPPD technology, calls in CFM IPPD method using improved QFD process, to determine customer requirements from reasonable selection of research objects and the methodology, furthermore, rough analytic hierarchy process is determined from QFD importance rating of customer requirement, so as to calculate compound basic importance rating of customer requirement for clutch friction material.

（2）Starting from customer requirement, and using improved QFD decomposition model, customer requirement is developed in part feature stage, process design stage and production control stage in sequence. From construction of the house of quality for customer requirements and part features, CFM part features are determined. Moreover, corresponding process control parameters are established based on process matrix, and finally, sample key process production control table of clutch friction material is established.

References

1. Dwivedi S.,Green J. The systems engineering approach to an IPPD implementation [J], International Journal of Computer Applications in Technology,2001,15.
2. John M.Usher,Utpal Roy,Hamid R.Parsaei.Integrated product and process development[M].John Wiley & Sons,Inc.
3. John R Hauser,Don Clausing.The house of quality [J],Harvard Business Review,1988.5-6
4. Shian.Xu,Fachao.Jiang.Automotive clutch [M]. Qinghua university press,2005 .

[a] Corresponding author: *tuguangping@126.com*

5. Schaeffler Friction Products GmbH.Our testing philosophy for clutch facings [M].Shanghai technical seminar.2010.5
6. Ronghua.Sen,Lin He.Friction material and products production technology [M].Beijing university press.2010
7. Edward B.Magrab.Integrated product and process design and development-the product realization process [M].Taylor & Francis Group,2009.73-110
8. Li Yanlai,Tang Jiafu,Yao Jianming.Mehtod of determining the importance ratings of customer requirements' improvements for house of quality.Chinese journal of mechanical engineering,2007,Vol.43(11)
9. Han C H,Kim J K,Choi S H.Prioritizing engineering characteristics in quality Function deployment with incomplete information: a linear partial ordering approach [J]. International journal of production economics, 2004,91(3).
10. X.T.Wang,W.Xiong,Rough AHP approach for determining the importance ratings of customer requirements in QFD [J]. Computer Integrated Manufacturing Systems, 2010, 16(4)
11. George E. Dieter,Linda C.Schmidt.Engineering Design [M],Fourth Edition.The McGraw-Hill Companies, Inc,2009

The FAC Experimental Research on Vapor-water Two-phase Flow Piping System of CAP1400 Secondary Loop

Shenhui Zhai[1], Simin Pang[1, a], Jie Shao[1], Hongqing Zhu[1], Xian Zhang[1] and Pengfei Liu[2]

[1]*State Nuclear Electric Power Planning Design & Research Institute Co., Ltd, Beijing 100095, China*
[2]*School of Nuclear Science and Engineering, Shanghai Jiao Tong University, Shanghai 200240, China*

Abstract. CAP1400 power plant is one of the advanced Third-generation Pressurized Water Reactor power plant. To ensure the safety and stability of the power plant secondary loop during operation, the corrosion characteristic of CAP1400 secondary loop should been investigated. This paper set up a test facility to study the vapor-water two-phase flow corrosion status, which could be seen in the piping system after the control valve of heater drain line. The key parameters, such as pressure drop, temperature, and surface topography have been observed and measured, and the different corrosion characteristic in straight pipe, elbow, tee, and reducer also have been acquired in the tests. The test results and analysis could provide the basis for the corrosion resistance design of the drain pipelines in the CAP1400 nuclear power plant.

1 Introduction

Nuclear power security is the top priority for the development of nuclear power. One of the most important factors in the design of conventional islands (CI) is flow accelerated corrosion (FAC) which can degrade the pipeline material even cause pipeline rupture [1-2]. Since late 1970s, several serious pipeline rupture accidents have occurred in the CI of some pressurized water reactor (PWR) nuclear power stations in the United States. These incidents have attracted the attention of the United States Nuclear Regulatory Commission and the nuclear industry. A special investigation team was established to investigate these incidents. The results of the investigation showed that that these pipeline failures were caused by FAC. Subsequently, many countries around the world also began conducting extensive research on the FAC. Unit 2 of the Arkansas Nuclear One Nuclear Power Plant in the US is a PWR unit owned by Combustion Engineering. In 1989, a rupture occurred at the welding place between the high-pressure steam extraction interface of the turbine and the extraction pipeline, as shown in Figure 1. The accident was the FAC damage that took place in the two-phase flow system since the Surry power station accident, which also reminded the nuclear power practitioners to strengthen their understanding of the destruction of two-phase flow FAC.

Chinese CAP1400 demonstration project will build two 1500MWe class PWR units; the design life is 60 years. CI adopts half-speed condensate Turbine-Generator unit. There are seven stages of regenerative feedwater heaters in secondary loop, which is consist of four stages LP heaters, two stages HP heaters and one deaerator. [3]

In the secondary loop of the CAP1400 nuclear power plant, a large number of vapor-water two-phase flow phenomenons exist. The pipe downstream of the heater drain control valve is a typical vapor-water two-phase flow pipeline. For the corrosion research and protection of the vapor-water two-phase flow pipelines in the secondary loop, no relevant regulations and rules can be followed in China, and the corrosion research of the vapor-water two-phase flow pipelines under the simulated actual working conditions is still in its infancy. This paper adopted experimental methods to study the FAC of the typical vapor-water two-phase flow pipelines in secondary loop in order to provide the basis for design and corrosion protection of the vapor-water two-phase flow pipelines of the subsequent CAP1400 nuclear power plants.

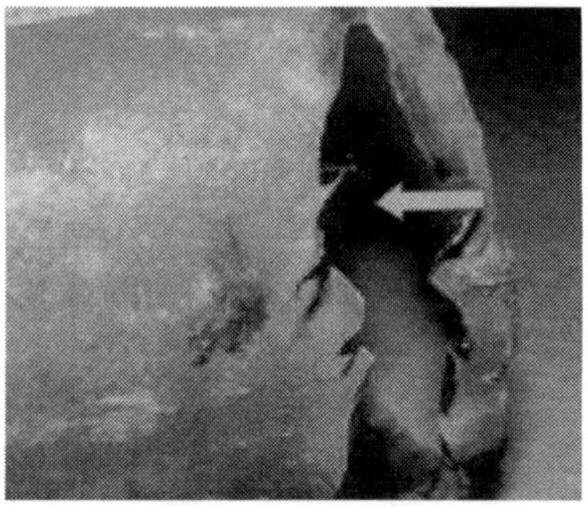

Figure 1. Cracked Welded Joint of Unit 2 of the Arkansas Nuclear One Nuclear Power Plant.

[a] Corresponding author: pangsimin@snpdri.com

© The Authors, published by EDP Sciences. This is an open access article distributed under the terms of the Creative Commons Attribution License 4.0 (http://creativecommons.org/licenses/by/4.0/).

FAC is a type of global corrosion, which is usually accelerated and deteriorated due to the medium flow. FAC is mainly formed by the interaction of physical and electrochemical processes. It is worth paying attention that FAC works through dissolution process rather than mechanical destruction of the oxide layer which is different from the common erosion/corrosion (EC) mechanism [4]. Moreover, the dissolution takes effect only when the medium is water. Therefore, the precondition of FAC is that the flowing medium in the pipeline is water or wet steam.

FAC formation is mainly divided into two processes: the first is the dissolution process of metal surface, and the second is the diffusion process of dissolved products.

Due to the oxidation reaction between iron atoms of the steel pipe and the water molecules of the flowing water in the pipe, oxide layers are formed which consist of an inner layer that grows inwards the metal matrix and a relatively loosely attached outer layer. Both layers are mainly composed of Fe_3O_4. When there are oxygen molecules in the water, Fe_2O_3 can also be formed which can strengthen the structure of the Fe_3O_4 oxide layer to make it denser. The formation of oxide layer can avoid direct contact between iron atoms of metal matrix and water molecules, thereby protecting the metal matrix. However, due to the oxide layer produced by natural reaction cannot form a dense structure that completely covers the surface of metal matrix, the pores and other discontinuous structures are inevitable in the formation of the oxide layer. These places become the weak points of the oxide layer, which makes the flowing medium and the metal matrix interacts continuously through the pores, resulting in the dissolution of the metal matrix and the oxide layer. The structure of the oxide layer formed on the metal surface is porous, therefore ferrous ions can diffuse through pores to the medium (water or wet steam). In three areas (main stream area, boundary layer, and oxide layer), the processes of dissolution and diffusion of ferrous ions are continuous. The ferrous ions produced by the corrosion of the metal matrix diffuse to the boundary layer outside the oxide layer through the pores, while the flowing water carries away these ferrous ions which are just diffused to the main stream area.

The FAC can cause a more obvious corrosion pattern on the inner surface of the metal pipeline. Due to the different state of the flowing medium in the pipeline, the corrosion forms are also different. In general, FAC can be divided into two types, single-phase flow FAC and two-phase flow FAC.

Under single-phase flow condition, FAC will cause horseshoe shaped pits on the metal surface, and the entire surface looks like an orange peel, as shown in Figure 2. In some areas where FAC rate is not very high, there are obvious dent features. If further investigation is conducted, it will be clear that these features have a certain direction and can be described as "serrated" or "horseshoe-shaped" pit "strings" whose tips point in the direction of flow [5]. As the corrosion becomes more severe, these serrated shapes gradually overlap. The most serious corrosion parts will create continuous rounded holes or orange-peel-shaped abrasions. Almost no oxide (Fe_3O_4) remains on the surfaces of these corrosion parts, and it can be seen on the metal profile that the oxide layer is very thin (about a few microns). Using a microscope scan with a higher magnification ratio, the structure of a typical rounded dent caused by FAC can be observed.

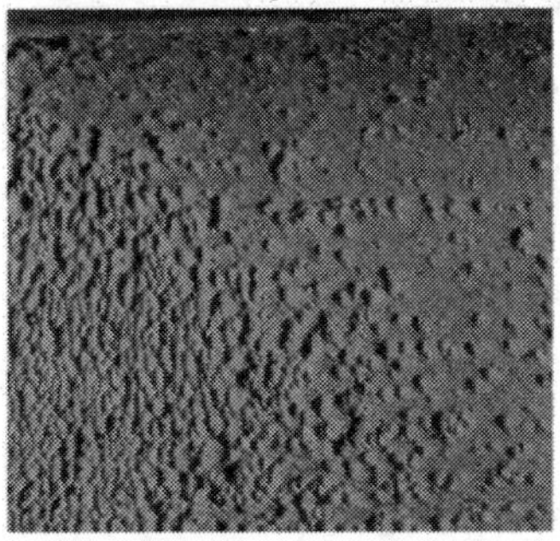

Figure 2. Corroded Metal Surface Caused by Single-phase FAC.

Under two-phase flow condition, FAC will cause tiger print shaped appearance on the metal surface, as shown in Figure 3. The black part of the figure is the area with relatively serious corrosion while the red part of the figure is the oxide layer on the metal surface. The two-phase FAC usually creates the wear marks on the metal surface which appear black and shiny, or even enamel-like. These marks usually have pit-like markings, which sometimes show serrated or horseshoe-like directional characteristics as the damage caused by single-phase flow.

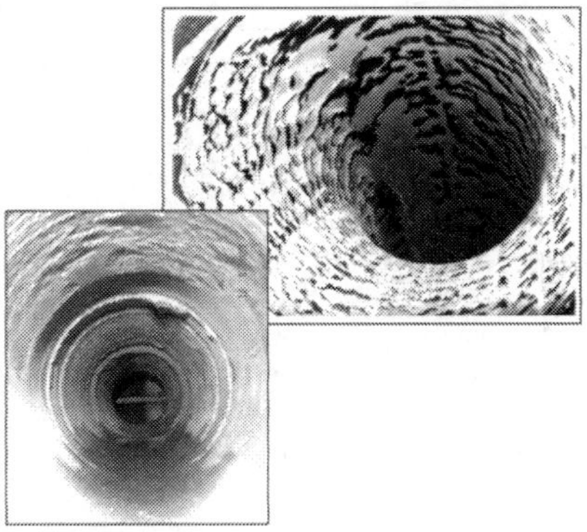

Figure 3. Corroded Metal Surface Caused by Two-phase FAC.

2 Experimental Methods

The research results on the influencing factors of FAC show that FAC has many influencing factors, such as temperature, pH value, velocity, and rate of vapor content of the medium etc. If the design condition of the secondary loop is fixed, some influencing factors are determined, such as pH valve, velocity, temperature, etc., which also brings convenience for experimental research of two-phase FAC. In this paper, the pipeline downstream of a heater drain control valve in the secondary loop of the CAP1400 nuclear power plant was taken as the test object. A two-phase flow erosion-corrosion facility (TPEC) was designed and built for experimental research, and the design parameters of the pipeline are shown in Table 1.

Table 1. Experimental Parameters of Vapor-water Two-Phase Flow in CAP1400 CI Pipeline.

Name	**Operating Pressure**	**Operating Temperature**	**Velocity**	**Piping Material After the Drain Valve**
Normal Drain Pipe of a High-Pressure Heater	Upstream of the Drain Control Valve: 1.867 MPa Downstream of the Drain Control Valve: 1.026 MPa	184.5℃	1.29 m/s	TP304L

The TPEC experimental device is a closed loop, which consists of a high-pressure (HP) tank (design pressure is 4MPa, and design volume is 2.5m3), a low-pressure (LP) tank (design pressure is 0.4MPa, and design volume is 10m3), a pump, a heat exchanger, in-tube heaters, test section, measurement and control system, and auxiliary system. The pressure control of the test section is jointly accomplished by the pump, the pressure control system of the HP tank, and the test section control valve: the pump with variable voltage and variable frequency (VVVF) system allows for coarse adjustment of pressure and flowrate; the pressure control system on the top of the HP tank can adjust and stabilize the pressure of the loop; the control valve before the test section is used for precise adjustment of the pressure before the test section; the downstream pressure of the pipeline is adjusted by the pressure control system of the LP tank. The temperature control of the loop is jointly accomplished by the in-tube heater, the in-tank heater of HP tank, and the preheater before the test section: the in-tube heater with a rated power of 600kW which can be adjusted is used for adjusting the inlet temperature of the HP tank; the in-tank heater with a rated power of 200kW is used for adjusting the outlet temperature of the HP tank; the preheater with a rated power of 50kW is used for precise adjustment of the inlet temperature of the test section to ensure the inlet temperature condition of the test section. A bypass pipeline of the test section is installed to regulate the flowrate and ensure the flow velocity in the test section. After the high-temperature and high-pressure water flashes to vapor-water mixture in the test section, it is pumped to the heat exchanger with a rated power of 430kW to be cooled down, then it flows back to the LP tank.

The specification of the pipe in the test is DN50. The material after the control valve is determined according to the test conditions, and the material of other parts of the test loop is 304 stainless steel.

The schematic diagram of TPEC test loop is shown in Figure 4.

Commonly used pipe fittings of the secondary loop include straight pipes, elbows, tees, and reducers, etc. Therefore, these typical pipe fittings were measured for the experiments. Considering the experimental conditions, the design parameters of the test section are shown in Table 2. The test period is designed to be 2 hours.

Table 2. Design Parameters of the Test Section.

Type	**Specification**	**Material**
Straight Pipe Test Section	Diameter: DN50; Wall thickness: 7mm; Length: 2.5m	No. 20 Steel/ TP304L
90° Elbow Test Section	Pipe Diameter: DN50; Bend Radius: 200mm	No. 20 Steel / TP304L
Reducer Test Section	Pipe Diameter before Control Valve: DN50; Pipe Diameter after Control Valve: DN80;	No. 20 Steel / TP304L
Tee Test Section	Pipe Diameter: DN50×DN50×DN50	No. 20 Steel / TP304L

2.1 Straight Pipe Test Section

In the test, a pipe with a nominal diameter of 50mm is adopted to perform surface analysis on the inner surface of the pipe. Measuring points and surface analysis test sections are set up at equal distance on the test pipe, which can observe and analyze the scour morphology on the pipe wall.

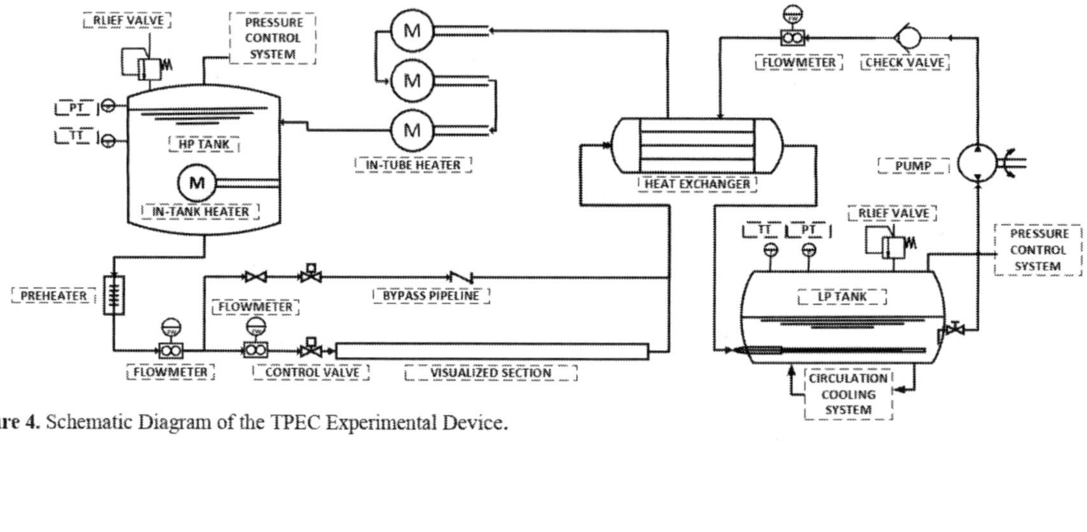

Figure 4. Schematic Diagram of the TPEC Experimental Device.

2.2 Elbow Test Section

The bend radius of the test section is 200mm, and the diameter of the pipe is 50mm. The test locations are designed at 5°, 25°, 45°, 65°, and 85°of the elbow. The test piece is fixed at the top of each test location. The test piece is replaceable, which is fixed to the measuring point by using the tighten bolt. The outer sleeve and the tighten bolts are threaded connected. There is a pressure measuring point and a temperature measuring point on the test section.

2.3 Reducer Test Section

Since the reducers are necessary components for connecting the pipes with different diameters. The existence of pipe expansion in the secondary loop of the nuclear power plant can also be one of the concerns of the two-phase flow scour test. In the test, the DN50×80 reducer is adopted, and the length of the reducer is 100mm. The scouring effect of the two-phase flow on the wall of the expansion part is observed. The reducer test section is designed to be replaceable, which is similar to the straight pipe test section. The test pieces are positioned through the positioning component and fixed by the threaded tighten pipe. The test pieces can be replaced by unscrewing the threaded tighten pipe in order to save the processing time of the test pieces.

2.4 Tee Test Section

Tee is one of the common components used in the piping system after which the flow pattern and velocity of the fluid change greatly. Therefore, the vapor-water two phase flow can also result in corresponding wear that different from other piping components on the tee. In the test, a standard DN50 equal tee is adopted, and the design of dismantling and replacement are also considered. The replaceable test pieces used in this test are with the same design as the pieces used in straight pipe test section.

3 Experimental Results and Analysis

Based on the parameters of the pipe after the heater drain control valve in the secondary loop of a CAP1400 nuclear power plant which is under construction, the two-phase flow corrosion experiments were conducted for typical piping components including straight pipe, elbow, tee, and reducer. The main experimental results are as follows.

3.1 Measurement of Thermodynamic Parameters of Two-Phase Flow

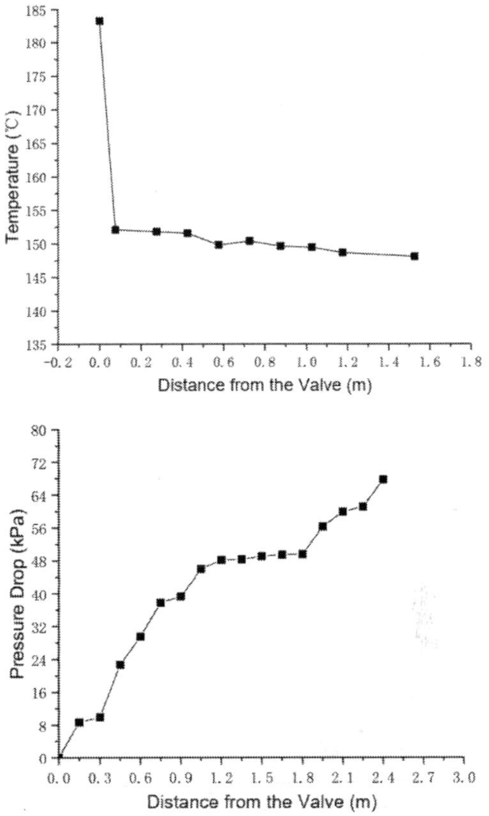

Figure 5. Curve of Temperature and Pressure Drop along Straight Pipe Section.

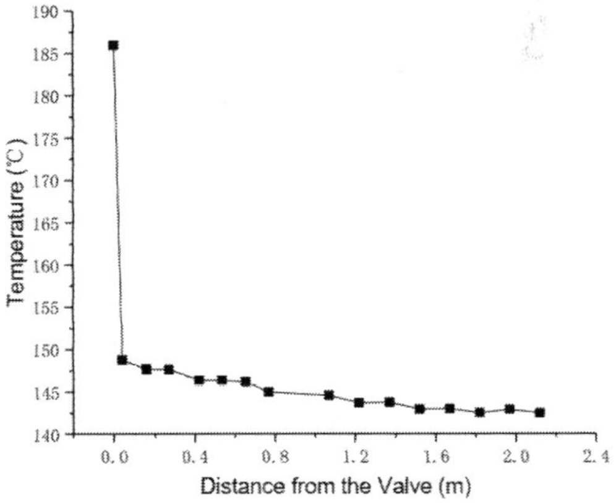

https://doi.org/10.1051/matecconf/201820704001

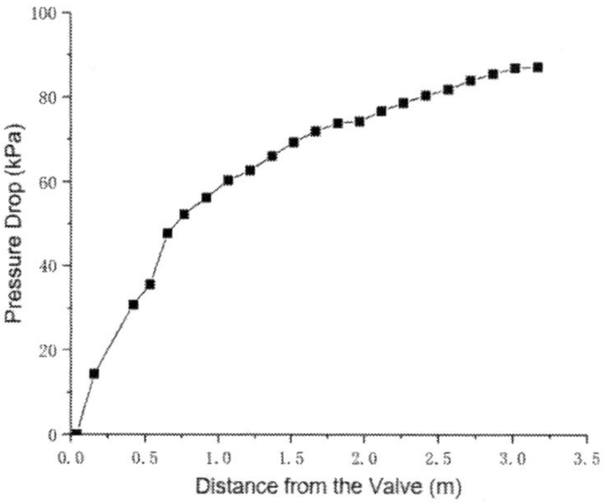

Figure 6. Curve of Temperature and Pressure Drop along Elbow Section.

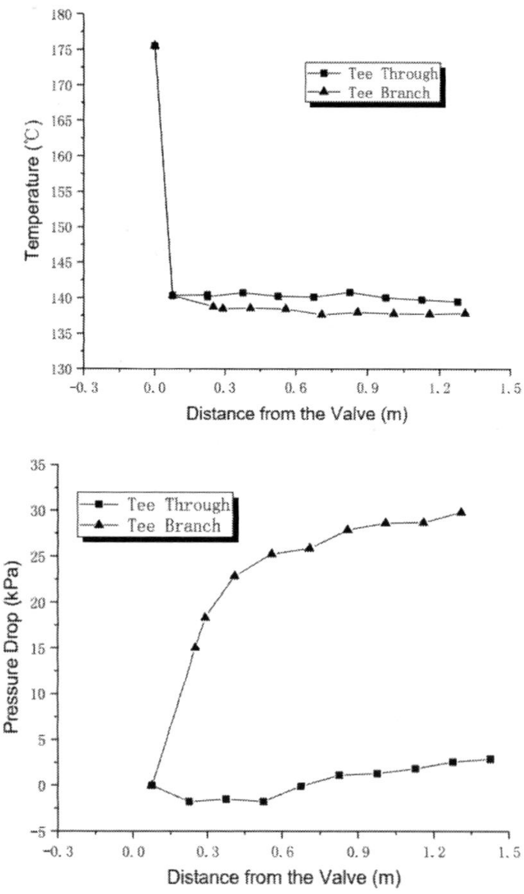

Figure 7. Curve of Temperature and Pressure Drop along Tee Section.

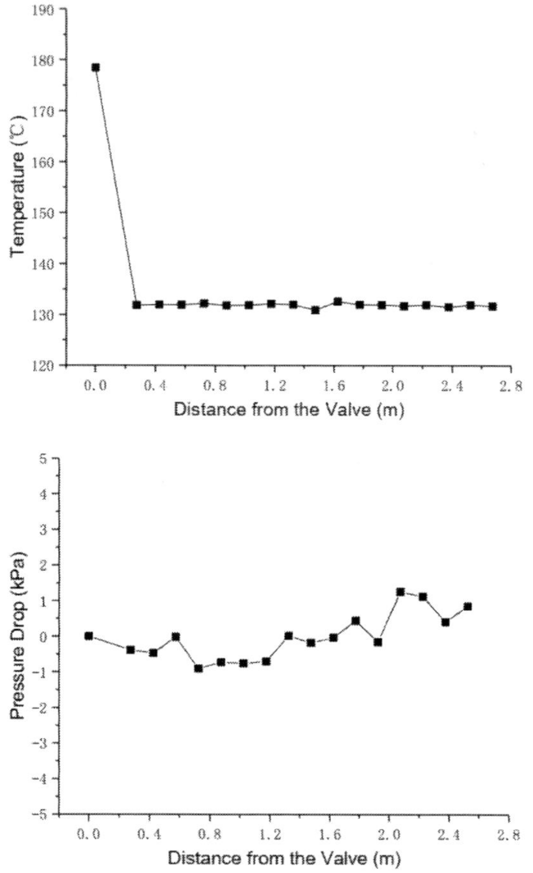

Figure 8. Curve of Temperature and Pressure Drop along Reducer Section.

The curves of temperature and pressure drop along straight pipe section, elbow section, tee section and reducer section under the condition listed in Table 1 were measured, as shown in Figure 5~ 8. As can be seen from the figure, after passing through the valve, due to the pressure drop, there will be a stepped steep drop in the temperature after the valve which is related to the form of the piping components. The temperature of the straight pipe section after the control valve is the highest while the temperature of the reducer section after the control valve is the lowest. The reason is that the straight pipe has the least influence on the flow pattern. However, for the reducer section, before entering to the pipe with larger diameter, the flow of the fluid is driven by the pressure difference, which means the fluid moves under favorable pressure gradient. When the fluid enters the area with a larger diameter, based on the mass continuity, the velocity of the fluid will decrease due to the increase of the flow area. According to the Bernoulli equation of incompressible fluid, the decrease of velocity will lead to the increase of static pressure, which will result in appearance of adverse pressure gradient in the local area on upper and lower sides of the reducer. The combined effect of adverse pressure gradient and wall viscous resistance will cause the fluid in the local area to decelerate continuously. Eventually, the inertial force cannot overcome the resistance and cause the flow to leave from the wall surface and generate vortex. Since the velocity of the viscous fluid does not change abruptly as the pipe shape, it changes gradually. This sudden change in the flow distribution results in severe turbulence in the local area, which will reduce the pressure in the local area and cause the temperature to drop rapidly. The temperature change of elbow and tee is between the straight pipe and reducer.

For the pressure drop along the distance, the change trend of straight pipe, elbow, and tee is basically the same. The pressure drop increases along with the distance, and it tends to be constant when the distance is long enough. According to the experimental results, the pressure drop of the elbow is larger than that of the straight pipe. Because of the diversion effect of the branch pipe, the flow of the tee cannot be compared with the straight pipe or the elbow directly. Therefore, it can be qualitatively considered that the pressure drop with flow pattern change is larger than that of the straight pipe. It can be seen from the figure that the pressure drop of the reducer along the distance tends to 0. It can be considered that the development area of the two-phase flow is extremely short and the interaction between two phases is small, therefore the flow pattern reaches the relatively stable state which is different from the straight pipe and elbow that need certain space for development to achieve a stable state.

According to the above experimental results, expanding the pipe diameter of the downstream pipe of the drain control valve by reducer is beneficial for the thermodynamic parameters of the fluid in the pipe after the control valve.

3.2 Morphology of the Surface Corroded by Two-Phase Flow

According to the sample size and surface scour marks, the scanning electron microscope (SEM) and energy dispersive spectrometer (EDS) were employed to scan the sample surface and the surface morphology of the samples were measured by a three-dimensional profilometer. In order to better observe the corrosion patterns in the experiments, samples made of different materials were used, one is made of No. 20 steel (carbon steel) and the other is made of TP304L (stainless steel). Since the corrosion effect of TP304L test samples is not obvious, the test results of carbon steel samples are listed only.

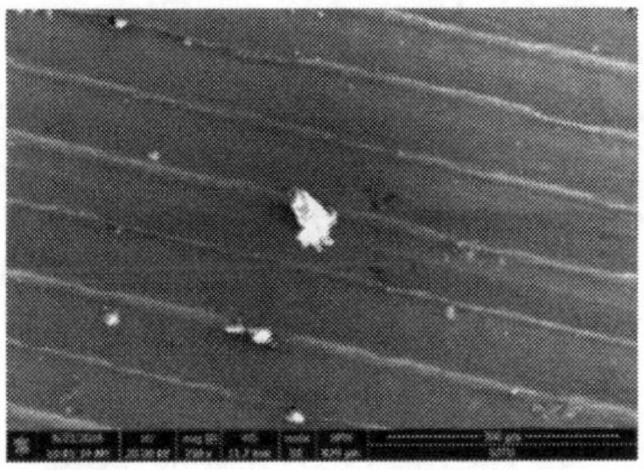

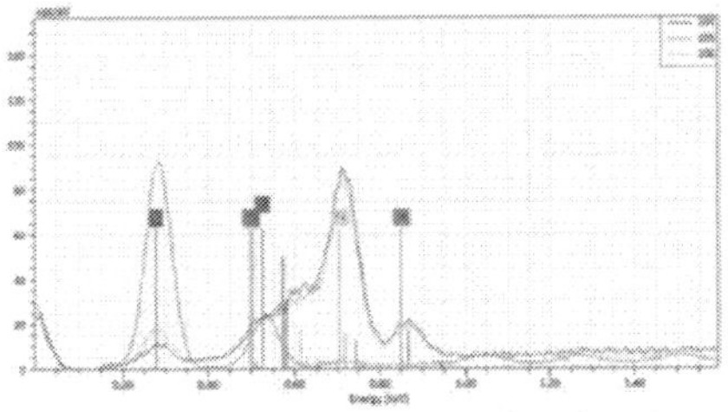

Figure 9. Surface Morphology and Element Distribution of Carbon Steel Sample in Straight Pipe Test Section.

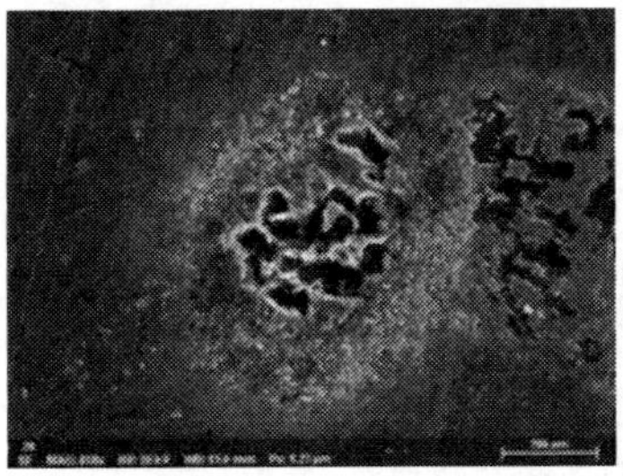

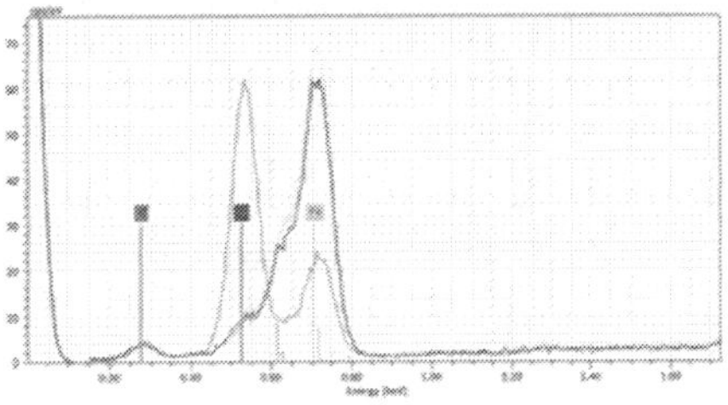

Figure 10. Surface Morphology and Element Distribution of Carbon Steel Sample in Elbow Test Section.

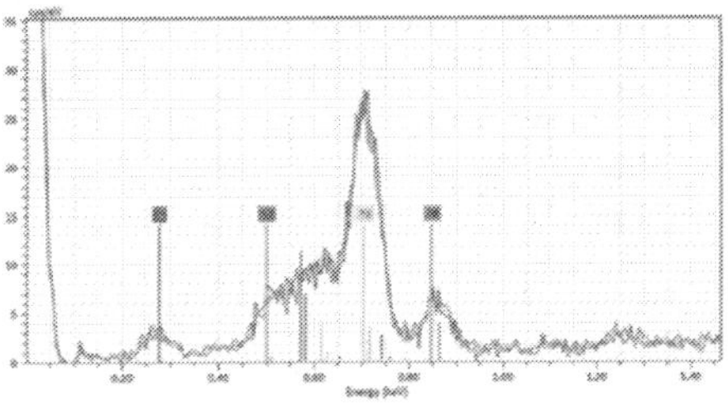

Figure 11. Surface Morphology and Element Distribution of Carbon Steel Sample in Tee Test Section.

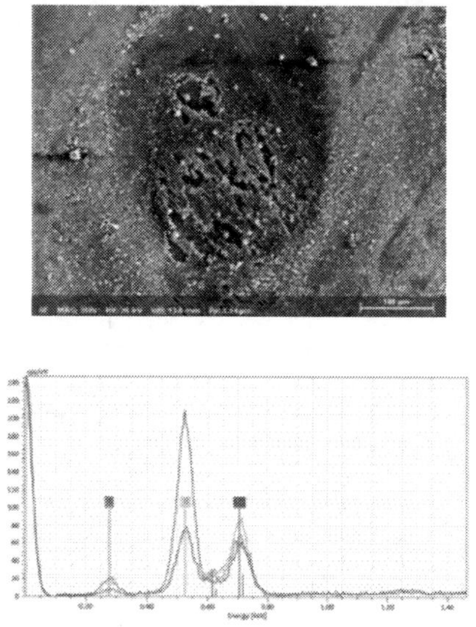

Figure 12. Surface Morphology and Element Distribution of Carbon Steel Sample in Reducer Test Section.

Figure.9 ~ Figure.12 show the surface morphology and elemental analysis results of carbon steel samples tested in each experiment of straight pipe, elbow, tee, and reducer. As can be seen from Figure 9, after the carbon steel samples scoured by two-phase flow in straight pipe section, the oxide concentration area can be observed. The concentration area is basically round, the size of it is ranging from 0.5~1mm. According to the EDS observation, no oxygen can be found out of the oxide concentration area. In the oxide concentration area, the proportion of oxygen is different in different area, the thickness of oxide layer is thicker in the center of the oxide concentration area than at the edge. The experimental results show that there is a significant concentration of oxygen element and a significant increase of carbon element in the destroyed area of the sample which means high temperature oxidation occurs at the place that result in embrittlement of the structure.

In Figure 10, it can be seen more clearly in the sample of elbow test that obvious round shape scour marks are showed on the surface of the No.20 steel sample, and a few dotted pits are generated inside these round shape scour marks. These pits are irregular in position, but mainly concentrated inside of the scour marks. The larger marks have more pits inside of them and the pit size is smaller in the smaller scour marks.

From Figure 11, it can be seen that the samples of tee test section can also be scoured, and there are round scour marks appear on the surface which looks less obvious than the scour marks on the samples of the straight pipe test and the reducer test. Considering the position of the samples, it can be considered that the two-phase flow area in the tee is relatively large, and thermodynamic parameters change significantly at the position of the tee (which due to the flow diversion caused by the branch pipe). Therefore, the two-phase effect is weakened, causing less corrosion.

As shown is Figure 12, the scour marks of the size between 0.1~0.3mm are also found on the surface of the samples of reducer section. The structural failure points are showed in the internal area of the scour marks. The failure points appear to be typical FAC morphology, and the material is scales off the surface. According to the EDS results, oxygen element is found in the damage point which indicates that oxidation also occurs at the damage point. And the oxide layer inside the scour marks appears the trend of thinning. It means that the dotted pits produced by the two-phase flow are generally dotted pits, which usually appears inside the round shape scour marks. The scour marks are created by the high temperature generated by the bubbles impact, which leads to the property change or thinning of the oxide layer. And the oxide layer is delaminated due to the bubble scouring or the mechanical force produced by the bubble collapse.

Under the same experimental conditions, no obvious scour marks are seen on the surface of the TP304L steel samples.

3.3 The Correlation between the Thinning of Piping Wall thickness and the Scouring Time

In order to observe the correlation between the thinning of the piping wall thickness and the time, the impact experiments of scouring time were conducted. Two sample points (point 1 is at the 0° of the elbow, point 2 is at the 30°of the elbow) were measured using a three-dimensional profilometer, and the test results are as follows.

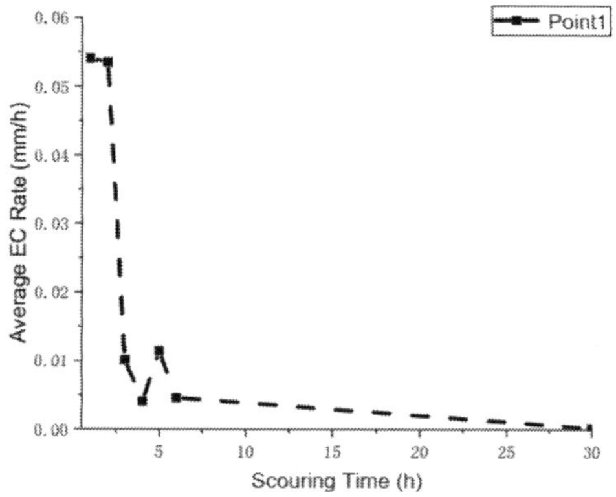

Figure 13. The Curve of the Average Piping Wall Thickness Reduction of the Samples at Point 1 of the Elbow Test Section with Time.

It can be seen from Figure 13~14, the erosion rate was larger during the initial 1h~2h, which was related to that the initial piping wall was already oxidized. The average EC rates are significantly different for the two sample points is due to the different initial oxidation degree. After the initial oxidation layer was scoured away, the erosion rate dropped rapidly, but it increased at 5 h. This is related to the process of erosion development. During the initial period of two-phase scouring, the pipe wall oxidized and generated the oxide layer in the water environment. Due to the low strength of the oxide layer, it was stripped away from the pipe surface by the two-phase flow bubbles and left several dotted pits. Since the peeling of the oxide layer exposed the underlying metal matrix which strength is relatively high and hard to be damaged by the two-phase flow bubbles, the erosion rate starts to decrease. With continuous oxidation of the metal matrix, the erosion rate increases slightly at later time.

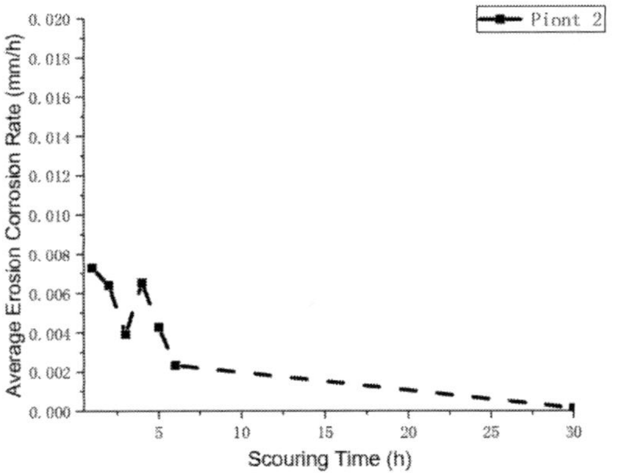

Figure 14. The Curve of the Average Piping Wall Thickness Reduction of the Samples at Point 2 of the Elbow Test Section with Time.

According to the experimental results, due to the reduction of the oxygen concentration of the fluid in the pipeline, the piping wall thickness reduction rate reached a relatively small steady value after a period of time, which was about 0.485μm/h.

4 Conclusions

In this paper, the TPEC experimental device was used to study the FAC effects of the pipeline after the drain valve in the CAP1400 secondary loop. The thermodynamic parameters of two-phase flow and the corrosion morphology of the pipe surface in different straight pipe sections, elbow sections, tee sections and reducer sections were measured. The following conclusions can be drawn from the experimental measurement and analysis:

(1) Due to the pressure drop, the high-temperature and high-pressure water will be vapor-water two-phase fluid after flowing through the drain valve. The thermodynamic parameters are related to the parameters before the drain valve. Under the same before-valve thermodynamic parameters, the energy loss of the straight pipe is smaller so that the temperature after valve of it is the highest; the flow development is faster and the interaction between two phases is relatively small in the reducer so that the temperature after valve of it is the lowest.

(2) The corrosion mechanism is the vapor phase of the vapor-water two-phase flow has an impact on the oxide layer which peels off the oxide layer from the pipe surface and exposes the metal matrix to the water to be further oxidized, therefore, the erosion-corrosion continues.

(3) According to the measurement results of the three-dimensional profilometer, the rate of the corrosion caused by vapor-water two-phase flow scouring follows a special law: at the beginning of the corrosion, the oxide layer becomes thicker due to the continuously oxidation, however, the impact from bubbles makes the oxide layer easier to be peeled so that the corrosion rate is faster at the initial period. With scouring continuing, the loose oxide layer is washed away and the corrosion rate decreases. However, the dotted pits created by the scouring expose the metal matrix, so that the oxidation proceeds and the corrosion rate increases. It can be seen that the erosion-corrosion and the metal oxidation interact on the pipeline together, therefore, no obvious erosion-corrosion marks were observed on the 304L stainless steel which has a strong resistance to oxidation

(4) At the initial period of the two-phase flow, the fluid can facilitate the formation of the oxide layer. After the oxide layer is formed, the oxide layer separates the metal matrix from the fluid so that the corrosion rate of the pipe wall tends to be stable.

References

1. Derek H. Lister, L. liu, A. Feicht, M. Khatibi, W. Cook, K. Fujiwara, and E. Kadoi. A Fundamental study of flow-accelerated corrosion in feed water systems, *15th International Conference on the Properties of Water and Steam Water, Steam, and Aqueous Solutions: Advances in Science and Technology for Power Generation*, ICPWS XV, Berlin, Sept. 8-11, (2008).
2. Recommendations for Controlling Cavitation, Flashing, Liquid Droplets Impingement, and Solid Particle Erosion in Nuclear Power Plant Piping Systems, *EPRI*, 1011231, 2004.
3. Chenge Lin, Zusheng Yu. Advanced passive pressurized water reactor technology. Beijing, *Atomic Energy Press*, 243-246, (2009).
4. Bryan Poulson, Complexities in predicting erosion corrosion, *Wear*: 497-507, (1999).
5. Barry Dooley, Flow accelerated Corrosion in Fossil and Combined Cycle/HRSG Plants: A summary. *Lyon: FAC2008 International conference*, 961-975 (2008).

New Nanocompoite Thermal Interface Materials Based on Graphene Flakes, Mesoscopic Microspheres and Polymers

Alex A Dmitriev, Alex S Dmitriev[a], Inna Mikhailova

NanoThermal Physics Laboratory, National Research University «Moscow Power Engineering Institute»
Russia, 111250, Moscow, Krasnokazarmennaya, 14

Abstract. In recent years, there has been a great interest in the development and creation of new functional energy materials, including for improving the energy efficiency of power equipment and for effectively removing heat from energy devices, microelectronics and optoelectronics (power micro electronics, supercapacitors, cooling of processors, servers and data centers). In this paper, the technology of obtaining new nanocomposites based on mesoscopic microspheres, polymers and graphene flakes is considered. The methods of sequential production of functional materials from graphene flakes of different volumetric concentration using epoxy polymers, as well as the addition of monodisperse microspheres are described. Data are given on the measurement of the contact angle and thermal conductivity of these nanocomposites with respect to the creation of thermal interface materials for cooling devices of electronics, optoelectronics and power engineering.

1 Introduction

Graphene research has attracted increasing interest from researchers in physics, materials science, chemistry, etc. Graphene consists of an unusual single-layer honeycomb lattice of C atoms, which differs from all other materials of this element. More specifically, the C atoms of graphene are sp^2-bonded and densely packed in a crystal lattice. This closely resembles the basic building block found in graphitic materials of different dimensionalities. It has been demonstrated that graphene exhibits some special characteristics, such as a high electron mobility of up to 230 000 $cm^2/V{\cdot}s$ at low temperatures, an exceptional thermal conductivity approaching 5000 $W/m{\cdot}K$ and superior mechanical properties with a large Young's modulus, exceeding 1.0 TPa the latter being comparable to those of single-walled nanotubes and even diamonds. With these intriguing physical and chemical properties, it is not surprising that graphene has been intensively investigated to exploit its functions as a promising material for future applications.

Graphene nanocomposites, in which graphene flakes are dispersed in a matrix (either polymer, ceramic or metal), have found numerous applications (energy, electronics and optoelectronics) that demand low weight with good mechanical as well as thermal performance [1-6]. The performance of composites is often limited by poor coupling between graphene flakes and matrix. For example, thermal transport through a composite can be dominated by thermal resistance at the interface that results primarily from phonon scattering due to mismatched phonon spectra in the two phases (graphene flakes and matrix) and from interfacial defects. The resistance is further increased by process-induced interfacial irregularities such as voids and interfacial stress, which result from graphene flakes surface roughness, poor adhesion between graphene flakes and matrix and dissimilarities in thermal expansion coefficient between graphene flakes and the surrounding matrix. Aside from the possibility of improving the overall mechanical properties of the composite, improved graphene flakes – matrix coupling would reduce interfacial thermal resistance and enhance heat conduction. Such an enhancement would have positive implications for a variety of applications, ranging from aircraft skins to lightweight heat exchangers.

In many cases, when discussing graphene thermal applications, we use the term graphene even when the actual material consists of a mixture of single layer graphene (SLG), bilayer graphene (BLG), and few-layer graphene (FLG). The latter is because for thermal applications the difference between SLG and FLG is not as important as for electronic applications. It is sometimes difficult to distinguish between FLG and graphite films or between FLG and graphite nano-platelets (GnP) or graphite nanoflakes (GnF) used in composite materials. The definition of SLG — a single atomic plane of sp^2-bound carbon — is strict.

It is important to note the possibility of using such hybrid nanocomposites as functional energy materials, for example, for lithium-ion batteries or materials with phase transitions. Development of high-power-density batteries, e.g., Li-ion batteries, enabled progress in mobile communications, consumer electronics, and

[a] Corresponding author: asdmitriev@mail.ru

© The Authors, published by EDP Sciences. This is an open access article distributed under the terms of the Creative Commons Attribution License 4.0 (http://creativecommons.org/licenses/by/4.0/).

 https://doi.org/10.1051/matecconf/201820704002

automotive industries [7,8]. Temperature rise beyond the normal operating range negatively affects Li-ion battery performance. If overheated, the battery can suffer thermal runaway, cell rupture or explosion. A conventional approach for thermal management of high-power-density-ion battery packs is based on the utilization of thermal phase change materials (PCMs). They reduce the temperature rise in the battery due to the latent heat storing and phase changes over a small temperature range [9, 10]. In this work, we will demonstrate new types of hybrid nanocomposites based on graphene flakes and metallic monodisperse microspheres. We investigate contact angles on the surface of such composites and measure the heat conductivity of several types of hybrid nanocomposites.

2 Material and methods

In the work, hybrid functional materials were studied in relation to energy problems, made of metal microspheres, polymer (epoxy resin) and graphene flakes. The method of obtaining composite materials is as follows. Graphene flakes produced by Graphene Star Ltd were used to prepare a nanocolloid solution (nanofluid), for which graphene flakes of various volume were added to pure distilled water and stirring was conducted for 20-30 minutes in a rotating centrifuge. Graphene flakes with a size of 5-10 microns with 3-10 layers of graphene (few-layer graphene, FLG) were obtained. After that, the microspheres from Sn and Er_3Ni (production of the Moscow Power Engineering Institute) were placed in a Petri dish and a nanocolloid solution of graphene flakes was poured into it. After that, the Petri dish was placed on the heater and the water from it was evaporated at different temperatures (from 60 to 100 °C). After evaporation of the water, the microspheres covered with graphene flakes were heated to a temperature of 250 ^{0}C for 1-1,5 hours, which led to the effect of quenching the microspheres with graphene flakes on the surface. A general view of the microspheres covered with graphene flakes after drying and the quenching effect is shown in Figure 1. It is not difficult to see that the microspheres were very homogeneous, with a good coating of graphene flakes. At the same time, the adhesion value of graphene flakes to metal microspheres was very high, which made it possible to mix microspheres without destroying the graphene coating.

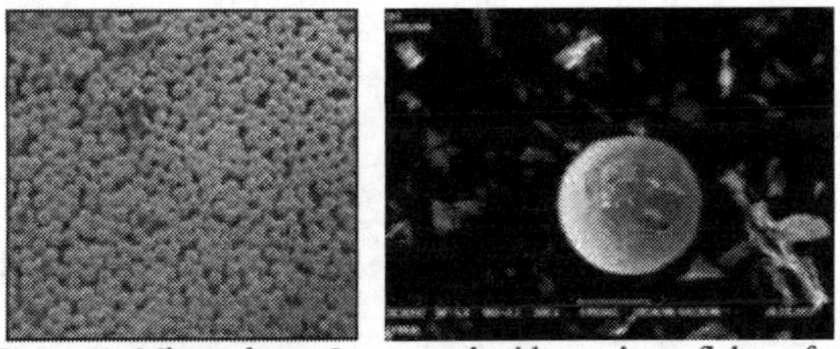

Figure 1. Microspheres Sn covered with graphene flakes after drying and the quenching effect

The structure of graphene nanocomposites is shown in the figure 2. Inside the composite, the metal microspheres are regularly located, the thickness of the coating of the microspheres by graphene flakes depends on their volume fraction, but does not exceed 20-25 μm.

Figure 2. Structure of graphene nanocomposites (SEM)

In addition to hybrid nanocomposites, tablets were made from graphene flakes of various thicknesses at various pressures. The production takes place in two stages: 1) the choice of a certain volume of graphene powder, on which the thickness of the obtained samples depends; 2) loading the selected volume of powder into the press for further pressing. The choice of a certain volume of powder is carried out by a conventional 20 ml medical syringe. To do this, the end of the syringe is carefully cut along the first line of the calibration. As a result, a large hole is obtained equal to the diameter of the syringe, with which it is already possible to easily obtain the required volume of graphene powder. The collected volume of graphene is pushed out by the plunger of the syringe into the mold, which is closed with a plug on top. Then, the press-fitting parameters are selected on the control panel of the press and then the process is started. At the end of the process, a cap is opened and a mold with a received graphene tablet is lifted using the control buttons. For this work, samples of graphene tablets were made at various press parameters and various powder volumes.

The measurement of quantity was carried out in volumetric units with the help of a syringe, which allowed to measure a certain amount and was convenient for loading the metered powder into the mold. Since this press is a highly specialized equipment, intended mainly for metallographic research, it constantly heats the material loaded in the mold. Therefore, the lowest possible temperature of 150 ^{0}C was used for manufacturing. As it was said above, 3 samples were prepared for each pressure value with powder volumes equal to 2, 5 and 10 ml respectively.

Different values of pressure and volume gave different thicknesses of the obtained samples (Figure 3).

In some experiments, isolated flakes were homogeneously dispersed in epoxy matrix and almost no large agglomerates or bundles were observed. The strong interfacial adhesion between oxygenic groups on the of GnF and epoxy molecular chains led to high embedding and tight binding of GnF in the matrix. For this reason, these wrinkled and isolated GNPs effectively prevents direct contact between graphite nanoparticles in the composite, which give advantages for forming a large microcapacitor network and suppressing the leakage current.

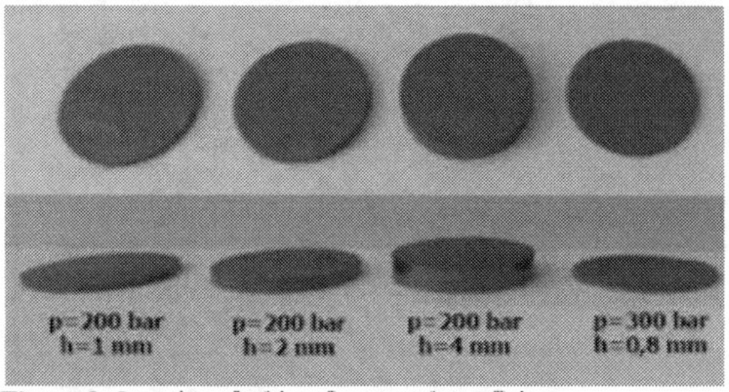

Figure 3. Samples of tablets from graphene flakes

3 Results and discussion

First, we studied the dynamic contact angles and evaporation rates of drops of distilled water on graphene tablets, which were obtained by pressing at various pressures (200 and 300 bar) of graphene flakes (Figure 4). Various properties of hybrid nanocomposites based on microspheres and graphene were investigated. In some cases, microspheres coated with graphene were placed in a polymer matrix. For the experiments, tablets were prepared in the form shown in Figure 3. For all types of samples, contact angles and thermal conductivity were studied. To determine the angle of wetting of drops on the surface of samples of a hybrid nanocomposite, Kruss EasyDrop measuring system was used with special heater and temperature control system.

Investigations of the electrical resistance of various samples were carried out under pressure variations during their preparation. Table 1 shows the results of these measurements.

Table 1. Electrical conductivity of graphene tablets

#Tablets	Thicknes s (mm)	Specific electrical resistance (Ohm / m)
P=200 bar	4	1375
P=300 bar	3.8	789.47

With an increase in pressure by 50%, the resistance dropped to 57%. This result is not accidental. As the pressure increases, the number of bonds for the passage of current along the sample increases. Thus, it can be argued that denser samples have a higher electrical conductivity.

We also investigated the rates of evaporation of droplets of distilled water on the surface of graphene tablets at different temperatures (Figure 5). To heat the substrates, the ITO was used, the measurements were made by a non-contact infrared thermometer and thermocouples.

Investigation of thermal conductivity of graphene hybrid nanocomposites includes experiments on the measurement of thermal conductivity by two methods: the method of thermal resistance of a layer and the method of laser flash (LFA-laser flash analysis) - Figure 6. The laser flash method is a more accurate method, so the results of these methods will be compared with each other. Measurements of the thermal conductivity of hybrid nanocomposites were carried out for dilute samples.

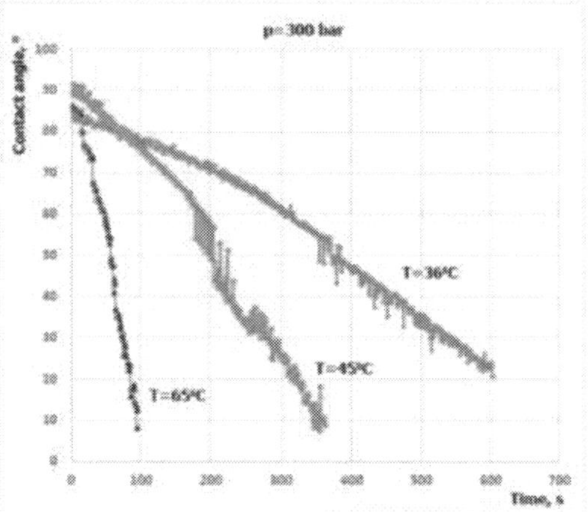

Figure 4. Changing the contact angle with time

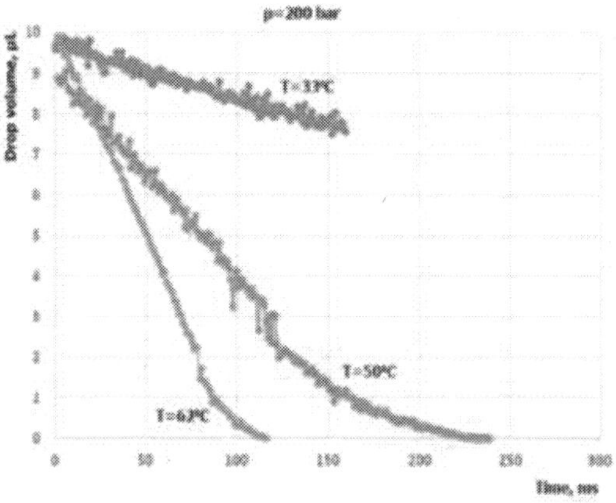

Figure 5. Evaporation rate of distilled water droplets on the graphene tablet surface

In the first series of experiments, samples based on a freely lying layer of microspheres coated with graphene flakes were investigated. In the second series of measurements, the thermal conductivity of densely packed microspheres coated with multilayer graphene was studied (Figure 2). It is important to note that the dependence of thermal conductivity on the volume fraction of graphene flakes has a sharp increase with a volume fraction of about 2.5-3% (Figure 7). Thus, the graphene layers near the microspheres play an important role in the process of heat transfer - with increasing amounts of graphene flakes, thermal conductivity sharply increases. This fact can be used to control heat transfer in various nanocomposites, in particular, used as thermal interface materials. The results of the thermal conductivity show that there is an optimal volume fraction of graphene flakes in the range 2,5 - 10%. For large values of the volume fraction, the thermal conductivity practically does not increase.

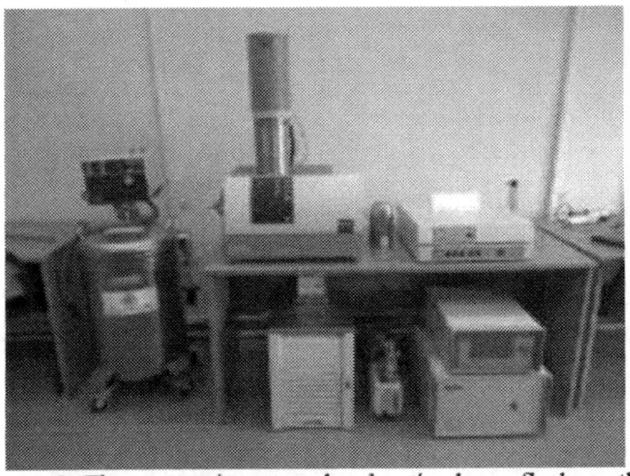

Figure 6. The measuring complex by the laser flash method NETZSCH LFA 457 MicroFlash

Apparently, physically this means that the observed percolation in this range of the mating fraction and the further increase in the amount of graphene flakes does not affect the mechanism of phonon heat transfer. The following series of experiments were carried out on the samples shown in Figure 8.

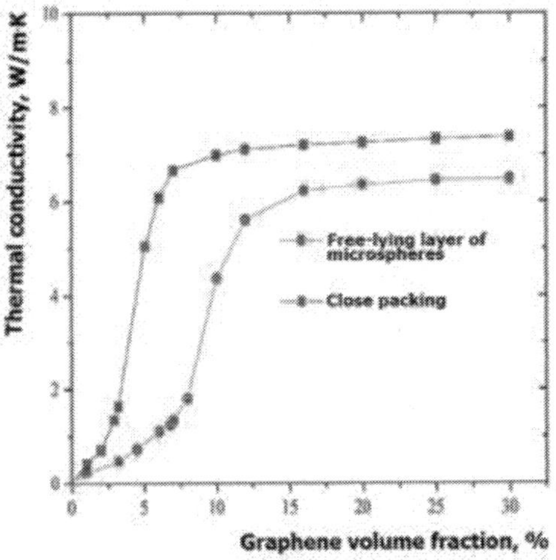

Figure 7. Thermal conductivity of microspheres coated graphene flakes

These samples were hybrid nanocomposites - the inclusion of graphene flakes in the polymer matrix of the epoxy, as well as the hybrid nanocomposite microspheres + graphene flakes inside the epoxy matrix. The dependences of the thermal conductivity of such hybrid nanocomposites are presented in Figure 8. In addition, the thermal conductivity of a polymer matrix (epoxy resin) is shown here.

With increasing temperature, the thermal conductivity of all samples decreases, which is associated with an increase in phonon scattering at the boundaries of graphene flakes. It is important to note that the use of graphene coated microspheres greatly increases the thermal conductivity. This indicates the possibility of using hybrid nanocomposites with microspheres as functional energy materials.

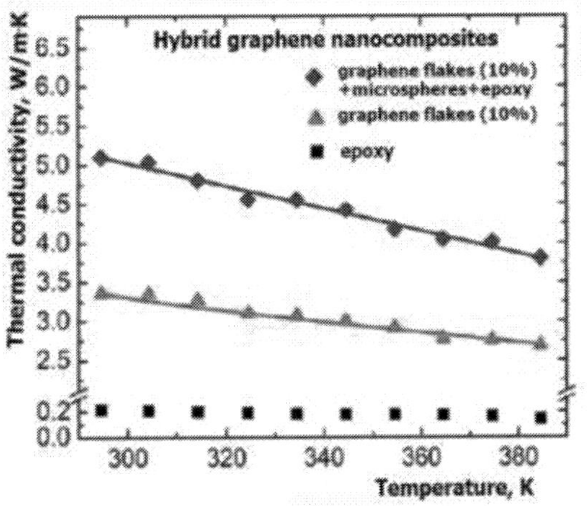

Figure 8. Thermal conductivity of hybrid nanocomposites

Conclusions

In this paper we presented the results of the properties of new hybrid nanocomposites based on graphene flakes, metal microspheres and epoxy polymers. The obtained data on the contact angles on the surface of such composites and their thermal conductivity at room temperature make it possible to draw conclusions about the potential use of these materials as functional energy materials (for example, as the thermal interface materials). In particular, for practical applications of graphene nanocomposites are outlined in an example using thermal phase change materials for thermal management of high-power battery packs.

Acknowledgments

The authors thank S Alekseev (Graphene Star Ltd, UK) for making graphene flakes. This work was supported by the Russian Science Foundation (project No. 17-19-01757).

References

[1]. J. Renteria, D. Nika and A. Balandin, Appl. Sci. **4**, 525-547 (2014)

[2]. A. Bar-Cohen, K. Matin and S. Narumanchi, J. Electronic Packaging **137** 040803-1 (2015)

[3]. A. Dmitriev, *Introduction to nanothermophysics*, (BINOM Press, Moscow, 2015)

[4]. A. Roy, B Farmer, V Varshney et al., ACS Appl. Mater. Interfaces **4** (2) 545–563 (2012)

[5]. S A Alekseev, A S Dmitriev, P G Makarov and I A Mikhailova, Journal of Physics: Conf. Series. **891** 012361 (2017)

[6]. F. Gardea, M. Naraghi and D. Lagoudas, ACS Appl. Mater. Interfaces **6** (2) 1061−1072 (2014).

[7]. A. Balandin, IEEE Spectr. 29 (2009) 35–39.

[8]. A.S. Dmitriev, A. R. Valeev, J. Phys.: Conf. Ser. **891** 012359 (2017)

[9]. F.Agyenim, N. Hewitt, P. Eames and S. Mervyn, Renew Sustain. Energy Rev. **14** 615–628 (2010)

[10]. V. Goyal and A Balandin, Appl. Phys. Lett. **100** 073113 (2012)

Laser Micromilling Technology as a Key for Rapid Prototyping SMD ceramic MEMS devices

Nikolay Samotaev[1,a], Konstantin Oblov[1] and Anastasia Ivanova[1]

[1]*National Research Nuclear University MEPhI (Moscow Engineering Physics Institute), 115409, Kasirskoe highway 31, Moscow, Russia*

Abstract. The flexible laser micromilling technology for ceramic MEMS producing of microhotplate in the surface mounted device (SMD) package for the metal oxide (MOX) gas sensors is describing. There are discusses technological and economic aspects of small-scale production of gas MOX sensors in comparison with classical clean room technologies using for mass production MEMS devices. The main technical factors affecting on using MOX sensors in various applications are presented. Current results demonstrate that using described technology possible to manufacturing all parts of MOX gas sensor in the SMD form-factor SOT-23 package type.

1 Introduction

The modern MOX sensor is a combination of several factors – a microhotplate (responsible for power consumption and temperature operation mode), MOX gas sensitive layer (responsible for sensitivity to target gases) and a package (responsible for the possible applications of MOX gas sensor) [1]. Last decades MOX gas sensors development there is a concentration of efforts, in the field of technologies allowing obtaining a cheap product for mass applications in the manner of indoor air quality (IAQ) monitoring. Similar solutions were found by the leading manufacturers of MOX sensors (Figaro, Sensirion, Bosch, AMS, SGX, etc.) [2-6]. Technological solutions that make it possible to obtain a sensor with satisfactory characteristics for such application consist in the use of a combination of silicon MEMS microhotplate [2-6] and plastic [2,3], metal-plastic [4] or metal-ceramic [5, 6] SMD package. But what need to do scientific groups which, at the moment there is no market or market still small for the return of investment in silicon MEMS technology and the production of specialized plastic or ceramic SMD package? In that case need using technology process close to 3D prototype philosophy – rapid, simple and cheap with ratio a single sample to total production cost.

After several experiments with wide spectres ceramics MEMS technologies and materials (ZrO_2 [7], Al_2O_3 [8], LTCC ceramics [9]) we find more flexible technology responsible to goals of 3D prototype philosophy. As a technology for manufacturing ceramic MEMS MOX sensor and SMD package for one, we propose to use combination of laser micromilling monolithic Al_2O_3 commercially available ceramics and screen-print/jet (aerosol) printing technologies to form platinum metallization on MEMS and package.

2 Experiment

The equipment involved in the described laser micromilling technology does not need clean rooms and is widely presented on the market. The software for technology is simple and accessible at the student level (3D printing software with open access). The only specialized product of the proposed technology is software for adaptation of machine vision to obtain the minimum possible size of the MEMS structure of the microheater and the deposition of the MOX gas sensitive layer and platinum metallization, as well as the translation program of the MEMS 3D models of 2D topology of metallization to a 4-axis laser micromilling facilities for automatic production (Photo of tech process present on fig.1).

Figure 1. Photos of micromilling by 4-axis CNC 1,064 µm wavelength fiber laser tool 250 µm thick Al_2O_3 substrate during fabrication microhotplates for MOX gas sensor present on fig. 4 and 5.

[a] Corresponding author: nnsamotaev@mephi.ru

© The Authors, published by EDP Sciences. This is an open access article distributed under the terms of the Creative Commons Attribution License 4.0 (http://creativecommons.org/licenses/by/4.0/).

Technological steps for our experimental works are follows:
- 3D modeling of MEMS microhotplate, top and bottom part of the package (Fig. 2a, 2b) in *.stl format (Autodesk Inventor or SolidWorks program) and 2D modeling topology of MEMS metallization in *.dxf format (AutoCAD program),
- Simulation of the MEMS microhotplate parameters in the COMSOL program for prediction approximate representations about the MOX sensor thermal characteristics (optional),

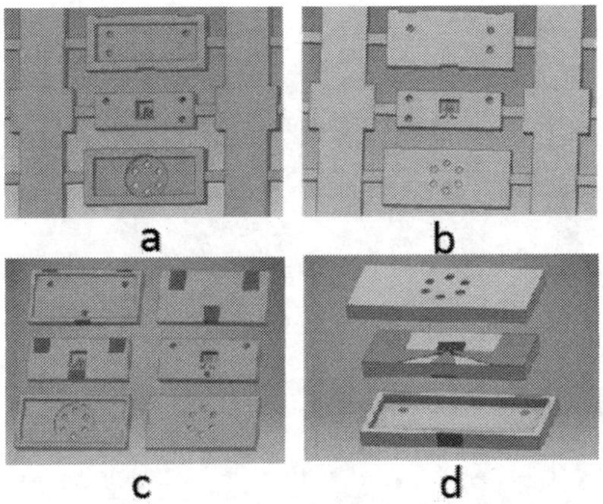

Figure 2. a) Bottom and b) Top view of 3D model of microhotplate and parts of SMD package as a task for laser micromilling. c) Scheme of metallization topology in 3D model. d) Scheme of assembling MEMS microhotplate in SMD package.

- Laser micromilling of monolithic ceramics on a 4-axis laser facility (Fig.1) by using 3D model of MEMS microhotplate, top and bottom part of the package (Fig 3),
- Deposition of platinum metallization using 2D model topology and technical annealing of metallization on MEMS according to specification on screen-print, jet or aerosol platinum materials (Fig. 2c),

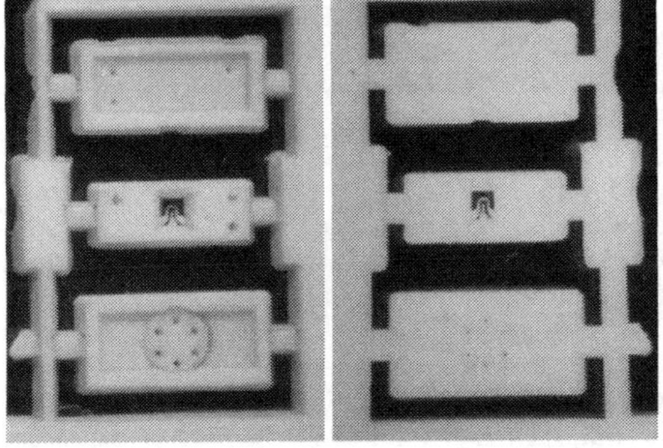

Figure 3. Photos of bottom (left) and top (right) views on 250 µm thick Al_2O_3 substrate with part of SOT-23 SMD package and microhotplate for MOX gas sensor after laser micromilling.

- Laser micromilling of metallization according 2D model (optional),
- Deposition and annealing of the MOX gas sensitive layer on the MEMS microhotplate,
- Assembling and conglutination by special glass single parts of sensor into a monolith package, like schematically represented for SOT-23 package on Fig. 2d.

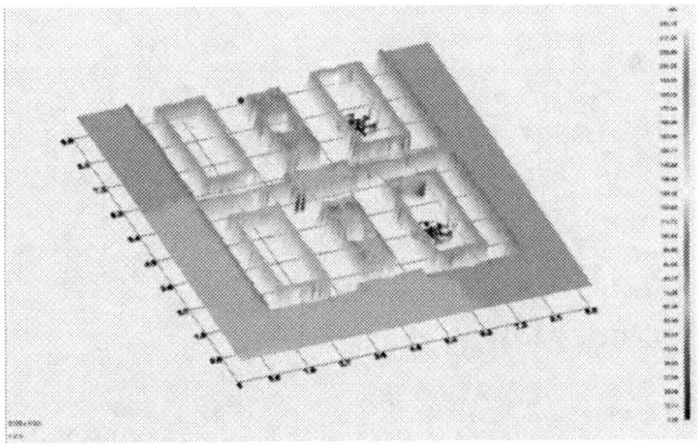

Figure 4. Dimensions parts of SMD package measured by optic noncontact profilometer. Standard SOT-23 packages dimension is 1,6 x 2,8 x1,0 mm.

Using present approach, experiments were carried out to fabricate a MEMS microhotplate, whose track width was 30 µm and a thickness of 20 µm (see Fig. 5). The power consumption of MEMS microhotplate at 450°C was approximately 250 mW (350 mW at ≈ 600°C – "burning" platinum temperature, see Fig. 6). Current results gives the prospect that manufacture MOX sensor in the SMD SOT-23 package type (max dissipating power by form-factor of package is 350 mW [10]) for surface mounting in a tape is available by using described technology.

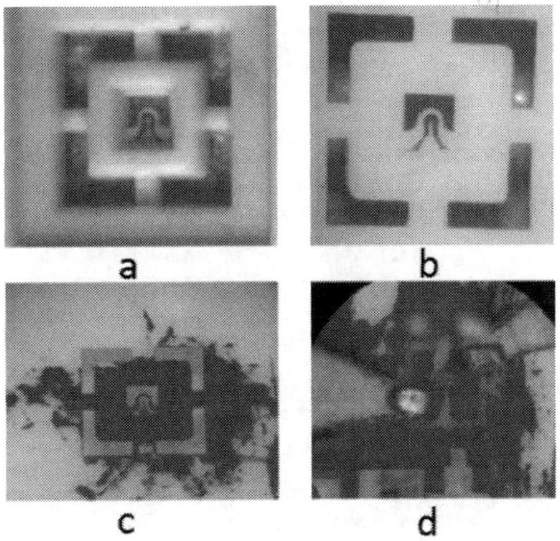

Figure 5. a) Bottom view of MEMS microhotpale after laser micromilling. Chip size 1,0 x 1,0 mm by border of internal frame. b) Top view of MEMS. Internal window size 0,3 x 0,3 mm. Central track width is 30 µm. c) Microheater track width 30 micron under cower Pt paste. d) Testing MEMS microhotplate under voltage load (pins). Hot area is ≈ 600°C.

 https://doi.org/10.1051/matecconf/201820704003

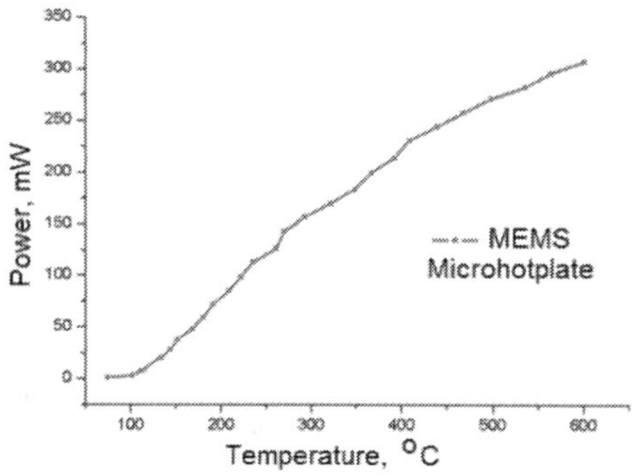

Figure 6. Power consumption of microhotplate fabricated by laser micromilling technology. Topology and SEM photo of MEMS microhotplate presents respectively on Fig. 7 and 8.

3 Discussion

Current experimental results gives prospects that manufacture MOX gas sensor in the SMD SOT-23 package type (max dissipating power by form-factor of package is 350 mW [10]) for surface mounting in a tape is available by using described above technology.

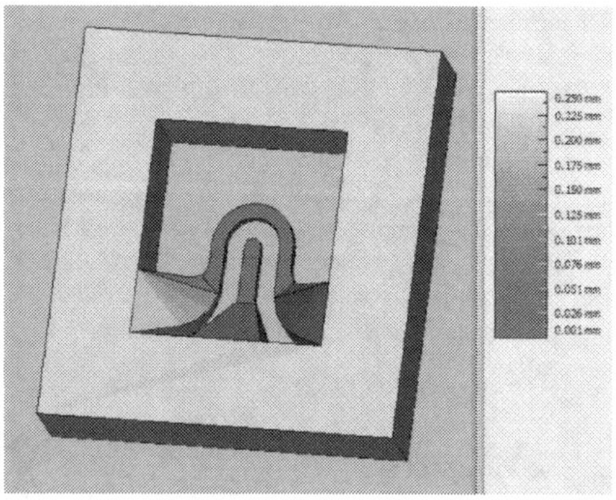

Figure 7. 3D model of MEMS microhotplate in *.stl format (Print Screen in Autodesk Inventor CAD program).

To control the accuracy of MEMS manufacturing, a scanning electron microscope was applied. On Fig.7 present 3D model of MEMS microhotplate as a print screen in Autodesk Inventor CAD program and for comparisons on Fig.8 present photo of the same MEMS microhotplate immediately after laser micromilling. From Fig.8 ensue that laser micromilling gives geometric dimensions withstand relatively well, opposite the roughness is the parameter on which depend from thickness.

The quality of the final MEMS structure strongly depends on its surface roughness. The minimal roughness will make the MEMS structure more durable, as well as allow for more qualitative metallization deposition processes on the chip surface. To obtain a minimum surface roughness, it is required to use a material whose thickness does not exceed the height of the obtained structure in order to avoid ablation of the volume of the material only in order to reduce its thickness. However, the cost of the final product depends heavily on the cost of the material, in our case, ceramic substrates Al_2O_3. The minimum cost have substrates with standard size. The standard sizes of the ceramic substrates being manufactured start at a thickness of 500 μm or higher, which, at a chip height of 250 μm, requires excess ablation of the material at a thickness of 250 μm. Operation of laser evaporation of excess material can be justified in the case of a high cost of manufacturing substrates of non-standard thickness and only in the case of a qualitatively selected laser micro-milling mode that does not lead to a strong increase in the surface roughness after removal of excess material thickness.

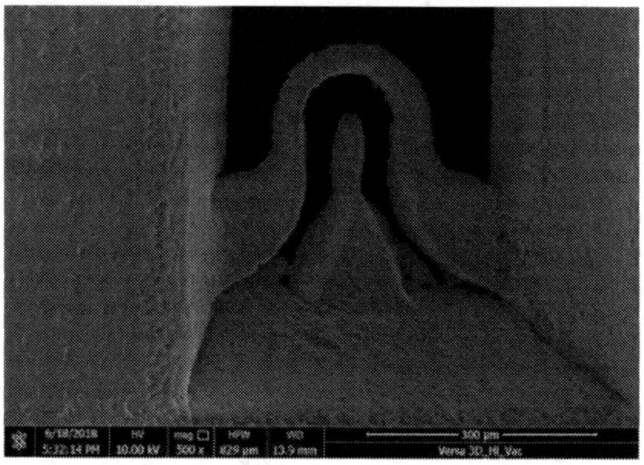

Figure 8. SEM photo MEMS microhotplate immediately after laser micromilling of 250 μm thick Al_2O_3 substrate using 3D model of MEMS microheater in *.stl format as a source CNC-code.

Laser radiation allows carrying out the process of Al_2O_3 ceramics ablation to a depth of 0.5 to 10μ in one pass. High radiation power and low speed of the laser beam accelerate the process of micro-milling, but lead to a higher surface roughness, the values of which can be more than 50 μm. Roughness is formed due to the fact that the intensity of laser radiation has a Gaussian distribution, which requires a very accurate selection of the laser beam overlapping region, when the entire material is engraved. On average, the surface roughness is 50 percent of the ablation depth per one pass of the laser beam. The minimum roughness was obtained by evaporation of the material in one pass at 1 μ and amounted to 2.3 μ.

The laser micro-milling process is a complex multifactor physical process, whose behaviour, depending on the type of material being processed, has a different character, which requires an empirical selection of the optimal surface treatment regime. Acceleration of the debugging process can be achieved by integrating the measuring equipment, able to produce a visual analysis of the surface obtained. After the ablation of each layer of material, such integration allows to analyze the resulting surface and correct the laser radiation regime by adjusting the radiation parameters or introducing additional cleaning passages aimed only at the surface roughness reducing. Suchlike machine learning is the task for specialized software developed within the framework of described technology.

Production time for laser micromilling of single MEMS structure was ~ 10 minutes, and 1 hour is necessary for group annealing of the MEMS metallization on substrate using muffle furnace. Our experience shows that, if technology is used by skilled personals, that gives fantastic design and manufacturing speed for ready-to-use microhotplates (about 2 hour)! The cost of materials (alumina substrate and platinum ink) for the manufacturing of single microhotplate with size of 1×1 mm is of about 0.50 Euro. The cost of a new set of equipment used in our experimental works is around 69k Euro.

4 Conclusion

We demonstrated that the application of laser micromilling technology for the fabrication of sensing MEMS devices is very prospective and assures important advantages in comparison with traditional expensive clean room technologies. These advantages are related first of all with a very considerable enlargement of the range of working temperatures (up to 800 ^{0}C) and of the range of technological treatment temperature of the sensor (up to almost 1000 ^{0}C compare with maximum 720 ^{0}C for silicon technology). This enlargement of temperature range results in the possibility to detect gases, not detectable with silicon based devices and to improve significantly the selectivity of gas sensors using temperature modulation regime of operation [10]. Another important result is the possibility to fabricate sensors operating under harsh environmental conditions including industrial and natural catastrophes using ceramic material stable for high temperatures and aggressive agents.

Acknowledgement

This work was supported by the Ministry Education and Science Russian Federation (Grant No. 14.584.21.0030 from 22.11.2017, unique identifier RFMEFI58417X0030) in frame of joint Russian-Korean project.

References

1. I. Simon, N. Bârsan, M. Bauer, U. Weimar, U. Sensors and Actuators, B, **73 (1)**, (2001).
2. https://www.sensirion.com/en/environmental-sensors/gas-sensors/
3. http://ams.com/eng/Products/Environmental-Sensors/Gas-Sensors/AS-MLV-P2
4. https://www.bosch-sensortec.com/bst/products/all_products/bme680
5. https://www.sgxsensortech.com/content/uploads/2014/08/1107_Datasheet-MiCS-2714.pdf
6. http://www.figarosensor.com/products/entry/tgs8100.html
7. A. Vasiliev, A. Lipilin, A. Mozalev, A.S. Lagutin, A. Pisliakov, N. Zaretskiy, N. Samotaev, A. Sokolov, Proc. of SPIE, **8066**, (2011).
8. A. Vasiliev, R. Pavelko, S. Gogish-Klushin, D. Kharitonov, O. Gogish-Klushina, A. Sokolov, A. Pisliakov, N. Samotaev, Sensors and Actuators, B, **132 (1)**, (2008).
9. A. Vasiliev, A. Pisliakov, A. Sokolov, N. Samotaev, S. Soloviev, K. Oblov, V. Guarnieri, L. Lorenzelli, J. Brunelli, A. Maglione, A. Lipilin, A Mozalev, A. Legin, Sensors and Actuators, B, **224**, (2016).
10. https://www.diodes.com/assets/Datasheets/ds18001.pdf
11. N. Samotaev, B. Podlepetsky, A. Vasiliev, A. Pisliakov, A. Sokolov, Automation and Remote Control, **74 (2)**, (2013).

Problems of sodium using in pulsating heat pipe made from fused silica

Radovan Nosek[1a], Tatiana Liptáková[2b], Libor Trško[3c], Zuzana Kolková[3d], Milan Malcho[1e] and Anna Kiijan[4f]

[1]*University of Žilina, Faculty of Mechanical Engineering, Department of Power Engineering, Univerzitná 8215/1, 01026 Žilina, Slovakia*
[2]*University of Žilina, Faculty of Mechanical Engineering, Department of Material Engineering, Univerzitná 8215/1, 01026 Žilina, Slovakia*
[3]*Research Centre of the University of Žilina, Univerzitná 8215/1, 01026 Žilina, Slovakia*
[4]*Silesian University of Technology, Gliwice, Poland*

Abstract. You Heat pipe is a high efficiency heat transfer element, depends on the evaporation, condensation and circulation of inside working fluid. The working fluid of a high temperature pulsating heat pipe is generally alkali metals, and sodium heat pipe can operate in range of 500-1100°C. In order to investigate terminal velocity of working fluid, the glass pulsating heat pipe was produced for experimental purposes. The experiment was carried out, in order to simulate real operating conditions in range of 500-1100°C. Sudden boiling of liquid sodium (b.p. = 883°C at 1 atm) inside the all quartz-made heat pipe results in high-temperature reaction of sodium vapour with the inner wall surface. The reaction became more aggressive with increasing vapour temperature and resulted in heat pipe explosion. The evaluation of damage character is analysed in this paper.

1 Introduction

Worldwide production of aluminium consumes about 3.5% of the total global electric power. During the aluminium production, 50% of the supplied energy is consumed by the chemical process, and 50% of the supplied energy is lost in form of heat. Heat losses are necessary to maintain a frozen side ledge to protect the side walls, so extra heat has to be wasted. As the energy costs account for 40% of the total production costs, it is clear that the heat recovery optimization offers a significant economical end environmental potential in the industry. In order to increase the energy efficiency of the process, it is necessary to significantly lower the heat losses dissipated by the furnace's external surface. Extracting the waste heat (capturing it in a heat transfer fluid) also requires adapted heat-exchanger designs. The technology based on the use of heat pipes, has been developed for utilization energy from the waste heat produced in the electrolytic process. The heat recovery system can control the ledge thickness and thermal energy can be utilized by other energy demanding processes and increase the aluminium production by 20% and still achieve thermal balance. With heat recovery system, 10-12% of the total energy consumption for the furnace can be recovered [1].

Heat pipe is very good thermal conductor and has the ability to transfer heat hundred times more than copper. The high and super high temperature heat pipes have emerged as interesting alternatives to conventional heat transfer technologies and have wide applications in space thermal control, solar storage power plant, nuclear thermal control, the constant industry stoves etc.[2].

The aim of the project was to develop an innovative heat to electricity converter without any moving parts, placed on the pulsating heat pipe (PHP). Device can generate electric power by means of the interaction of a moving fluid and a magnetic field. One of the tasks was to measure the velocity of working fluid in the glass heat pipe and then analyse the efficiency of converter.

2 Experimental measurements

The experimental setup for the energy utilization of waste heat consists of PHP, which is a two phase heat transfer device with outstanding effective thermal conductivity. It is a vacuum tight device consisting of an envelope (here made of quartz glass) and flowing fluid (liquid sodium), which transfers heat under isothermal conditions at vaporization.

Sodium is a very reactive element because of his electron configuration in the valence shell where it is a single electron. That it readily donates, creating a positively charged ion - the Na+ cation. To achieve nonreactive condition, the pulsating heat pipe was cleaned with isopropyl alcohol 3 times. Subsequently, the pipe was heated up to 120°C, evacuated with turbomolecular pump and filled with argon (Purity 6). This procedure was repeated 3 times, in order to eliminate concentration of oxygen and impurities.

After cleaning the system and filled with argon, the pulsating heat pipe was evacuated for seven days and then filled with cleaned sodium. In order to prevent the

[a] Corresponding author: [a]radovan.nosek@fstroj.uniza.sk, [b]tatiana.liptakova@fstroj.uniza.sk, [c]libor.trsko@rc.uniza.sk, [d]zuzana.kolkova@rc.uniza.sk, [e]milan.malcho@fstroj.uniza.sk, [f]anna.kiljan@polsl.pl

© The Authors, published by EDP Sciences. This is an open access article distributed under the terms of the Creative Commons Attribution License 4.0 (http://creativecommons.org/licenses/by/4.0/).

solidification of Na in the system, the entire heat pipe and also filling system were heated up to 120°C.

The evacuation was stopped in this phase, valve to the vacuum pump was closed and liquid sodium has been released through the filling system to the heat pipe. After filling of PHP, the silicon tube was closed by two tubing clamps. The experimental setup is shown on Fig. 1.

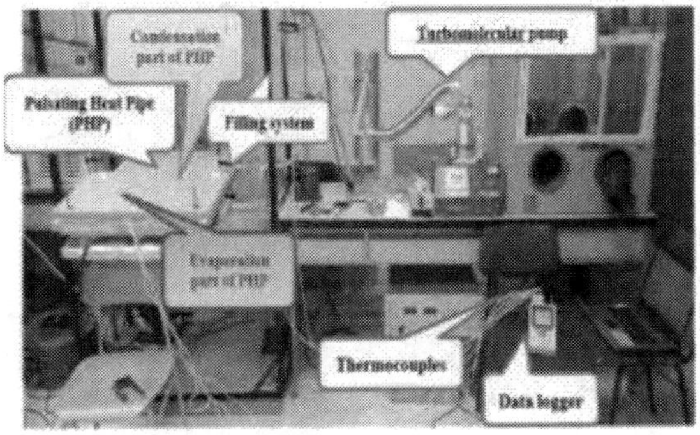

Figure 1. Experimental setup

3 Evaluation of the failure reason

Sodium was used in experimental apparatus as transport medium because of its convenient physical properties (conductivity, low melting and evaporating temperature, low weight etc.). It was supposed that its high chemical aggressiveness will be restricted by high vacuum and keeping of constant working temperature [3]. But in two hours the experimental device exploded in the evaporation part. It can be seen (Fig. 2.) that character of compounds inside of PHP is in various parts different and is changed with temperature. The temperature profiles of the PHP are presented in Fig. 3 and 4. To identified composition of black substance on the pipe walls in the part of explosion (evaporation part of black colour) microscopic and XPS methods were applied, in order to analyse the factors causing the explosion of heat pipe.

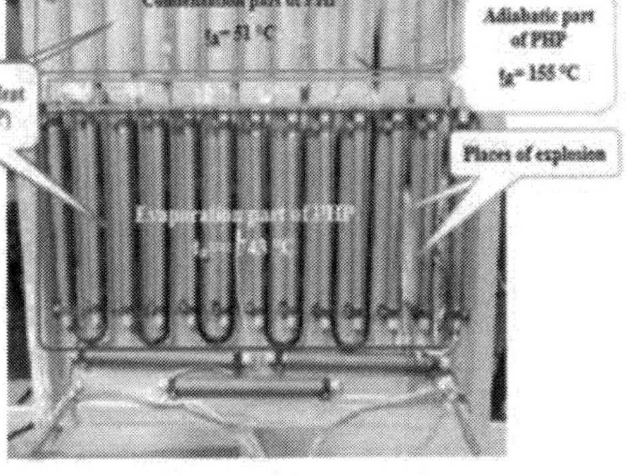

Figure 2. Detail of damaged pulsating heat pipe

The XPS analysis shows the presence of the pure carbon in high concentration on the quartz pipe wall (Tab. 1).

Table 1. Composition of the black products adherent on the pipe walls

Element	Weight %	Atomic %
Carbon C	81.11	87.79
Oxygen O	9.9	8.04
Silicone Si	9.0	4.17

According to the analysis, the explosion was caused by the high temperature followed devitrification of the fused silica pipe by sodium [4] and heterogeneity of reactions inside of experimental system. Increasing of diffusion through the weakening walls and created failures were followed by chemical reaction of sodium with components of ambient atmosphere.

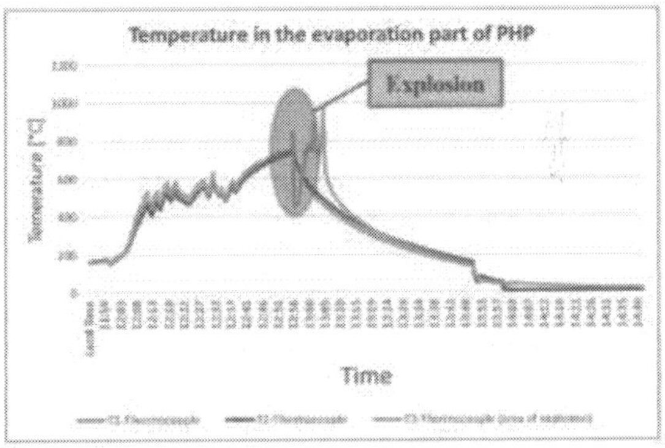

Figure 3. Profile of temperature in the evaporation of pulsating heat pipe

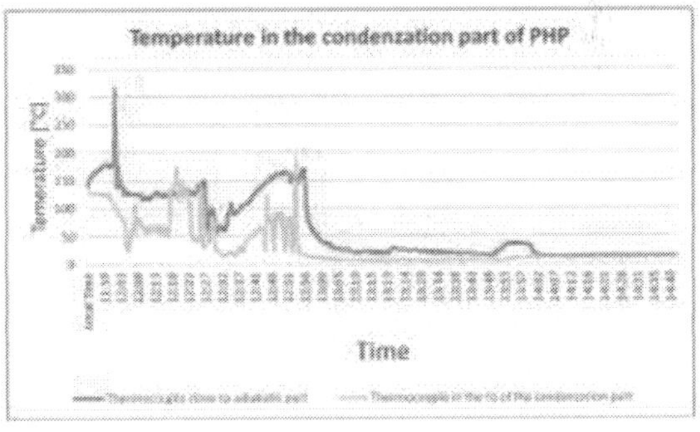

Figure 4. Profile of temperature in the condensation part of pulsating heat pipe

4 Summary and discussion

By thermal influence, the devitrified areas tend to weaken the PHP mechanically and change its properties and compatibility. Also, these devitrified areas which provide grain boundaries in the previously amorphous structure encourage diffusion by forming easy diffusion paths for environmental impurities to molten sodium [5]. The mechanism of devitrification as enhanced by sodium involves diffusion of sodium into the fused quartz. The sodium is thought to diffuse quickly through the open structure of the silica via strings of interconnected interstices [6]. The sodium sets up nucleation sites for crystallization. One can view a piece of a fractured glass pipe from a wet oxidation that has failed due to devitrification. It will be observed that the heavily devitrified region is inside of the pipe wall too. This is because water acts to greatly enhance the rate of devitrification [7].

The character of the pipe walls damage at various temperatures in the experimental apparatus as shown in Fig. 5 and is in the agreement with the published knowledge. As it can be seen the great differences are observed in intensity and nature of silica glass damage of the PHP. The Fig. 5a shows the expressive traces of surface etching as well as cracks caused by interaction of sodium with fused glass at high temperature and interactions with environment in the evaporation part. In the transit part, the temperature rapidly dropped and destruction of pipe was lower as it is presented in Fig. 5b where only slight attack of pipe surface was noticed. It is interesting that in the condensation part in spite of lower temperature the pipe surface damage was more intensive (Fig. 5c) which is probably due to different ambient conditions. In Fig. 5d, the original pipe surface is shown for comparison of various characters and intensity of attack in the heat pipe.

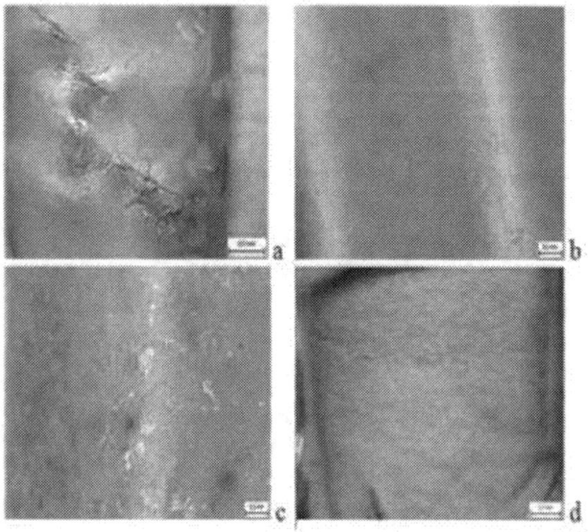

Figure 5. Character of fuse quartz pipe walls damage a) evaporation zone, b) transit zone, c) condensation zone, d) original pipe

The intensity of wall damage was affected by temperature very strong. The dimensions and quantity of disturb localities subjected reaction intensity as well as various characters of reaction products. Reactions of infused reactants with boiling sodium are very inhomogeneous as it can be seen from character of products especially in evaporation part (Fig. 6).

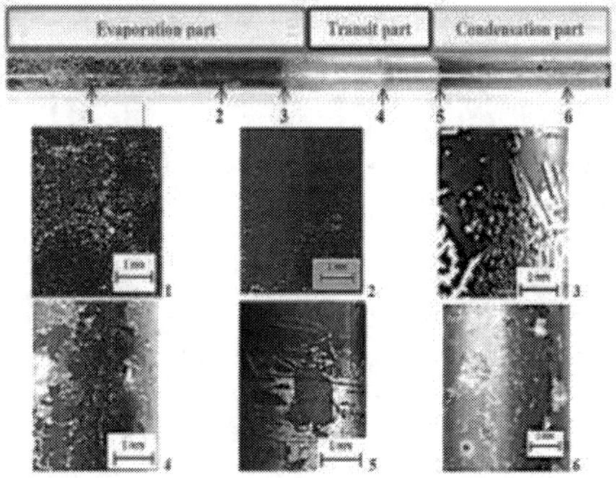

Figure 6. Character of the products in the fused glass pipe in different parts of experimental equipment differing in temperature

This fact made possibility to evoke different tensions and chemical composition in various localities of evaporation part and consequent explosion of the quartz pipe. The differences in reaction products are identified in transit and condensation parts. The variabilities of reactions become evident by various colour of reaction products [8].

This can be affected by different ambient and diffusion conditions in the condensation part from atmosphere since the transit part of the PHP is insulated and fixed in a solid holder.

5 Conclusions

- The visualization process of the liquid sodium flow in the PHP is not possible to investigate in fused silica tubes because the sodium can penetrate inside of pipe in very short time at temperature higher than 700°C and destroy it.
- The destruction of the fused silica pipe is influenced by penetration of sodium to pipe walls as well as with reaction compounds penetrated inside to tube through created defects.
- As the result can be concluded, that visualization in this form is not applicable and using of the fused silica pipe is possible only in the condensation part.

Acknowledgement

This article is supported by the financial assistance of the project KEGA No. 042ŽU-4/2016" Cooling by physical and chemical processes" and Visegrad Scholarship Fund 2017/18.

References

1. R. Nosek, et al., *Utilization of waste heat from aluminium melting furnace* (AIP Conference Proceedings, Volume **1917**, PIM, 2017)

2. Q.Wei, *Progress Works of High and super high temperature heat pipes* (Developments in Heat Transfer, Chapter 25, ISBN 978-953-307-569-3, pp. 504-512, 2011)
3. J.Krätsmár-Šmogrovič et al., *Všeobecná a anorganická chémia (in Slovak)* (II. edition, Osveta ISBN 8080632458, 2007)
4. K. B. Richter, *Prevention of sodium diffusion through furnace tube quartz* (Lehigh University-Lehigh Preserve Theses and Dissertations, 1984)
5. P.F. Schmidt, *Contamination-Free High Temperature Treatment of Silicon or other materials* (J. Electrochem. Soc.,Vol. **130**, 1983)
6. S. J. Rothman, *Diffusion of Alkali Ions in Vitreous* (Silica,11 J. Am. Ceram. Soc.T, Vol. **65**, 1982)
7. W. Heng-yu and W. Jia-fan, *A study on the devitrification of silica glass tube the diffusion furnace* (Silicate Journal, Vol. **9**, 1981)
8. K. D. Kihm, D. M. Pratt and A. D. Swanson, *Condensation of sodium and high-temperature reaction with quartz pore inner surface* (J. Heat Transfer 131(8), 2009)

MATEC Web of Conferences **207**, 04005 (2018) https://doi.org/10.1051/matecconf/201820704005
ICMMPM 2018

Welding Process Optimization of WELDOX960 High Strength Steel

Min Hu

Jiangsu province Xuzhou technician Institute, Xuzhou Jiangsu 221151,China

Abstract: This paper studies WELDOX960 high strength steel, analysis of the welding ability of WELDOX960 high strength steel. Analyze the weld ability of WELDOX960 high-strength steel materials, and study the influence of process parameters such as welding current, welding voltage, and welding speed on penetration depth and weld width in the automated welding process. Through this test, the welding process is optimized to ensure the weld quality. The results show that WELDOX960 high-strength steel adopts multi-layer and multi-pass welding to form better welds.

INTRODUCTION

With the continuous development of large-tonnage crane products, the use of high-strength steel plate is large, and the strength of the jib is also improved. The welding process is the main process for manufacturing the metal structure of the hoisting machinery. The welding quality and assembly process are directly related to the quality of the hoisting machinery. As the main force part of heavy equipment, boom is mainly made of imported high strength steel, and the welding technology of boom is the key and difficult point in manufacturing process. According to market research, this kind of low alloy high strength steel produced by Sweden is usually made of large boom. This kind of low alloy high strength steel is usually achieved by welding process. Although some mature welding techniques are adopted in the process of production, the same kind of welding process is not exactly the same in different environment, and the welding workers with different construction environment of the same kind of material are welded. Art is also different. Compared with the traditional alloy structural steel, the yield strength of WELDOX960 high strength steel is greater than that of 960MPa. It has high strength and good plasticity, and has good advantages in welding and mechanical properties. However, experts and scholars have proved through practice that under the steel alloying principle, the strength of low-alloyed steels is increased, and the carbon content and alloying elements in the chemical composition are also increased. In the welding robot welding process, brittleness of the welded joints is likely to occur. Welding defects, such as welding cracks, destroy the structure and performance of welded structural parts. Therefore, it is particularly important to study WELDOX960 high-strength steel weld properties, improve the weld ability of low-alloy high-strength steel, and reduce the welding defects.

1 WELDOX960 HIGH STRENGTH STEEL WELD ABILITY

Low alloy quenched and tempered has good weld ability and is close to low carbon steel. With the increase of alloy elements in the steel, the strength grade is increased, the hardenability is great and the

© The Authors, published by EDP Sciences. This is an open access article distributed under the terms of the Creative Commons Attribution License 4.0 (http://creativecommons.org/licenses/by/4.0/).

weld ability of steel is becoming worse and worse. The problems mainly include two aspects, one is the crack sensitivity, and the carbon content of the secondary high strength steel is lower, generally below. The high strength and good mechanized performance are improved by adding a variety of improvement on the low carbon base. Hardenability alloying elements and quenched and tempered heat treatment obtained low carbon lath martensite and some lower bainitic structure with high strength and toughness. Low strength low-alloy high-strength steel, such as grade I, is the mechanical property of welding heat affected area due to the low content of alloy element in steel. How to ensure the welding quality of welded components and obtain the proper strength and good toughness is the key of low alloy high strength steel welding. Therefore, in the process of welding low-alloy high-strength steel, it is necessary to improve the strength and impact toughness of weld metal and welding heat affected zone while ensuring the quality of weld. Due to the excellent weld ability and mechanical properties of high strength steel, the steel plate thickness between mm and mm can be applied to all conventional welding methods[1]. Such as gas solid wire, flux-cored wire welding, manual arc welding, self-protection flux cored wire, submerged arc welding, can be used in practice, the selection of welding methods are often based on the structure, construction conditions, welding quality, efficiency and the level of cost accounting and other comprehensive consideration, often use the selection principles for under the lower economic cost accounting, labor condition security friendly, can efficiently complete the high quality of weld. In these welding methods, gas shielded welding is the first choice. Reasonable selection of welding material can guarantee the weld composition and comprehensive performance, but it is more important to choose the proper welding process specification. This paper adopts KUKA welding robot to weld WELDOX960 high strength steel plate butt weld, optimize welding process and ensure qualified weld joint[2].

2 WELDING MATERIAL

In this test, the thickness of the WELDOX960 low alloy high strength steel is three specifications: 12mm, 16mm and 20mm. Table1.1 the mechanical performance of weldox960 and Table1.2 the chemical compositions of weldox960(wt%)

Table1.1 the mechanical performance of weldox960

Material、thickness	σ_b	σ_s	δ	AKV(J)10			RT\180°
	(MPa)	(MPa)	(%)	0°	−20°	−40°	R=3t
Weldox960 t=20mm	980/1150	960	12	35	30	27	Perfect

Table1.2 the chemical compositions of weldox960(wt%)

Material、thickness	C	S_i	M_n	P	S	N_b	V	N_i
Weldox 960 t=20mm	0.20	0.50	1.6	0.02	0.01	0.04	0.06	2.00
	C_u	C_r	M_0	T_i	Al	B	N	
	0.30	0.70	0.70	0.04	0.018	0.005	0.015	

The KUKA CO2 gas shielded welding robot was welded with gas content of 20%CO2+80%Ar and gas pressure of 15~20Mpa. The choice of welding material is the solid core wire produced by the Swedish company of diameter 1.2mm, the type is E-FK1000 welding wire. The chemical composition and mechanical properties are shown in table 1.3

Table1.3 Chemical composition of wire

Welding wire	C	S_i	M_n	P	S	C_r	M_0	N_i
E-FK1000 ϕ1.2	0.13	0.81	1.73	0.012	0.024	0.024	0.59	2.27

3 WELDING TECHNOLOGY

In view of the characteristics of the welded material, the actual structure or the form of the joint, the welding process should be used to weld bottom welding, fill welding, cover welding and angle welding when welding with different welding methods. In this paper, the process measures include preheating before welding, cooling after welding and multi-layer multi-channel welding. Proper selection of process parameters is the key to ensure welding quality of joints. In the design of welding process, the arc voltage, welding current and welding speed three welding process parameters is the key factor affecting the quality of welding process parameters, the reasonable selection directly affects the size of the welding line energy, in order to meet the design requirements of welding joint, should be within a certain range, adjust the arc voltage, welding current and welding speed. Whether the welding current size is suitable or not will directly affect the final quality of welding. In the design of welding process, the arc voltage, welding current and welding speed three welding process parameters is the key factor affecting the quality of welding process parameters, the reasonable selection directly affects the size of the welding line energy, in order to meet the design requirements of welding joint, should be within a certain range, adjust the arc voltage, welding current and welding speed[3]. Whether the welding current size is suitable or not will directly affect the final quality of welding. If the welding current is too large, though it can increase the penetration depth, it is easy to have the defects such as nibble welding tumor, and it is difficult to control the welding pool in the welding operation of vertical welding. With the increase of arc length, produce the bite edge welding current will decrease greatly, while the penetration depth decreases, and very easy to appear incomplete fusion, slag inclusion, fusion defects such as bad, low-alloy and high strength steel with thick plate welding, the recommended injection in order to achieve the transition to an appropriate welding current.

In the welding process, reasonable control of the arc length is a key factor to ensure the stability of the welding quality. If the arc is too long, the protection of the molten metal will deteriorate, and harmful gases such as oxygen and nitrogen in the air will invade easily, making the weld seam easy to produce. Air holes, while reducing the mechanical properties of the weld metal. However, the arc length should not be too short, otherwise it will cause sticking. At the same time, the surface pressure of the welding arc on the molten pool is too large, which affects the gas and slag floating velocity in the molten pool of the molten pool, which can cause slag inclusions, pores, and other defects. Therefore, before welding, the arc voltage matching the welding current should be adjusted to obtain higher welding quality. In the test, the welding current is 80~100A, the welding voltage is 20~22V, the welding speed is 220~260mm/min, the welding current is 240~260A, the welding voltage is 28~30V, the welding speed is 400~450mm/min, and the cover welding The welding current is 280~300A, the welding voltage is 28~30V, the welding speed is 400~450mm/min; the dry length of the welding wire is 12~18mm. In the welding process, in order to minimize the residual stress and deformation of the welded structure, it is necessary to pay attention to the welding from the most rigid position to the free end of the unconstrained welding; avoid the beginning and end of the weld in the high stress area Under the premise of satisfying the performance requirements of the weld, try to use low-strength welding material with good toughness; perform symmetrical welding; minimize the cross-section of the weld and the cross-section to ensure minimum; consider the geometry of the joint; reduce the joint Deformation and residual stress measures.

4 WELDING PROCESSING

WELDOX960 low-alloy high-strength steel has excellent weld ability, and generally does not require post-heating and post-weld heat treatment. However, for thick slabs with a tendency to cold cracks, where the stress is concentrated on the weld, if the post weld heat treatment cannot be performed immediately after the welding work is stopped, postheating should be performed. After the welding work is stopped, the weldment is heated to a certain temperature immediately. 200°C ~ 400°C, heat for a certain period of time, an hour, so that the weldment slowly cool down to accelerate the escape of hydrogen in a welding heat treatment process. For high-strength steel parts with high degree of restraint or deposition, in order to avoid welding cracks, asbestos can be used for heat preservation immediately after welding. The holding time is not less than 0.5 hour for each 25mm of the composite sheet thickness, and the total holding time must not be less than 1 Hours to determine, reach the holding time should be slowly cooled to room temperature. Under the guidance of the technical staff, it is also possible to add a post-heating process for hydrogen elimination treatment, and use a flame heater or the like for heating. The heating temperature is controlled at 200°C to 250°C. After the welding work is completed, the weldment is heated to a temperature below the phase change temperature of the material, AC1, for a certain period of time to allow the weldment to slowly cool down to improve the microstructure and properties of the weld joint or to eliminate residual stresses. Welding heat treatment process. High-strength steel should be post-weld heat treatment only if there are special requirements in the design rules.

5 SUMMARY

WELDOX960 low-alloy high-strength steel welding joint quality of the main problem is to ensure that the impact of the prevention of embrittlement and softening, high overall performance of the weld, to ensure that the fusion zone and heat-affected zone does not occur cold and hot cracks, and a certain plastic, toughness. Through the above tests, it is proved that the multi-pass multi-pass automatic welding process parameters are set reasonably, the weld seam is well formed, and it can meet the production needs of enterprises.

REFERENCE

1 Dong Mingshui, Chen Ming, Qin Zhu Feng. Welding technology of low alloy high strength steel WELDOX700 and WELDOX900 [J]. welding, 2006 (7): 45-46.

2 Fang Naiwen, Li Liansheng, Chen Mo, and so on. Weldox900 steel welding no crack technology research [C]// China Mechanical Engineering Society Welding Society eighteenth national welding academic conference paper - S03 welding technology and equipment. 2013.

3 Lu Anjun, Wu Yanming. Research on weldability and application of WELDOX900E high strength steel [J]. technology innovation and application, 2014 (15): 29-30.

Horizontal welding quality control of the CO2 gas shielded welding

Zhi-ling Wang

Jiangsu Province Xuzhou Technician Institute, Xuzhou 221151, China

Abstract: In the process of horizontal butt welding of CO2 gas shielded welding, the molten metal will squat under the effect of its own weight. Therefore, the upper part of the weld seam is very easy to produce undercuts, and the lower part is prone to defects such as welding and unwelding. If the problem is serious, it will cause welding. Seam cannot pass the weld quality test. This article is based on the welding skill training topic "CO2 gas shielded welding transverse welding". Through trial and error of preparations before welding, selection of welding process parameters, and welding operation process, Weld seam quality is well controlled, weld seams are beautifully formed, and relevant experience is promoted in practical training.

INTRODUCTION

Transverse welding means that the weld angle is 0 to 180 degrees, and the weld angle is 0 to 180 degrees. n this paper, the teaching links of CO2 gas shielded welding transverse welding, such as pre welding preparation, selection of welding parameters and welding operation process, are carried out repeatedly, and the weld quality is well controlled by the practice process[1].

1PRE WELDING PREPARATION

1.1Welding parts.

The weldment we chose is Q235 low carbon steel plate, with a thickness of 12mm, a length of 300mm and a width of 150mm. We use the cutting machine to cut down the material and use the planer to form the V type 60 degree slope. The root clearance is b=2.0 to 3, and the blunt side is P=0.5 to 1. The technical drawings of the welding parts are shown in Figure 1.

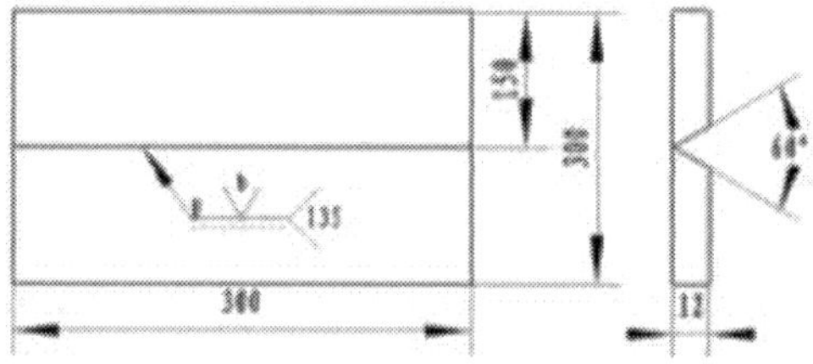

Figure 1 Technical drawings for welding parts

1.2 welding wire.

Welding wire specifications: H08Mn2Si, diameter 1.2.

1.3 Cleaning of welding parts.

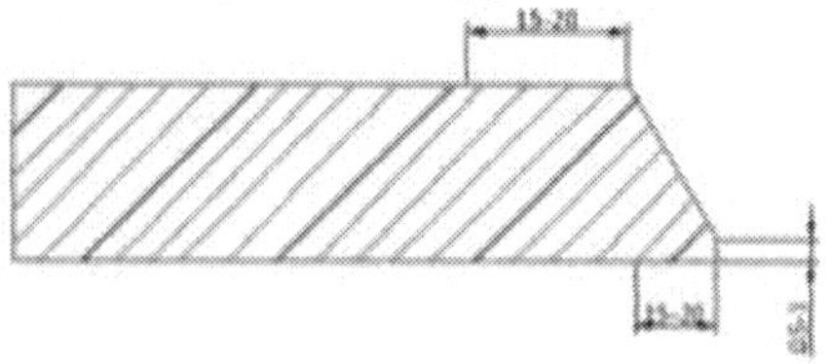

Figure 2. cleaning range on welding parts

We use sand turbine to clean the oil, rust and dirt in the range of 15 ~ 20mm on both sides of the test plate and the edge of the slope, to prevent and reduce the production of non fusion. The cleanup scope is shown in Figure 2. Then, we use the fitter tvice to repair the blunt edge of the groove on the weld, and we guarantee that the blunt side

© The Authors, published by EDP Sciences. This is an open access article distributed under the terms of the Creative Commons Attribution License 4.0 (http://creativecommons.org/licenses/by/4.0/).

size is 0.5 to 1mm, the blunt side of the straightness is in line with the standard.

1.4 Our welding machine.

We choose the NBC - 350 type CO2 gas shielded welding machine, using argon rich mixed gas, CO2 gas content 14%, Ar gas content 86%. Before welding, we check the power supply, wire feeding mechanism, welding gun, gas cylinder and pressure reducing flow regulator respectively, and adjust the flowmeter to the required flow, and then carry out the test welding to do the whole inspection of the welding parts[2]. Before welding, we must check and clean the spout on the nozzle, and spray the nozzle with silicone oil.

1.5Assembly positioning before welding.

The assembly gap of the welded test plate should not be too large or too small. If the gap is too large, the back weld will be too high and drop down. If the clearance is too small, the welding will not be able to penetrate. Theoretically, we choose the assembly gap of the initial welding end to be about 3.0mm, and the final welding end assembly gap is about 4mm.But in actual operation, we usually set the gap smaller, the welding end 2.5mm, the final welding end 3.0mm[3]. The weld pool is easier to control and the back weld is beautiful.

The length of positioning welding is 10 ~ 15mm. As shown in Figure 3, the welding quality requirement of the positioning weld is the same as that of the formal weld.

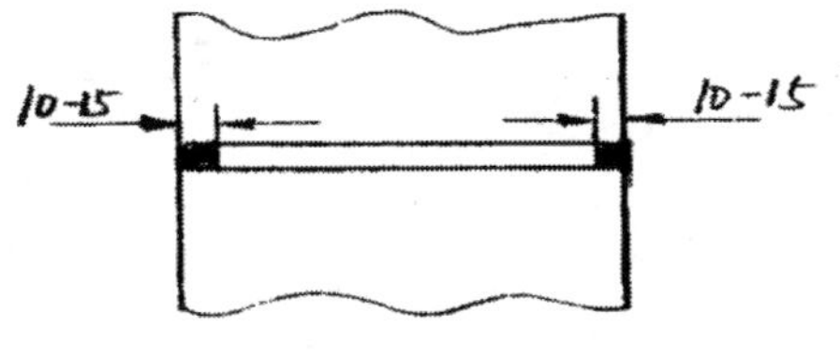

Figure 3. positioning weld

At the same time, the angle of the reverse deformation is about 3 degrees. We usually draw materials from under two boards with an electrode of 3.2. The assembly is accurate and convenient. The specific assembly method is shown in Figure 4.

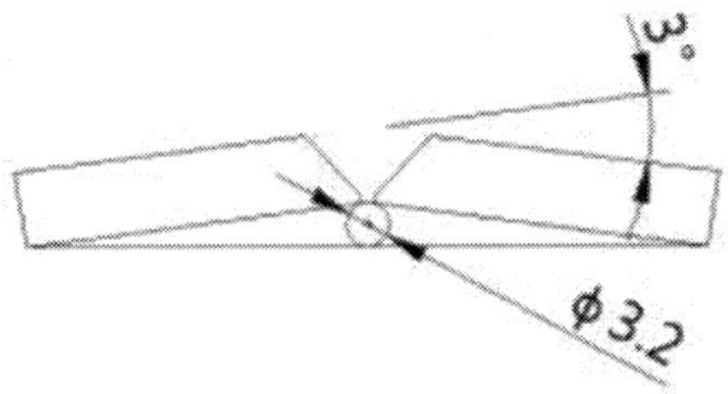

Figure 4. Reverse deformation of a workpiece

1.6Auxiliary tools and measuring tools.

We used CO2 gas flowmeter, CO2 gas cylinder, grinding wheel grinding machine, slag hammer, weld universal gauge and so on.

2Welding parameters

Which the selection of welding parameters is correct or not affects the shape, size, welding quality and the productivity of the weld directly. Therefore, the selection of appropriate welding parameters is a very important problem in the welding production. The welding process parameters of CO2 gas shielded welding mainly affect welding quality, such as welding current, arc voltage, welding speed, length and diameter of welding wire, gas flow and so on. The selection of the recommended parameters in this topic is shown in Table 1.

Table 1 Welding process parameters

Number of welds	**wire diameter (mm)**	**welding current (A)**	**welding current (V)**	**welding speed (m/h)**	**Gas flow (L/min)**
Bottom seam weld		100 ~ 110	20 ~ 22		
Filling layer	1.2	130 ~ 150	20 ~ 22	20 ~ 22	
Cover layer		130 ~ 150	20 ~ 24		15 ~ 20

3 Welding operation process

The plate thickness of the test plate is 12mm, which is usually welded 3 layers and 6 channels, the first layer is the bottom layer, the second layer is filling layer (two weld), and the third layer is welded with 3 welds of the cover layer. The weld height will not be too high, and the deformation will be reduced and the angular deformation can be reduced.

3.1 bottom welding

The test piece is assembled on the welding workstation frame, and the left side welding is adopted, and the one with the smaller gap is placed at the initial welding end of the right side. The reference welding parameters of backing welding are 100 to 110A, and the voltage is 20 to 22V. The angle between the welding torch and the weld is from 85 to 90 degrees, and the angle between the welding torch and the surface of the specimen is 75 to 85 degrees.

In the welding, due to the small gap, we should try to adjust the current and voltage to a large point, current 115A, voltage 20V, so as to ensure the penetration and prevent the production of unfusion. The arc movement is a small skew saw tooth, so that the welding is narrow and fast, and the two sides are very beautiful. The oblique circle can also be used. In the welding process, we should carefully observe the shape of the weld pool and the weld hole, keep the size of the weld pool moderate and uniform, and the welding speed should be uniform. When the welding torch nozzle swings in the groove gap, the residence time at the blunt edge of the upper slope should be slightly longer than that at the edge of the lower groove, so as to prevent the falling of the molten metal.

Welding should be done at once to reduce the joint not to break the test plate welding. It should be welded at same time to reduce the joint, thereby reducing the defect. If the arc is broken during the welding process, the arc should be re drawn from 15mm after the arc breaking, and the welding gun is swinging with a small amplitude, and when welding to the edge boundary of the weld hole, the welding gun should press forward, make a pause, and then resume the oblique sawtooth swing forward to weld, complete the welding of the bottom layer, weld the specimen arc, extinguish arc and welding. The gun can not be removed immediately until the weld pool is solidified before the welding torch can be removed to prevent the blowhole in the arc protection area.

3.2 Filled welding After.

Cleaning the spatter of the weld surface, the welding current we choose is 130 to 150A and the voltage is 20 to 22V. During welding, we should maintain the angle between the welding torch and the horizontal plane at 5~10 degrees. When we weld the first weld, the welding gun is 0 to 10degree angle, and the arc is swinging in the horizontal oblique circle with the lower edge of the bottom of the soldering track to ensure the good fusion. The welding path is 1/2 to 2/3 before the welding. When the second road is filled, the welding gun is 0 to 10degree angle. The arc is centered on the upper edge of the bottom welding path, and it swings between the first and the upper slope of the filling layer to ensure good fusion, and the welding path 1/2 to 2/3 before overlap.

In the operation, the oblique circle method is used. In fact, the trajectory of the motion is not necessarily a skew circle, it may be a slanted L, and the height of each weld and the forming of the weld are controlled. And the use of such a gun method can ensure good fusion on both sides of the slope, and avoid the phenomenon of groove between the welds, so as to avoid the production of non fusion, control the height of the weld seam and the forming of the weld. The welding of the filler layer should be staggered. The welding speed is slower, the thickness of the filling layer is less than 1.5 to 2mm, and the edge angle of the edge of the groove can not be melted to help the welding of the surface. Otherwise, it is not only unfavorable to welding, but also directly affects the appearance of the weld. If the pores are filled or the droplets can not be cleaned up, the welding holes and droplets must be thoroughly removed before the welding can be carried out.

3. 3Cover welding.

After cleaning the spatter on the weld and the groove, the matching of the welding current and voltage is 130~140 A and the voltage is 20 to 24V. The current and voltage matching between the two front welds is current 135A and voltage 21V. In the last welding line, the current voltage is smaller than that of the filler layer. Usually we tune it to 130A, and the voltage is adjusted to 20V. The angle of the welding gun is about 0~10 degrees. If the last electric current is larger, the weld is very easy to bite. At the same time, the last layer is high and the pool is easy to droop. So try to adjust the current and change the angle of the welding gun to about 15 degrees. The key to the first weld of the cover surface is not only to weld straight, but also to form a smooth transition of the weld shape. The nozzle of the welding torch is drawn circle or straight line reciprocating strip. The welding speed is slowed down, the residence time at the edge of the pool is increased, and the arc force drives the metal flow upward. During welding, the fusion of both sides of the weld should be ensured, so the weld pool edge is better than the groove edge 1.5 ~ 2mm. The welding layers are lapped to 1/2 to 2/3 to prevent the appearance of overlapping groove on the welding layer, which will affect the appearance of the weld surface.

While the V shaped groove butt welding is easy to operate, the weld surface is easy to operate, but the weld surface is not easy to be symmetrical, so the weld pool should be as small as possible. In addition, multi pass welding method is used to adjust the appearance of the weld. Finally, the symmetry of the weld appearance is obtained. The angle deformation of the specimen in the transverse welding is large. It is related to the welding parameters, the number of welds, the number of welds per layer and the stay time between the welding tracks. The larger the melting pool, the longer the residence time, the high interlayer temperature, the larger angle deformation, and vice versa. Most of the non fusion occurs in the filling layer, which is common outside the fusion line and between the two channels. The main reason is that there are oxide film or rust skin on the surface of the weld area, insufficient heat input, too large weld pool, improper welding technology and unreasonable joint design. ical industry press, May 2018.

Undercut is also common in welding process, especially when welding the last surface of the transverse plate cover. The main reason is that the welding speed is too high, the arc voltage is too high, the current is too large, the residence time is not enough, and the angle of the welding gun is not correct.

Through the above improvements, the angular deformation of the weld is controlled, and the non fusion, undercut and welding seam are reduced, and the weld shape is also beautiful.

4 Conclusion

Through the study of the welding process parameters and the operation process of CO2 gas shielded welding in horizontal welding, we draw the following conclusions: the use of positioning technique before welding (such as using the head of 2.5mm and 3.2mm welding rod at the blunt edge of the mouth of the test plate and fixing two specimens.) The operation techniques of each layer of the weld (using small oblique serrated wire) and the selection of the technology can improve the welding operation level and the fine rate of the welding parts faster and better, and greatly improve the quality of the weld.

Reference

1 Zhang Wenyue, welding metallurgy (basic principles) Machinery Industry Press, February 2014.

2 Shen Hui Tang "welding technology and brilliant idea" Machinery Industry Press 2012.

3 Fang Hongyuan "welding structure", mechan

Analysis and study on the thermodynamic performance of S-CO_2 simple Brayton cycle

Lei Sun[1], Lin Tao[1], Yonghui xie[1,a], Yuanjian Dang[2], Yongqing Wang[2]

[1]*MOE Key Laboratory of Thermo-Fluid Science and Engineering, School of Energy and Power Engineering, Xi'an Jiaotong University, Xi'an, China*
[2]*State Grid Shaanxi Electric Power Research Institute, Xi'an, China*

Abstract. The S-CO_2 Brayton cycle system has many characteristics such as low cost and compact structure, and is one of the hotspots in the fields of waste heat utilization, new energy and so on. In this paper, a mathematical model of the S-CO_2 simple Brayton cycle is constructed, and the cyclic characteristics are analysed. The relationship between the component parameters of the system and the calculation parameters of the cycle such as the cycle efficiency and the output net work are obtained under the design conditions. Meanwhile, calculation models of turbine and compressor under off-design conditions are established, and the characteristics of the whole cycle are analyzed. Thus, the characteristics of off-design conditions of the simple cycle are obtained. This study provides some reference for the related research of the S-CO_2 Brayton cycle and the practical engineering application.

1 Introduction

The supercritical carbon dioxide Brayton cycle system is a thermodynamic system using supercritical carbon dioxide as the working substance, which transforms the heat of the heat source into mechanical energy and finally outputs electric energy. CO_2 has good thermal stability, physical properties and safety, it is non-toxic, non-combustible and cheap. The critical temperature is 31.2 ℃, and the critical pressure is 7.38 MPa [1], which is far below the critical point of water, thus it is easy to achieve supercritical state. When CO_2 is in a supercritical state (S-CO_2), it has the density similar to liquid and the fluidity similar to gas, and there is no phase change during the cycle. This way, it can greatly improve the working capacity in turbine machineries, and under the same power generation capacity, the system volume is much smaller than other turbine machineries. In particular, there is no need for a large number of boiler piping equipment for steam making [2].

At present, research institutions in many countries or regions have carried out the research and prototype manufacturing of the S-CO_2 Brayton cycle system. The Sandy National Laboratory (SNL) has built a recompression Brayton cycle experimental system using two-stage compression, two regenerative structure, and a compact printed circuit plate heat exchanger (PCHE). The heat source power is 260kW in the initial design in 2010, and has been raised to 780W at present [3]. Tokyo Institute of Technology, Japan (TIT) also carried out the research on the S-CO_2 closed loop Brayton cycle. They improved the simple Brayton cycle by adding the process of flow rate, intermediate compression, and intermediate cooling, in order to improve the cycle efficiency [4]. On this basis, a complete set of supercritical carbon dioxide Brayton cycle experimental system was built by KAERI, in which the output power is 10kW, the single stage centrifugal compressor's pressure ratio is less than 2, and the turbine inlet temperature is 500 ℃ [5].

Iverson et al. studied the supercritical carbon dioxide Brayton cycle solar power generation system in the experimental system of 780kW. The results show that the cycle efficiency of the system can be improved by supercritical carbon dioxide Brayton cycle, especially when the turbine inlet refrigerant temperature is higher than 600℃ [6]. Sienicki et al. proposed a conceptual design of the supercritical carbon dioxide Brayton cycle system of 100MWe sodium cold fast reactor and pointed out that this system provides a cycle efficiency 1% or more higher than that of conventional steam circulation systems, and the turbine and reactor size ratio are also smaller [7]. In addition, Dyreby [8], Muto [9], Bae [10], Jeong [11], Moullec [12] and Ahn [13] et al. carried out detailed studies on the supercritical carbon dioxide Brayton cycle and achieved valuable results.

However, most of these studies are about the design of systems or components, there are few records on the research of variable conditions of the system. In the actual situation, the changes of the residual heat parameters are objective existence. Therefore, it is particularly important to study the change of the heat source of the system. In this paper, the design condition calculation model and the variable condition calculation

[a] Corresponding author: yhxie@mail.xjtu.edu.cn

© The Authors, published by EDP Sciences. This is an open access article distributed under the terms of the Creative Commons Attribution License 4.0 (http://creativecommons.org/licenses/by/4.0/).

model of S-CO_2 simple Brayton cycle are studied and constructed, and by changing the temperature of the heat source, the results are obtained and then compared and discussed.

2 Mathematical models and calculation process

The schematic diagram of a supercritical CO_2 Brayton cycle using afterheat is shown in Figure 1. The heat is provided by the gas turbine exhaust and transferred by the heat exchanger and some assumptions are made to simplify the mathematical models.

1、The system could reach a steady state in a short time.

2、The heat transfer between equipment and environment and the pressure drop in system are neglected.

3、The CO_2 temperature at the outlet of the pre-cooler can keep constant by changing the mass flow rate of the environmental water or air.

The system mainly consists of three kinds of equipment: turbines, compressors, and heat exchangers and the mathematical models are given as follows.

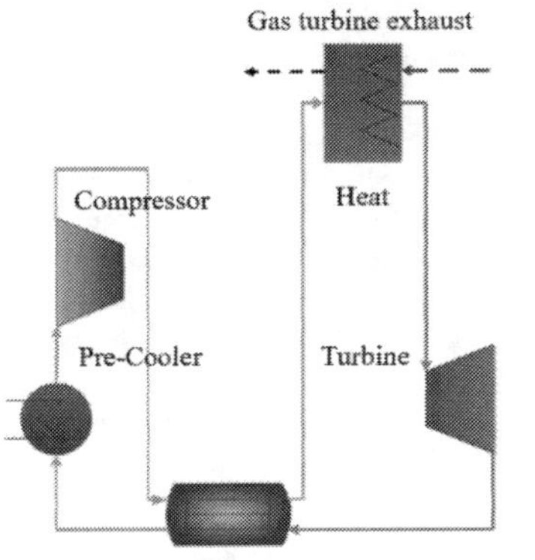

Figure 1. Supercritical CO2 simple cycle layout.

2.1 Turbine

Under the design conditions, an isentropic efficiency is used to describe the non-isentropic expansion process of turbines. The formula used follows:

$$\eta_{t\_d} = \frac{H_{t\_in}\text{-}H_{t\_out}}{H_{t\_in}\text{-}H_{t\_out,s}} \tag{1}$$

The power output of turbines is:

$$W_{t\_d} = m_{t\_d}(H_{t\_in} - H_{t\_out}) \tag{2}$$

When it operates under off-design conditions, the Stodola's ellipse method [14], which is established upon the flow characteristics of turbine, is employed to calculate the parameters of turbines:

$$\phi_{off} = m_{t\_off} \frac{\sqrt{T_{t\_in\_off}}}{P_{t\_in\_off}} \tag{3}$$

From the above two equations, it can be obtained the mass flow rate of working fluid through turbine under the off-design conditions once the turbine inlet pressure, turbine inlet temperature and turbine back pressure are determined.

$$m_{t\_off} = \frac{P^2_{t\_in\_off} - P^2_{t\_out\_off}}{T_{t\_in\_off} Y_d} \tag{4}$$

Where:

$$Y_d = \frac{P^2_{t\_in\_d} - P^2_{t\_out\_d}}{P^2_{t\_in\_d}\phi_d^2} \tag{5}$$

The turbine isentropic efficiency under the off-design conditions changes and could be determined by the following equation [14]:

$$\eta_{t\_off} = \eta_{t\_d} \sin\left[0.5\pi\left(\frac{m_{t\_off}\rho_{t\_in\_d}}{m_{t\_d}\rho_{t\_in\_off}}\right)^{0.1}\right] \tag{6}$$

2.2 Compressor

Under the design conditions, an isentropic efficiency is also used to describe the non-isentropic compression process of compressors. The formula used follows:

$$\eta_{c\_d} = \frac{H_{c\_out,s}\text{-}H_{c\_in}}{H_{c\_out}\text{-}H_{c\_in}} \tag{7}$$

The power compressed by compressor is:

$$W_{c\_d} = m_{t\_c}(H_{t\_out} - H_{t\_in}) \tag{8}$$

When it operates under off-design conditions, the performance curve, which is illustrated in Figure 2, is employed to calculate the parameters of compressors:

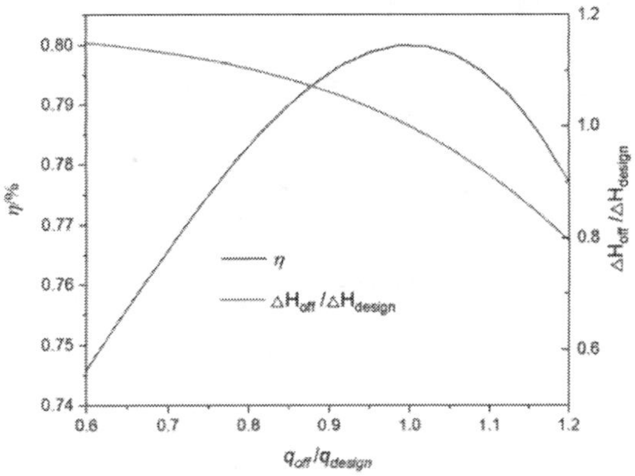

Figure 2. Compressor performance curve.

2.3 Heat exchanger

Since the heat exchangers are the largest components in the cycle, their careful design is an important issue. In order to reduce the total volume of heat exchangers, printed circuit heat exchangers(PCHE) are chosen in this article. The heat exchanger was divided into several axial nodes to simply the model.

The heat exchanger performance calculation can start from either the hot or cold end, therefore either hot or cold side operating conditions must be known. The heat exchanger calculation proceeds from the known end to the other one by sequentially evaluating the performance of all nodes.

The heat transferred in a node was calculated from:

$$q_i = U_i A_i (T_h - T_c) \tag{9}$$

The overall heat transfer coefficient is calculated from:

$$U_i = \left(\frac{1}{U_{i\_h}} + \frac{1}{U_{i\_c}}\right)^{-1} \tag{10}$$

U_{i_h} and U_{i_c} are the heat transfer coefficients on hot and cold sides respectively which are similarly to calculate. U_{i_h} is calculated from:

$$U_{i\_h} = \frac{\frac{f_h}{8}(\mathrm{Re}_h - 1000)\mathrm{Pr}_h \frac{L_h}{PD_h}}{1 + 12.7(\mathrm{Pr}_h^{2/3} - 1)(\frac{f_h}{8})^{0.5}(1 + \frac{PD_h}{L})^{2/3}} \tag{11}$$

Where:

$$f_h = \left(\frac{1}{1.8\lg(\mathrm{Re}_h) - 1.5}\right)^2 \tag{12}$$

$$PD_h = \frac{4\pi D^2 / 8}{\pi D/2 + D} \tag{13}$$

In the heat, the correlations above are used again to calculate the parameters of heat.

2.4 Calculation process

After the mathematical models above are built, Figure 3 illustrates the flow chart of the calculation process of the system under off-design conditions. The calculation contains two iteration loops to determine CO_2 turbine inlet pressure and mass flow rate of water in the heater, respectively.

Once the above two thermodynamic parameters are obtained, all other parameters can be calculated by the mathematical models above, and then the off-design performance of the system is obtained.

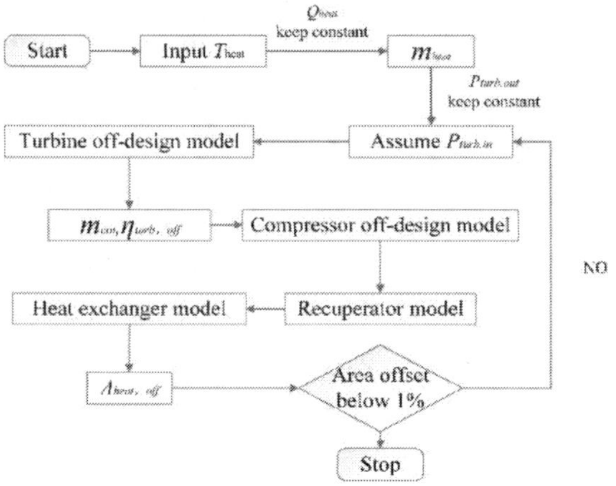

Figure 3 Flow chart of calculation process of the system under off-design conditions.

3 Results and discussion

3.1. Design condition results

The calculation models, including design and off-design conditions are both built by MATLAB software and the physical properties of carbon dioxide are directly retrieved by the database of National Institute of Standards and Technology. Before discussing the results under off-design conditions, several system settings and operation parameters under the design condition are provided to be compared to the results under the off-design conditions. Table 1 shows the operation parameters and some main thermodynamic parameters under the design condition. Table 2 illustrates corresponding system performance under the design condition.

Table 1. Operation and thermodynamic parameters of the system under design condition

Term	Value	Unit
Heat source temperature	500	℃
Heat source mass flow rate	80	$kg\cdot s^{-1}$
Heat source pressure	0.1018	MPa
Mass flow rate of CO_2	73.82	$kg\cdot s^{-1}$
Turbine inlet temperature	480	℃
Turbine inlet pressure	20	MPa
Turbine isentropic efficiency	85	%
Compressor inlet temperature	35	℃
Compressor inlet pressure	8	MPa
Compressor isentropic efficiency	80	%

Table 2. System performance under design condition

Term	Value	Unit
Power output of turbine	7.58	MW
Power consumption of compressor	2.12	MW
Net power output	5.46	MW
Exergy input	17.01	MW
Exergy efficiency	32.08	%

3.2 Off-design condition results and discussion

Figure 4 and Figure 5 illustrate the variations of off-design parameters and performance of the system with respect to the heat source temperature. It is noted that in each figure, the values of black points represent the values of design condition.

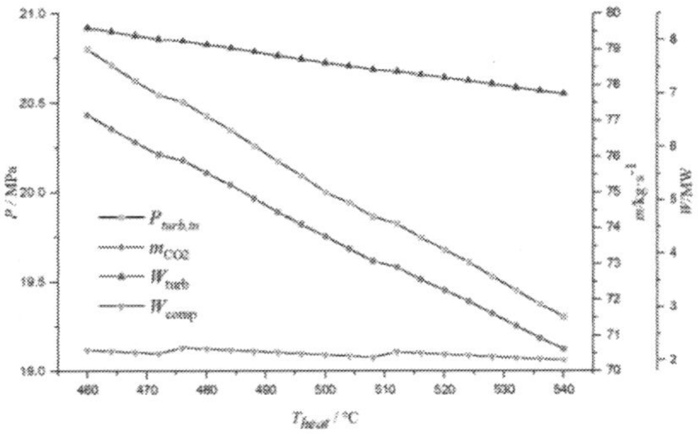

Figure 4. Variation of system parameters with respect to heat source temperature.

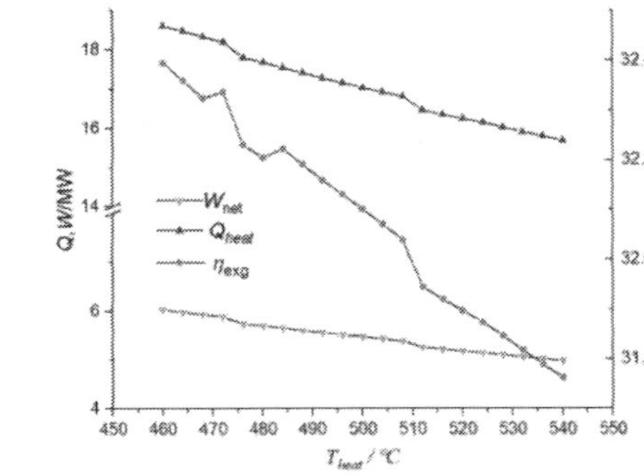

Figure 5. Variation of system performance with respect to heat source temperature.

Figure 4 shows the variation of some parameters with respect to the heat source temperature. Firstly, the heat source mass flow rate decreases as the temperature increases because the quantity of heat keeps constant. It can be seen that the turbine inlet pressure should decreases from 20.8 MPa to 19.30 MPa by regulation as the heat source temperature increases from 460 °C to 540 °C. With the decreased heat mass flow rate, the CO_2 mass flow rate decreases from 77.18 kg·s-1 to 70.63 kg·s-1. Similarly, since the turbine inlet pressure decreases while the outlet pressure keeps constant and the CO_2 mass flow rate decreases, power output of turbine decreases from 8.24 MW to 7.00 MW. But the power consumption of compressor is not changed basically because the inlet temperature of compressor changes correspondingly.

Figure 5 presents the variation of total system performance with respect to the heat source temperature. The net power output of system decreases from 6.03 MW to 4.98 MW when the heat source temperature increases from 460 °C to 540 °C. The quantity utilized by the system decreases from 18.61 MW to 15.69 MW. Finally, it is obtained that with the increase of heat source temperature, the system exergy efficiency decreases from 32.40% to 31.76%. It can be deduced easily that the heat source mass flow rate plays a more influential role than the heat source temperature in the system.

4 Conclusions

In this article, the off-design performance analysis of a supercritical CO_2 simple cycle was carried out. An off-design mathematical model for the system was established to examine the variation of system performance with the variations of heat source and temperature. Some main conclusions are summarized as follows.

(1) When the heat source temperature increases, the power output of turbine and the net power of whole system, and the system exergy efficiency decrease because of the heat source mass flow rate decreases.

(2) It can be deduced that the heat source mass flow rate plays a more influential role than the heat source temperature in the system by the results.

(3) Though the system parameters and system performance change when the heat source heat source temperature increases, they can be kept to be a certain range. So, we can come to the conclusion that supercritical CO_2 simple cycle is a good way to utilize the waste heat.

References

1. Tanaka H, Nishiwaki N, Hirata M, et al. FORCED CONVECTION HEAT TRANSFER TO FLUID NEAR CRITICAL POINT FLOWING IN CIRCULAR TUBE[J]. International Journal of Heat and Mass Transfer, 1971, 14(6): 739-+.
2. H Gil L, Otin S.F, Embid J.M, et al . Experimental setup to measure critical properties of pure and binary mixtures and their densities at different pressures and temperatures Determination of the precision and uncertainty in the results[J]. Journal of Supercritical Fluids, 2008, 44(2):123-138.

3. Conboy T, Pasch J, Fleming D. Control of a Supercritical CO2 Recompression Brayton Cycle Demonstration Loop[J]. Journal of Engineering for Gas Turbines and Power-Transactions of the Asme, 2013, 135(11).
4. Kato Y, Nitawaki T, Yoshizawa Y. Carbon dioxide partial condensation direct cycle for advanced gas cooled fast and thermal reactors[J]. Bulletin of the Research Laboratory for Nuclear Reactors, 2001, 25: 91-92.
5. Cho J, Choi M, Baik Y.J, et al. Development of the turbomachinery for the supercritical carbon dioxide power cycle[J]. International Journal of Energy Research, 2016, 40(5): 587-599.
6. Iverson B D, Conboy T M, Pasch J J, et al. Supercritical CO2 Brayton cycles for solar-thermal energy. Applied Energy, 2013, 111:957-970.
7. Sienicki J, Moisseytsev A, Krajtl L. Utilization of the Supercritical CO2 Brayton cycle with sodium-cooled fast reactors. Proceeding of the 4th International Symposium-Supercritical CO2 Power Cycles. Pittsburgh, Pennsylvania, USA: Argonne National Laboratory, 2014.
8. Dyreby J J , Klein S A, Nellis G F, et al. Design considerations for supercritical carbon dioxide Brayton cycles with recompression. Journal of Engineering for Gas Turbines and Power, 2014, 136(10): 101701.
9. Muto Y, Ishiyama S, Kato Y, et al. Application of supercritical CO2 gas turbine for the fossil fired thermal plant. Journal of Energy and Power Engineering,2010, 4(9): 7-15.
10. Bae S J, Lee J, Ahn Y. Preliminary studies of compact Bryton cycle performance for small modular high temperature gas-cooled reactor system. Annals of Nuclear Energy,2015,75:11-19.
11. Jeong W S, Lee J I, Jeong Y H. Potential improvements of supercritical recompression CO2 Brayton cycle by mixing other gases for power conversion system of a SFR. Nuclear Engineering and Design, 2011,241(6): 2128-2137.
12. Moullec Y L. Conceptual study of a high efficiency coal-fired power plant with CO2 capture using a supercritical CO2 Brayton cycle. Energy, 2013,49: 32-46.
13. Ahn Y, Lee J, Kim S G, et al. Studies of supercritical carbon dioxide Brayton cycle performance coupled to various heat sources. Proceedings of the ASME 2013 Power Conference, Boston, Massachusetts, USA, 2013.
14. Calise, F., C. Capuozzo, A. Carotenuto, and L. Vanoli. 2014. Thermoeconomic analysis and off-design performance of an organic Rankine cycle powered by medium-temperature heat sources. Solar Energy 103:595–609.

MATEC Web of Conferences **207**, 04008 (2018) https://doi.org/10.1051/matecconf/201820704008
ICMMPM 2018

The Thermal Hydraulic Performance of Wavy PCHE with Different Materials and Geometric Parameters

Yunzhu Li[1,a], Yonghui Xie[1] and Di Zhang[1]

[1] *Shaanxi Engineering Laboratory of Turbomachinery and Power Equipment, School of Energy and Power Engineering, Xi'an Jiaotong University, Xi'an, China*

Abstract. The printed circuited heat exchanger (PCHE) contain several different channel configurations, such as straight channel, zigzag channel and wavy channel. The wavy channel has better thermal performance than the straight channel and better hydraulic performance than the zigzag channel. This paper explores the thermal hydraulic performance of wavy channel PCHE. The numerical analysis of the PCHE in different materials and geometric parameters are conducted by computational fluid dynamics (CFD) tool. The materials applied in simulations involve Alloy617, Titanium Grade 3, Carlson 2205, UNS S30400 and Sandvik 253A. The results show that the materials have little effect on the thermal-hydraulic performance. The geometric parameters include channel degree varying from 10° to 50°, channel amplitude varying from 1mm to 5mm and the radium of hot/cold channel varying from 0.4mm to 2.0mm. It is found that the larger radium of hot channel results out the lower Nusselt number and lower fanning-friction factor while the higher radium of cold channel produces the higher Nusselt number and lower fanning-friction factor. The larger channel amplitude indicates the higher fanning-friction factor and lower Nusselt number. The larger channel degree indicates the higher fanning-friction factor, and higher Nusselt number.

1 Introduction

The PCHEs are more and more popular in many areas, including nuclear energy, electronic cooling and ocean engineering. Compared with the conventional shell and tube heat exchanger, the PCHEs can reduce 85% weight and only take place one third room. Besides, the diffusion bonding technique allows the PCHEs enduring a wide range of temperature and pressure. Due to plenty of advantages, researchers pay much attentions on PCHE and make abundant studies around the thermal-hydraulic performance. Justin et al.[1] have investigated the thermal and hydraulic performance for straight channel PCHEs in Very High Temperature Reactors(VHTRs). Kim et al.[2] proposed a newly PCHE based and explored the local friction factor with different hydraulic diameters and inclination degrees in lower Reynolds. Ting Ma et al.[3] identifies the local thermal-hydraulic performance of zigzag channel PCHE at high temperature of 900℃. The study implies that the heat transfer enhancement with inclined angles should consider the operation conditions. Jeon et al.[4] outlined an innovative type of PCHE, the heterogeneous and considered the effect of channel sizes, the spacing between channels and the channel cross-sectional shape on the thermal-hydraulic performance. Aneesh et al.[5] clarified the heat transfer rate and thermo-hydraulic performance of trapezoidal, sinusoidal and triangular channel PCHE, and they expounded the trapezoidal channel have the highest heat transfer rate and optimal thermo-hydraulic performance. The helium test loop and numerical simulations were conducted by In Hun Kim et al.[6, 7], they obtained the fanning-friction factor and Nusselt number correlations with the operation condition parameters. Sung et al.[8] interpreted the effect of tangled channels on the thermal-hydraulic performance in PCHE with the Reynolds number at 1000 to 3000. Tsuzuki et al.[9, 10] proposed a new configuration PCHE with S-shaped fin configuration, and reported the heat transfer and pressure drop of S-shaped fin PCHE with different geometrics.

The academic community has extensively explored the thermal-hydraulic performance for PCHEs. However, little research has been conducted to show the effect of the materials and geometric parameters for the sinusoidal channel PCHE. The study explores thermal-hydraulic performance of sinusoidal channel PCHE with different materials and geometric parameters

2 SYSTEM DESCRIPTION AND NUMERICAL MODEL

The traditional PCHE system as illustrated in Figure1 right contains a stack of photochemical etching planets with many semicircular fluid passages. The sinusoidal channel PCHEs are investigated and the reduced models are used in three-dimensional simulations. In this work, the radium of the hot channel and cold channel change from 0.4mm to 2.0mm, the amplitude A ranges from

[a] Corresponding author: author@e-mail.org

© The Authors, published by EDP Sciences. This is an open access article distributed under the terms of the Creative Commons Attribution License 4.0 (http://creativecommons.org/licenses/by/4.0/).

1mm to 5mm and the angle varies from 10° to 50° as illustrated in Figure1 left. The cold and hot SCO2 are utilized as the working fluid, the inlet temperature and the outlet pressure in cold fluid are set as 381K and 8.5MPa, while those in hot fluid are set as 553K and 2.6MPa respectively.

In this work, the hexahedral mesh as shown in Figure 2 is applied by the ANSYS ICEM CFD, the first cell height is set as 0.02mm. The grid independence verification as shown in Table 1 is done to guaranteed accuracy. The results show that the solutions of the chose model has less than 1% different from the last model.

ANSYS FLUENT is used to numerical simulate the PCHEs and the finite volume method is taken to describe the governing equations as shown in Eqs. (1-3). These equations are the continuous equation, momentum equation and energy equation used for heat and mass transfer in fluid domain. Eq. 4 is the energy equation applied for the solid domain.

Figure 1. The configuration of PCHE.

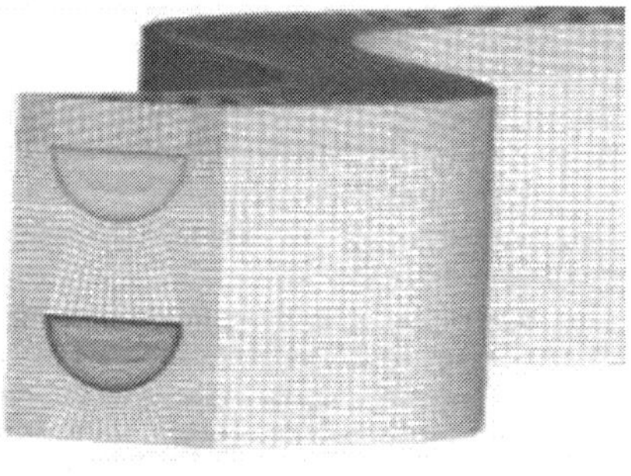

Figure 2. A small section of wavy PCHE for general mesh representation.

Table 1. Grid independence verification.

Model	Mesh (104)	$T_{h,out}$ (K)	Error (%)	Δp_c (Pa)	Error (%)
1	160	424.34	1.41	3467	1.31
2	309	420.16	0.41	3490	0.65
3	442	419.59	0.28	3504	0.26

$$\frac{\partial(\rho u_i)}{\partial x_i} = 0 \quad (1)$$

$$\rho\frac{\partial}{\partial u_i}(u_i u_j) = -\frac{\partial p}{\partial x_i} + \mu\frac{\partial}{\partial x_i}\left(\frac{\partial u_j}{\partial x_i} + \frac{\partial u_i}{\partial x_j} - \frac{2}{3}\delta_{ij}\frac{\partial u_k}{\partial x_k}\right) \quad (2)$$

$$\frac{\partial}{\partial x_i}\left[u_i(\rho E + p)\right] = \frac{\partial}{\partial x_i}\left[\left(k_f + k_t\right)\frac{\partial T}{\partial x_i}\right] \quad (3)$$

$$\frac{\partial}{\partial x_i}(k_s\frac{\partial T}{\partial x_i}) = 0 \quad (4)$$

The RNG k-ε model is adopted to model the turbulence. The mass flow inlet and pressure outlet boundary conditions are specified on the cold channel and hot channel correspondingly. The periodic boundary condition is applied on the up and down walls and the adiabatic boundary condition is set for the front walls, back walls and side walls. . The NIST real gas model is applied to estimate the thermal properties of the SCO2. The thermal-hydraulic characteristic considered include the heat transfer coefficient, Nusselt number and fanning friction factor as illustrated in Eqs.(5-8).

The hydraulic diameter of the channel is

$$D_h = \frac{\pi R}{(\pi/2+1)} \quad (5)$$

where the R is the channel diameter. The heat transfer coefficient is described as followed

$$h = \frac{q}{T_b - T_w} \quad (6)$$

where the T_b is the volume average temperature of fluid, T_w is the area weighted temperature of the channel walls. The Nusselt number is shown as followed

$$Nu = \frac{hD_h}{\lambda} \quad (7)$$

where the λ is the heat conductivity of fluid. The fanning friction factor is expressed as followed

$$f = \frac{\Delta p \cdot D_h}{2\rho L u_{in}^2} \quad (8)$$

where the Δp is the pressure different between the inlet and outlet., ρ is the fluid density, L is the total length of channels and the u_{in} is the inlet velocity.

Fluent 16.0 provides many turbulent models to simulate the actual situations, including Standard k-ε, Realizable k-ε, RNG k-ε, Standard k-ω, SST k-ω. In this paper, the data in research[11] is taken as the standard, the boundary conditions are completely same and the physical model is basically consistent. The results showed in Table 3 demonstrated that the numerical model with RNG k-ε is closest to the standard. In this paper, the turbulent model of RNG k-ε is chose as the calculate method.

Table 3. Validation of different turbulent models.

Variable	Ref [11]	Realizable *k-ε*	Standard *k-ω*
Nu	69.8	78.9	94.8
$error_{Nu}$	0	0.1304	0.3582
f	0.0513	0.0569	0.0608
$error_f$	0	0.1092	0.1852
Variable	Ref [11]	Standard k-ε	RNG k-ε
Nu	69.8	76.2	73.6
$error_{Nu}$	0	0.0917	0.0544
f	0.0513	0.0544	0.0544
$error_f$	0	0.0604	0.0604

3 Results and Discussion

3.1. Effect of Materials

The Table 2 shows the properties of five different materials including Alloy617, Titanium Grade 3, Carlson 2205, UNS S30400 and Sandvik 253A. Under the same boundary conditions, the PCHEs have similar thermal and hydraulic performance as shown in Figure 2. Take the results with Alloy 617 as a standard, the maximum error of Nusselt number of hot fluid and cold fluid is 0.59% and 1.03%, while the maximum error of Fanning friction factor is 3.54% and 1.93%.

Table 2. The properties of five different materials.

Number	Material	Density (kg/m3)	Specific heat capacity (J/kg·°C)	Thermal conductivity (W/m·K)
1	Sandivk 253 MA	7800	0.540	16.000
2	UNS S30400	8000	0.500	17.525
3	Titanium Grade3	4500	0.560	18.100
4	Carlson 2205	7820	0.418	19.000
5	Alloy 617	9240	0.465	16.300

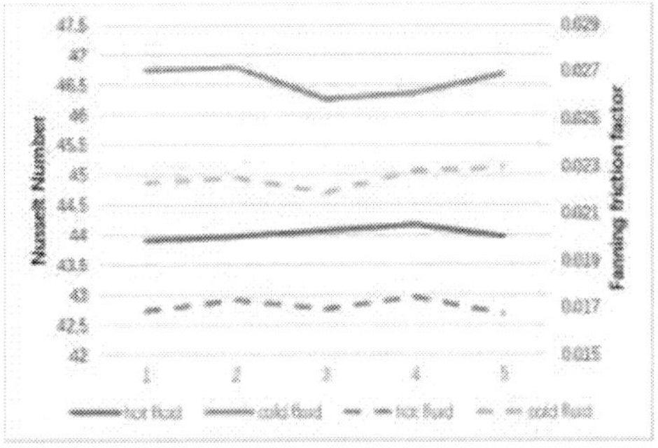

Figure 3. The effect of materials on the Nusselt number (denoted by solid lines) and fanning friction factor (denoted by dotted lines)

3.2 Effect of Geometric Parameters

Plenty of study about the effect of the hydraulic radium have been done, but the radium of hot channel equals to that of cold channel. In this work, the effect of hot and cold channel radium is investigated respectively.

Figure 4 shows the effect of the radium of hot channel on the Nusselt number and fanning friction factor. The results prove that the Nusselt number as well as the fanning friction factor decrease with the radium. It is obvious that the bigger the radium is the lower the velocity is, so that the pressure drop is relatively lower and the fanning friction factor declines. Moreover, the reduction of the heat transfer coefficient is much larger than increasement of hydraulic radium, which results out the lower Nusselt number.

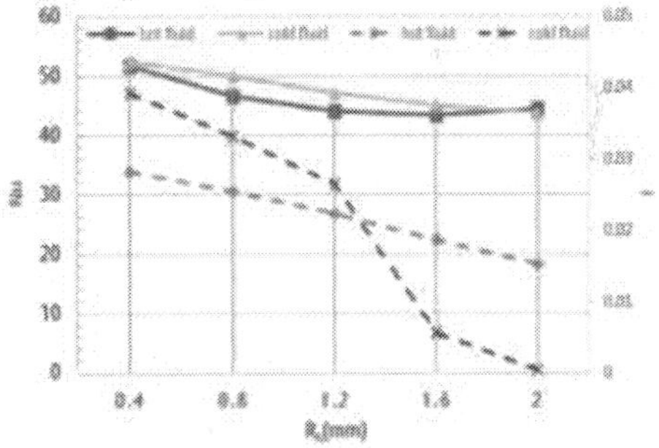

Figure 4. The effect of radium of the hot channel on the Nusselt number (denoted by solid lines) and fanning friction factor (denoted by dotted lines)

The effect of the radium of hot channel on the Nusselt number and fanning friction factor is shown in Figure 5. The results prove that the Nusselt number increases while the fanning friction factor decreases with the radium. The reduction of fanning friction factor has the same reason with that of hot channel radium. Although the heat transfer coefficient has a slightly drop therefore that the enlarged radium of cold channel and the reduced fluid dominate. According to eq.(7), the Nusselt number increases.

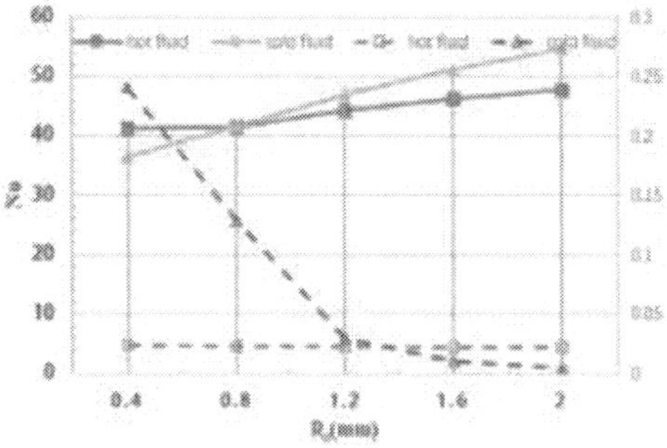

Figure 5. The effect of radium of the cold channel on the Nusselt number (denoted by solid lines) and fanning friction factor (denoted by dotted lines)

The important geometry parameters controlling the shape of channel include the amplitude and bend angle of the channel. Figure6 and Figure7 show the effect of the amplitude and angle on the Nusselt number and fanning friction factor respectively. In Figure 6, the amplitude changes from 1mm to 5 mm in step of 1mm. It is indicated that the Nusselt number decreases with the amplitude, and the reason is that the lower heat exchange coefficient with higher amplitude. It is also indicated that fanning friction factor increases with the amplitude and the reason is that the higher amplitude leads to a longer flow length and larger mixing areas.

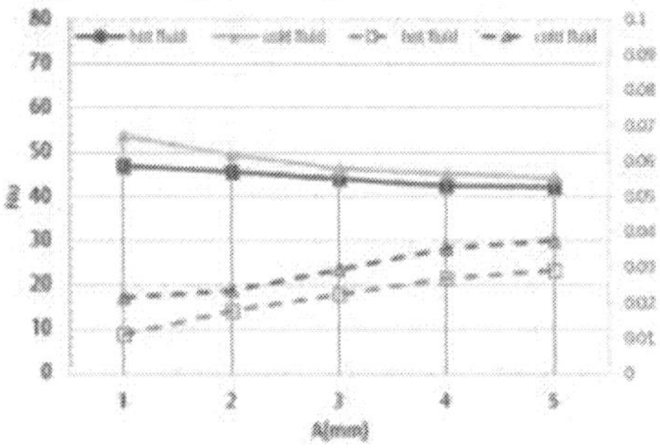

Figure 6. The effect of the amplitude of the channel on the Nusselt number (denoted by solid lines) and fanning friction factor (denoted by dotted lines)

In Figure 7, the pitch length changes from 10mm to 50 mm in step of 10mm. It can be see that the Nusselt number increases with the growth of angle, which due to the higher mixing around the corners of channel. The Nusselt number of hot fluid is slightly less than that of cold fluid because of the lower Prandtl number in cold fluid. As for the fanning friction factor, it also shows a positive proportional to bend angle and the slope is much higher for cold fluid which causes that the friction factor of cold fluid becomes high than the hot one. It can be concluded to the higher mixing and the lower flow length.

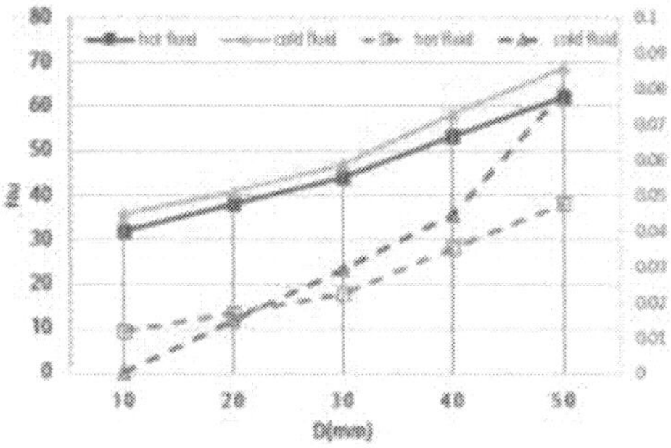

Figure 7. The effect of channel angles on the Nusselt number (denoted by solid lines) and fanning friction factor (denoted by dotted lines)

4 Conclusions

In this work, the five different materials are applied in sinusoidal channel PCHEs. The thermal-hydraulic performance and the structural analysis are investigated for those PCHEs with different materials. The thermal-hydraulic performance for various channel diameters, channel amplitudes and channel angles. Based on the numerical results, the conclusions are drawn as followed.

(1) The materials have little effect on the thermal-hydraulic performance, the maximum error between the Nusselt numbers and fanning friction factors are less than 4%.

(2) The Nusselt numbers of sinusoidal channel PCHEs increase with the channel angle increase, the channel amplitude reduces, the channel diameter of cold fluid increase and the radium of hot fluid declines. Especially the channel angle has the maximum effect.

(3) The fanning friction factors of sinusoidal channel PCHEs increases with the channel angle increases, the channel amplitude increases, the channel diameter declines Especially the channel angle has the maximum effect.

References

1 J. Figley, X. Sun, S. K. Mylavarapu and B. Hajek, Progress in Nuclear Energy **68,** 89-96(2013)

2 J. H. Kim, S. Baek, S. Jeong and J. Jung, Applied Thermal Engineering **30,** 2157-2162(2010)

3 T. Ma, L. Li, X.-Y. Xu, Y.-T. Chen and Q.-W. Wang, Energy Conversion and Management **104,** 55-66(2015)

4 S. Jeon, Y.-J. Baik, C. Byon and W. Kim, International Journal of Heat and Mass Transfer **102,** 867-876(2016)

5 A. M. Aneesh, A. Sharma, A. Srivastava and P. Chaudhury, International Journal of Heat and Mass Transfer **118,** 304-315(2018)

6 I. H. Kim and H. C. No, Applied Thermal Engineering **31,** 4064-4073(2011)

7 I. H. Kim, H. C. No, J. I. Lee and B. G. Jeon, Nuclear Engineering and Design **239,** 2399-2408(2009)

8 J. Sung and J. Y. Lee, International Journal of Heat and Mass Transfer **115,** 647-656(2017)

https://doi.org/10.1051/matecconf/201820704008

9 N. Tsuzuki, Y. Kato and T. Ishiduka, Applied Thermal Engineering **27,** 1702-1707(2007)
10 N. Tsuzuki, Y. Kato, K. Nikitin and T. Ishizuka, Journal of Nuclear Science and Technology **46,** 403-412(2009)
11 A. Meshram, A. K. Jaiswal, S. D. Khivsara, J. D. Ortega, C. Ho, R. Bapat and P. Dutta, Applied Thermal Engineering **109,** 861-870(2016)

Investigation of vibration characteristics for bladed-disks with dry friction nonlinearity

Tianyuan Liu[1,a], Yonghui Xie[1] and Di Zhang[1]

[1]*MOE Key Laboratory of Thermo-Fluid Science and Engineering, School of Energy and Power Engineering, Xi'an Jiaotong University, 710049 Xi'an, China*

Abstract. This paper focuses on the vibration characteristics of the bladed-disk subjected to the dry friction damping under periodic excitation. Firstly, the multi-harmonic equation basing on the frequency-domain analysis is established to predicate the steady response of the bladed-disk. Then, the algorithm to solve the nonlinear multi-harmonic balance equation is given step by step. In the numerical simulation, a simplified lumped parameter model of a turbine bladed-disk as well as the elastic Coulomb friction model between the root and disk contacting surfaces are applied. The normal load level of the friction interfaces, which are the focal points of the bladed-disk design, is analysed for the nonlinear vibration characteristics of the blades. The results show that vibration response of bladed-disk is affected significantly by normal load, and there exists an optimal value of the normal load under the operating condition of the turbomachinery, which can provide a qualitative assessment for the design practice of friction dampers.

Introduction

In operating conditions, the turbomachinery undergoes many sources of excitation, making vibration mitigation a challenge of great interest in bladed-disk design. Friction dampers such as snubbers, shrouds and under-platform dampers, which turn out to be the most effective method to reduce vibration, are widely used in modern turbomachinery.

Due to the strong nonlinearity of the friction dampers, it is difficult to predict the response of the bladed disks under periodic excitation. Especially, it is a big problem to design the dampers to reduce the vibration of the structure. For the approximate linear model widely adopted in industrial, Wu and Yuan [1] applied the equivalent linearization of the stiffness and damping coefficients of the friction dampers, which can be regarded as single harmonic method. However, this method has been proved with low accuracy in hazardous conditions. Recently, analysis of the vibrations of bladed discs with friction dampers is usually performed in the frequency-domain owing to the high computational efficiency of this approach. Petrov [2] derived the analytical formulation for the vectors of contact forces and the stiffness matrix of the non-linear friction contact interface based on multi-harmonic vibrations in the frequency-domain, and developed this method for high-fidelity models of different bladed disks [3, 4]. Laxalde et al. [5] proposed an alternating frequency-time domain method for the calculation of implicit non-linear forces. Joannin et al. [6] investigated the phenomenology of turbomachinery bladed disk subjected to dry friction nonlinearities by means of a lumped-parameter model. These investigations expand the scope of multi-harmonic methods in the design of bladed-disk.

This paper is organized as follows: 1. The dynamic equation for bladed disk is established in accordance with the multi-harmonic balance method; 2. The algorithm to solve the nonlinear equation is given step by step; 3. The elastic Coulomb friction model with variable norm load is performed to the root-disk friction interface of the bladed disk. At last, a simplified lumped parameter model is applied in the numerical simulation to investigate how the contact norm load of the dampers impact on the response of the bladed-disks.

1 Multi-harmonic balance equations

The dynamic equation for forced vibration of a bladed-disk system with nonlinear interactions can be can be written in the form:

$$\boldsymbol{Kx} + \boldsymbol{C\dot{x}} + \boldsymbol{M\ddot{x}} = \boldsymbol{fn}(\boldsymbol{x}, \dot{\boldsymbol{x}}, \ddot{\boldsymbol{x}}) + \boldsymbol{fe}(t) \qquad (1)$$

Where, $\boldsymbol{x}, \dot{\boldsymbol{x}}, \ddot{\boldsymbol{x}}$ are vectors of displacements, velocities and acceleration for all degrees of freedom (DOFs) in the bladed-disk structure, respectively. $\boldsymbol{M}$, $\boldsymbol{C}$, $\boldsymbol{K}$ are structural stiffness, damping and mass matrices. $\boldsymbol{fe}(t)$ is a vector of excitation forces and $\boldsymbol{fn}(\boldsymbol{x}, \dot{\boldsymbol{x}}, \ddot{\boldsymbol{x}})$ is the vector of nonlinear forces which, for a general case considered here, can be explicitly dependent on $\boldsymbol{x}, \dot{\boldsymbol{x}}, \ddot{\boldsymbol{x}}$. It should be noticed that the excitation forces of the bladed disk in equation (1) are usually considered as the periodic forces, which satisfy $\boldsymbol{fe}(t)=\boldsymbol{fe}(t+T)$, where vibration period T can be written as $T=2\pi/\omega$.

[a] Corresponding author: tianyuannn@163.com

© The Authors, published by EDP Sciences. This is an open access article distributed under the terms of the Creative Commons Attribution License 4.0 (http://creativecommons.org/licenses/by/4.0/).

Then time variation of displacements $\boldsymbol{x}$, the nonlinear forces or the periodic excitation forces $\boldsymbol{f}$ for the steady-state periodic regimes by a restricted Fourier series:

$$x(t) = X_0 + \sum_{i=1}^{Nh} (X_i^c \cos \omega_i t + X_i^s \sin \omega_i t) \quad (2)$$

$$f(t) = f_0 + \sum_{i=1}^{Nh} (f_i^c \cos \omega_i t + f_i^s \sin \omega_i t) \quad (3)$$

For multi-harmonic equations, the frequencies ω_i of the harmonic components can be expressed as $i\omega$, and Nh is the number of harmonic. Let $\boldsymbol{B}$ denotes the transformation matrix and $\boldsymbol{X}$, $\boldsymbol{F}$ denote the vectors of harmonic coefficients for displacements, nonlinear forces and excitation forces respectively. The term $\boldsymbol{X}$, $\boldsymbol{F}$ in the frequency-domain can be expressed as:

$$x = BX = B\{X_0, X_c^1, \cdots, X_s^{Nh}\}^T \quad (4)$$

$$f = BF = B\{F_0, F_c^1, \cdots, F_s^{Nh}\}^T \quad (5)$$

The multi-harmonic balance procedure of the time-domain Eq. (1) gives the following nonlinear equation with respect to vector X.

$$AX = Fn + Fe \quad (6)$$

$$A = \bar{K} + \bar{C}T_C + \bar{M}T_M \quad (7)$$

Where $\bar{K} = diag(K, \cdots, K)$, $\bar{C} = diag(C, \cdots, C)$, $\bar{M} = diag(M, \cdots, M)$ and the matrices T_C and T_M in Eq. (7) can be written as:

$$T_C = diag(0, \omega_1 \begin{bmatrix} 0 & I \\ -I & 0 \end{bmatrix}, \cdots, \omega_{Nh} \begin{bmatrix} 0 & I \\ -I & 0 \end{bmatrix})$$
$$T_M = diag(0, -\omega_1^2 \begin{bmatrix} I & 0 \\ 0 & I \end{bmatrix}, \cdots, -\omega_{Nh}^2 \begin{bmatrix} I & 0 \\ 0 & I \end{bmatrix}) \quad (8)$$

2 Arc-length solver for nonlinear equations

The Newton-Raphson method is usually applied to solve the nonlinear equation Eq. (6), however, it always causes difficulty in convergence near the resonance frequency. In order to efficiently deal with the convergence problem, an arc-length continuation was used to trace the loci of the solutions [6]. An equation is required to supplement Eq. (6) :

$$ds = \sqrt{(\omega - \omega_0)^2 + \|X - X_0\|^2} \quad (9)$$

Where an unknown frequency ω is introduced. The variables ω_0 and X_0 are the last converged solutions and the radius is ds. Let $Y=\{\omega, X^T\}^T$ denote the new unknown vector. The step by step descriptions of the arc-length solver for the multi-harmonic balance equation are as follows:

1). Obtain m initial solutions of the equation in weak nonlinear region by Newton-Raphson iteration and compute arc-length control variable s_q (q=2, ..., m) of these m solutions, and set s_1=0.

$$s_q = \sum_{p=1}^{q-1} \|Y_{p+1} - Y_p\| \quad (10)$$

2). Start outer iteration (j=m+1, m+2, m+3, ...). Compute new arc-length control variable s_j

$$s_j = s_{j-1} + ds \quad (11)$$

3). Start inner iteration (k= 1, 2, 3, ...). The Lagrange extrapolation predictor [7] based on the last m computed solutions is thus used to initialize the vector Y.

$$Y_j^{(1)} = \sum_{p=j-m}^{j-1} \prod_{\substack{q=j-m \\ q \neq p}}^{j-1} (\frac{s_j - s_q}{s_p - s_q}) Y_p \quad (12)$$

4). Compute the Jacobian matrix $\boldsymbol{J}$ of the nonlinear system [8] at $Y_j^{(k)}$.

$$J = \begin{bmatrix} \frac{(\omega_j^{(k-1)} - \omega_{j-1})}{\|Y_j^{(k-1)} - Y_{j-1}\|} & \frac{(X_j^{(k-1)} - X_{j-1})^T}{\|Y_j^{(k-1)} - Y_{j-1}\|} \\ \frac{\partial A}{\partial \omega} X_j^{(k-1)} - \frac{\partial Fn}{\partial \omega} & A - \frac{\partial Fn}{\partial X} \end{bmatrix} \quad (13)$$

5). Compute the residual vector $R = \{R_1, R_2^T\}^T$. If $\|R\| <$ err, then stop inner iteration, let j=j+1, go to step 2); else go to step 6).

$$\begin{cases} R_1 = \|Y_j^{(k-1)}\| - ds \\ R_2 = AX_j^{(k-1)} - Fn^{(k-1)} - Fe \end{cases} \quad (14)$$

6). Solve the linear equation.

$$\{\Delta\omega, \Delta X^T\}^T = J^{-1} \cdot R \quad (15)$$

7). Update $Y_j^{(k)}$, let k=k+1, go back to step 4).

$$Y_j^{(k)} = Y_j^{(k-1)} - \{\Delta\omega, \Delta X^T\}^T \quad (16)$$

3 Dry friction model

The non-linear restoring force between the contacting surfaces of a bladed-disk system, such as root-disk joints or shrouds, are usually modelled by means of an elastic Coulomb friction model [5]. It should be noticed that, due to vibration, the normal load $N(t)$ is inevitably non-constant in practical bladed-disks. So during relative motion of the surfaces three states are possible: stick, slip or separation.

$$f_n = \begin{cases} f_0 + k_x \cdot (x - x_0) & stick \\ \xi_0 \mu N(t) & slip \\ 0 & seperation \end{cases} \quad (17)$$

Where μ and k_x are the friction and stiffness coefficient of the Coulomb model. ξ_0 is a sign function of the friction force at the time instant of slip state initiation; x_0 and f_0 are values of the relative tangential displacement and the interaction force at the beginning of the stick state. However, the constants ξ_0, x_0 and f_0 are history-dependent, and the transition states for the periodic set should be analysed in this model [9].

The Jacobi matrix in Eq. (12) of this friction model can be calculated by DFT in the time-domain, as shown in Eq. (18), where $g(t)$ is the harmonic basis function. It should be noticed that the force harmonic coefficient $\boldsymbol{Fn}$ not only depends on the relative displacement x but also x_0 and f_0. The detailed derivation process can be seen in [2], which has been proved can overcome difficulties in convergence of the Newton-Raphson method.

$$\frac{\partial Fn}{\partial X} = \frac{2}{T} \int_0^T (\frac{\partial fn}{\partial x} \frac{\partial x}{\partial X} + \frac{\partial fn}{\partial x_0} \frac{\partial x_0}{\partial X} + \frac{\partial fn}{\partial f_0} \frac{\partial f_0}{\partial X}) \cdot g(t) \mathrm{d}t \quad (18)$$

4 Lumped parameter model for bladed disk

In this paper, a one-dimensional lumped parameter model (LPM) has been devised to study the phenomena of a bladed disk system. The high-fidelity model consists of 30 symmetric sectors as shown in the left Fig. 1. Each sector is simplified as 4 DOFs, including the blade tip, middle root and disk. Two adjacent sectors are coupled by the disk DOFs and the boundary conditions are considered by the stiffness and damping of ground. All the values of structural parameters calculated according to actual dimensions and material properties of bladed disk are reported in Table 1.

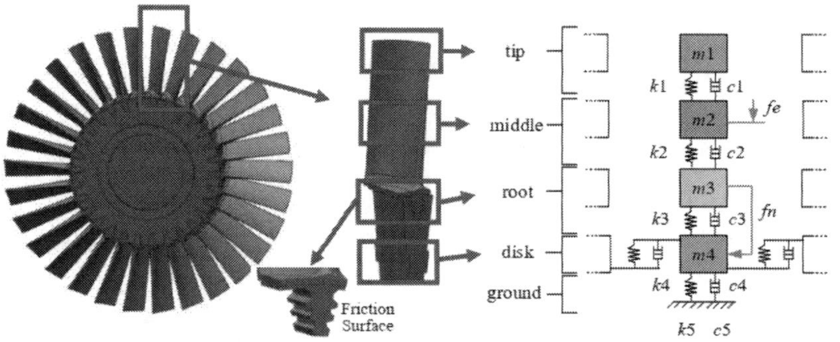

Fig 1. LPM for the bladed disk with dry friction nonlinearity

Table 1. Parameters values of LPM.

Parameter (unit)	Tip	Middle	Root	Disk	Ground
m (kg)	0.02	0.1	0.055	0.1	—
c (N•s/m)	0.4	0.4	0.4	0.4	0.4
k (N/m)	5×10^5	2×10^6	1×10^8	5×10^7	10^7

In a turbomachine, the external excitation acted on the blade disk is mainly gas exciting force, which can be simplified to a circular periodic excitation. Here the excitation force is expressed as a cosine traveling wave with one nodal diameter $fe(t)=E_0\cos(\omega t+2\pi j/Nb)$, while the force is applied on the middle of jth blade as shown in Fig. 1. The nonlinear dry friction force fn is applied on the disk and the root of the blade to simulate the coupling of two surfaces, which is the function of the relative displacement between the disk and root DOFs. The parameters k_x and μ of the Coulomb friction model is set to 2×10^8 and 0.3 respectively. The normal $N(t)$ load force between the contacting surfaces is also regarded as one nodal diameter wave expressed as Eq. (19).

$$N(t) = N_0 + A_0\cos(\omega t + \frac{2\pi j}{Nb}) \tag{19}$$

5 Results and discussion

The vibration characteristics of the LPM bladed disk with different normal loads N_0 and A_0 are presented and discussed in this section. The residual condition in step 5) $err=10^{-9}$ is used to check the convergence of the iteration process. The natural frequencies of the first family modes of the linear system are calculated for two special cases: bonded and frictionless interfaces of the root–disk, while frequencies of first nodal diameter are 555.13Hz and 552.37Hz respectively.

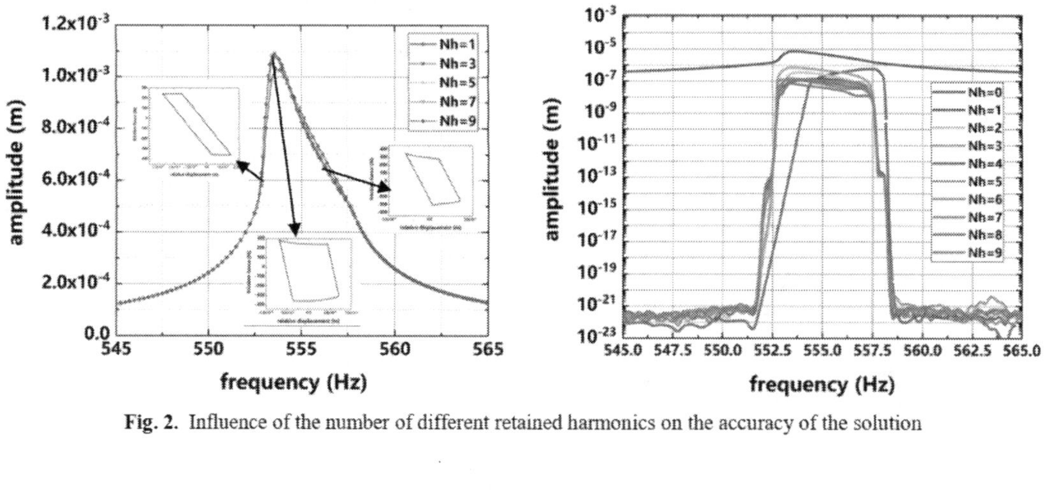

Fig. 2. Influence of the number of different retained harmonics on the accuracy of the solution

First of all, the forced response with different numbers *Nh* of the retained harmonics curves are compared. As suggested in [6], a sufficient number of harmonics should be kept to account for the stick–slip transitions occurring at the interface when the nonlinearity is activated. It can be clearly seen that as *Nh* increasing from 1 to 9, the response curves can converge gradually, and the hysteresis loops illustrate that the stick–slip transitions can be described accurately. It should be pointed out that, more accurate than approximate linear model, which is equivalent to the single harmonic method (*Nh*=1), the responses near resonance frequency are obviously higher. Further, the harmonic spectrum of the multi-harmonic vibration is shown quantitatively in the right side of Fig. 2. To reconcile the computational cost and the accuracy, 7 harmonics are kept in this study.

In this section, the excitation amplitude E_0 is set to 5N, with an excitation direction perpendicular to the blade surface. Different normal loads ranging from 600N to 3000N are selected to be the contact pressure between interfaces, as shown in Fig. 3. It can be seen that, there is an optimal load at 1100N (dotted line), which makes the forced vibration response amplitude at the blade tip to be a minimum. The effect of dry friction damping is obvious that the maximal vibration amplitude of whole circle blades under the normal load of 1100N is 0.107mm. However, if the normal load becomes an inappropriate value, which means the contact interfaces of root-disk joints are too loose or tight, the response may improve significantly, especially the normal load is designed for a smaller value.

The resonance frequency is equal to that of the underlying linear system (555.13Hz) with bonded DOFs at the interface, which means the contact interfaces is too tight. Once the normal load decreases to certain extent, the friction interface steps into slip state completely and the resonance frequency will be close to frictionless linear model (552.37Hz). The phenomenon reveals that the resonance frequency gradually shifts from the bonded model toward the frictionless model.

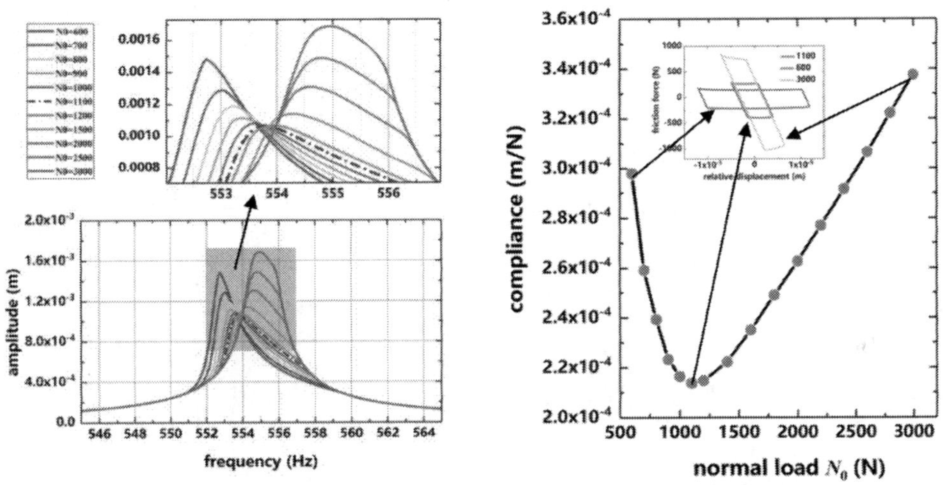

Fig. 3. Response amplitude curves under different normal loads N_0

Further, the compliance of the blade tip at the resonant peak frequency with different normal loads is given by the right graph in Fig. 3, which can be expressed as the ratio of the maximum displacement amplitude to the excitation level E_0. It can be observed clearly that with an optimum in terms of damping for a normal load N_0 of about 1100 N noticeable in the graph. Meanwhile, due to the nonlinear friction damping characteristics, hysteresis curves of dry friction damped could be influenced significantly by the magnitude of normal load under the same excitation, which lead to resonance amplitude and resonance frequency out of original state.

These results indicate that for a specific blade disk structure, there exist a suitable contact normal load design under the excitation force of general operating conditions. This analysis method proposed in this paper can accurately predict this optimal normal load. In this numerical example, the optimal displacement amplitude of the blade tip can be reduced by 28.26% compared with that of normal load at 600N, which means better vibration performance of the bladed-disk.

Last but not least, the sensitive of normal load at the optimal value is investigated, with the normal load fluctuation A_0 varying from 0 to 1000N. It should be emphasized that the contact surfaces have no separation during the periodic time. The hysteresis loop of the friction surface degenerates into the form of general Coulomb friction for A_0=0N, which is a parallelogram. As the A_0 increases, the hysteresis loop develops into a closed curve, meanwhile the relative displacement between the contact surfaces expands slowly. However, the incensement is so small that the vibration amplitude of blade tip has no much difference (from 0.106mm to 0.119mm) as shown in Fig. 4. It appears that the vibration performance of system is insensitive to variation of normal load fluctuations.

[a] Corresponding author: tianyuannn@163.com

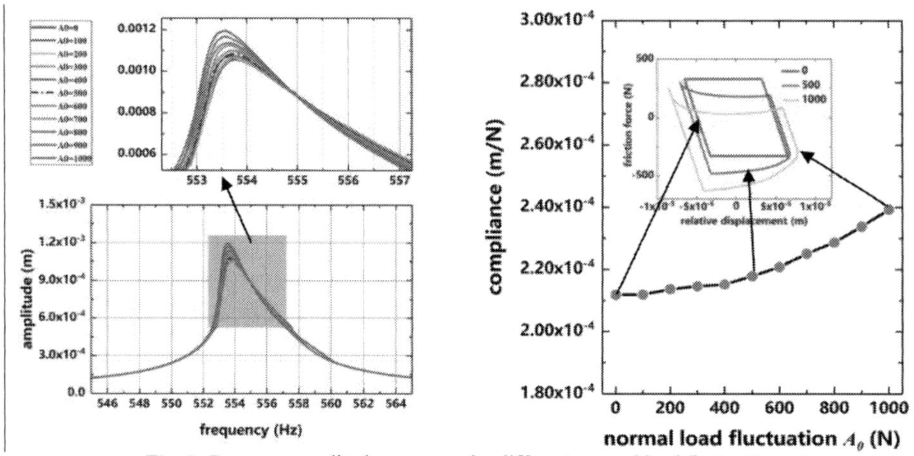

Fig. 4. Response amplitude curves under different normal load fluctuations A_0

Conclusions

This paper focuses on the vibration characteristics of the bladed-disk subjected to the dry friction damping under periodic excitation. A lumped parameter model is applied to investigate the influence of normal load of the blade root-disks on the vibration response. Firstly, the forced response with different numbers of the retained harmonics curves are compared, which reveals that the approximate linear model shows poor accuracy in the resonance region. Then the forced response with different normal loads are performed to seek an optimal norm load for the bladed-disk design. It appears that the resonance frequency gradually shifts from the tight normal load to the loos normal load and the optimal value can be obtained by the qualitative analysis of the nonlinear vibration characteristic. At last, the influence of the normal load fluctuations for the friction interfaces has also been addressed. The results show that the vibration performance of system is insensitive to variation of normal load fluctuations, which can be a supplement to the robustness evaluation at the optimal normal load of the bladed-disk design. The investigation provides a qualitative assessment for the design practice of friction dampers.

References

1. Jun W, Ruishan Y, Zhenwu H, et al. Experimental Study on Dry Friction Damping Characteristics of the Steam Turbine Blade Material with Nonconforming Contacts[J]. Advances in Materials Science & Engineering, 2015, 2015: 1-15.
2. Petrov E, Ewins D. Analytical Formulation of Friction Interface Elements for Analysis of Nonlinear Multi-Harmonic Vibrations of Bladed Discs[J]. Journal of Turbomachinery, 2003, 125(2):364-371.
3. Petrov, E, and Ewins, D, 2006, Effects of Damping and Varying Contact Area at Bladed-Disk Joints in Forced Responses Analysis of Bladed Disk Assemblies, ASME J Turbomach., 128(2), pp. 403–410.
4. Petrov, E., Zachariadis, Z., and Beretta, A., 2012, A Study of Nonlinear Vibrations in a Frictionally Damped Turbine Bladed Disk with Comprehensive Modeling of Aerodynamic Effects, ASME J. Gas Turbines Power, Vol.3, pp. 1239–1251.
5. Laxalde, D, and Thouverez, F, 2009, "Complex Non-Linear Modal Analysis for Mechanical Systems: Application to Turbomachinery Bladings with Friction Interfaces, Journal of Sound Vibration[J], 322(4–5), pp. 1009–1025.
6. Joannin, C., Chouvion, B., and Thouverez, F., 2015, Nonlinear modal analysis of mistuned periodic structures subjected to dry friction, ASME J. Eng. Gas Turbines Power[J], Vol.7
7. Cheung Y K. Application of the incremental harmonic balance method to cubic non-linearity systems. Journal of Sound & Vibration[J], 1990, Vol. 140, No 2, pp. 273-286.
8. Ferreira J V, Serpa A L. Application of the arc-length method in nonlinear frequency response". Journal of Sound & Vibration[J], 2005, Vol. 284, No 1, pp. 133-149.
9. Yang, B.D., Chu, M.I. and Menq, C.H., Stick-slip-separation analysis and non-linear stiffness and damping characterization of friction contacts having variable normal load, Journal of Sound & Vibration, Vol. 210, No 4, 1998, pp. 461-481.

[a] Corresponding author: tianyuannn@163.com

AUTHOR INDEX

AUTHOR INDEX